Wildland Fire, Forest Dynamics, and Their Interactions

Wildland Fire, Forest Dynamics, and Their Interactions

Special Issue Editors

Marc-André Parisien
Enric Batllori
Carol Miller
Sean A. Parks

MDPI • Basel • Beijing • Wuhan • Barcelona • Belgrade

MDPI

Special Issue Editors
Marc-André Parisien
Northern Forestry Centre
Canadian Forest Service Natural Resources
Canada

Enric Batllori
CREAF (Center for Ecological Research and Forestry Applications)
Spain

Carol Miller and Sean A. Parks
Aldo Leopold Wilderness Research Institute
Rocky Mountain Research Station
US Forest Service
USA

Editorial Office
MDPI
St. Alban-Anlage 66
Basel, Switzerland

This edition is a reprint of the Special Issue published online in the open access journal *Forests* (ISSN 1999-4907) from 2017–2018 (available at: http://www.mdpi.com/journal/forests/special_issues/ Wildland_Fire_Forest_Dynamics_Inte ractions).

For citation purposes, cite each article independently as indicated on the article page online and as indicated below:

LastName, A.A.; LastName, B.B.; LastName, C.C. Article title. *Journal Name* **Year**, Article number, Page Range.

ISBN 978-3-03897-099-6 (Pbk)
ISBN 978-3-03897-100-9 (PDF)

Cover image courtesy of Carol Miller.

Contents

About the Special Issue Editors

Marc-André Parisien, research scientist. Marc-André Parisien is a research scientist at the Canadian Forest Service, Northern Forestry Centre, where he has been working with the fire research group since 2000. He was trained as a forest ecologist and holds a BSc from McGill University, a MSc from l'Université du Québec à Rimouski, and a PhD from the University of California, Berkeley. His research on wildland fire is focused on understanding biophysical controls on fire regimes at landscape, regional, and continental scales, mostly within the boreal biome of North America. To study large-scale fire patterns, he uses a variety of quantitative methods including empirical, as well as process-based simulation modeling, and has developed tools and techniques for mapping wildfire risk.

Enric Batllori post-doctoral researcher and adjunct lecturer. Enric Batllori is a post-doctoral researcher at the Center for Ecological Research and Forestry Applications (CREAF) and adjunct lecturer at the University of Barcelona (UB). Trained as a forest ecologist, he holds a MSc and PhD from the UB and was a research associate specialist at University of California, Berkeley. His research focus is to understand the mechanisms governing the response and resilience of terrestrial ecosystems to environmental change from local to global scales, especially in relation to disturbance regimes. His investigations integrate field work, environmental classifications, and ecological modelling to assess fire-vegetation-climate relationships and the effects of compound disturbance regimes, mostly within Mediterranean-type ecosystems.

Carol Miller, research scientist. Carol has been a research ecologist with the US Forest Service since 2001. She earned a MS in Forest Science and a PhD in Ecology at Colorado State University where she developed and used a simulation model to study the interactions among climate, fire, and forest pattern in the national parks in the Sierra Nevada of California. After a postdoctoral appointment at the University of Montana, she joined the staff at the Aldo Leopold Wilderness Research Institute in 2001 to lead its fire research program. She is particularly interested in fire as an agent of landscape pattern formation and the scientific value of wilderness as a natural benchmark for change. Her program of research at the Leopold Institute seeks to help land managers understand how to include wildland fire as an ecological process to landscapes.

Sean A. Parks, research scientist. Sean has worked in various capacities for the research branch of US Forest Service since 2002 and achieved the position of research scientist in 2015. Sean earned a BS in Environmental Biology and Management and an MA in Geography from the University of California, Davis; he earned a PhD in Forestry from the University of Montana. Sean's research interests include the spatial interactions between past wildland fire and subsequent fire events and in better understanding how climate shapes fire regimes, the latter of which is particularly relevant given that climate change will inevitably result in altered fire regimes. He also actively studies the drivers and distribution of fire regime characteristics at spatial extents ranging from individual protected areas to the North American continent. Sean heavily relies on satellite imagery, geographic information systems, and statistical modeling to conduct his research.

Preface to "Wildland Fire, Forest Dynamics, and Their Interactions"

As a consequence of changing climates and human pressure on fire-prone systems, wildfire regimes are changing over much of the globe. The same environmental change that is altering the nature of fire in our forests is likewise affecting various aspects of system dynamics (e.g., productivity, reproduction, insects and disease) that, in turn, influence both the outcome and the drivers of fire regimes. Forests may thus face stresses to which they may be resilient or which may degrade their ecological integrity and ultimately undermine their ability to provide necessary ecosystem services.

This book, derived from a special issue published in Forests, includes 18 articles that examine diverse aspects of wildland fire and forest dynamics in many fire-prone parts of the world. Several articles offer new perspectives on basic fire and forest interactions and surmise how altered dynamics may arise in the future. Similarly, articles in this book address the growing concern of cumulative impacts of multiple natural disturbances (drought, insect outbreaks) whose incidence is also changing, sometimes drastically. Also represented in the book is a topic that is—rightfully—gaining recognition across the globe: the impact of humans on wildfire regimes.

Whereas the above mentioned articles shed light on fundamental (and often mutating) first-order effects and interactions in fire-prone forests, other papers in this book provide insights for increased resilience to fire-induced change. That is, they offer ecologically based advice and guidance to land managers, forest practitioners, and forest conservationist alike. Overall, the collection of articles in this special issue reinforce the notion that wildfire fires do not act in isolation in a given forest system: they constantly interact with an environment that is dynamic and complex.

Marc-André Parisien, Enric Batllori, Carol Miller, Sean A. Parks

Special Issue Editors

Review

Revisiting Wildland Fire Fuel Quantification Methods: The Challenge of Understanding a Dynamic, Biotic Entity

Thomas J. Duff [1],*, **Robert E. Keane [2]**, **Trent D. Penman [3]** and **Kevin G. Tolhurst [3]**

[1] School of Ecosystem and Forest Sciences, Faculty of Science, University of Melbourne, Burnley 3121, Australia

[2] Missoula Fire Sciences Laboratory, Rocky Mountain Research Station, US Forest Service, 5775 Highway 10 West, Missoula, MT 59808, USA; rkeane@fs.fed.us

[3] School of Ecosystem and Forest Sciences, University of Melbourne, Creswick 3363, Australia; trent.penman@unimelb.edu.au (T.D.P.); kgt@unimelb.edu.ai (K.G.T.)

*   Correspondence: tjduff@unimelb.edu.au; Tel.: +61-418-552-726; Fax: +61-353-214-166

Received: 24 August 2017; Accepted: 13 September 2017; Published: 18 September 2017

Abstract: Wildland fires are a function of properties of the fuels that sustain them. These fuels are themselves a function of vegetation, and share the complexity and dynamics of natural systems. Worldwide, the requirement for solutions to the threat of fire to human values has resulted in the development of systems for predicting fire behaviour. To date, regional differences in vegetation and independent fire model development has resulted a variety of approaches being used to describe, measure and map fuels. As a result, widely different systems have been adopted, resulting in incompatibilities that pose challenges to applying research findings and fire models outside their development domains. As combustion is a fundamental process, the same relationships between fuel and fire behaviour occur universally. Consequently, there is potential for developing novel fuel assessment methods that are more broadly applicable and allow fire research to be leveraged worldwide. Such a movement would require broad cooperation between researchers and would most likely necessitate a focus on universal properties of fuel. However, to truly understand fuel dynamics, the complex biotic nature of fuel would also need to remain a consideration—particularly when looking to understand the effects of altered fire regimes or changing climate.

Keywords: bushfire; grassfire; flammability; forest fire; quantitative methods; wildland fire; vegetation dynamics

---

1. Introduction

Fire behaviour is the product of the weather, topography, human intervention and, importantly, the fuel properties at the time a fire occurs [1,2]. In the case of wildland fires, this consists of vegetative matter, both living and dead [3]. Wildland fires, while essential to ecosystem processes, impose costs on societies including the loss of life, productivity, property, infrastructure, and ecosystem services [4–7]. The management of the landscape to minimise these costs requires that fire and, by necessity, fuel, be understood [8–11].

Fuels have particular importance to managers as they are the only element of the landscape that can be modified to influence the behaviour of future fires [10–12]. Substantial efforts are put into the treatment of fuel for risk reduction [9–11,13,14] and parameterisations of fuel are a core component of fire prediction systems [12,15–17]. Dead fine fuels in particular, have long been a focus of fire managers and researchers as they respond to weather over short time scales [18,19] and so are important determinants of fire occurrence and behaviour [3,20–22].

Effective fire management before, during and after fire events demands an understanding of the properties of fuel that will contribute the greatest hazard to values of interest, and methods to quantify and represent these spatially [23,24]. While parameterisations of fuel for risk assessment and modelling purposes have been a chief focus of land managers over recent decades, recognition of the dynamic, biotic nature of fuel is also increasing [25–27] due to the magnitude of effects that changing vegetation composition can have on fire behaviour (e.g., [28,29]), particularly in the face of a changing climate [6,30–33].

The development of methods to describe, quantify and map fuels has occurred relatively independently between regions, leading to a wide diversity of approaches and standards, including multiple ways of describing the same fuel properties. In this paper, we provide a critical review of current approaches for wildland fuel description, summarization and mapping in use worldwide. To conclude, we make recommendations on future directions in methods for the evaluation of fuel that have the potential to increase accuracy, utility and our understanding of fuel dynamics.

2. Quantifying Fuel

At a fundamental level, wildfires are uncontrolled and sustained combustion reactions that spread between organic fuel elements in the landscape [3,34]. These elements have intrinsic and extrinsic properties that influence the occurrence, rate and intensity of combustion of fires. These properties include chemical composition, particle density, size, shape, arrangement (both vertical and horizontal) and moisture content [16]. Here, we refer to fundamental fuel properties as 'attributes' and measured abstractions used for modelling as 'parameters' sensu Hollis et al. [35]. The actual values used in models are referred to as 'arguments'. We use the term 'fuelbed' to refer to the entire live and dead fuel complex at a site including surface, shrub and canopy sensu Riccardi et al. [36].

The behaviour of a fire is a function of the components of a fuelbed, and fuelbed is a function of the vegetation community at a site, including species composition, condition, and structure [21,27,29]. The vegetation community itself is a function of complex processes including climate, geology, herbivory and disturbance [37–39]. Methodologies for representing fuelbed properties have predominantly been driven by a need to forecast and manage fire impacts rather than understand dynamic processes [3].

Forecasting the progression of fires requires that methods be developed to describe, measure, summarise and map fuelbeds across the landscape. The methods selected to quantify and map fuel fundamental properties can have consequences on the applicability, accuracy, precision and compatibility of the modelled outcomes [40–45]. Creating fuel maps is a multi-stage process; it requires (A) having defined and measureable fuel parameters; (B) a method for assessment of parameters in the field; (C) a method to summarise or convert information to conform to model input argument requirements; and (D) a method for mapping summarised units [3]. These four steps and the implications of various approaches are discussed separately below.

2.1. Parameterising Fuel

Due to the need to manage fire, there is a long history of the assessment of fuels in wildland landscapes (e.g., [46] and [47]). However, a particular driver for the development of new fuel description and quantification methods was the advent and development of wildfire modelling in the 20th century [3], in which numerous models were created for a range of vegetation types, fuel conditions and regions [17,48,49]. To predict fire behaviour, it is necessary to parameterise the fuel attributes that are most influential over fire behaviour. However, the combustion of vegetation is a complex process [34,50] and there is no universal set of parameters common to all models. Fire behaviour is strongly determined by the properties of vegetation and consequently, features that are important in one system may be absent in another. Additionally, any parameterisation requires a degree of abstraction of the real world into something measurable; the degree of abstraction can vary, resulting in fuel parametrizations that vary along a spectrum from those thought to be fundamental to fire behaviour processes (as in the Rothermel Model [16]), to representations of vegetation type linked to fire behaviour through empirical observation (as in the Canadian Fire Danger Prediction System [51]). Some examples of operationally used models and the diversity of their key fuel input parameters are presented in Table 1. Further details of the contrasting inputs for the Australian models are presented in [52]. Although methods of quantification vary greatly, there are commonalities between approaches; operational fire models invariably include some form of consideration of the amount, physical characteristics and spatial configuration of fine fuels (<6 mm diameter [53]—the fuels that readily ignite in a flaming fire front).

Early fire models provided estimates of fire rate of spread for a defined set of conditions—they were inherently aspatial. To predict fire spread, their outputs had to be interpreted and mapped by hand [54–56]. To achieve this, maps of fuel were necessary to select the appropriate model to use and obtain the necessary fuel arguments. More recently, driven in-part by increasing computational power, models have been developed to be spatially explicit. Fire behaviour simulators are now routinely used operationally to solve large-scale real-time fire prediction problems to provide emergency decision support, e.g., FARSITE [57] and PHOENIX RapidFire [58,59]. Additionally, the applications of fire models are increasingly being extended, including applications such as strategic risk assessment [60,61], the assessment of ecological fire regimes [62,63] and carbon accounting [6]. In addition to modelling, fuel maps are also important for strategic purposes to enable managers to visualise fuels across the landscape relative to topography and vulnerable assets.

The development of spatial fire models has substantially increased demand for high quality maps of input arguments. Models developed for the management of fire risk typically require that predictions be made faster-than-realtime so wildfire spread can be forecast as they occur. As fires can be very large (i.e., 10's of square kilometres), this has influenced the practicality of data collection and affected the precision adopted in parametrising fuel. However, with increases in computer processing power, there has also been development of complex physical models that, while generally slower than real time, allow insight into the physical processes within fires, e.g., WRF-Fire [64], FIRETEC [65] and the Wildland Fire Dynamics Simulator [66]. The development of such models of fire poses additional challenges to fuel quantification as physical models require that the physio-chemical properties of fuel elements be known at the scale of the processes being emulated—these scales are typically much smaller than used in empirically models [49]. Furthermore, as empirical models are statistically fit, the fitting process can somewhat compensate errors in measurements—a luxury not afforded to physical models. Physical models are crucial to understanding fundamental combustion processes, so being able to accurately quantify fuels in the field to allow their verification and validation against real-world fire outcomes remains important.

To date, the development of fuel quantification and mapping systems has predominantly focused on providing arguments for specific fire models rather than representing the fundamental properties of fuel important to fire behaviour [67,68]. This means that the information collected is highly regional and focused on the limited number of parameters and methods specific to local vegetation types (e.g., Eucalyptus forests [69] or grasslands [70]).

One attempt to reduce this model-centric focus has been the development and implementation of the 'Fuel Characteristic Classification System (FCCS)' in the USA. Within this system, fuel beds are described in great detail with the aim of being able to provide inputs to a wide variety of models that operate at different scales and for different purposes [71].

Table 1. Selection of fire models used for operational faster-than-real-time fire behaviour prediction by landscape managers, and the fuel input arguments required for their computation *. The models presented utilise unique functions for deriving fire behaviour from fuel. Modelling systems that utilise these functions are not considered here.

Model	Region of Use	Intended Vegetation	Fuel Arguments
Anderson shrublands [1]	Australia, Europe	Shrublands	Vegetation height
Buttongrass model [2]	Australia	Buttongrass plains	Cover Fuel load % dead
Canadian FFDPS [3]	Canada, New Zealand	Various	Fuel type Grass curing
CSIRO Grass [4]	Australia	Temperate grasslands	Grassland structure Grass curing
CSIRO Tropical grass [5]	Australia	Tropical grasslands	Grassland type Grass curing
Mallee-Heath model [6]	Australia	Mallee Heath	Vegetation height Vegetation cover Near surface fuel load
McArthur [7]	Australia	Southern Australian forests	Fine fuel load Soil dryness / fuel availability
PHOENIX Rapidfire [8]	Australia	Various	Surface fine fuel load Near surface fine fuel load Bark fuel fine fuel load Shrub fine fuel load Grassland structure Grass curing Wind reduction factor
Rothermel [9]	USA, Europe	Various	Fuel load by size class and category Surface area: volume by class and category Fuelbed depth Dead fuel extinction moisture content Heat content of live and dead fuels
Vesta [10]	Australia	Southern Australian forests	Surface fine fuel load Near surface fine fuel load Shrub fine fuel load Bark fuel fine fuel load

* Short-term dynamic fuel properties (e.g., moisture content) are computed separately using weather data. [1] [72]; [2] [73]; [3] [51]; [4] [74]; [5] [75]; [6] [76]; [7] [12]; [8] [58]; [9] [16]; [10] [15].

2.2. Assessing Fuel Attributes in the Field

The effective spatial representation of fuel requires some level of assessment or verification in the field [77]. Extensive vegetation surveys are expensive, so invariably some form of sampling is required [78,79]. In designing a fuel inventory, the questions of what to measure within a sampling unit and how units should be sampled (including number and stratification) need to be resolved [3]. An ideal method for sampling within measurement units is one that can be completed efficiently and accurately with minimal expertise. As some fire model arguments are not easily measurable outside of a laboratory (e.g., fuel element energy, oil and mineral content) and others are time consuming to measure directly (e.g., bulk density and surface area to volume ratio), an alternative has been to undertake a number of simple measurements combined with visual estimates. This commonly involves textual descriptions combined with photos, keys and simple measurements (e.g., [77,80]) to approximate parameter arguments (or groups of parameter arguments) from a limited number of classes. Such class-based approaches can greatly increase the efficiency of field surveys; however, there is a cost in terms of the degree of accuracy and precision [81,82]. Additionally, error can be introduced due to variation in the way assessors interpret classification guidelines [83,84].

To understand fire behaviour processes from a scientific point of view, the ideal field assessments of fuel within a site would be comprehensive evaluations that quantify fuel element attributes in three dimensions to allow virtual fuelbed reconstruction. In addition, non-fuel details such as species composition, canopy cover and soil type would also be recorded as they can provide insight into the dynamics that result in particular fuel configurations [27,85]. Apart from the FCCS, such intensive fuel audits are rare outside research. However, recent developments in technology have the potential to improve the efficiency, accuracy and precision of highly detailed field assessments, in particular terrestrial LiDAR [86,87] and photogrammetry [88]. These enable the rapid quantification of structure in three dimensions, enabling sites to be digitally represented at extremely fine scales.

Fuels can have high levels of spatial variation [25] which can be important determinants of fire behaviour and impacts [43,44]. The capture of such variation necessitates a large number of sampling plots, resulting in trade-offs between the level of detail measured at a sampling unit and the number of sampling units that can be collected. To resolve this requires an understanding of the sensitivities of fire models to the relevant inputs (e.g., [89,90]), although ideally this would be driven by fundamental fire theory [91].

2.3. Summarizing Fuel to Develop Maps

The process of summarizing measured fuel attributes at a site level and developing mapping methodologies is often concurrent, as site level classes are typically used as mapping units. During a site fuel survey, a diversity of attributes is independently considered. However, it is rare to map each attribute directly—values are usually first summarised using a single, exclusive site-level class. Attributes are given values that apply to the entirety of the assigned class. An example is the use of Fire Behaviour Fuel Models in the US to represent fuel loading, depth and moisture of extinction [92]. When assigning classes, there are three approaches that are used: association (using existing vegetation classifications), classification by fuel fundamental properties (using statistical or descriptive methods), and abstraction (grouping fuels based on a common secondary property such as fire behaviour). These approaches are comprehensively summarised in Keane [41].

Regardless of classification approach, the summarization of measurements into site level classes results in a loss of information if sites that have properties of more than one class are forced into a single class [93]. This effectively compresses information, resulting in approaches that do not represent the heterogeneity or potential range of values present in these systems. There is also an assumption that the site attributes consistently co-vary—i.e., that bulk density and crown base height are at consistent ratios for a particular vegetation class. This assumption may not be always valid as natural systems often have gradients of change [94] and high levels of independent variation occur in space and time in both species composition and fuel attributes [25,27,38,95]. The importance of considering this variation is particularly evident at the interface between wildlands and urban environments where vegetation is heavily modified (resulting in novel fuel configurations that are not well represented by existing classifications) and there are high concentrations of values at risk (so there are potentially greater consequences for errors) [96].

Variation within classes can be accounted for with the addition of intermediate classes [67,97]; however, large numbers of classes can provide additional challenges, such as difficulty in identifying or verifying them in the field [41]. This is a particular issue where fuels change rapidly post fire—fixed classifications have limited potential to represent the continuum of change that occurs as a forest recovers. One method that has been used to account for this is the adjustment of class attribute values to account based on other landscape properties. This approach is applied in Australia in systems where the forest overstorey typically survives fires and vegetation (and consequently fuel) re-accumulates after fire following a negative exponential pattern [27,53,98]. This pattern is used to moderate fuel loading from class equilibria based on time since last fire [59]. While this approach is unique to Australia, such patterns of recovery are not (e.g., [99,100]). Furthermore, with variation in post fire conditions [27] or fire severity [101,102] having the potential to influence vegetation recovery, using time since fire as the sole moderator of fuel properties may not necessarily deliver outcomes that meet manager's expectations. Additionally, fire is only one of many potential disturbances that can impact fuels—it may also be important to recognise other disturbances such as timber harvesting or drought.

The continuous and dynamic nature of vegetation through space and time means that high within-class heterogeneity and independent variation of attributes will remain a challenge with any fuel classification, necessitating monitoring or biophysical modelling to maintain reliability [3].

2.4. Creating Maps of Fuel

Mapping fuels at large scales faces challenges typical of mapping vegetation; practicality limits the proportion of the landscape that can be measured directly and high inherent heterogeneity limits the potential for interpolating between measured sites [103,104]. For broad-scale fuel mapping, there are three main approaches that can be applied; direct (where methods directly measure properties of interest—such as measuring canopy structure with LiDAR), indirect (where methods use the direct measurement of a proxy for the properties of interest—such as using images to create classes based on overstorey tree species as a proxy for fuel structure) or derived (where values are derived statistically from a range of sources including combinations of biophysical variables and indirect measurements—such as modelling fuel loading using climatic and vegetation community data) [23,105,106]. The methods available for mapping fuel are highly dependent on the ways fuel has been sampled and classified. Many of the parameters used in fire behaviour models (e.g., bulk density of fine fuels or surface fuel depth) are impractical to quantify with direct measurement so their values must be determined through other means.

Indirect assignation of classes, in particular assigning estimated fuel attributes to existing classifications, has been common as it allows managers to apply existing maps—often of vegetation type—as fuel maps, reducing the need for extensive surveys or mapping programs [41]. However, the value of such maps will be dependent on (1) how well they represent existing vegetation type classes (as the accuracy of the derived fuel map cannot be greater than the vegetation map it is derived from); (2) how representative the existing classifications are of fuel attributes in space and time; and (3) how internally consistent the units are. Additionally, having a fuel map based on extant classifications means there is limited flexibility in adjusting values where there are known inconsistencies, such as those resulting from changing abundances of particular species that have unusual flammability properties (e.g., [28,29]).

Where there are site level classifications of fuel that can be discriminated aerially, remote sensing approaches can be used to directly assess and classify them [107]. While obscuration by tree canopies has provided a challenge for directly measuring many fuel properties [23], in recent years there have been rapid developments in technologies that allow the measurement of sub-canopy fuel properties, including airborne LiDAR [108], hyper and multi-spectral imagery [109], and radar [110]. These have the potential to yield detailed measurements of attributes that have been difficult to measure over large areas, in particular vertical and horizontal structure. Additionally, remote sensing approaches can now provide information on the status of fuels, including the degree of curing [111] and live moisture status [112–114].

Derived approaches are becoming increasingly available to allow attributes that are not so readily measurable remotely to be estimated using statistical approaches [115]. They have the strength of being able to use modelling to combine disparate sources of data to predict attributes in a parsimonious manner [23,27,116–118]. Advantages include the ability respond to dynamic changes (such as incorporating observations [119]) as well as being able to spatially quantify uncertainty around attribute values. Understanding uncertainty can be important for prioritizing the collection of data and for Monte Carlo style fire risk analysis [120].

The accuracies of fuel maps reflect the approaches used in their creation. There are a number of sources of error that may contribute to poor results. These include (1) inappropriate fuel sampling methods and designs; (2) improper classifications; (3) errors in the application of methods; (4) improper geo-registration; and (5) scale incompatibilities (both between fuel attributes at a site and between sampling scale and mapping scale) [3,95]. The level of error in using classes can be high: a review of the LANDFIRE fuel mapping products found that correlation between mapped units and fuel properties was relatively low (ranging between 5% and 85% correct, regardless of mapping approach) due to scale and resolution mismatches and the possible insensitivity of the attributes used [121].

3. Future Directions, Opportunities and Needs

3.1. Parameterising Fuel

It is important that the quality of fuel data is commensurate with the gravity of the decisions being made using them. Fuel maps are a key input in wildfire modelling systems; such systems are becoming increasingly important to land managers. Despite this, there are no universal standards used for quantifying and representing fuel worldwide. Single purpose methodologies are widespread, but incompatibilities in the parameters that are represented limits the ease at which models can be applied outside their development localities. This is because where one model is used operationally, the appropriate measurements for alternative models are rarely collected, necessitating unit conversion and approximation. The adoption of a more universal system would increase the applicability of fire models and research findings, foster collaboration and reduce research duplication by allowing findings to be generalised across regions [35,68,122].

While there is a great diversity of ecosystems prone to wildland fire worldwide, the fundamental processes behind combustion and fire propagation are common to all. As a result, fuel quantification systems that have a basis in fundamental fire properties will have a degree of universality by default. The adoption of a hierarchical system could provide for abstraction while allowing for base level fuel attributes to be reconstituted [25,123]. Such a hierarchy could be considered in terms of:

- Primary attributes; those that can be directly linked to fire behaviour (e.g., fuel element dimensions, chemistry, moisture content and spatial configuration);
- Secondary attributes; those that can measured in the field but require transformation to be linked to the primary attributes (e.g., plant species may be used as a proxy for element chemical composition);
- Tertiary attributes; those that summarise primary and secondary attributes (e.g., vegetation type may be used to describe the likely properties at a site) and can be used for mapping;
- Accessory attributes; those that are not directly related to fuel, but are important for understanding processes, such as species composition, site age and soil properties.

Due to the diversity in vegetation community properties worldwide, the development of a practical and functional system is a great challenge. However, by considering primary attributes as directly as possible and ensuring that any secondary attributes can be readily transformed into primary attributes, a basis for commonality can be maintained. A sample of measurable secondary fire behaviour attributes, their related primary attributes, and their effect on fire behaviour is presented in Table 2. One thing that is immediately evident from this table is the complexity of the problem—each secondary attribute may influence multiple primary attributes.

Increasing detail in the parameterisation of fuel is likely to exacerbate the issue where the standard site level classifications currently used for mapping are too coarse to represent the known variation between components of the fuel bed. It is regressive to discard detailed information (such as from LiDAR) to constrain fuel information to a fixed classification. An alternative could be to treat fuel attributes as independent continuous variables. While separate maps of each fuel parameter of interest may cause difficulties in human interpretation, simulation models should be able to process the values directly.

Table 2. Some commonly measured fuel attributes that are assessed at a site level (secondary attributes), the associated (primary) attributes of these that affect fire behaviour, and the fundamental fire behaviour processes they influence [16,34,50,77]. Processes may be associated with more than one primary attribute.

Secondary Attributes	Primary Attributes	Associated Fire Behaviour Processes *
Fuel element geometry	Size Shape Surface area to volume ratio	Heat transfer (including cooling) Ignitability Residence time
Fuel type (species) and condition	Stratum particle density Stratum bulk density Stratum packing ratio Species composition Moisture content Fuel availability Chemistry (Fats, Salts, Ash content, Carbohydrates, Sugars and other extractives) Proportion dead Decomposition state	Ignitability Energy balance Air: fuel mixture Reaction chemistry Heat transfer H_2O Latent heat absorption Combustible air: fuel mixture Heat conductivity Residence time Combustion efficiency Smoke production Proportion of fuel remaining unburnt
Horizontal continuity fuel continuity	Distance between fuel elements Distance between fuel clumps	Connectivity/sustainability thresholds (i.e., wind and flame properties) Heat transfer efficiency Combustible air: fuel mixture
Mass and location of fuel in different strata	Fuel element spatial configuration Stratum particle density Stratum bulk density Stratum packing ratio Wind adjustment factor Wind profile and turbulence Overall fuel load	Flame height/depth Energy output Ignitability Preheating of fuel Residence time Spread rate
Firebrand potential	Mass of loose material Nature of loose material Location of loose material	Number of viable embers produced Aerodynamic properties of embers Likelihood of lofting Sustainability of embers

Ideally, fuel quantification would be purely directed by fundamentals; however, areas of ambiguity remain as fire science is not settled. There is not yet a fundamental framework describing the process of wildfire spread [124], and there are clear challenges in transferring the concepts of flammability from the laboratory to landscape scales, as fire is more complex than a spreading flame front [125–128]. For example, the different dimensions of flammability (for example, ignitability and combustibility) take on different meanings at different scales, each of which may require particular fuel information in order to be understood [126]. Other processes, such as the spread of fire through spotting (considered in Australian fire models due to the nature of Eucalyptus bark) incorporate firebrand generation, transport and spot fire ignition [129]—this cannot be replicated in totality in a laboratory. Despite these issues, there are a number of attributes that are already currently common components of fire models including fuel element size, amount, spatial distribution and status (live or dead) that are already quantified and mapped in various forms. A review of these would be a potential starting point for considering a more universal system.

The adoption of a new set of universal model parameters would require unit conversion for the majority of existing fire models. Ideally, models would be updated to process primary attributes without the use of intermediate units—or alternatively, novel models could be developed to supersede the current ones. It is unlikely, due to the complexity of natural systems and the vastly different scales of processes (i.e., from molecular decomposition to terrain wind channelling), that any single model (or fuel quantification system) will meet all needs at all scales. However, in principle, a universal fuel quantification system could support the development of a universally applicable fire model. There are substantial benefits that could be realised from this—in particular, increased leverage of research and development, and greater availability of wildfire data for testing.

3.2. Cooperative Development

Many parts of the world subject to wildfire are likely to have fuel quantification systems currently in place based on contemporary fire models, as evidenced by the Canadian and US field assessment systems [130,131]. As moving to a new system would require investment, a compelling case needs to be made as to what the benefits would be. These are likely to include:

- The ability to share research and apply models developed elsewhere;
- The ability to adopt new systems as science progresses;
- The ability to combine fire behaviour and fire effects systems.

Furthermore, increasing the breadth and applicability of fuel information has the potential to increase efficiency and reduce costs by avoiding duplication between localities and providing for research leverage. This is particularly important when considering the research of rare events, such as extreme fire behaviour, where small sample sizes are an issue.

Any move towards universality in fuel quantification systems would require the cooperation of a broad range of users in multiple jurisdictions to ensure all needs are considered. Unless a system is able to meet the majority of needs of potential users, there is the risk of merely introducing an additional competing system [132]. Ideally, such a system would proceed as part of broader fire management information sharing agreements, allowing ecological, fire behaviour and operational data to be pooled internationally [133]. Such a process would require consensus on how to quantify various attributes, data formats, minimum levels of precision and accuracy, and units of measurement to allow interoperability between jurisdictions. Open ended standards have the benefit over set specifications of allowing higher quality information to be integrated where available so they do not impede improvement as technology advances. For example, this issue is already apparent with recent developments in remote sensing—we are beginning to have more detailed data (e.g., describing the nature of ladder fuels to the canopy using LiDAR [134]) than existing fire models can utilise. The operational fire simulation models discussed in this paper (FARSITE, PHOENIX RapidFire and Prometheus) are all based on point rate-of-spread models that were developed in the previous century [57,59,135], and so are not able to directly utilise more detailed information as it becomes available. These models were constrained by the processing and informational limitations at the time. Ideally, as improved fuel information becomes available, so too does the potential to develop new fire behaviour models that can process such data directly.

There is precedence for multijurisdictional cooperative development in fire sciences—for example, within Europe, the Paradox project [136] and within the US the Joint Fire Science Program [137]. There are also examples of multidisciplinary approaches to model development—for example, the FIREX climate and air study [138]. Ideally, such programs could be used to provide a framework for developing a broader framework for unifying approaches in localities with wildfire problems worldwide.

While it would be expected that the initial focus would be on the subset of attributes currently being used for fire models, it would be ideal to agree on protocols for as broad a set of attributes as possible. Such an attribute set would provide for the development of new, improved models, would allow integration with other ecological modelling systems and would allow broader uses of the data such as the analysis of ecological processes and spatial patterning in three dimensions [123]. An enduring challenge with the development of such a system is that there are multiple needs that require the quantification of fuels, in particular:

- The need for quantifying the fundamental properties of fuel that contribute to fire behaviour;

- The need for estimating fire effects such as smoke, carbon loss or watershed impacts;
- The need to have methods for evaluating fuel hazard and model verification in the field; and
- The need for understanding how fuel properties relate to vegetation, climate, and environmental variation.

These needs have different requirements (Table 3) and the levels of detail required for each are not the same. For example, simplicity and efficiency are priorities when conducting field fuel hazard assessments; however, the data collected are unlikely to have suitable resolution, accuracy or precision for developing landscape fuel dynamics models. Currently, no system is available that is suited to all phases of fire management [41]. Due to the diversity of fire prone ecosystems worldwide, the assessment of secondary and tertiary attributes may require different assessment methods and no 'one-size-fits-all' approach is likely to be feasible for all uses. A fundamental fire basis for fuel quantification will greatly help understand *what* the current conditions are. To understand *how* and *why* they will change, we need to continue to develop our understanding of the ecological processes behind fuel development.

Table 3. Uses of fuel quantifications and key features required to fulfil desired use.

Use of Fuel Quantification	Features Required for Efficacy
Field identification of fuel hazard	Limited number of classes to select from Potential for rapid assessment with limited expertise Distinctive classes that can be field identified Ability to provide dichotomous keys
Modelling of fire behaviour	Element moisture content Element arrangement (vertically and horizontally) Element dimensions Element load (in relation to spatial arrangement) Element chemical composition Element bulk density
Modelling of fire effects	Fuel element fundamental properties (as above) Expected fire/fuel interaction (fire behaviour outputs) Fuel/impact relationships (e.g., fuel type/sediment flow) Properties of less flammable components (e.g., duff, logs)
Spatio-temporal fuel/vegetation models	
Spatial information	Species abundances and properties Community dynamics (co-occurring species, dominance other interactions) Species—fuel relationships Seasonal variation
Temporal information	Fuel condition (e.g., current status) Live: dead ratio or curing properties Life cycle properties Fire responses
Accessory attributes	Disturbance history (e.g., landuse, fire) Biophysical attributes (e.g., soil, climate)

3.3. Rethinking Fuel–Fuel as an Ecological Entity

While fuels can be parameterised solely in terms of their potential contribution to fire behaviour, in order to understand their properties through time, it is important to also recognise that they are biological products that are a product of complex and dynamic processes [3,27,123]. To date, there has been a tendency to consider fuel separately from the vegetation it is derived from; however, to be truly understood, the biotic nature of fuel needs to be taken into consideration. Importantly, what is thought of as 'fuel' by land managers is, in essence, potential fuel—it only acts as fuel when it is involved with combustion; otherwise, it is vegetable matter. At broad scales, the occurrence of wildfires is dependent on a suitable combination of climate, weather, vegetation and ignitions [139–141]. Furthermore, climate is a key driver of the composition of plant species at a particular location (combined with other environmental tolerances, competition and disturbance [142]). With a changing climate, range shifting species and communities have the potential to alter fuel properties at a landscape level, resulting in changes in the relative distribution of fuel hazard through space and time by altering flammability [33,126,143]. Additionally, altered fire regimes driven by increased fire weather have the potential to cause abrupt shifts in vegetation communities, potentially resulting in rapid changes [39,144,145]. Even within communities, changing abundances of individual species may result in changes to flammability at the landscape scale [28,146,147]. The ecological aspects of wildland fuels are also strongly evident in the way fuel recovers after fire or other disturbances. The rate of vegetation recovery and the composition of a community is a function of the weather conditions before, during and after a fire—weather affects both the severity of a fire and resources available for growth [27,30,32,101]. The severity of a fire could also be considered in terms of the fuels that do not burn in a fire—understanding the availability of the lesser flammable fuels (logs, duff, soil etc.) to burn under particular conditions is important for predicting how a system recovers after fire in terms of fuel and important ecosystem services (carbon storage, faunal habitat, water quality). Other non-fuel properties of vegetation communities can also influence short-term fuel dynamics, for example, the overstorey of a forest plays a role in defining the understorey microclimate, influencing the water available for both plant growth and fuel moisture dynamics [148,149]. In the face of changing climates, understanding the interactions between plant ecology, fuel properties and fire regimes [150–153] will be critical for understanding future fire. A focus on processes can provide insight into fuel properties as they exist today and provide an indication of what may change with different forms of disturbance [145,153,154] or changing environmental conditions [155,156].

Due to ecosystem complexity, finding the best way to incorporate ecological processes and fuel quantification methods is likely to remain an enduring challenge. To begin to understand such relationships, the first step would be to begin to consider fuel data collection in a holistic manner and ensure that information about ecosystem properties are collected in conjunction with fuel surveys (for example, including assessing species abundances, their structural roles and site properties under which they occur). While such information may not add immediate value to a survey intended to provide a snapshot of the current fuel status, ultimately, consideration of ecosystem processes (i.e., looking at fuel types and components through an ecological lens) can both assist in the development of more appropriate and accurate sampling techniques and support the development of dynamic fuel models that improve estimates of fuel properties through time [41].

4. Conclusions

There is currently a wide variety of practices used in measuring wildland fuels worldwide. This has resulted in challenges in applying research findings and models outside of their development regions, limiting collaboration and resulting in duplicated efforts. Methods could potentially be focused in a hierarchical manner using the universal fundamental physical processes of wildfire behaviour as a basis. Additionally, it remains important to appreciate that fuel is of biotic origins—while it can be described in terms of fundamental fire properties, it can only be understood by ensuring that the complex biological processes are also recognised.

The movement towards a more universal approach to fuel quantification would require a deliberate concerted effort from many parties. A new system would be disruptive to many existing management systems; however, the benefits could be expected to be substantial. There have been regional scale multijurisdictional and multidisciplinary programs in fire science—the challenge now is to gain support for such an approach internationally.

Acknowledgments: This research was partially funded by a grant by the Department of Environment, Land Water and Planning, Victoria, Australia as part of the integrated Forest Ecosystem Research project (iFER). We gratefully thank Alen Slijepcevic and the contribution of our anonymous reviewers to this document. A fellowship from the Churchill Trust funded by the Lord Mayor of Sydney also contributed to this work.

Author Contributions: Thomas J. Duff and Kevin G. Tolhurst conceived the manuscript. Thomas J. Duff was responsible for writing with assistance from Robert E. Keane; Robert E. Keane, Kevin G. Tolhurst and Trent D. Penman contributed additional material and participated in the drafting and review process.

Conflicts of Interest: The authors declare no conflict of interest. The founding sponsors had no role in the design of the study; in the collection, analyses, or interpretation of data; in the writing of the manuscript, and in the decision to publish the results.

References

1. Byram, G.M. Combustion of forest fuels. In *Forest Fire: Control and Use*; Davis, K.P., Ed.; McGraw Hill Book Company Inc.: New York, NY, USA, 1959; pp. 61–89.

2. Fuller, M. *Forest Fires: An Introduction to Wildland Fire Behaviour, Management, Firefighting and Prevention*; John Wiley & Sons, Inc.: Hoboken, NJ, USA, 1991.

3. Keane, R.E. *Wildland Fuel Fundamentals and Applications*; Springer: New York, NY, USA, 2015; p. 183.

4. Mason, C.L.; Lippke, B.R.; Zobrist, K.W.; Bloxton, T.D.; Ceder, K.R.; Comnick, J.M.; McCarter, J.B.; Rogers, H.K. Investments in fuel removals to avoid forest fires result in substantial benefits. *J. For.* **2006**, *104*, 27–31.

5. Gorte, J.K.; Gorte, R.W. *Application of Economic Techniques to Fire Management—A Status Review and Evaluation*; Forest Service, U.S. Department of Agriculture: Ogden, UT, USA, 1979.

6. Weise, D.R.; Wright, C.S. Wildland fire emissions, carbon and climate: Characterizing wildland fuels. *For. Ecol. Manag.* **2014**, *317*, 26–40. [CrossRef]

7. Blanchi, R.; Leonard, J.; Haynes, K.; Opie, K.; James, M.; de Oliveira, F.D. Environmental circumstances surrounding bushfire fatalities in Australia 1901–2011. *Environ. Sci. Policy* **2014**, *37*, 192–203. [CrossRef]

8. Bradstock, R.A.; Cary, G.J.; Davies, I.; Lindenmayer, D.B.; Price, O.F.; Williams, R.J. Wildfires, fuel treatment and risk mitigation in Australian eucalypt forests: Insights from landscape-scale simulation. *J. Environ. Manag.* **2012**, *105*, 66–75. [CrossRef] [PubMed]

9. Vaillant, N.M.; Fites-Kaufman, J.A.; Stephens, S.L. Effectiveness of prescribed fire as a fuel treatment in Californian coniferous forests. *Int. J. Wildland Fire* **2009**, *18*, 165–175. [CrossRef]

10. Fernandes, P.M.; Botelho, H.S. A review of prescribed burning effectiveness in fire hazard reduction. *Int. J. Wildland Fire* **2003**, *12*, 117–128. [CrossRef]

11. Thompson, M.P.; Vaillant, N.M.; Haas, J.R.; Gebert, K.M.; Stockmann, K.D. Quantifying the potential impacts of fuel treatments on wildfire suppression costs. *J. For.* **2013**, *111*, 49–58. [CrossRef]

12. McArthur, A.G. *Fire Behaviour in Eucalypt Forests*; Forestry and Timber Bureau; Athur, A.J., Eds.; Commonwealth Government Printer: Canberra, Australia, 1967.

13. Penman, T.D.; Collins, L.; Price, O.F.; Bradstock, R.A.; Metcalf, S.; Chong, D.M.O. Examining the relative effects of fire weather, suppression and fuel treatment on fire behaviour—A simulation study. *J. Environ. Manag.* **2013**, *131*, 325–333. [CrossRef] [PubMed]

14. Gorte, R.W. *The Rising Cost of Wildfire Protection*; Headwaters Economics: Bozeman, MT, USA, 2013.

15. Gould, J.S.; McCaw, L.; Cheney, N.P.; Ellis, P.; Matthews, S. *Project Vesta: Fire in Dry Eucalypt Forest: Fuel Structure, Fuel Dynamics and Fire Behaviour*; Ensis-CSIRO, Canberra, Australian Capital Territory, and WA Department of Environment and Conservation: Perth, Australia, 2007.

16. Rothermel, R.C. *A Mathematical Model for Predicting Fire Spread in Wildland Fuels*; Forest Service, U.S. Department of Agriculture: Ogden, UT, USA, 1972.

17. Sullivan, A.L. Wildland surface fire spread modelling, 1990–2007. 3: Simulation and mathematical analogue models. *Int. J. Wildland Fire* **2009**, *18*, 387–403. [CrossRef]

18. Matthews, S. Dead fuel moisture research: 1991–2012. *Int. J. Wildland Fire* **2013**, *23*, 78–92. [CrossRef]

19. Viney, N. A review of fine fuel moisture modelling. *Int. J. Wildland Fire* **1991**, *1*, 215–234. [CrossRef]

20. Morvan, D. Numerical study of the effect of fuel moisture content (FMC) upon the propagation of a surface fire on a flat terrain. *Fire Saf. J.* **2013**, *58*, 121–131. [CrossRef]

21. Schunk, C.; Wastl, C.; Leuchner, M.; Menzel, A. Fine fuel moisture for site- and species-specific fire danger assessment in comparison to fire danger indices. *Agric. For. Meteorol.* **2017**, *234*, 31–47. [CrossRef]

22. Rossa, C.G. The effect of fuel moisture content on the spread rate of forest fires in the absence of wind or slope. *Int. J. Wildland Fire* **2017**, *26*, 24–31. [CrossRef]
23. Keane, R.E.; Burgan, R.E.; van Wagtendonk, J. Mapping wildland fuels for fire management across multiple scales: Integrating remote sensing, GIS and biophysical modelling. *Int. J. Wildland Fire* **2001**, *10*, 301–319. [CrossRef]
24. Loveland, T.R. Toward a national fuels mapping strategy: Lessons from selected mapping programs. *Int. J. Wildland Fire* **2001**, *10*, 289–299. [CrossRef]
25. Keane, R.E.; Gray, K.; Bacciu, V. *Spatial Variability of Wildland Fuel Characteristics in Northern Rocky Mountain Ecosystems*; Rocky Mountain Reseach Station, Forest Service, U.S. Department of Agriculture: Missoula, MT, USA, 2012.
26. Rollins, M.G.; Keane, R.E.; Parsons, R.A. Mapping fuels and fire regimes using remote sensing, ecosystem simulation, and gradient modeling. *Ecol. Appl.* **2004**, *14*, 75–95. [CrossRef]
27. Duff, T.J.; Bell, T.L.; York, A. Predicting continuous variation in forest fuel load using biophysical models: A case study in south-eastern Australia. *Int. J. Wildland Fire* **2012**, *22*, 318–332. [CrossRef]
28. Rossiter, N.A.; Setterfield, S.A.; Douglas, M.M.; Hutley, L.B. Testing the grass-fire cycle: Alien grass invasion in the tropical savannas of northern Australia. *Divers. Distrib.* **2003**, *9*, 169–176. [CrossRef]
29. Baeza, M.; Raventós, J.; Escarré, A.; Vallejo, V. Fire risk and vegetation structural dynamics in Mediterranean shrubland. *Plant Ecol.* **2006**, *187*, 189–201. [CrossRef]
30. Penman, T.D.; York, A. Climate and recent fire history affect fuel loads in *Eucalyptus* forests: Implications for fire management in a changing climate. *For. Ecol. Manag.* **2010**, *260*, 1791–1797. [CrossRef]
31. Montenegro, G.; Ginocchio, R.; Segura, A.; Keeley, J.E.; Gomez, M. Fire regimes and vegetation responses in two Mediterranean-climate regions. *Rev. Chil. Hist. Nat.* **2004**, *77*, 455–464. [CrossRef]
32. Zhang, C.; Tian, H.; Wang, Y.; Zeng, T.; Liu, Y. Predicting response of fuel load to future changes in climate and atmospheric composition in the Southern United States. *For. Ecol. Manag.* **2010**, *260*, 556–564. [CrossRef]
33. Pausas, J.G.; Paula, S. Fuel shapes the fire-climate relationship: Evidence from Mediterranean ecosystems. *Glob. Ecol. Biogeogr.* **2012**, *21*, 1074–1082. [CrossRef]
34. Sullivan, A.L. Inside the Inferno: Fundamental Processes of Wildland Fire Behaviour. Part 1: Combustion chemistry and heat release. *Curr. For. Rep.* **2017**, *3*, 132–149. [CrossRef]
35. Hollis, J.J.; Gould, J.; Cruz, M.G.; Doherty, M.D. *Scope and Framework for an Australian Fuel Classification*; Featherstone, G., Ed.; Australasian Fire and Emergency Services Council (AFAC) and the Commowealth Science and Industrial Research Organisation (CSIRO): East Melbourne, Australia, 2011.
36. Riccardi, C.L.; Ottmar, R.D.; Sandberg, D.V.; Andreu, A.; Elman, E.; Kopper, K.; Long, J. The fuelbed: A key element of the Fuel Characteristic Classification System. *Can. J. For. Res.* **2007**, *37*, 2394–2412. [CrossRef]
37. Haslem, A.; Kelly, L.T.; Nimmo, D.G.; Watson, S.J.; Kenny, S.A.; Taylor, R.S.; Avitabile, S.C.; Callister, K.E.; Spence-Bailey, L.M.; Clarke, M.F.; et al. Habitat or fuel? Implications of long-term, post-fire dynamics for the development of key resources for fauna and fire. *J. Appl. Ecol.* **2011**, *48*, 247–256. [CrossRef]
38. Duff, T.J.; Bell, T.L.; York, A. Managing multiple species or communities? Considering variation in plant species abundances in response to fire interval, frequency and time since fire in a heathy Eucalyptus woodland. *For. Ecol. Manag.* **2013**, *289*, 393–403. [CrossRef]
39. Bowman, D.M.J.S.; Murphy, B.P.; Neyland, D.L.J.; Williamson, G.J.; Prior, L.D. Abrupt fire regime change may cause landscape-wide loss of mature obligate seeder forests. *Glob. Chang. Biol.* **2014**, *20*, 1008–1015. [CrossRef] [PubMed]
40. Cary, G.J.; Keane, R.E.; Gardner, R.H.; Lavorel, S.; Flannigan, M.D.; Davies, I.; Li, C.; Lenihan, J.M.; Mouillot, F. Comparison of the sensitivity of landscape-fire-succession models to variation in terrain, fuel pattern, climate and weather. *Landsc. Ecol.* **2006**, *21*, 121–137. [CrossRef]
41. Keane, R.E. Describing wildland surface fuel loading for fire management: A review of approaches, methods and systems. *Int. J. Wildland Fire* **2013**, *22*, 51–62. [CrossRef]
42. Bachmann, A.; Allgower, B. Uncertainty propagation in wildland fire behaviour modelling. *Int. J. Geogr. Inf. Sci.* **2002**, *16*, 115–127. [CrossRef]
43. King, K.J.; Bradstock, R.A.; Cary, G.J.; Chapman, J.; Marsden-Smedley, J.B. The relative importance of fine-scale fuel mosaics on reducing fire risk in south-west Tasmania, Australia. *Int. J. Wildland Fire* **2008**, *17*, 421–430. [CrossRef]

44. Loudermilk, E.L.; O'Brien, J.J.; Mitchell, R.J.; Cropper, W.P.; Hiers, J.K.; Grunwald, S.; Grego, J.; Fernandez-Diaz, J.C. Linking complex forest fuel structure and fire behaviour at fine scales. *Int. J. Wildland Fire* **2012**, *21*, 882–893. [CrossRef]

45. Thaxton, J.M.; Platt, W.J. Small-scale fuel variation alters fire intensity and shrub abundance in a pine savanna. *Ecology* **2006**, *87*, 1331–1337. [CrossRef]

46. Hornby, L.G. *Fire Control Planning in the Northern Rocky Mountain Region; Progress Report No. 1*; Northern Rocky Mountain Forest and Range Experiment Station, Forest Service, U.S. Department of Agriculture: Ogden, UT, USA, 1936.

47. Cochrane, G.R. Vegetation Studies in Forest-fire Areas of the Mount Lofty Ranges, South Australia. *Ecology* **1963**, *44*, 41–52. [CrossRef]

48. Sullivan, A.L. Wildland surface fire spread modelling, 1990–2007. 2: Empirical and quasi-empirical models. *Int. J. Wildland Fire* **2009**, *18*, 369–386. [CrossRef]

49. Sullivan, A.L. Wildland surface fire spread modelling, 1990–2007. 1: Physical and quasi-physical models. *Int. J. Wildland Fire* **2009**, *18*, 349–368. [CrossRef]

50. Sullivan, A.L. Inside the Inferno: Fundamental Processes of Wildland Fire Behaviour. Part 2: Heat transfer and interactions. *Curr. For. Rep.* **2017**, *3*, 150–171. [CrossRef]

51. Fire Danger Group. *Development and Structure of the Canadian Forest Fire Behavior System*; Forestry Canada Science and Sustainable Development Directorate: Ottawa, ON, Canada, 1992.

52. Cruz, M.G.; Gould, J.; Alexander, M.E.; Sullivan, A.L.; McCaw, L.; Matthews, S. *A Guide to Rate of Fire Spread Models for Australian Vegetation*; Australasian Fire and Emergency Service Authorities Council Ltd.; Commonwealth Scientifc and Industrial Research Organisation: East Melbourne, Ausralia, 2015.

53. Gould, J.S.; McCaw, L.W.; Cheney, P.N. Quantifying fine fuel dynamics and structure in dry eucalypt forest (*Eucalyptus marginata*) in Western Australia for fire management. *For. Ecol. Manag.* **2011**, *262*, 531–546. [CrossRef]

54. Andrews, P.L. Methods for predicting fire behavior-you do have a choice. *Fire Manag. Notes* **1986**, *47*, 6–10.

55. Cheney, N.P. Predicting fire behaviour with fire danger tables. *Aust. For.* **1968**, *32*, 71–79. [CrossRef]

56. Rothermel, R.C. *How to Predict the Spread and Intensity of Forest and Range Fires*; Forest Service, U.S. Department of Agriculture: Boise, ID, USA, 1983.

57. Finney, M.A. *FARSITE: Fire Area Simulator—Model Development and Evaluation*; Rocky Mountain Reseach Station, Forest Service, U.S. Department of Agriculture: Missoula, MT, USA, 2004.

58. Tolhurst, K.G.; Shields, B.; Chong, D. PHOENIX: Development and application of a bushfire risk management tool. *Aust. J. Emerg. Manag.* **2008**, *23*, 47–54.

59. Paterson, G.; Chong, D. Implementing the Phoenix fire spread model for operational use. In Proceedings of the Surveying and Spatial Sciences Biennial Conference, Wellington, New Zealand, 21–25 November 2011; New Zealand Institute of Surveyors and the Surveying and Spatial Sciences Institute: Wellington, New Zealand.

60. Penman, T.D.; Bradstock, R.A.; Price, O.F. Reducing wildfire risk to urban developments: Simulation of cost-effective fuel treatment solutions in south eastern Australia. *Environ. Model. Softw.* **2014**, *52*, 166–175. [CrossRef]

61. Ager, A.A.; Vaillant, N.M.; Finney, M.A. Integrating fire behavior models and geospatial analysis for wildland fire risk assessment and fuel management planning. *J. Combust.* **2011**, *2011*. [CrossRef]

62. Pausas, J. Simulating Mediterranean landscape pattern and vegetation dynamics under different fire regimes. *Plant Ecol.* **2006**, *187*, 249–259. [CrossRef]

63. He, H.S.; Shang, B.Z.; Crow, T.R.; Gustafson, E.J.; Shifley, S.R. Simulating forest fuel and fire risk dynamics across landscapes–LANDIS fuel module design. *Ecol. Model.* **2004**, *180*, 135–151. [CrossRef]

64. Coen, J.L.; Cameron, M.; Michalakes, J.; Patton, E.G.; Riggan, P.J.; Yedinak, K.M. WRF-Fire: Coupled weather-wildland fire modeling with the weather research and forecasting model. *J. Appl. Meteorol. Climatol.* **2013**, *52*, 16–38. [CrossRef]

65. Linn, R.; Reisner, J.; Colman, J.J.; Winterkamp, J. Studying wildfire behavior using FIRETEC. *Int. J. Wildland Fire* **2002**, *11*, 233–246. [CrossRef]

66. Morvan, D.; Hoffman, C.; Rego, F.; Mell, W. Numerical simulation of the interaction between two fire fronts in grassland and shrubland. *Fire Saf. J.* **2011**, *46*, 469–479. [CrossRef]

67. Ottmar, R.D.; Sandberg, D.V.; Riccardi, C.L.; Prichard, S.J. An overview of the Fuel Characteristic Classification System–Quantifying, classifying, and creating fuelbeds for resource planning. *Can. J. For. Res.* **2007**, *37*, 2383–2393. [CrossRef]

68. Sandberg, D.V.; Ottmar, R.D.; Cushon, G.H. Characterizing fuels in the 21st century. *Int. J. Wildland Fire* **2001**, *10*, 381–387. [CrossRef]

69. Gould, J.S.; McCaw, W.L.; Cheney, N.P.; Ellis, P.F.; Matthews, S. *Field Guide–Fuel Assessment and Fire Behaviour Prediction in Dry Eucalypt Forest*; Ensis-CSIRO, Canberra, Australian Capital Territory, and WA Department of Environment and Conservation: Perth, Australia, 2007.

70. Country Fire Authority. *Grassland Curing Guide*; Country Fire Authority: Burwood East, Australia, 2015.

71. Riccardi, C.L.; Prichard, S.J.; Sandberg, D.V.; Ottmar, R.D. Quantifying physical characteristics of wildland fuels using the Fuel Characteristic Classification System. *Can. J. For. Res.* **2007**, *37*, 2413–2420. [CrossRef]

72. Anderson, W.R.; Cruz, M.G.; Fernandes, P.M.; McCaw, L.; Vega, J.A.; Bradstock, R.A.; Fogarty, L.; Gould, J.; McCarthy, G.; Marsden-Smedley, J.B.; et al. A generic, empirical-based model for predicting rate of fire spread in shrublands. *Int. J. Wildland Fire* **2015**, *24*, 443–460. [CrossRef]

73. Marsden-Smedley, J.B.; Catchpole, W.R. Fire behaviour modelling in Tasmanian buttongrass moorlands. I. fuel characteristics. *Int. J. Wildland Fire* **1995**, *5*, 203–214. [CrossRef]

74. Cheney, N.P.; Gould, J.S.; Catchpole, W.R. Prediction of fire spread in grasslands. *Int. J. Wildland Fire* **1998**, *8*, 1–13. [CrossRef]

75. Cheney, N.P.; Sullivan, A.L. *Grassfires: Fuel, Weather and Fire Behaviour*; CSIRO Publishing: Collingwood, Australia, 1997.

76. Cruz, M.G.; McCaw, W.L.; Anderson, W.R.; Gould, J.S. Fire behaviour modelling in semi-arid mallee-heath shrublands of southern Australia. *Environ. Model. Softw.* **2012**, *40*. [CrossRef]

77. Hines, F.; Tolhurst, K.G.; Wilson, A.G.; McCarthy, G.J. *Overall Fuel Hazard Assessment Guide*, 4th ed.; Department of Sustainability and Environment Victoria: Melbourne, Australia, 2010.

78. Bonham, C.D. *Measurements for Terrestrial Vegetation*; John Wiley & Sons: New York, NY, USA, 1989.

79. Benson, J.S. Sampling, strategies and costs of regional vegetation mapping. *Globe* **1995**, *43*, 18–28.

80. Fischer, W.C. *Photo Guide for Appraising Downed Woody Fuels in Montana Forests: Interior Ponderosa Pine, Ponderosa Pine-Larch-Douglas-Fir, Larch-Douglas-Fir, and Interior Douglas-Fir Cover Types*; Forest Service, U.S. Department of Agriculture: Ogden, UT, USA, 1981.

81. Sikkink, P.G.; Keane, R.E. A comparison of five sampling techniques to estimate surface fuel loading in montane forests. *Int. J. Wildland Fire* **2008**, *17*, 363–379. [CrossRef]

82. Gopal, S.; Woodcock, C.E. Theory and methods for accuracy assessment of thematic maps using fuzzy sets. *Photogramm. Eng. Remote Sens.* **1994**, *60*, 182–188.

83. Gosper, C.R.; Yates, C.J.; Prober, S.M.; Wiehl, G. Application and validation of visual fuel hazard assessments in dry Mediterranean-climate woodlands. *Int. J. Wildland Fire* **2014**, *23*, 385–393. [CrossRef]

84. Watson, P.J.; Penman, S.H.; Bradstock, R.A. A comparison of bushfire fuel hazard assessors and assessment methods in dry sclerophyll forest near Sydney, Australia. *Int. J. Wildland Fire* **2012**, *21*, 755–763. [CrossRef]

85. Reich, R.M.; Lundquist, J.E.; Bravo, V.A. Spatial models for estimating fuel loads in the Black Hills, South Dakota, USA. *Int. J. Wildland Fire* **2004**, *13*, 119–129. [CrossRef]

86. Rowell, E.M.; Seielstad, C.A.; Ottmar, R.D. Development and validation of fuel height models for terrestrial lidar–RxCADRE 2012. *Int. J. Wildland Fire* **2016**, *25*, 38–47. [CrossRef]

87. Loudermilk, E.L.; Hiers, J.K.; O'Brien, J.J.; Mitchell, R.J.; Singhania, A.; Fernandez, J.C.; Cropper, W.P.; Slatton, K.C. Ground-based LIDAR: A novel approach to quantify fine-scale fuelbed characteristics. *Int. J. Wildland Fire* **2009**, *18*, 676–685. [CrossRef]

88. Korpela, I.; Tuomola, T.; Välimäki, E. Mapping forest plots: An efficient method combining photogrammetry and field triangulation. *Silva Fenn.* **2007**, *41*, 457–469. [CrossRef]

89. Clark, R.E.; Hope, A.S.; Tarntola, S.; Gatelli, D.; Dennison, P.E.; Moritz, M.A. Sensitivity analysis of a fire spread model in a chaparral landscape. *Fire Ecol.* **2008**, *4*, 1–13. [CrossRef]

90. Benali, A.; Ervilha, A.R.; Sá, A.C.L.; Fernandes, P.M.; Pinto, R.M.S.; Trigo, R.M.; Pereira, J.M.C. Deciphering the impact of uncertainty on the accuracy of large wildfire spread simulations. *Sci. Total Environ.* **2016**, *569*, 73–85. [CrossRef] [PubMed]

91. Finney, M.A.; Cohen, J.D.; McAllister, S.S.; Jolly, W.M. On the need for a theory of wildland fire spread. *Int. J. Wildland Fire* **2013**, *22*, 25–36. [CrossRef]

92. Anderson, H.E. *Aids to Determining Fuel Models for Fire Behavior*; Forest Service, U.S. Department of Agriculture: Ogden, UT, USA, 1982.
93. Woodcock, C.E.; Gopal, S. Fuzzy set theory and thematic maps: Accuracy assessment and area estimation. *Int. J. Geogr. Inf. Sci.* **2000**, *14*, 153–172. [CrossRef]
94. Austin, M.P.; Gaywood, M.J. Current problems of environmental gradients and species response curves in relation to continuum theory. *J. Veg. Sci.* **1994**, *5*, 473–482. [CrossRef]
95. Keane, R.E. Spatiotemporal variability of wildland fuels in US Northern Rocky Mountain forests. *Forests* **2016**, *7*, 129. [CrossRef]
96. Mell, W.E.; Manzello, S.L.; Maranghides, A.; Butry, D.; Rehm, R.G. The wildland–urban interface fire problem–Current approaches and research needs. *Int. J. Wildland Fire* **2010**, *19*, 238–251. [CrossRef]
97. Parresol, B.R.; Scott, J.H.; Andreu, A.; Prichard, S.; Kurth, L. Developing custom fire behavior fuel models from ecologically complex fuel structures for upper Atlantic Coastal Plain forests. *For. Ecol. Manag.* **2012**, *273*, 50–57. [CrossRef]
98. Tolhurst, K.G.; Kelly, N. *Effects of Repeated Low Intensity Fire on Fuel Dynamics of a Mixed Eucalypt Foothill Forest in South-Eastern Australia*; Forest Science Centre, University of Melbourne, Creswick: Melbourne, Australia, 2003.
99. Terrier, A.; Paquette, M.; Gauthier, S.; Girardin, P.M.; Pelletier-Bergeron, S.; Bergeron, Y. Influence of fuel load dynamics on carbon emission by wildfires in the clay belt boreal landscape. *Forests* **2017**, *8*, 9. [CrossRef]
100. Chiono, L.A.; O'Hara, K.L.; De Lasaux, M.J.; Nader, G.A.; Stephens, S.L. Development of vegetation and surface fuels following fire hazard reduction treatment. *Forests* **2012**, *3*, 700–722. [CrossRef]
101. Coppoletta, M.; Merriam, K.E.; Collins, B.M. Post-fire vegetation and fuel development influences fire severity patterns in reburns. *Ecol. Appl.* **2016**, *26*, 686–699. [CrossRef] [PubMed]
102. Ferster, J.C.; Eskelson, N.B.; Andison, W.D.; LeMay, M.V. Vegetation mortality within natural wildfire events in the Western Canadian boreal forest: What burns and why? *Forests* **2016**, *7*, 187. [CrossRef]
103. Keane, R.E.; Rollings, M.G.; McNicoll, C.H.; Parsons, R.A. *Integrating Ecosystem Sampling, Gradient Modelling, Remote Sensing and Ecosystem Simulation to Create Spatially Explicit Landscape Inventories*; U.S. Department of Agriculture, Forest Service, Rocky Mountain Research Station: Fort Collins, CO, USA, 2002.
104. Benson, D. Mapping vegetation. *Globe* **1995**, *41*, 40–44.
105. Ferrier, S. Mapping spatial pattern in biodiversity for regional conservation planning: Where to from here? *Syst. Biol.* **2002**, *51*, 331–363. [CrossRef] [PubMed]
106. Thomas, P.B.; Watson, P.J.; Bradstock, R.A.; Penman, T.D.; Price, O.F. Modelling surface fine fuel dynamics across climate gradients in eucalypt forests of south-eastern Australia. *Ecography* **2014**, *37*, 827–837. [CrossRef]
107. Arroyo, L.A.; Pascual, C.; Manzanera, J.A. Fire models and methods to map fuel types: The role of remote sensing. *For. Ecol. Manag.* **2008**, *256*, 1239–1252. [CrossRef]
108. Jakubowksi, M.K.; Guo, Q.; Collins, B.; Stephens, S.; Kelly, M. Predicting surface fuel models and fuel metrics using lidar and CIR imagery in a dense mountenous forest. *Photogramm. Eng. Remote Sens.* **2013**, *79*, 37–49. [CrossRef]
109. Mutlu, M.; Popescu, S.C.; Stripling, C.; Spencer, T. Mapping surface fuel models using lidar and multispectral data fusion for fire behavior. *Remote Sens. Environ.* **2008**, *112*, 274–285. [CrossRef]
110. Saatchi, S.; Halligan, K.; Despain, D.G.; Crabtree, R.L. Estimation of forest fuel load from radar remote sensing. *IEEE Trans. Geosci. Remote Sens.* **2007**, *45*, 1726–1740. [CrossRef]
111. Newnham, G.J.; Verbesselt, J.; Grant, I.F.; Anderson, S.A.J. Relative Greenness Index for assessing curing of grassland fuel. *Remote Sens. Environ.* **2011**, *115*, 1456–1463. [CrossRef]
112. Yebra, M.; Chuvieco, E.; Riaño, D. Estimation of live fuel moisture content from MODIS images for fire risk assessment. *Agric. For. Meteorol.* **2008**, *148*, 523–536. [CrossRef]
113. Danson, F.M.; Bowyer, P. Estimating live fuel moisture content from remotely sensed reflectance. *Remote Sens. Environ.* **2004**, *92*, 309–321. [CrossRef]
114. Chladil, M.A.; Nunez, M. Assessing grassland moisture and biomass in Tasmania—The application of remote-sensing and empirical-models for a cloudy environment. *Int. J. Wildland Fire* **1995**, *5*, 165–171. [CrossRef]
115. Hudak, A.T.; Dickinson, M.B.; Bright, B.C.; Kremens, R.L.; Loudermilk, E.L.; O'Brien, J.J.; Hornsby, B.S.; Ottmar, R.D. Measurements relating fire radiative energy density and surface fuel consumption–RxCADRE 2011 and 2012. *Int. J. Wildland Fire* **2016**, *25*, 25–37. [CrossRef]

116. Poulos, H.M. Mapping fuels in the Chihuahuan Desert borderlands using remote sensing, geographic information systems, and biophysical modeling. *Can. J. For. Res.* **2009**, *39*, 1917–1927. [CrossRef]
117. Fernandes, P.; Luz, A.; Loureiro, C.; Ferreira-Godinho, P.; Botelho, H. Fuel modelling and fire hazard assessment based on data from the Portuguese National Forest Inventory. *For. Ecol. Manag.* **2006**, *234*, S229. [CrossRef]
118. Fernandes, P.M. Combining forest structure data and fuel modelling to classify fire hazard in Portugal. *Ann. For. Sci.* **2009**, *66*, 415. [CrossRef]
119. García, M.; Chuvieco, E.; Nieto, H.; Aguado, I. Combining AVHRR and meteorological data for estimating live fuel moisture content. *Remote Sens. Environ.* **2008**, *112*, 3618–3627. [CrossRef]
120. Cechet, B.; French, I.A.; Kepert, J.D.; Tolhurst, K.G.; Meyer, M. *Fire Impact and Risk Evaluation*; Bushfire Cooperative Research Centre: Melbourne, Australia, 2013.
121. Keane, R.E.; Herynk, J.M.; Toney, C.; Urbanski, S.P.; Lutes, D.C.; Ottmar, R.D. Evaluating the performance and mapping of three fuel classification systems using Forest Inventory and Analysis surface fuel measurements. *For. Ecol. Manag.* **2013**, *305*, 248–263. [CrossRef]
122. McCaw, L.W. Measurement of fuel quantity and structure for bushfire research and management. In *Conference on Bushfire Modelling and Fire Danger Rating Systems*; Cheney, N.P., Gill, A.M., Eds.; CSIRO: Canberra, Australia, 1998; pp. 147–155.
123. Krivtsov, V.; Vigy, O.; Legg, C.; Curt, T.; Rigolot, E.; Lecomte, I.; Jappiot, M.; Lampin-Maillet, C.; Fernandes, P.; Pezzatti, G.B. Fuel modelling in terrestrial ecosystems: An overview in the context of the development of an object-orientated database for wild fire analysis. *Ecol. Model.* **2009**, *220*, 2915–2926. [CrossRef]
124. Finney, M.A.; Cohen, J.D.; Forthofer, J.M.; McAllister, S.S.; Gollner, M.J.; Gorham, D.J.; Saito, K.; Akafuah, N.K.; Adam, B.A.; English, J.D. Role of buoyant flame dynamics in wildfire spread. *Proc. Natl. Acad. Sci. USA* **2015**, *112*, 9833–9838. [CrossRef] [PubMed]
125. Pérez, Y.; Pastor, E.; Àgueda, A.; Planas, E. Effect of wind and slope when scaling the forest fires rate of spread of laboratory experiments. *Fire Technol.* **2011**, *47*, 475–489. [CrossRef]
126. Pausas, J.G.; Keeley, J.E.; Schwilk, D.W. Flammability as an ecological and evolutionary driver. *J. Ecol.* **2016**, *105*, 289–297. [CrossRef]
127. Gill, A.M.; Zylstra, P. Flammability of Australian forests. *Aust. For.* **2005**, *68*, 87–93. [CrossRef]
128. Fernandes, P.M.; Cruz, M.G. Plant flammability experiments offer limited insight into vegetation–Fire dynamics interactions. *New Phytol.* **2012**, *194*, 606. [CrossRef] [PubMed]
129. Koo, E.; Pagni, P.J.; Weise, D.R.; Woycheese, J.P. Firebrands and spotting ignition in large-scale fires. *Int. J. Wildland Fire* **2010**, *19*, 818–843. [CrossRef]
130. Scott, J.H.; Burgan, R.E. *Standard Fire Behavior Fuel Models: A Comprehensive set for Use With Rothermel's Fire Spread Model*; Forest Service, U.S. Department of Agriculture: Fort Collins, CO, USA, 2005.
131. Taylor, S.W.; Pike, R.G.; Alexander, M.E. *Field Guide to the Canadian Forest Fire Behaviour Prediction (FBP) System, FRDA Handbook 012*; Natural Resources Canada, Canadian Forest Service and the BC Ministry of Forests: Pacific Forestry Centre, Victoria, BC, Canada, 1996.
132. Monroe, R.P. Standards. *XKCD*. Available online: http://xkcd.com/927/ (accessed 24 August 2017).
133. Duff, T.J.; Chong, D.M.; Cirulis, B.A.; Walsh, S.F.; Penman, T.D.; Tolhurst, K.G. Gaining benefits from adversity: The need for systems and frameworks to maximise the data obtained from wildfires. In *Advances in Forest Fire Research*; Viegas, D.X., Ed.; Imprensa da Universidade de Coimbra: Coimbra, Portugal, 2014; pp. 766–774.
134. Kramer, A.H.; Collins, M.B.; Kelly, M.; Stephens, L.S. Quantifying ladder fuels: A new approach using LiDAR. *Forests* **2014**, *5*, 1432–1453. [CrossRef]
135. Tymstra, C.; Bryce, R.W.; Wotton, B.M.; Taylor, S.W.; Armitage, O.B. *Development and Structure of Prometheus: The Canadian Wildland Fire Growth Simulation Model*; Canadian Forest Service: Edmonton, AB, Canada, 2010.
136. Fernandes, P.M.; Rego, F.C.; Rigolot, E. The FIRE PARADOX project: Towards science-based fire management in Europe. *For. Ecol. Manag.* **2011**, *261*, 2177–2178. [CrossRef]
137. Clark, B. Congress Funds Joint Fire Science Program. *Fire Manag. Notes* **1998**, *58*, 29.
138. Warneke, C.; Roberts, J.M.; Schwarz, J.P.; Yokelson, R.J.; Pierce, B. *Fire Influence on Regional and Global Environments Experiment (FIREX) The Impact of Biomass Burning on Climate and Air Quality: An Intensive Study of Western North America Fires*; National Oceanic & Atmospheric Administration: Boulder, CO, USA, 2014.

139. Krawchuk, M.A.; Moritz, M.A. Constraints on global fire activity vary across a resource gradient. *Ecology* **2011**, *92*, 121–132. [CrossRef] [PubMed]
140. Parisien, M.A.; Moritz, M.A. Environmental controls on the distribution of wildfire at multiple spatial scales. *Ecol. Monogr.* **2009**, *79*, 127–154. [CrossRef]
141. Bradstock, R.A. A biogeographic model of fire regimes in Australia: Current and future implications. *Glob. Ecol. Biogeogr.* **2010**, *19*, 145–158. [CrossRef]
142. Austin, M.P.; Smith, T.M. A new model for the continuum concept. *Plant Ecol.* **1989**, *83*, 35–47. [CrossRef]
143. Bond, W.J.; Keeley, J.E. Fire as a global 'herbivore': The ecology and evolution of flammable ecosystems. *Trends Ecol. Evol.* **2005**, *20*, 387–394. [CrossRef] [PubMed]
144. Marlon, J.R.; Bartlein, P.J.; Walsh, M.K.; Harrison, S.P.; Brown, K.J.; Edwards, M.E.; Higuera, P.E.; Power, M.J.; Anderson, R.S.; Briles, C.; et al. Wildfire responses to abrupt climate change in North America. *Proc. Natl. Acad. Sci. USA* **2009**, *106*, 2519–2524. [CrossRef] [PubMed]
145. Fletcher, M.S.; Wood, S.W.; Haberle, S.G. A fire-driven shift from forest to non-forest: Evidence for alternative stable states? *Ecology* **2014**, *95*, 2504–2513. [CrossRef]
146. Murray, B.R.; Hardstaff, L.K.; Phillips, M.L. Differences in Leaf Flammability, Leaf Traits and Flammability-Trait Relationships between Native and Exotic Plant Species of Dry Sclerophyll Forest. *PLoS ONE* **2013**, *8*, e79205. [CrossRef] [PubMed]
147. Dimitrakopoulos, A.P. A statistical classification of Mediterranean species based on their flammability components. *Int. J. Wildland Fire* **2001**, *10*, 113–118. [CrossRef]
148. Cawson, J.G.; Duff, T.J.; Tolhurst, K.G.; Baillie, C.C.; Penman, T.D. Fuel moisture in Mountain Ash forests with contrasting fire histories. *For. Ecol. Manag.* **2017**, *400*, 568–577. [CrossRef]
149. Walsh, S.F.; Nyman, P.; Sheridan, G.J.; Baillie, C.C.; Tolhurst, K.G.; Duff, T.J. Hillslope-scale prediction of terrain and forest canopy effects on temperature and near-surface soil moisture deficit. *Int. J. Wildland Fire* **2017**, *26*, 191–208. [CrossRef]
150. Clarke, P.J.; Knox, K.J.E.; Wills, K.E.; Campbell, M. Landscape patterns of woody plant response to crown fire: Disturbance and productivity influence sprouting ability. *J. Ecol.* **2005**, *93*, 544–555. [CrossRef]
151. Pausas, J.G.; Ribeiro, E. The global fire-productivity relationship. *Glob. Ecol. Biogeogr.* **2013**, *22*, 728–736. [CrossRef]
152. Pausas, J.G.; Bradstock, R.A. Fire persistence traits of plants along a productivity and disturbance gradient in mediterranean shrublands of south-east Australia. *Glob. Ecol. Biogeogr.* **2007**, *16*, 330–340. [CrossRef]
153. Penman, T.D.; Binns, D.L.; Brassil, T.E.; Shiels, R.J.; Allen, R.M. Long-term changes in understorey vegetation in the absence of wildfire in south-east dry sclerophyll forests. *Aust. J. Bot.* **2009**, *57*, 533–540. [CrossRef]
154. Dantas, V.D.L.; Batalha, M.A.; Pausas, J.G. Fire drives functional thresholds on the savanna–Forest transition. *Ecology* **2013**, *94*, 2454–2463. [CrossRef]
155. Krawchuk, M.A.; Moritz, M.A.; Parisien, M.A.; Van Dorn, J.; Hayhoe, K. Global pyrogeography: The current and future distribution of wildfire. *PLoS ONE* **2009**, *4*, 1–12. [CrossRef] [PubMed]
156. Matthews, S.; Sullivan, A.L.; Watson, P.; Williams, R.J. Climate change, fuel and fire behaviour in a eucalypt forest. *Glob. Chang. Biol.* **2012**, *18*, 3212–3223. [CrossRef] [PubMed]

![forests logo] *forests*

MDPI

Article

Environmental Influences on Forest Fire Regime in the Greater Hinggan Mountains, Northeast China

Qian Fan [1], Cuizhen Wang [1,3,*], Dongyou Zhang [2] and Shuying Zang [3]

[1] Department of Geography, University of South Carolina, Columbia, SC 29208, USA; fanq@email.sc.edu
[2] College of Geography Sciences, Harbin Normal University, Harbin 150025, China; zhangdy@163.com
[3] Key Laboratory of Remote Sensing Monitoring of Geographic Environment, College of Heilongjiang Province, Harbin Normal University, Harbin 150025, China; zsy6311@163.com
* Correspondence: cwang@mailbox.sc.edu; Tel.: +1-803-777-5867

Received: 14 June 2017; Accepted: 26 September 2017; Published: 30 September 2017

Abstract: Fires are the major disturbances in the Greater Hinggan Mountains, the only boreal forest in Northeast China. A comprehensive understanding of the fire regimes and influencing environmental parameters driving them from small to large fires is critical for effective forest fire prevention and management. Assisted with satellite imagery, topographic data, and climatic records in this region, this study examines its fire regimes in terms of ignition causes, frequencies, seasonality, and burned sizes in the period of 1980–2005. We found an upward trend for fire occurrences and burned areas and an elongated fire season over the three decades. The dates of the first fire in a year did not vary largely but those of the last fire were significantly delayed. Topographically, spring fires were prevalent throughout the entire region, while summer fires mainly occurred at higher elevations under severe drought conditions. Fall fires were mostly human-caused in areas at lower elevations with gentle terrains. An ordinal logistic regression revealed temperature and elevation were both significant factors to the fire size severity in spring and summer. Other than that, environmental impacts were different. Precipitation in the preceding year greatly influenced spring fires, while summer fires were significantly affected by wind speed, fuel moisture, and human accessibility. An important message from this study is that distinct seasonal variability and a significantly increasing number of summer and fall fires since the mid-1990s suggest a changing fire regime of the boreal forests in the study area. The observed and modeled results could provide insights on establishing a sustainable, localized forest fire prevention strategy in a seasonal manner.

Keywords: Greater Hinggan Mountains; boreal forest; fire regime; fire season; ordinal logistic regression

1. Introduction

Forests are important natural resources and play a significant role in regulating climate and the carbon cycle. Boreal forests, also known as Taiga in high northern latitudes across North America and Eurasia, account for 29% of the world's forests, and store 37% of global terrestrial carbon [1,2]. Forest fire is primarily a natural process in boreal ecosystems [3]. With a low decomposition rate, the post-fire productivity of boreal forests could decline for up to 80 years before the organic leaf litter layer is reestablished [4]. Under the pressure of climate warming and accelerated human activities, fire behavior in boreal forests has been found to be undergoing dramatic changes [5]. It is crucial to understand these changes of fire characteristics and to identify the driving factors for sustainable forest management.

Fire regime defines the combined characteristics of fire in terms of its frequency of occurrences, size, intensity, seasonality, cause, and severity. Instead of considering a forest fire as a singular random event, fire regime treats it as a landscape-level spatial process, which helps us understand the forest fire and its causal factors at a larger spatial extent in a climate change context [6,7]. The interaction of top-down and bottom-up factors governs forest fire regimes over a range of spatial and temporal scales. The bottom-up controls usually act at fine scales by regulating fire physics and behavior [8]. For instance, fire propagation is mainly controlled by weather, local terrain plus fuel load, moisture content, and fuel continuity. Topographic factors (i.e., elevation, slope, and aspect) also strongly influence the forest environment in aspects of potential incident radiation and temperature. On the other hand, climate acts as a top-down control, which impacts fire occurrence through intra- and inter-annual climatic variations. Studies have shown that the impact of intra-annual precipitation variability on fire frequency is greater than the total annual precipitation in forests of the eastern United States [9]. It is not clear whether the top-down or bottom-up factors are leading factors. In years of extreme drought, climate would create weather and fuel conditions to overtake the bottom-up controls, allowing fires to cross natural barriers like streams or roads. Controlling factors vary in different biophysical scenarios and, sometimes, are a combination of multiple factors [10]. Anthropogenic forces also play a significant role in influencing forest fire regime. It is reported that more human-induced fires in Russian boreal forests have occurred due to the lack of control and ineffectual fire management policies since the creation of the Russian Federation [11]. In Northeast China, extensive logging increases the forest vulnerability to future burning and the half-century fire suppression policy has greatly altered its fire patterns [12]. It is challenging to understand how these factors interact to regulate the fire regime.

Boreal forests of China are mainly distributed in the Greater Hinggan Mountains that are located at the southern end of Siberian boreal forest. Fire regimes vary spatially across the region due to different species compositions, physiographic conditions, climate characteristics, and characteristics of the local economies. Intensive studies have been conducted to examine the controlling factors on fires in this region. For instance, Wu et al. found that climate was the primary factor influencing fire occurrence, while human activities were the secondary control [13]. Another study from Hu et al. reported that climatic factors were dominant drivers for lightning-caused fires, but not for human-caused ones [14]. Three fire environment zones were identified in this area through spatial clustering of environmental variables [15]. Chang et al. utilized a binary logistic regression to predict the fire occurrence patterns and to assess fire risks in Heilongjiang Province, China [16]. Forest fire regime and the surrounding environments usually exhibit dramatic seasonal variations; however, few studies have examined it from this perspective.

Forest fires in the Greater Hinggan Mountains have been analyzed in a seasonal manner, with spring season from March to June, summer season from July to August, and fall season starting in September and generally lasting to October when it begins to snow [17]. Moreover, extremely large fires, sometimes named mega fires, are catastrophic and their impacts to the landscape are complex and far reaching [18]. Usually a small number of large fires constitute the majority of burned areas [19]. Studies have also shown that fire burning sizes varied with environmental conditions such as vegetation, topography, and weather [20,21]. It is necessary to examine how these environmental factors regulate the fires in terms of fire sizes in different seasons, which could be of great help for effective fire control in this remote, boreal forest.

However, there exist some challenges to carry out such quantitative fire studies at the landscape scale. One is the data availability. Taking fuel conditions as an example, it is difficult to obtain actual in-field fuel conditions when a fire occurs. Remote sensing imagery becomes a promising data source for its frequent updating and synoptic coverage. Studies have shown that vegetation index is correlated with fuel moisture content. For example, the Normalized Difference Vegetation Index (NDVI) products from the Advanced Very High Resolution Radiometer (AVHRR) [22] and Moderate Resolution Imaging Spectroradiometer (MODIS) [23] imagery have been successfully used to estimate fuel moisture content. Therefore, vegetation index could serve as a good proxy for fuel moisture at the landscape scale. Another challenge is the difficulty of quantifying human impacts on fires.

In limited studies, distance to the most nearby road was used to approximate the accessibility to a fire location [24]. Road network density could be an indicator of the intensity of human activities.

The primary goal of this study is to identify the forest fire regimes in the Greater Hinggan Mountains, and to characterize the controlling environmental factors in spring, summer, and fall seasons. Integrating multiple sources of data sets, this study analyzes how these factors regulate the fire severity through a statistical analysis approach in this boreal region.

2. Materials and Methods

2.1. Study Area

The Greater Hinggan Mountains, covering an approximate area of 7.3 million ha in Northeast China (Figure 1), is one of the largest national forests of China. It lies between the Inner Mongolia Plateau and the Northeast China Plain, covering a large geographic area between 51°30′–53°33′N and 121°10′–127°08′E. It comprises about 10% of the boreal ecosystems in the Northern Eurasia region [25]. Located within the sub-arctic climatic zone, winters of the study area are long, dry, and cold. The annual average temperature ranges from −4 °C to −2 °C and annual precipitation ranges from 400 to 500 mm, with almost half of the precipitation falling in summer, especially in July and August [25]. As a southern extension of Eurasia's boreal ecosystem, vegetation is dominated with deciduous coniferous tree species. From the 1:1,000,000 China Vegetation Atlas at the *Environmental and Ecological Science Data Center for West China* (http://westdc.westgis.ac.cn), larch (*Larix gmelini*) covers 55.4% of the study area. Other tree species include evergreen coniferous such as Mongolian pine (*Pinus sylvestris* var. *mongolica*) and spruce (*Picea koraiensis*), and deciduous broadleaved trees such as birch (*Betula platphylla*) and aspen (*Populus tremuloides*). Non-forest land covers are limited, mostly in forms of herbaceous grasses and shrubs in the valleys and croplands in the plains at lower elevations.

The study area is one of the major timber production bases in China. It is composed of five counties. From northwest to southeast are Mohe, Tahe, Huzhong, Xinlin, and Huma counties (as marked in Figure 1). Mohe has the highest forest cover of 93.3% across the county, while Huma has the least. Huma is the only county relying on an agriculture-based economy. The population in the study area is approximately 500,000 according to the 2010 census from the *National Bureau Statistics of China*. Tahe has the largest population, followed by Mohe and Huma.

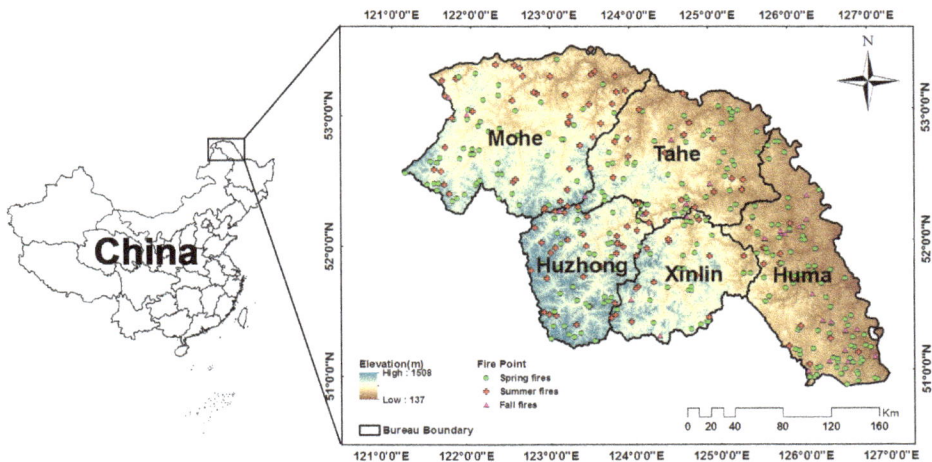

Figure 1. The study area and historic fire points in 1980–2005.

2.2. Data Sets

2.2.1. Historical Fire Records

The 26-year fire data (1980–2005) in the Greater Hinggan Mountains were obtained from the *Forest Fire Prevention Office, China Forest Bureau*. A total of 404 fires were recorded in 1980–2005. Points of the recorded fires are marked in Figure 1. These fires were real-time observations by field staff at *China Forest Bureau*. Attributes of a fire record include: fire location, time and date of ignition (start date) and extinction (end date), burned size (burned area of a fire), and cause of ignition. Complying with strict fire prevention policies in Northeast China, fires were often rapidly extinguished. The average time to put out a fire was 26.33 h in the study area, although there was an exception of the "Black Dragon" fire on 6 May–2 June 1987, the largest fire in this area that burned 1.3 million ha of forests in 26 days. Among all fire records, the lightning-caused fires accounted for 63% (254 fires); 29% (117 fires) were caused by human activities such as smoking, debris burning, equipment usage, and short circuit of power lines. A small portion (8%) of the records were fires with unknown causes (34 fires).

2.2.2. Topography and Road Network

The digital elevation model (DEM) was acquired from the Shuttle Radar Topography Mission (SRTM) global products at 1 arc-second cell size (approximately 30 m). As shown in Figure 1, the terrain of the study area is mountainous, rising from east to west at elevations ranging from 130 to 1500 m with an average of 560 m. Slopes and aspects of terrain surfaces were derived from the DEM. Road infrastructure is limited in this natural forest. Provincial-level and township-level road data were obtained from the *National Geomatics Center of China*.

2.2.3. Meteorological Data

The meteorological data were downloaded from the U.S. National Oceanic and Atmospheric Administration (*NOAA*) *Earth System Research Laboratory*. The gridded climate data, with a $0.5°$ (approximately 50 km) cell size, include monthly mean temperature and monthly total precipitation in the period of 1901 to 2014, and long-term mean of the monthly mean temperature and that of monthly total precipitation. Wind speed was obtained from the *China Meteorological Administration*. Five weather stations are evenly distributed in the study area, and have monitored daily wind speed (km/h) since the 1950s. Adopting the Beaufort Scale [26], wind speeds were transformed to the categorical scales from 1 to 5, representing breeze, moderate, strong, very strong, and stormy winds, respectively.

2.2.4. Vegetation Data

Fuel moisture condition is an essential control of forest fires. Studies have shown that vegetation index is correlated with fuel moisture content [27]. The bi-weekly NDVI product of the Global Inventory Modeling and Mapping Studies (GIMMS), namely the AVHRR GIMMS NDVI3g with a pixel size of 8 km, were obtained from the U.S. National Aeronautics and Space Administration (*NASA*) *Earth Exchange* (NEX) platform. At each fire point, the accumulative NDVI in the snow-free growing season (May–September) of the preceding year was computed to serve as a surrogate for fuel moisture condition (representing the organic layers). It was referred to as $\Sigma NDVI_{preceding}$ in this study, with a range of $[-10, 10]$ accumulated in the 6-month period.

2.3. Approaches

2.3.1. Data Processing

Fire records in the study area were grouped into categories of spring (March–June), summer (July–August), and fall (September–October) fires according to their igniting dates. The burned area of each fire was recorded in the fire data. Since the degrees of fire severity were not recorded in this historical data set, here we took the burned area as a measure of fire size severity, or FSS. Note that it is different from the terms fire severity and burn severity which are often interchangeably used based on the loss of soil and aboveground organic matter [28]. Following the standards of the *Chinese Forest Fire Prevention Office*, fires were assigned into four ordinal levels on basis of burned areas, i.e., $\leq$1 ha, $\leq$100 ha, $\leq$1000 ha, and >1000 ha, corresponding to low, moderate, moderate/high, and high FSS, respectively. With these FSS data, the causes of ignition (lightning vs. human) and seasonality (spring, summer, fall) of fire regimes in the study area were examined.

With the data sets described in Section 2.2, environmental parameters were extracted to assess the environmental influences on fires in the study area (Table 1). Considering fire as a natural process that behaves in a spatial extent, environmental effects on a fire are a spatial representation within this extent. For this reason, parameters in Table 1 are retrieved from an areal buffer instead of merely at a single fire point. For the fire records in this study, the average burned area was 4864 ha/fire. To extract the environmental parameters for each fire, we approximated each fire as a circular buffer centered at the fire point with a radius of 4000 m. For each fire, the environmental parameters in Table 1 are calculated as the average values within this buffer. Although the spatial coverage of the burned area of each fire was not available, the spatial average within such a buffer fairly represented the environmental variables when this fire broke out.

Table 1. Environmental parameters used in this study.

Data Category	Abbreviation	Parameter	Format	Unit	Cell Size
Topography		Slope	Continuous	Degree	30 m
		Elevation	Continuous	m	30 m
		Aspect	Continuous	Unitless	30 m
Climate	MAT	Mean Annual Temperature	Continuous	Celsius	0.5°
	$TAP_{current}$	Total Annual Precip. (current year)	Continuous	mm	0.5°
	$TAP_{preceding}$	Total Annual Precip. (preceding)	Continuous	mm	0.5°
	MTP	Monthly Temperature Percentage	Continuous	%	0.5°
	MPP	Monthly Precipitation Percentage	Continuous	%	0.5°
	Wind Speed	Daily Mean Wind Speed	Categorical	1 to 5	0.5°
Vegetation		$\Sigma NDVI_{preceding}$	Continuous	Unitless	8 km
Road		Distance to Nearest Road	Continuous	m	/
		Road Density	Continuous	km/km^2	/

For topographic data, the average values of elevation, slope, and aspect within each buffer represented the three topographic parameters at this fire point. For vegetation data, the average GIMMS NDVIg3 value accumulated in May–September in the preceding year was the $\Sigma NDVI_{preceding}$ at this fire point.

Using the province- and township-level road network, the distance from any fire point to the nearest road was extracted. We used the distance to the nearest road as a proxy of human accessibility at a given pixel. A shorter distance indicated higher accessibility and therefore higher possibility of human-induced fires. Local dirt roads and pathways were rare in this remote boreal forest with low population. Fire behavior such as fire spread at a landscape was not considered in this study. The road density map in the study area was generated using the kernel density tool in ArcGIS (Figure 2). Road density at each fire point was thus extracted.

For meteorological data, the annual climatic variables, mean annual temperature (MAT) and total annual precipitation (TAP), were extracted at each fire point. Also, studies have shown that meteorological conditions in the preceding year could directly affect fire risks for the coming spring [29]. For example, several severe spring fires broke out in Huma County in 2003, which directly followed the prolonged drought in 2002 when the annual precipitation was reduced by 50–90% in comparison to normal years [30]. Hence, in addition to the precipitation of the current year ($TAP_{current}$) when a fire broke out, a variable of total precipitation over the past year ($TAP_{preceding}$) was also analyzed.

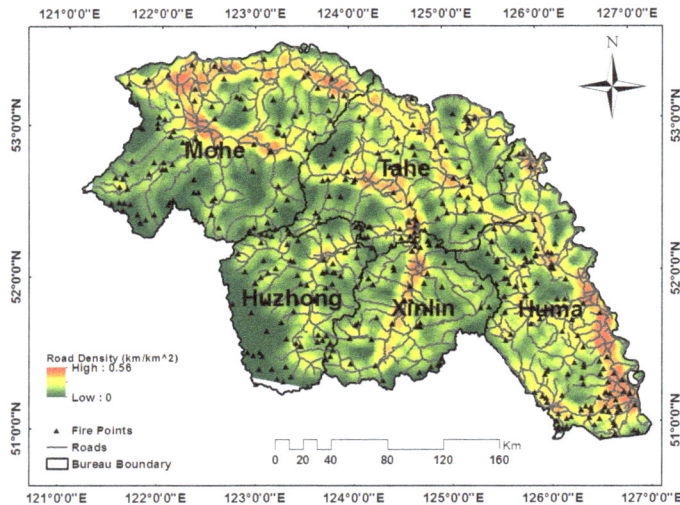

Figure 2. Road density map in km/km^2.

To examine the weather abnormality of a fire, we calculated the ratios of monthly precipitation and temperature over their long-term means, respectively. This process effectively alleviated the spatial and temporal bias of fire points across the study area in the 26-year period. All temperature measures were converted to the unit of Kelvin. Given the LTM(T) and LTM(P) as the long-term monthly mean temperature and monthly total precipitation, the ratios at a fire point can be calculated as:

$$\text{Monthly Temperature Percentage (MTP)} = \frac{\text{monthly mean temperature}}{\text{LTM(T)}} \tag{1}$$

$$\text{Monthly Precipitation Percentage (MPP)} = \frac{\text{monthly total precipitation}}{\text{LTM(P)}} \tag{2}$$

With Equations (1) and (2), the climatic data are standardized to represent local variations at the same scale. The standardization also reduces the spatial correlation between these explanatory variables and therefore, benefits the logistic modeling in the next section.

The wind speed at each fire point was assigned as the daily mean wind speed recorded at the nearest station on the ignition date.

2.3.2. Analytical Approaches

Descriptive statistics were implemented to explore the fire regimes and their decadal trends from 1980 to 2005 in terms of ignition causes, fire occurrence frequencies, fire burned areas (sizes), and seasonality. The Welch's ANOVA tests were applied to examine the variations of the explanatory variables, including topographic factors, weather/climate conditions, human impacts, as well as fuel conditions in different seasons. For the categorical variable (wind speed), the non-parametric Kruskal-Wallis test was applied.

An ordinal logistic regression was developed to quantitatively examine the driving factors that impact the fire size severity (FSS) in each season. Defining the FSS as the response variable, y, and the environmental parameters as the set of independent explanatory variables, $\mathbf{X}$, the logistic regression is described as [31]:

$$\text{logit}[P(y \le j)] = \log\left[\frac{P(y \le j)}{P(y > j)}\right] = \alpha_j + \beta * \mathbf{X}, \tag{3}$$

where $j = 1, 2, \cdots, c - 1$, with $c = 4$ in this study (the four FSS levels). Each cumulative logit uses all FSS levels.

Equation (3) is an ordinary logit model for a binary response in which categories 1 to j form one outcome and categories $j + 1$ to j form the other [31]. The Pearson's r and rank-based Spearman's rho (for categorical data) are used to identify the correlation among all explanatory variables.

The rule of thumb for the sample size in a logistic regression is that there are at least 10 events for each explanatory variable [32]. Only 30 fall fires were recorded in our study period, which was not a sufficient number for model establishment with the 12 explanatory variables listed in Table 2. Therefore, the ordinal logistic analysis was only conducted for spring and summer fires.

Table 2. The mean and standard deviation values of the extracted environmental parameters for fires in each season and the ANOVA tests for their seasonal differences.

Parameters	Statistics	Spring	Summer	Fall	Welch's ANOVA p-Value
Elevation (m)	Mean	538.8	641.0	440.7	$p < 0.0001$ [#]
	Std. Dev.	211.2	208.1	189.6	
Slope (°)	Mean	6.35	8.15	4.61	$p < 0.0001$ [#]
	Std. Dev.	3.13	3.49	2.77	
Aspect	Mean	−0.03	−0.02	−0.01	$p = 0.5108$
	Std. Dev.	0.15	0.14	0.14	
MTP (%)	Mean	100.39	100.36	100.30	$p = 0.5630$
	Std. Dev.	0.53	0.24	0.43	
MPP (%)	Mean	71.33	59.11	79.89	$p = 0.0005$ [#]
	Std. Dev.	41.95	22.15	46.62	
MAT (°C)	Mean	−3.58	−3.40	−2.22	$p < 0.0001$ [#]
	Std. Dev.	1.60	1.26	1.53	
$TAP_{current}$ (mm)	Mean	454.8	325.2	399.2	$p < 0.0001$ [#]
	Std. Dev.	82.3	58.3	62.3	
$TAP_{preceding}$ (mm)	Mean	423.9	425.6	444.5	$p = 0.3283$
	Std. Dev.	83.28	63.55	69.36	
Wind Speed (Beaufort scale 1–5)	Mean	2.6	2.3	2.4	$p < 0.0001$ [*,#]
	Std. Dev.	0.65	0.49	0.67	
$\Sigma NDVI_{preceding}$	Mean	7.89	7.87	7.78	$p = 0.4682$
	Std. Dev.	0.39	0.37	0.44	
Distance to road (m)	Mean	2694.3	3400.2	2358.4	$p = 0.0751$ [#]
	Std. Dev.	3122.1	3504.6	1883.0	
Road density (km/km^2)	Mean	0.174	0.147	0.182	$p = 0.0478$ [#]
	Std. Dev.	0.111	0.095	0.095	

[*] Wind Speed was tested with the Kruskal Wallis test; [#] indicated significant differences.

When implementing the ordinal logistic regression in the SAS package, the response variable (FSS) was entered in a descending manner. In this way, the resulted positive coefficient of each explanatory variable (environmental parameter) represents a positive influence of this parameter, i.e., the increased value of a specific parameter produces higher odds of larger fires, and vice versa.

The interpretation of the results is as follows. Assume the coefficient for a parameter in the logistic model is β_1. For a continuous variable such as elevation, given that all other parameters in the model are held stable, with 1 m increase of elevation, the odds of a larger fire would be computed as e^{β_1} of a smaller fire. For categorical parameters (e.g., wind speed), a base level is required. In this study, the wind speed scale 1 (i.e., breeze wind) was chosen as the base level. With 1 scale increase of wind speed, the odds of a larger fire would be e^{β_1} over breeze wind.

A significance level $\alpha = 0.1$ was set in the Wald Chi-square Test to examine the significance of the environmental parameters in the logistic model. With the ordinal logistic model, the environmental parameters that play a significant role in spring and summer fires were thus identified.

3. Results

3.1. Characteristics of Fire Regimes in the Study Area

The 26-year variations of fire occurrences are plotted against three causal factors: lightning, human-induced, and unknown (Figure 3). Total occurrences showed relatively stable counts in years before 2000 and an obvious increase after then, especially in 2000, 2002–2003, and 2005.

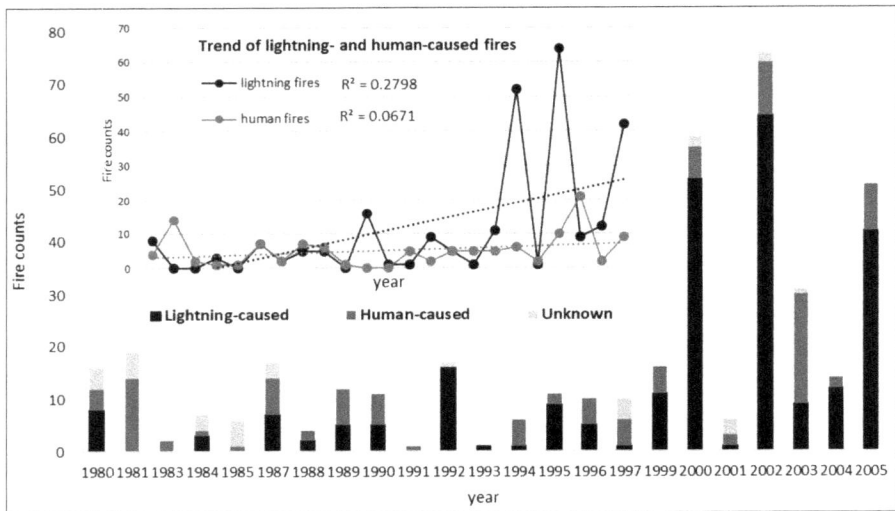

Figure 3. The occurrences of lightning- and human-caused fires in 1980–2005.

Lightning fires were dominant in the study area, accounting for ~63% of total occurrences. For the two different causal factors, an apparent increase of lightning fires was observed ($r = 0.53$, $p = 0.009$). As shown in the inset of Figure 3, the counts of human-induced fires did not display a statistically significant change in 26 years ($p = 0.23$).

Fire season length in each year was calculated as the duration between the start date of the first fire and the end date of the last fire in this year. In Figure 4, the first fire date did not show a statistically significant trend ($p = 0.58$), with most outbreaks having occurred in late April. On the contrary, the last fire date showed a significantly increasing trend, revealing a prolonged fire season length in past decades ($r = 0.60$, $p = 0.003$). An apparent change to fire season length was caused by fall fires (Day of Years (DOY) > 240). In the 1980s to early 1990s, there were no fall fires except in 1989. After 2000, however, fall fires occurred every year except 2003.

Fire seasonality (i.e., the season when a fire broke out) in the study area was analyzed with all fire records in 1980–2005. Figure 5a fairly reflects the seasonal categorization of this study which groups all fire records into spring, summer, and fall fires. Spring fires (March–June) accounted for the largest proportion (64%) of all fire counts, followed by summer fires (July–August) at 29% and fall fires (September–October) at 7%. The causal factors of fire ignition showed apparent seasonal variations. For fire counts, spring fires were fairly split between lightning-caused (54%) and human-caused (35%). Oppositely, almost all summer fires (96%) were lightning-caused, and most fall fires (74%) were human-caused. More specifically, lightning-caused fires mainly occurred from spring through summer (May to August), while human-caused fires were split between early spring (April–May) and fall (September–October). In Figure 5b, spring fires had the largest burned areas, followed by fall fires (October). The extremely high burned areas in May came from the catastrophic "Black Dragon" fire in 1987. The burned areas of summer fires were limited, probably because of ground wetness in peak growing season.

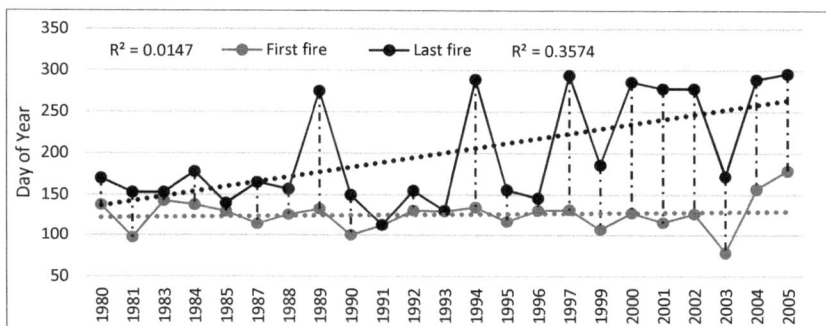

Figure 4. The 26-year variations of fire season length (the first and last fires in a year).

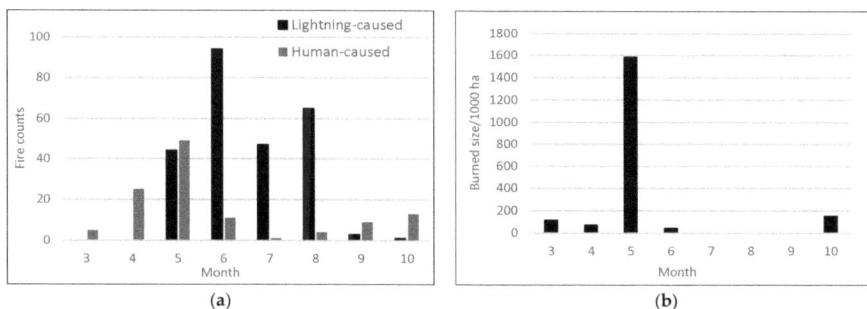

Figure 5. Fire counts (**a**) and burned areas (**b**) by month.

In addition to the increased fire occurrences and extended fire season length, the annual total burned area also showed an upward trend ($r = 0.55$, $p = 0.007$). In Figure 6, the area (ha) on the y-axis is transformed to logarithmic form for better visualization of the plot. Before 1994, burned areas were predominantly from spring fires. In later years, areas burned from fall fires dramatically increased. Summer fires were rare in the 1980s to 1990s, but burned large areas in 1999, 2002, and 2004–2005. While areas burned from spring fires remained relatively stable, more areas were burned from summer and fall fires in recent years, contributing to a significant increase in total burned areas. It is therefore reasonable to assume that the fire regime in the study area has changed in comparison to past decades.

The total burned area in 1980–2005 was about 1.97 million ha, and a small number of severe fires disproportionately burned excessive areas. Among the 404 fire records, 27 severe fires with high FSS (>1000 ha) composed 98.8% of the total burned area. The majority of these high FSS fires took place in spring (for example, the most catastrophic fire in 1987). From Figure 5, spring fires accounted for the largest number of fires as well as the most burned areas. No high FSS fire (>1000 ha) broke out in summer during our studied period.

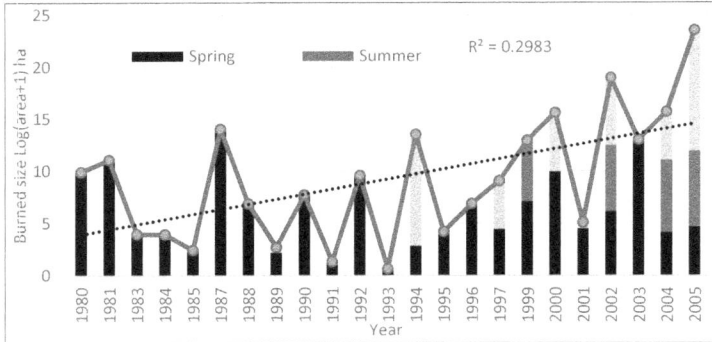

Figure 6. The 26-year variations of total burned areas by season.

The kernel densities of fire occurrences in three seasons are extracted in Figure 7. The density maps highlight the fire hotspots in spring (Figure 7a), summer (Figure 7b), and fall (Figure 7c). Both spring and fall fires were common in Huma County, which had the most agricultural lands in the study area. Summer fires were mostly located in Huzhong County at higher elevations. While fall fires mostly occurred in the agriculture-based Huma County, spring and summer fires spread across the forested mountains in other counties.

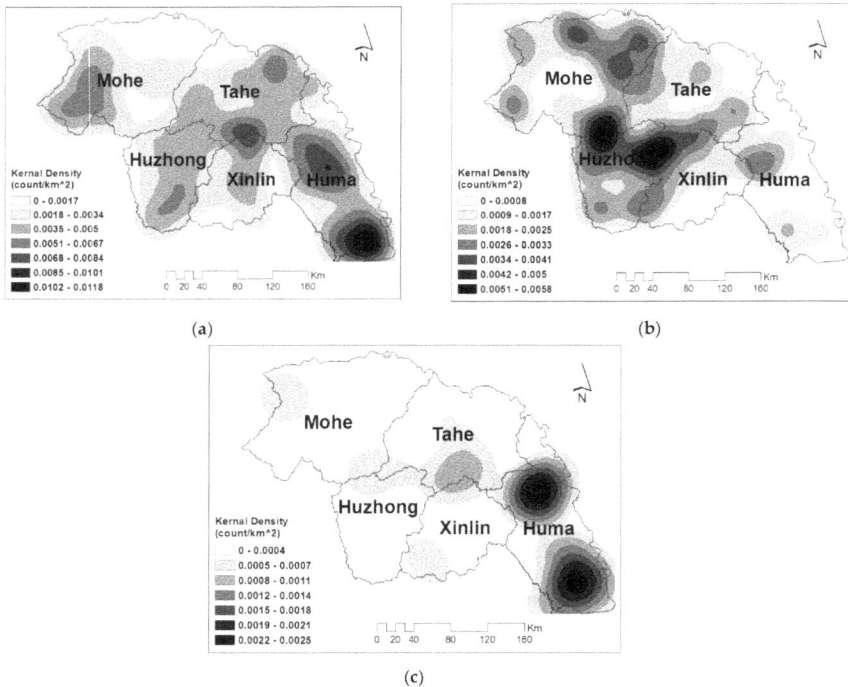

Figure 7. Kernel densities of fire occurrences in spring (**a**); summer (**b**); and fall (**c**).

3.2. Seasonal Variations of the Explanatory Variables

Descriptive statistics of the environmental parameters in Table 1 are summarized (Table 2). The Welch's ANOVA tests were performed to examine if a parameter showed significant differences in the three different seasons. A significant difference indicated that this parameter played an active role on the seasonality of fire occurrences. For topographic parameters, elevation and slope showed significant impacts on fire seasonality ($p < 0.0001$), while aspect was irrelevant. As shown in Figure 7a, spring fires were distributed across the whole study area, from Mohe at higher elevations to Huma at lower elevations. In Figure 7b, summer fires exhibited a higher density in Huzhong at higher elevations (average = 641 m) and steeper slopes (average = 8.15°) than spring and fall fires. There was a higher tendency for lightning strikes at higher elevations, which explained how lightning mainly caused summer fires (as revealed in Figure 5). Around two-thirds of fall fires occurred in Huma County, which had more cultivated lands and higher populations in plain areas. Therefore, topography (elevation and slope) had different impacts on fire occurrences in different seasons in the study area.

For meteorological parameters, the monthly temperature percentage (MTP%) was slightly higher than 100.0% in all three seasons. This indicated that temperature in each season had been slightly increasing from 1980 to 2005. However, this inter-annual increase of temperature was not seasonally different in the ANOVA test. Oppositely, the MPP% was much lower than 100% (in a range of 59–80%), indicating that there was dramatically decreased precipitation in this period. The ANOVA test confirmed that the inter-annual decrease of precipitation was seasonally different ($p = 0.0005$). In other words, the decreased precipitation casted a significant impact on fire seasonality. There also existed significant seasonal variations for the mean annual temperature (MAT) and total annual precipitation (TAP$_{current}$) (both with $p < 0.0001$). Considering both MTP% and MPP%, it was reasonable to assume that fire seasonality could be related to seasonal temperature and precipitation in current years as well as precipitation reduction from the preceding year. For example, fall fires were often accompanied by higher mean annual temperature while summer fires were associated with much lower precipitation. While some studies indicated the effects of precipitation in the preceding year [29], this study found that precipitation in the preceding year did not cast a significant effect on fire seasonality ($p = 0.3283$). Wind speed was a categorical variable. The Kruskal Wallis test was performed to examine its seasonal differences in Table 2 ($p < 0.001$). Statistics also showed that spring and fall fires suffered more severe wind conditions than summer. The strong to stormy (scale 3 to 5) winds composed 53.7% of all wind scales for spring fires and 46.7% for fall fires, while all wind speeds were within the category of strong wind (scale 3) for summer fires.

For vegetation, no significant variations of $\Sigma NDVI_{preceding}$ were found ($p = 0.4682$), indicating that fuel moisture conditions were not significantly different among the three seasons.

Regarding human impacts, the road density had a strong significant impact on fire seasonality ($p < 0.048$). The distance to roads had a weaker impact ($p < 0.075$). Summer fires exhibited the longest distance to roads and lowest road density in comparison to spring and fall fires. Oppositely, fall fires held the highest road density and shortest distance, while spring fires were in middle. These results were consistent with the distributions of ignition causes in Figures 5a and 7, revealing that summer fires occurred in more remote areas at higher elevations. Fall fires were in more populated areas at lower elevations. Spring fires featured both aspects.

In short, forest fires exhibited distinct seasonal variability in terms of topographical, meteorological, and human-related conditions in this area. Specifically, summer fires mainly occurred in drier conditions by lightning and were located at high elevations and remote areas. Fall fires were in relatively flat areas and were mostly human-caused fires. Spring fires took place across the whole region. In the following analysis, we simulated how these environmental parameters regulate the fire size severity in each season.

3.3. Environmental Influences with the Ordinal Logistic Regression

The collinearity analysis showed that two topographic factors, elevation and slope, were highly correlated in both spring and summer fire datasets ($\rho > 0.8$, $p < 0.001$). For forest fires in mountainous areas, the critical slope threshold is 25 degrees beyond which the burning behavior changes [33]. The maximum slope for all fires was <20 degrees in the study area. Therefore, slope was not used in our model. There was no significant correlation among meteorological variables ($|\rho| < 0.45$, $p > 0.05$) in spring fires. However, in summer fires the monthly precipitation was highly correlated with annual precipitation ($\rho = 0.75$, $p < 0.001$), probably because summer was the primary rain season in this region. The Spearman's correlation showed that monthly precipitation had higher correlation with FSS than annual precipitation, hence annual precipitation was not used. Variables of human influence proxy (road density and distance) did not show significant correlation ($|\rho| < 0.2$, $p \approx 0.05$). The environmental parameters used in the model are listed in Table 3.

Table 3. Ordinal logistic regression of spring and summer fires.

	Parameter	Coefficient	Wald Chi-Square	p-Value
Spring fires	Elevation	−0.00014	5.5775	0.0182
	MTP(%)	0.4991	4.4273	0.0354
	TAP$_{preceding}$	−0.00834	25.5820	<0.0001
Summer fires	Wind Speed (scale 2)	0.1999	0.1632	0.6863
	Wind Speed (scale 3)	1.7719	8.1004	0.0044
	Elevation	−0.00359	5.1185	0.0237
	NDVI	−0.00285	11.1797	0.0008
	MAT (°C)	0.4287	3.1402	0.0764
	Distance to road	0.00014	3.4953	0.0615

Both final modes in Table 3 are converged and the Wald Chi-square tests for the proportional odds assumption are significant ($p < 0.1$). For spring fires, the three significant environmental parameters were elevation, monthly temperature, and total annual precipitation of the preceding year. The interpretation of the results was as follows. Given that all other parameters in the model were held stable, with 1 mm more precipitation in the preceding year, the odds of a larger fire were calculated as $e^{-0.00834} = 99.2\%$ of a smaller fire. Similarly, a 1 m increase of elevation lowered the odds of larger fires by $(1 - e^{-0.00014}) = 0.01\%$. It was particularly noteworthy that when monthly temperature increases by one percent over long-term monthly mean temperature the odds for a larger fire were $e^{0.4991} = 164.7\%$ of a smaller fire. The coefficient value of each parameter revealed its relative importance. For spring fires, the most influencing parameters were monthly temperature, followed by the precipitation of the preceding year, and lastly elevation. Other environmental parameters examined in this study did not significantly affect the spring fires.

For summer fires, five influencing parameters were identified: wind speed, elevation, NDVI, mean annual temperature, and distance to road. The distance to road, mean annual temperature, and wind speed positively affected the fire size severity. Overall, wind speed was a significant variable that affected the fire size severity. A change of wind speed from scale 1 to 2 (breeze to moderate wind) did not significantly affect the fire size severity ($p = 0.6863$). However, when wind speed increased from scale 1 to 3 (breeze to strong wind), the odds of larger fires were $e^{1.7719} = 588.2\%$ of smaller fires, thus greatly increasing the chance of large fires. For the mean annual temperature, a 1 °C increase would increase the odds for larger fires by 1.538 times. The impacts of distance to road, NDVI, and elevation were limited.

4. Discussion

4.1. Fire Regime Changes

This study carried out a comprehensive analysis of fire regimes in the study area. Increasing trends for fire occurrences and burned areas were found in 1980–2005, which were greatly attributed to there being more summer and fall fires after the mid-1990s, thus resulting in a prolonged fire season length. Fire prevention in the study area has been one of the most important management activities of the *National Forest Bureau of China*. In the past decades, spring and fall have received the primary attention because it was not easy to form large fires in summer due to the high wetness from precipitation and leaf moisture of green canopies. The significant upward trend of summer fire occurrences since the mid-1990s indicates that more attention should be given to the summer season. While high FSS fires (burned area > 1000 ha) were not recorded in summers of 1980–2005, an increased number of moderate fires (100–1000 ha) have been observed in recent years. Under certain circumstances—for example, a stronger wind speed or insufficient fire fighting forces—these moderate-sized fires could possibly develop into larger conflagrations. Therefore, our study raised a sound alarm that summer fires cannot be ignored for effective fire prevention.

Causes of these fire regime changes could be twofold. Firstly, great efforts of fire prevention and management have been enforced in the Greater Hinggan Mountains since the catastrophic "Black Dragon" Fire in 1987 [34]. Through a rich set of evenly distributed lookout towers, an expanded fire monitoring network was established for field staff to detect and report wildfires in a timely manner (Mr. Huadong Wu, Vice Director, Fire Prevention Office, Tuqiang Forest Bureau, personal communication on 5 June 2015). Improved community learning programs and strict fire management policies led to reduced human-caused fires. On the other hand, the increased lightning fires, especially in summer, could be strongly impacted by climate change. For all fires in 1980–2005 in the study area, apparent weather anomalies were observed when fires broke out, with a trend that the mean annual temperature increased by 0.672 °C and the total annual precipitation decreased by 32.38 mm. The prolonged fire season length observed in this study was also inter-related with climate change. Studies have projected that the fire season will be prolonged by 20 to 30 days under the Intergovernmental Panel on Climate Change (IPCC) A2 and B2 scenarios of climate change in Northeast China [35].

4.2. Environmental Impacts on Fire Seaonality

Spatial patterns of fire points in different seasons varied with topography and road accessibility. Both spring and fall fires prevail in Huma County, a more populated and agriculture-heavy area at lower elevations that connects to the Northeast Plain of China. Therefore, fire prevention management should be always vigilant from spring to fall, especially in regard to human-induced fires. On the other hand, the dominancy of summer lightning fires at higher elevations in Huzhong County indicated that this area should be given higher attention during summer fire watch.

Fires in all three seasons were affected by similar meteorological changes, such as significantly higher temperature and lesser precipitation. However, they still varied in some aspects; for instance, summer fires usually occurred in dry years with dramatically reduced precipitation. Fall fires erupted when there was much higher annual temperature.

4.3. Environmental Influences on Fire Size Severity

Elevation had a negative impact on both spring and summer fires in the Greater Hinggan Mountains. At lower elevations, local temperature tended to be higher, which promoted dry fuel accumulation and led to an enlarged combustion area. Good accessibility to road network at lower elevations could facilitate the occurrences of human-caused fires in spring.

It is counterintuitive that wind speed was an influencing factor in summer but not in spring. Spring is prone to large forest fires due to the rising temperature and accumulated dry leaves on the ground, accompanied with prevailing strong wind speeds. Wind speed is expected to be one of the most influential factors for fire spread and consequently larger burned area [19]. This study found that about half of spring fires were accompanied with strong to stormy winds (scale 3 to 5, respectively). In contrast, wind condition was much milder in summer, all within the scale of 1 to 3 which represents breezes to strong winds, respectively. The logistic regression results revealed that wind speed was a significantly positive factor facilitating the formation of larger fires in summer. In other words, although wind speed was still the key factor affecting large fire formation, summer fires were more sensitive to wind speed than spring fires at the landscape scale. Wind not only directly propagated the fire spread, but affected the moisture content of surface fuels more rapidly.

Fuel moisture significantly influenced summer fires. As shown in Table 3, NDVI has a negative coefficient in the logistic model. Summer fires were often caused by lightning—in most cases by dry thunderstorms, occurring in dense forests at higher elevations, where the high leaf moisture in peak growing season restricted the occurrences of large fires. Moreover, distance to nearest roads acted as a constraint to developing large fires in summer, since the distant location of fire incidences hindered firefighters to extinguish the fires in time. Even worse, if accompanied with strong wind, the situation became more difficult to control. In contrast, distance to nearest roads may act as a driving factor of fire occurrences in other seasons. In spring and fall, for example, high accessibility to road network could facilitate the occurrences of human-caused fires in Huma County.

Two other significant factors for spring fires were the monthly temperature (MTP against long-term mean) and the precipitation in the preceding year. Usually, fine fuels like fallen leaves and forest litter accumulate in spring. High temperature would expedite the evaporation and drying of fire fuel, resulting in a high flammability and propagation speed. A dry preceding meteorological condition created a vulnerable condition for fires in the coming spring, consequently posing an elevated risk for larger fires. Taking the preceding meteorological factors into consideration helps us develop more appropriate measures for fire prevention in the coming spring.

4.4. Limitations and Future Work

Some limitations remain in this study. The fire records that we could access are only available for a relatively short period (26 years), and fall fire is not examined in this study, given its low number of fire records. This study indicates that more fall fires would occur with a prolonged fire season. When longer data series are available in the future, fall fires could be better studied, thus allowing for a more comprehensive understanding of the fire regime and its evolution in the long run. This study also implies the promising application of integrating remote-sensed data, such as NDVI, into fire studies. In the future, more remote sensing products, for example, land surface temperature and active fire data from coarse-resolution satellite imagery, could be applied to substitute the limited in-field observations in fire studies.

Human impacts on fires in boreal forests were not deeply examined in this study. The distance to the nearest road and road density surrounding a fire point were simply extracted to approximate the anthropogenic disturbances. In the past decades, human activities in the study area have been accelerating (e.g., logging, planting, sawmills, and wood product transportation). In the early 2000s, the "Tian Bao" Project was enforced to permanently prohibit logging activities in the Greater Hinggan Mountains for natural forest protection [34]. It is therefore necessary to integrate these human activities and consequences of policies into fire studies in order to establish a better understanding of human impacts in this unique boreal forest region.

5. Conclusions

This study explored the statistical characteristics of fire records in boreal forests of the Greater Hinggan Mountains in the period of 1985–2006, analyzed the impacts of different environmental parameters on fire seasonality, and performed an ordinary logistic regression in order to identify the influencing environmental parameters on the fire regime in this region. It was found that spring fires accounted for the largest proportion of fire occurrences as well as the most burned areas, and fall fires were more related to anthropogenic activities in harvesting season. Summer fires were mostly lightning-caused and were rare before the mid-1990s. However, the increased summer fires in recent years, together with prolonged fire season length, deserve higher attention as a result of the possibly changing fire regime of the region. Different sets of significant environmental factors were identified: elevation, temperature, and precipitation in the preceding year for spring fires; wind speed, elevation, temperature, NDVI, and distance to road for summer fires. Spatial distributions and densities of fires in different seasons varied across the study area. The spatially and seasonally specific fire patterns extracted from this study could help to develop more localized fire prevention strategies for sustainable forest management.

Acknowledgments: This study was financially supported by the National Natural Science Foundation of China (No. 41371397).

Author Contributions: Qian Fan performed data analysis and drafted the manuscript. Cuizhen Wang completed and revised the manuscript and was in charge of communication with co-authors for revisions and comments to improve the manuscript. Dongyou Zhang collected the fire data and performed pre-processing for spatial and statistical analysis. Shuying Zang supervised and revised the manuscript in aspect of the fire regime change and environmental influences.

Conflicts of Interest: The authors declare no conflicts of interest.

References

1. Goldammer, J.G.; Furyaev, V.V. *Fire in Ecosystems of Boreal Eurasia*; Springer Science & Business Media: Berlin, Germany, 2013.
2. Kuusela, K. *The Dynamics of Boreal Coniferous Forests*, 1st ed.; SITRA: Helsinki, Finland, 1990.
3. De Groot, W.J.; Cantin, A.S.; Flannigan, M.D.; Soja, A.J.; Gowman, L.M.; Newbery, A. A comparison of Canadian and Russian boreal forest fire regimes. *For. Ecol. Manag.* **2013**, *294*, 23–34. [CrossRef]
4. Ward, C.; Pothier, D.; Paré, D. Do Boreal Forests Need Fire Disturbance to Maintain Productivity? *Ecosystems* **2014**, *17*, 1053–1067. [CrossRef]
5. Kelly, R.; Chipman, M.L.; Higuera, P.E.; Stefanova, I.; Brubaker, L.B.; Hu, F.S. Recent burning of boreal forests exceeds fire regime limits of the past 10,000 years. *Prec. Natl. Acad. Sci. USA* **2013**, *32*, 13055–13060. [CrossRef] [PubMed]
6. Liu, Z.; Yang, J.; Chang, Y.; Weisberg, P.; He, H. Spatial patterns and drivers of fire occurrence and its future trend under climate change in a boreal forest of Northeast China. *Glob. Chang. Biol.* **2012**, *18*, 2041–2056. [CrossRef]
7. Bergeron, Y.; Leduc, A.; Harvey, B.D.; Gauthier, S. Natural fire regime: A guide for sustainable management of the Canadian boreal forest. *Silva Fenn.* **2002**, *36*, 81–95. [CrossRef]
8. Parks, S.A.; Parisien, M.A.; Miller, C. Spatial bottom-up controls on fire likelihood vary across western North America. *Ecosphere* **2012**, *3*, 1–20. [CrossRef]
9. Lafon, C.; Quiring, S. Relationships of Fire and Precipitation Regimes in Temperate Forests of the Eastern United States. *Earth Interact.* **2012**, *16*. [CrossRef]
10. Falk, D.A.; Heyerdahl, E.K.; Brown, P.M.; Farris, C.; Fule, P.Z.; McKenzie, D.; Swetnam, T.W.; Taylor, A.H.; Van Horne, M.L. Multi-scale controls of historical forest-fire regimes: New insights from fire-scar networks. *Front. Ecol. Environ.* **2011**, *9*, 446–454. [CrossRef]
11. Mollicone, D.; Eva, H.; Achard, F. Ecology—Human role in Russian wild fires. *Nature* **2006**, *440*, 436–437. [CrossRef] [PubMed]
12. Chang, Y.; He, H.; Hu, Y.; Bu, R.; Lia, X. Historic and current fire regimes in the Great Xing'an Mountains, northeastern China: Implications for long-term forest management. *For. Ecol. Manag.* **2008**, *254*, 445–453. [CrossRef]
13. Wu, Z.W.; He, H.S.; Yang, J.; Liu, Z.H.; Liang, Y. Relative effects of climatic and local factors on fire occurrence in boreal forest landscapes of northeastern China. *Sci. Total Environ.* **2014**, *493*, 472–480. [CrossRef] [PubMed]
14. Hu, T.Y.; Zhou, G.S. Drivers of lightning- and human-caused fire regimes in the Great Xing'an Mountains. *For. Ecol. Manag.* **2014**, *329*, 49–58. [CrossRef]

15. Wu, Z.; He, H.S.; Yang, J.; Liang, Y. Defining fire environment zones in the boreal forests of northeastern China. *Sci. Total Environ.* **2015**, *518*, 106–116. [CrossRef] [PubMed]

16. Chang, Y.; Zhu, Z.L.; Bu, R.C.; Chen, H.W.; Feng, Y.T.; Li, Y.H.; Hu, Y.M.; Wang, Z.C. Predicting fire occurrence patterns with logistic regression in Heilongjiang Province, China. *Landsc. Ecol.* **2013**, *28*, 1989–2004. [CrossRef]

17. Wu, Z.; Chang, Y.; He, H.; Hu, Y. Analyzing the spatial and temporal distribution characteristics of forest fires in Huzhong area in the Great Xing'an Mountains. *Guangdong Agric. Sci.* **2011**, *5*, 189–193. (In Chinese)

18. Romme, W.H.; Everham, E.H.; Frelich, L.E.; Moritz, M.A.; Sparks, R.E. Are large, infrequent disturbances qualitatively different from small, frequent disturbances? *Ecosystems* **1998**, *1*, 524–534. [CrossRef]

19. Dimitrakopoulos, A.; Gogi, C.; Stamatelos, G.; Mitsopoulos, I. Statistical analysis of the fire environment of large forest fires (>1000 ha) in Greece. *Pol. J. Environ. Stud.* **2011**, *20*, 327–332.

20. Fang, L.; Yang, J.; Zu, J.; Li, G.; Zhang, J. Quantifying influences and relative importance of fire weather, topography, and vegetation on fire size and fire severity in a Chinese boreal forest landscape. *For. Ecol. Manag.* **2015**, *356*, 2–12. [CrossRef]

21. Wu, Z.W.; He, H.S.; Liang, Y.; Cai, L.Y.; Lewis, B.J. Determining Relative Contributions of Vegetation and Topography to Burn Severity from LANDSAT Imagery. *Environ. Manag.* **2013**, *52*, 821–836. [CrossRef] [PubMed]

22. Chuvieco, E.; Cocero, D.; Riano, D.; Martin, P.; Martinez-Vega, J.; de la Riva, J.; Perez, F. Combining NDVI and surface temperature for the estimation of live fuel moisture content in forest fire danger rating. *Remote Sens. Environ.* **2004**, *92*, 322–331. [CrossRef]

23. Yebra, M.; Chuvieco, E.; Riano, D. Estimation of live fuel moisture content from MODIS images for fire risk assessment. *Agric. For. Meteorol.* **2008**, *148*, 523–536. [CrossRef]

24. Chuvieco, E.; Aguado, I.; Yebra, M.; Nieto, H.; Salas, J.; Martin, M.P.; Vilar, L.; Martinez, J.; Martin, S.; Ibarra, P.; et al. Development of a framework for fire risk assessment using remote sensing and geographic information system technologies. *Ecol. Model.* **2010**, *221*, 46–58. [CrossRef]

25. Xu, H. *Forest in Great Xing'an Mountains of China*; Science Press: Beijing, China, 1998. (In Chinese)

26. Beaufort Scale. Wikipedia. Available online: https://en.wikipedia.org/w/index.php?title=Beaufort_scale& amp;oldid=746213871 (accessed on 12 June 2017).

27. Glenn, E.; Huete, A.; Nagler, P.; Nelson, S. Relationship between remotely-sensed vegetation indices, canopy attributes and plant physiological processes: What vegetation indices can and cannot tell us about the landscape. *Sensors* **2008**, *8*, 2136–2160. [CrossRef] [PubMed]

28. Keeley, J.E. Fire intensity, fire severity and burn severity: A brief review and suggested usage. *Int. J. Wildland Fire* **2009**, *18*, 116–126. [CrossRef]

29. Liu, X.; Yan, Z. Analysis of Pre-meterological conditions of forest spring fires in Greater Hinggan Mountain in 2003. *Heilong Jiang Meteorol.* **2003**, *4*, 29–30. (In Chinese)

30. Zhang, A. Forest Fires in Great Hinggan Mountains. *China News*, 29 March 2003. (In Chinese)

31. Agresti, A.; Kateri, M. *Categorical Data Analysis*; Springer: Berlin/Heidelberg, Germany, 2011.

32. Peduzzi, P.; Concato, J.; Kemper, E.; Holford, T.R.; Feinstein, A.R. A simulation study of the number of events per variable in logistic regression analysis. *J. Clin. Epidemiol.* **1996**, *49*, 1373–1379. [CrossRef]

33. Butler, B.; Anderson, W.; Catchpole, E. Influence of slope on fire spread rate. In *Proceeedings of the Fire Environment—Innovations, Management, and Policy*; USDA Forest Service, Rocky Mountain Research Station: Fort Collins, CO, USA, 2007; pp. 75–82.

34. Wang, J.; Wang, C.; Zang, S. Assessing re-composition of Xing'an larch in boreal forests after the 1987 fire, Northeast China. *Remote Sens.* **2017**, *9*, 504. [CrossRef]

35. Tian, X.; Shu, L.; Zhao, F.; Wang, M.; McRae, D.J. Future impacts of climate change on forest fire danger in northeastern China. *J. For. Res.* **2011**, *22*, 437–446. [CrossRef]

forests

MDPI

Article

Effects of Linear Disturbances and Fire Severity on Velvet Leaf Blueberry Abundance, Vigor, and Berry Production in Recently Burned Jack Pine Forests

Charlotte A. Dawe, Angelo T. Filicetti and Scott E. Nielsen *

Department of Renewable Resources, University of Alberta, Edmonton, AB T6G 2H1, Canada;
cadawe@ualberta.ca (C.A.D.); filicett@ualberta.ca (A.T.F.)
* Correspondence: scottn@ualberta.ca; Tel.: +1-780-492-1656

Received: 29 September 2017; Accepted: 13 October 2017; Published: 18 October 2017

Abstract: There is limited information on how velvet leaf blueberry (*Vaccinium myrtilloides* Michx.) responds to fires and existing small forest gaps associated with narrow linear disturbances. We measured the effects of narrow forest linear gaps from seismic lines used for oil and gas exploration versus adjacent (control) forests across a fire severity (% tree mortality) gradient on the presence, abundance (cover), vigor (height), and berry production of *Vaccinium myrtilloides* in recently (five-year) burned jack pine (*Pinus banksiana* Lamb.) forests near Fort McMurray, Alberta. Presence was greatest in forests that experienced low to moderately-high fire severities with declines at high fire severity. Abundance did not differ among seismic lines or adjacent forest, nor did it differ along a fire severity gradient. In contrast, vigor and berry production were greater on seismic lines compared to adjacent forests with fire severity positively affecting berry production, but not plant vigor. After controlling for changes in plant cover and vigor, berry production still increased with fire severity and within seismic lines compared with adjacent forests. Our findings suggest that narrow gaps from linear disturbances and fire severity interact to affect the fecundity (berry production) and growth (height) of *Vaccinium myrtilloides*. This has important implications for assessing the ecological effects of fire on linear disturbances associated with energy exploration in the western boreal forest.

Keywords: fire; severity; seismic line; disturbance; jack pine; production; vigor; abundance; presence; Ericaceae; velvet leaf blueberry; *Vaccinium myrtilloides*

1. Introduction

Fire is a common element in the boreal forest [1] with many of its dominant plants having adaptations that allow their long-term persistence, even under high fire frequencies [2–4]. For example, jack pine (*Pinus banksiana* Lamb.) is a common overstory species that dominates drier sites of the boreal forests of North America whose serotinous cones open and release seeds following fire [5,6]. Many shrub and herbaceous species are similarly adapted to disturbance, including fire, but instead through vegetative regeneration from underground rhizomes [7,8]. Fires also alter site conditions that favor understory plants by increasing light availability in the understory [9] and by reducing total plant cover, thereby reducing competition [10]. Fire severity also plays an important role in affecting the composition of vegetation both directly and indirectly through changes in below-ground processes [11]. For example, high severity fires in jack pine stands have lower species richness and cover, while lower severity fires have the highest species richness and cover, even when compared to unburned stands [12]. A common emphasis of forest fire studies is in understanding changes in tree composition and density [13–16]. Much less is known, however, about understory responses despite having the potential to influence the direction of post-fire succession [17].

Ericaceous plants are a common understory shrub of fire-prone forests that often respond positively to fire [18]. For instance, the germination rate of *Erica umbellate* L. increases following fire [19], while greater light availability post-fire increases the sexual reproduction and vigor of *Gaultheria shallon* Pursh [20]. Likewise, *Kalamia augustifolia* L. and *Vaccinium* species regenerate post-fire because their rhizomatic roots often escape the mortality of fires resulting in vegetative resprouting [7,21], although responses in these species may be negative when fire intensities are high due to extreme heat [19,21]. Although post-fire communities often exhibit lower levels of competition immediately following fire, dense tree regeneration can reduce understory cover, including ericaceous shrubs, due to direct competition for light and resources. For example, *Vaccinium angustifolium* Aiton and *Vaccinium myrtilloides* Michx. both responded negatively post-fire to competition with other rapidly colonizing species [22], including competition with tree species [23]. Thus, any positive responses following fire may be short lived once tree recruitment dominates a site.

Although fire is a common natural disturbance to boreal forests [24], anthropogenic disturbances have recently become more widespread [25]. In Alberta, Canada, common anthropogenic disturbances that create gaps in the forest canopy include forest harvest clear-cuts and seismic lines. Seismic lines are used to predict subsurface properties of the Earth, such as oil and gas reserves, and have been used in the exploration of oil and gas since 1924 [26]. Unlike clear-cuts, seismic lines are long, narrow, and linear forest gap features. As canopy gap size increases, the mean and variability of light levels within the gap also increases [27], affecting the amount of sunlight received by understory foliage and thus stimulating its growth [23]. For instance, *Gaultheria procumbens* responds positively to clear-cutting by stimulating below-ground vegetative growth and the release of new shoots from the existing network of rhizomes [8]. Open canopy gaps have also been shown to benefit *Vaccinium myrtillus* [28].

Although previous research has examined responses in *Vaccinium myrtilloides* to large forest openings from clear-cut forest harvesting [23], little is known about the effects of smaller forest canopy gaps, including the narrow linear seismic line disturbances, and how these small gaps interact with fires. Seismic line disturbances are individually narrow in their footprint (~3–12 m wide), but their densities are high (mean density of 1.77 km/km^2) [29], making them the most abundant linear features in Alberta's boreal forest. Many of these seismic lines are failing to recover decades after their initial disturbance [30,31], thus altering groundlayer vegetation composition [32]. Understanding specific responses to individual species, such as *Vaccinium myrtilloides*, is important for determining the overall effects of these linear disturbances on the boreal forest community and how that may change following natural disturbances of wildfires. The objectives of this study were to examine the responses of *Vaccinium myrtilloides* to small canopy gaps created by seismic lines (including variation in their width) and to test whether fire interacts with these disturbances along a fire severity gradient. Specifically, we examine changes in *Vaccinium myrtilloides* presence, abundance (cover), vigor (height), and berry production across a fire severity gradient five-years post-fire on narrow linear disturbances (corridors) used for seismic exploration and paired control sites in jack pine forests in northeast Alberta.

2. Materials and Methods

2.1. Study Area

The study area is located in the northeast part of Alberta, Canada, within an area known as the Richardson and McClelland Lake areas on the Athabasca Sand Plain approximately 115 km north of Fort McMurray. Elevation of the area is ~300 m above sea level and characterized by dry, sandy, gently sloping terrain dominated by jack pine (*Pinus banksiana*) forests that have among the highest observed fire frequencies in Canada's boreal forest [24].

In 2011, the Richardson fire burned 707,648 hectares of forest making it among Alberta's largest recorded fire [33]. Fire severity (% tree mortality) was, however, highly variable with tree mortality ranging from 0 to 100 percent. Jack pine stands in the area varied in age, but mostly represented mature forests of jack pine that were often characterized as having semi-open woodland conditions where tree density was moderate to low. Common shrub and understory species included velvet leaf blueberry (*Vaccinium myrtilloides*), pin cherry (*Prunus pensylvanica* L.), saskatoon (*Amelanchier alnifolia* Nutt.), bearberry (*Arctostaphylos uva-ursi* L.), reindeer lichen (*Cladina* spp.), and rose (*Rosa* spp.). Some sites that were slightly more mesic contained a minor element of trembling aspen (*Populus tremuloides* Michx.) and green alder (*Alnus crispa* Aiton).

2.2. Experimental Design and Field Measures

2.2.1. Site Selection

Sample sites were selected (stratified) across a range of fire severity classes (Figure 1), which were defined by percent tree mortality (tree mortality rates of: 0–25%; 26%–50%; 51%–75%; 76%–100%) to ensure the representation of all levels of fire severity (≥8 replicate sites per strata with 2 paired plots per site). Fire severity was measured within stands adjacent to seismic lines based on percent tree mortality within the surrounding area. In total, 66 sites were sampled in the summer of 2016 (from 5 July to 15 August) representing conditions five-years post-fire. At each site, a pair of plots was established in areas with a similar fire severity and stand age with one plot positioned in the middle of the seismic line (treatment) and the other 25 m into the adjacent forest (control). The total number of plots equaled 132 (66 paired plots or sites). Direction from the seismic line for the control plot was randomly assigned to being left or right of the seismic line using a coin flip. To avoid edge effects, all plots were at least 30 m from any other forest types, forest edges, or another seismic line, and at least 90 m from clear-cuts, roads, or major trails. Each plot was represented as a 30-m long transect that followed the orientation of the seismic line and *Vaccinium myrtilloides* characteristics were measured along the transect. Additional forest measures were made in the adjacent forest stand (e.g., basal area, tree stem density, stand age), but were not further considered here since they had little additional effect on variation in *Vaccinium myrtilloides* measures.

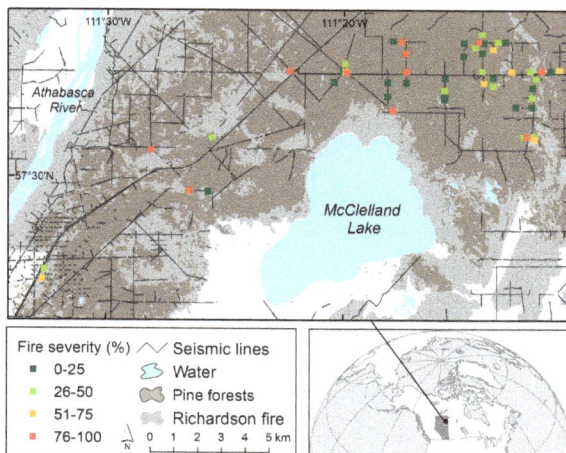

Figure 1. Location of study site in northeast Alberta, Canada (inset map), with the main map depicting the location of the 2011 Richardson fire (gray), unburned areas (white), pine forests (dark brown where burned and in a few places light brown where unburned), seismic lines, and sample sites symbolized by four levels of fire severity. Each site represents two-paired plots (a seismic line and an adjacent forest) with sites restricted to areas with no forest harvesting (not shown, but extensive in the area). Note that nearly all pine forests were burned and that some sites were within ~200 m of each other (not shown in this map) when stand and fire severity levels differed or there were gaps in the seismic line disturbance.

2.2.2. Measured Responses in *Vaccinium myrtilloides* and Treatment Variables

Vaccinium myrtilloides presence, abundance (cover), vigor (maximum height), and berry production were measured along each 30-m transect in 1-m wide belts (i.e., a 30-m^2 plot). Presence of *Vaccinium myrtilloides* within each 30-m^2 plot was first recorded and if present the abundance (cover) of *Vaccinium myrtilloides* plants was measured within ten sequentially spaced sub-plots (quadrats), each having a 3 × 1 m size (3 m^2). Within each quadrat, cover was estimated using 10 ordinal cover classes following the Carolina Vegetation Survey protocol that is similar to the Domin and Daubenmire techniques [34] with final cover estimates for each plot estimated as the average of midpoint cover values in each quadrat. Plant vigor was measured in each plot as the average maximum height of *Vaccinium myrtilloides* plants among the 10 sampled quadrats. Maximum height was used as a measure of vigor since it has been known to be an important predictor of fruit production [35], particularly after recent disturbances. Finally, berry production in *Vaccinium myrtilloides* was recorded after fruit set and before full ripening in each 3-m^2 quadrat and summed to estimate total berry production on a per 30 m^2 (1 × 30 m) basis.

Treatment variables used to describe responses in *Vaccinium myrtilloides* presence, abundance, vigor, and berry production included fire severity and location (seismic lines vs. adjacent forests). Fire severity was measured as either ordinal categories (0%–25%, 26%–50%, 51%–75%, or 76%–100%) for simple comparisons and graphs, or original continuous values for linear models. The seismic line and adjacent forest (location) factor was measured as a binary variable with seismic lines being coded as 1 and adjacent forest plots being coded as 0 (reference comparison).

2.3. Data Analysis

2.3.1. Responses in *Vaccinium myrtilloides* on Seismic Lines and Adjacent Forests

For each response measure of *Vaccinium myrtilloides* (presence, abundance, vigor, and berry production), we first summarized and graphed average values by treatment and fire severity class and then compared seismic line treatments to adjacent forests within each fire severity class using paired *t*-tests for continuous data (abundance, vigor, and berries) and a McNemar's χ^2 test for presence and absence data. Second, we used mixed effect regression models (logistic family for presence/absence and Gaussian family for all others) to examine responses across the full fire severity gradient (original linear values from 0 to 100% tree mortality), location (seismic line treatment; seismic line = 1; adjacent forests = 0), and their interaction. A random effect on site ID was used in all models to account for the paired nature of the sampling design (paired plots per site). All plots were used for presence data, while abundance, vigor, and berries were only assessed where plants were present at the site level. A $\log_{10}$ transformation with a constant of 1 was used for berry counts in all statistical analyses to normalize their distribution, while plant cover and the height of plants were kept in their original scale since their distribution was approximately normal. Responses to fire severity (tree mortality) were fit by both linear and quadratic terms to consider possible non-linear responses. Model selection followed the development of a global model that included all variables (fire severity, location treatment of seismic line, and their interaction) with the subsequent removal of non-significant ($p < 0.1$), non-treatment variables. Fire severity and location treatment were thus retained in all models regardless of their significance given that it was the main part of the study design. Model variance explained (R^2) was used to assess the overall model fit.

2.3.2. Effects of Seismic Line Width (Forest Gap Size) on *Vaccinium myrtilloides*

As well as more generally comparing responses between seismic lines and adjacent forests across a fire severity gradient, we also assessed whether forest gap size, based on seismic line width (~4–10 m), affected responses in *Vaccinium myrtilloides*. We did this by comparing responses by seismic line width using simple linear regression including reporting on model fit assessed with adjusted R^2 and model error assessed with Root Mean Square Error (RMSE). Here, only the seismic line treatments (plots) were included in the analyses since we were only comparing differences between seismic line disturbance widths. Collinearity between seismic line width and fire severity was low ($r = 0.24$) with a good representation of different line widths across the fire severity gradient. Scatter plots were used to help visualize univariate relationships. Models included all seismic line plots regardless of the presence of *Vaccinium myrtilloides* ($n = 66$), as well as only those plots where *Vaccinium myrtilloides* plants were present on seismic line plots ($n = 41$). The restriction to plots where plants were present allowed an evaluation of the changes in berry production due to forest gap size, while accounting for general changes in plant abundance (cover) and vigor (height) that would be expected to change under open forest gap conditions. Thus, we asked whether differences in gap size (seismic line width) increased the fecundity (berry production) of plants on a per capita basis through increases in resources (e.g., light) that would boost flowering and/or pollinator activity that could affect fruit set. All analyses were performed in STATA/SE 13.1 (STATA Corp., College Station, TX, USA).

3. Results

3.1. Presence

Of the 66 sites and 132 plots sampled, *Vaccinium myrtilloides* was present in 41 (62.1%) seismic line plots versus 43 (65.2%) adjacent forest plots (Figure 2) with no significant difference between treatments ($p = 0.148$; Table 1). Fire severity had, however, a significant, non-linear effect on presence ($p < 0.001$, Table 1) with increases occurring in both the seismic line and the adjacent forest stand up to a moderately-high fire severity (51%–75%; Figure 2). Thereafter, high fire severity (76%–100%) resulted in decreases in presence (Figure 2). The mixed effects linear model had a significant negative interaction between fire severity and seismic line treatment ($p = 0.014$; Table 1), demonstrating the reduced presence of *Vaccinium myrtilloides* on seismic lines compared to adjacent forests at moderate to high fire severity (Table 1; Figure 1). Overall model fit was high ($R^2 = 0.41$; Table 1).

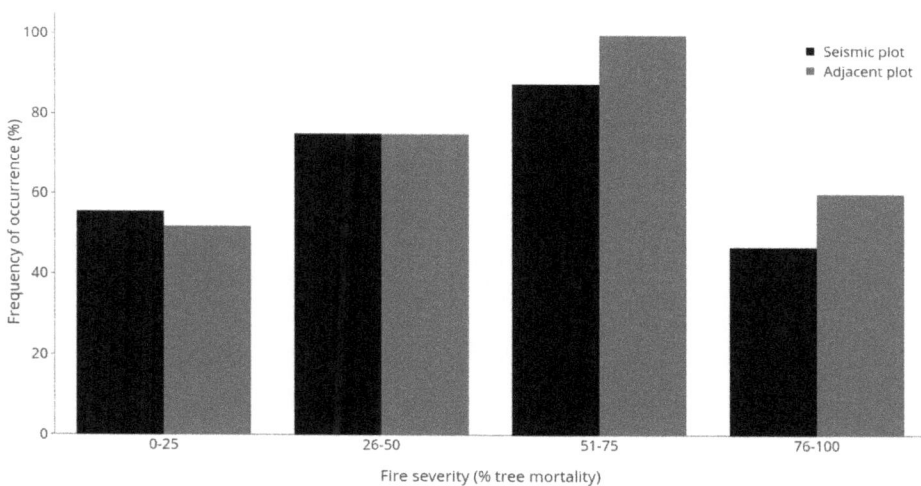

Figure 2. Frequency (presence) of *Vaccinium myrtilloides* plants in 30-m^2 plots on seismic lines versus adjacent forests across four fire severity classes. Comparisons between seismic and adjacent plots within individual fire severity classes using McNemar's χ^2 tests revealed no significant ($p < 0.1$) differences.

Table 1. Mixed effect regression results relating the presence, abundance (cover), and vigor (maximum height) of *Vaccinium myrtilloides* plants as a function of location (seismic line versus adjacent forest), fire severity, and the interaction of fire severity (% tree mortality) and location. Model coefficients (Coef.), Standard Error (S.E.) of coefficients, and significance (p) reported. Main treatment variables were retained regardless of significance, while a non-linear effect for fire severity and interaction term between fire severity and seismic lines were included only if significant.

Variable	Presence			Abundance (Cover)			Vigor (Max. Height)		
	Coef.	S.E.	p	Coef.	S.E.	p	Coef.	S.E.	p
Seismic line	4.167	2.88	0.148	0.323	0.428	0.451	4.516	1.357	0.001
Fire severity	1.010	0.175	<0.001	0.004	0.008	0.642	−0.006	0.028	0.832
Fire severity^2	−0.007	0.001	<0.001						
Seismic × Fire	−0.105	0.043	0.014						
Constant	−9.446	3.935	0.016	1.509	0.524	0.004	19.255	1.754	<0.001
Model fit, R^2		0.41			0.01			0.09	

3.2. Abundance (Cover)

Abundance (cover) where *Vaccinium myrtilloides* was present within either plot at the site level did not significantly differ ($p = 0.451$; Table 1) between seismic lines (Mean cover = 2.0%, S.E. = 0.33) and adjacent forest stands (Mean cover = 1.7%, S.E. = 0.36). Likewise, there was no significant difference in cover across the fire severity gradient ($p = 0.642$; Table 1, Figure 3). A non-linear fire gradient response was not supported, nor was an interaction between location (seismic line) and fire severity. Model fit was low overall ($R^2 = 0.01$; Table 1). This suggested no direct increase in abundance relative to differences in forest gap size (seismic line) or fire severity. Although variation was high and no significant responses were detected, there were weak trends towards greater cover in sites with moderate (26%–75%) fire severity, especially in adjacent control forests (Figure 3).

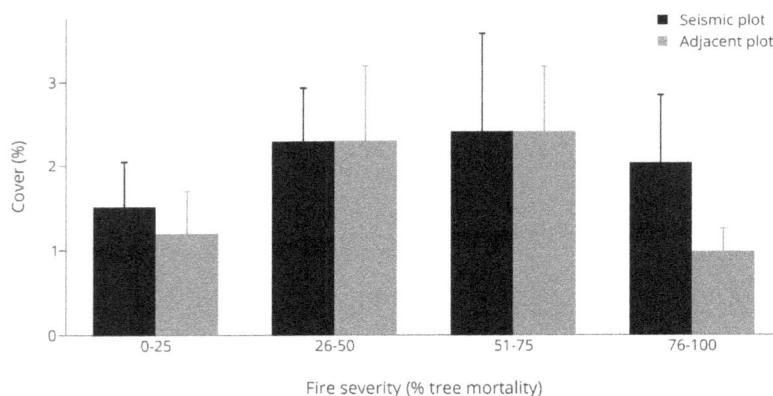

Figure 3. Mean cover of *Vaccinium myrtilloides* when present for seismic lines and adjacent forests across a fire severity gradient. Comparisons between seismic and adjacent plots within fire severity classes were tested using paired *t*-tests revealing no significant ($p < 0.1$) differences.

3.3. Vigor (Maximum Height)

Seismic line disturbances increased plant vigor, when present in plots, as measured by the maximum height of plants when compared to adjacent forest stands ($p = 0.001$; Table 1). Maximum heights in seismic line forest gaps averaged 23.5 cm (S.E. = 1.40), while being 19.0 cm (S.E. = 0.82) in the adjacent forest. Plant vigor did not vary across fire severity classes ($p = 0.832$), nor was there a significant interaction between seismic lines and fire severity (Table 1), although vigor on seismic lines did decrease under the highest fire severity class with a t-test revealing significant location effects within individual fire severity classes up to 75% tree mortality (Figure 4). Model fit explaining plant vigor was low overall ($R^2 = 0.09$; Table 1) demonstrating a high amount of unexplained variability.

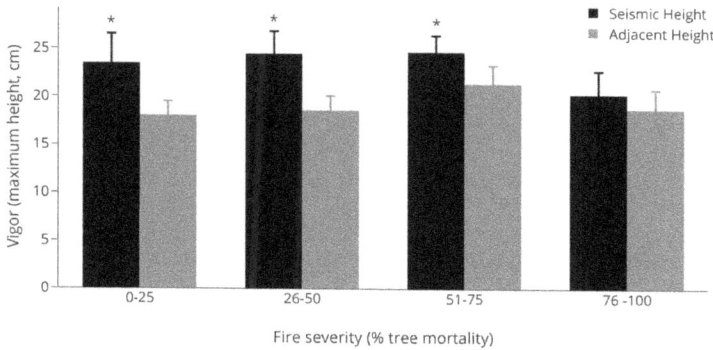

Figure 4. Average maximum height of *Vaccinium myrtilloides* when present, across a fire severity gradient (four classes) for seismic lines (black) and adjacent forests (gray). Significance among location by fire severity class was assessed by paired *t*-tests (* = $p < 0.1$).

3.4. Berry Production Across All Sites

Average berry production for sites where *Vaccinium myrtilloides* was present in at least one of the paired plots was significantly higher ($p < 0.001$) in seismic lines (36.9 berries per 30 m^2, S.E. = 8.5) compared with adjacent forest stands (11.4 berries per 30 m^2, S.E. = 2.7; Table 2). Fire severity positively affected berry production ($p = 0.009$; Table 2, Figure 5), although there was a significant negative interaction between seismic line treatment and fire severity ($p = 0.032$; Table 2) demonstrating that increases in berry production across fire severity classes was primarily related to changes in adjacent forest stands with berry production being more constant in seismic lines (Figure 4). Indeed, when testing seismic lines versus adjacent forests for each fire severity class using paired t-tests, berry production was significantly higher on seismic lines in the low (0%–25%) and low-moderate (26%–50%) fire severity classes, but not the moderately high (51%–75%) to high (76%–100%) fire severity classes (Figure 4), where tree mortality was high and thus canopy cover low and light levels high. Model fit explaining berry production was moderately-low overall ($R^2 = 0.15$; Table 2).

Table 2. Mixed effect linear regression model testing differences in *Vaccinium myrtilloides* berry production in response to location (seismic line versus adjacent forest), fire severity, and the interaction of fire severity and location. Note that results are based on a log$_{10}$ transformed count of berries.

Variable	Number of Berries (Plot)		
	Coef.	S.E.	p
Seismic line (binary)	0.753	0.213	<0.001
Fire severity (tree mortality)	0.007	0.003	0.009
Seismic × Fire severity	−0.008	0.004	0.032
Constant (intercept)	0.391	0.163	0.016
Model fit, R^2		0.15	

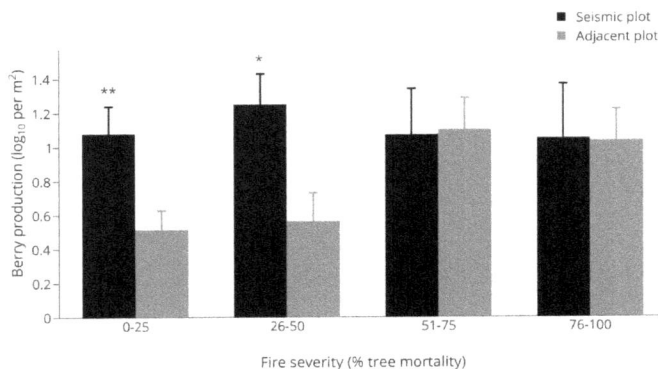

Figure 5. *Vaccinium myrtilloides* berry production ($\log_{10}$ scale) by fire severity class and location (seismic line versus adjacent forest). Significance among location by fire severity class was assessed by paired *t*-tests (* = $p < 0.1$; ** = $p < 0.01$).

3.5. Responses in Berry Production When Plants Were Present

Further analysis of berry production was conducted using data only from those plots (transects) with *Vaccinium myrtilloides* plants present in order to assess changes in berry production after controlling for the effects of shrub cover and vigor. Not surprisingly, plots that contained taller *Vaccinium myrtilloides* plants ($p < 0.001$) and those with increased cover ($p < 0.001$) produced more berries (Table 3). More interesting was that berry production on seismic lines was still significantly greater than in the adjacent forest stand ($p = 0.002$) after accounting for differences in plant cover and vigor (Table 3). Likewise, fire severity also increased berry production ($p < 0.001$) after controlling for cover and vigor with a weak significant effect of the interaction of seismic line treatment and fire severity ($p = 0.097$), suggesting a diminishing effect on berry production on seismic lines that experienced higher fire severity (Table 3). Model fit explaining berry production where plants were present was high overall although much of this was likely due to the effects of changes in cover and vigor ($R^2 = 0.59$; Table 3).

Table 3. Mixed effect linear regression results relating the berry production of *Vaccinium myrtilloides* to location (seismic line versus adjacent forest), fire severity, interaction of fire severity and location, abundance (cover), and vigor.

Variable	Number of Berries (Where Present)		
	Coef.	S.E.	*p*
Seismic line (binary)	0.422	0.134	0.002
Fire severity (tree mortality)	0.006	0.002	0.001
Seismic × Fire severity	−0.004	0.002	0.097
Constant (intercept)	−0.426	0.165	0.010
Shrub abundance (cover)	0.105	0.024	<0.001
Shrub vigor (max. height)	0.037	0.008	<0.001
Model fit, R^2		0.59	

3.6. Effect of Forest Gap Width from Seismic Lines

Vaccinium myrtilloides cover and vigor was positively and significantly ($p = 0.003$) related to seismic line forest gap width with model fit being higher for vigor ($R^2 = 0.19$, RMSE = 2.02) than for cover ($R^2 = 0.12$, RMSE = 8.09) (Table 4, Figure 6). Likewise, berry production was positively related to seismic line width (p <0.001, R^2 = 0.16, RMSE = 0.71; Table 4, Figure 6), although this appeared to be due to seismic lines with larger widths having significantly higher cover and vigor since berry production on sites with plants present was related to plant cover (p = 0.004) and vigor (p = 0.018), but not line width (p = 0.298, R^2 = 0.57, RMSE = 0.43; Table 5). Regardless, the presence of seismic lines still increased berry production above and beyond that expected by increases in cover and vigor (Table 3), but not due to changes in line width (gap size), suggesting an indirect effect of seismic line presence on berry production.

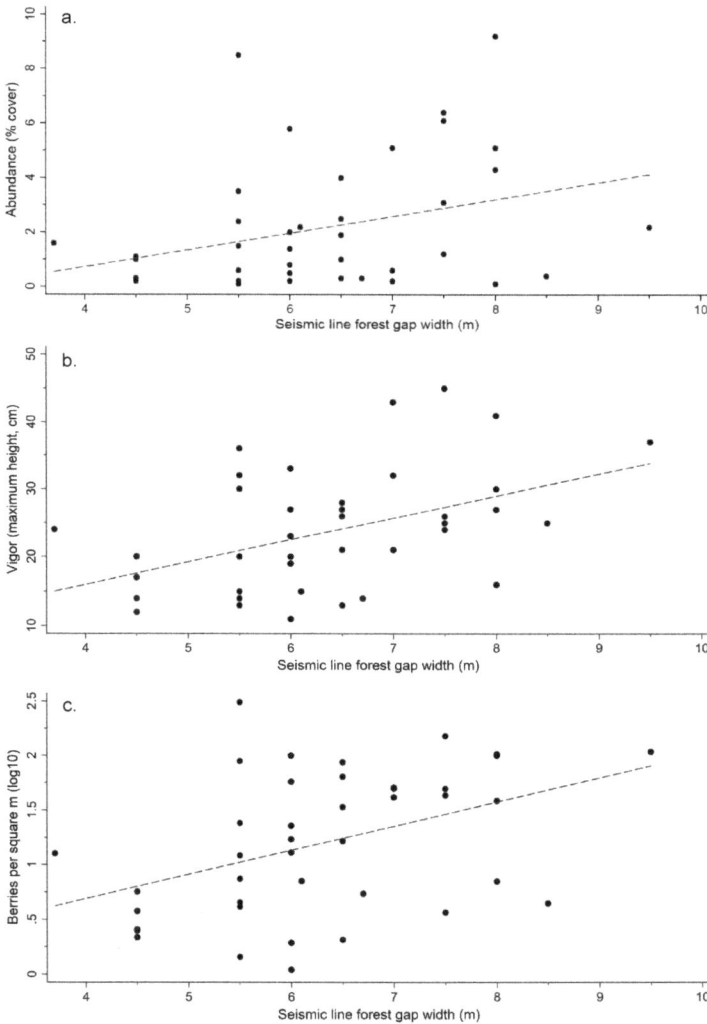

Figure 6. Scatterplot of seismic line width and *Vaccinium myrtilloides* cover (**a**), vigor (**b**), and berry production (**c**) on seismic lines where plants were present. Dashed line represents the fit of a linear regression relationship with overall model fit as follows: abundance (cover) R^2 = 0.12 (**a**), vigor (height) R^2 = 0.19 (**b**), and berry production R^2 = 0.16 (**c**).

Table 4. Linear regression results relating forest gap (seismic line) width to abundance and vigor of *Vaccinium myrtilloides*.

Variable	Abundance (Cover)			Vigor (Max. Height)		
	Coef.	S.E.	p	Coef.	S.E.	p
Seismic line width	0.676	0.219	0.003	3.244	1.019	0.003
Constant (intercept)	−2.774	1.357	0.045	3.010	6.549	0.648
Model fit, R^2 (RMSE)	0.12 (2.02)			0.19 (8.09)		

Table 5. Linear regression results relating forest gap (seismic line) width to berry production ($\log_{10}$ transformed) in *Vaccinium myrtilloides*.

Variable	Number of Berries (All Sites)			Number of Berries (Where Present)		
	Coef.	S.E.	p	Coef.	S.E.	p
Seismic line width	0.285	0.077	<0.001	0.064	0.061	0.298
Shrub abundance (cover)				0.115	0.038	0.004
Shrub vigor (max. height)				0.027	0.011	0.018
Constant (intercept)	−0.993	0.478	0.042	−0.076	0.358	0.833
Model fit, R^2 (RMSE)	0.16 (0.71)			0.57 (0.43)		

4. Discussion

4.1. Responses to Fire

We found that the presence of *Vaccinium myrtilloides* in recently burned (five-years) jack pine forests depended on fire severity with low to moderately-high severity (0%–75% tree mortality) increasing the presence in both seismic line forest gaps and adjacent forest stands. These results support prior research where low to moderate severity fires triggered positive responses to ericaceous shrubs by increasing light availability from canopy openings [17], stimulating the germination rate [19], and decreasing initial competition among understory plants [10]. In contrast, fires of a higher severity (>76% tree mortality) decreased the presence of *Vaccinium myrtilloides* on seismic lines and adjacent forest stands. This suggests that high severity fires can cause direct lethal effects to ericaceous shrubs most likely by the elimination of rhizomes in the soil layer [21]. Interestingly, the cover and height of *Vaccinium myrtilloides* was not affected by fire severity suggesting that its main effect is related to initial mortality and potentially to fecundity (berry production). Indeed, we found fire severity to be positively related to berry production in *Vaccinium myrtilloides*. The reproduction of ericaceous and *Vaccinium* species is known to be stimulated post-fire by increases in available light resources [20]. Fire not only increases light availability, but also nutrients with nitrogen, phosphorus, potassium, calcium, and magnesium potentially increasing following fire which can further stimulate growth and reproduction [36,37].

Our results demonstrate important interactions between forest gaps (seismic lines) and fire severity. Berry production in low to low-moderate fire severity (0%–50% tree mortality) was significantly higher on seismic lines than adjacent forest stands. Under low fire severity, canopy in the adjacent forest remains largely intact in contrast to local open canopy gaps associated with seismic lines. Increased light availability on seismic lines should therefore increase in berry production. Likewise, light availability increases in forests exposed to higher severity fires so light should be similar here to seismic line forest gaps, thereby resulting in similar increases in berry production.

4.2. Response to Seismic Lines

Small forest canopy gaps associated with seismic line disturbances did not affect the presence or abundance (cover) of *Vaccinium myrtilloides*, but it did increase plant vigor (maximum height) and berry production after fire disturbance. Past studies examining the responses of ericaceous species to clear-cut forest harvesting have also suggested that plant vigor is coupled to increases in light resources [38], although it may also be partly related to releases of allelochemcials [39]. Small forest canopy gaps affected berry production in *Vaccinium myrtilloides* with seismic lines producing ~30% more berries than the adjacent forest stand. Because seismic lines did not affect the abundance (cover) of *Vaccinium myrtilloides*, increases in reproduction are associated with increases in resource availability (light and nutrients) and/or increases in insect pollination.

Research in forest clear-cuts demonstrated similar changes in the reproductive strategies of ericaceous species with reproduction being more strongly influenced by irradiance than shoot density [20]. Past research has suggested that only large plants were capable of sexual reproduction and therefore light indirectly influences sexual reproduction through effects on plant size [35]. However, we found that after accounting for abundance, vigor (height), and fire severity, berry production was still greater on seismic lines compared with adjacent forests. This may be due to the larger canopy opening of seismic lines allowing greater light availability to the understory and thus higher rates of photosynthesis that can support the development of more fruit structure [40].

The second possibility is that seismic lines are increasing abundance and behaviour of insect pollinators, thus affecting fruit set. *Vaccinium* species are known to be pollinated by bees [41] with fruit production, size, and seediness of *Vaccinium* species being significantly higher in plants exposed to natural pollination [42]. However, *Vaccinium* species have also been shown to have higher flowering densities in sites with an open canopy [42] and because we did not measure flower production we cannot distinguish between these factors. In fact, it is most likely to be a combination of greater flower production due to photosynthetic resources and increases in pollinator activity. If the effect is due to pollinators, then visitation rates by pollinators on seismic lines should be higher than in the adjacent stand. Future studies should examine these interactions by measuring flower production, pollinator visitation rates, and fruit set.

4.3. Management Implications: Fire and Seismic Lines

Jack pine forests have evolved with fire as evidence of their aerial seedbanks of serotinous cones that require heat for the opening and release of seeds [13]. Post-disturbance boreal forest stands are often dominated by an understory of ericaceous species [4]. This flush in ericaceous productivity is an important phase of forest succession, providing a valuable habitat for wildlife. For instance, increasing the foraging opportunities of black bears [43] that are known to prefer seismic lines over forest stands, particularly in mid-to-late-summer due to increased berry production [44]. Indigenous peoples also benefit from post-fire resources with fire historically used to maintain a mosaic of successional stages that increase the diversity and abundance of forest resources, including berries from *Vaccinium* species [45]. Indeed, the gathering of *Vaccinium* berries is one of the most important aspects of indigenous traditions [46]. Therefore, management should consider the natural role that fire plays in jack pine forests and how changes in fire regimes may affect cultural values.

A number of considerations must be taken into account when considering seismic lines. These disturbances are known to affect wildlife by changing their behaviour and predator-prey relationships. In particular, woodland caribou in Northeastern Alberta are listed as threatened under the Federal Species at Risk Act [47] with seismic line restoration being a top priority [48] since these linear disturbances affect hunting and predation efficiency on caribou [49,50]. Black bear distribution in the forest is also affected by seismic line disturbances [44], possibly increasing opportunities for the human harvesting of black bear, as well as the predation of caribou calves by bears in recent disturbances [51]. Seismic lines also displace songbirds, although disturbances of smaller widths (canopy gaps) have less effect [52]. Regardless, there is general support for directly or indirectly promoting tree growth on seismic lines that fail to recover trees. Although the management of seismic lines is not directed towards *Vaccinium myrtilloides*, we expect such restoration actions that promote tree growth on seismic lines to reduce vigor and berry production similar to that of adjacent forests. Secondly, natural fires on these lines can interact to promote higher berry production. This may further affect wildlife, such as black bears, and the value of sites for harvesting by indigenous peoples.

5. Conclusions

Our study demonstrates that vigor and berry production in *Vaccinium myrtilloides* increases in narrow forest gaps associated with linear (seismic line) disturbances compared to adjacent forest stands after forest fires. We also found that as canopy gap (seismic line) width increases from 4 to 10 m, the cover, vigor, and berry production of *Vaccinium myrtilloides* also increases. This suggests that plants respond to changes in light and pollinators even within these small (4–10 m) forest gaps. Future research should manipulate these factors experimentally to further understand the mechanisms involved. Relative to wildfire, *Vaccinium myrtilloides* responded positively, except for high severity fires where declines were apparent. The management of seismic lines should consider the risks and benefits associated with restoration actions and wildfires on the habitat for *Vaccinium myrtilloides* if attempting to manage this species or species that utilize it, including wildlife (e.g., black bears) and indigenous peoples.

Acknowledgments: This work was conducted as part of the Boreal Ecosystem and Recovery Assessment (BERA) project with research and publication fees supported by grants from industry partners Cenovus Energy, Conoco Phillips Canada, and Alberta-Pacific Forest Industries, as well as matching funding from the Natural Sciences and Engineering Research Council of Canada (NSERC). Additional support was provided by the Alberta Biodiversity Conservation Chairs program funded by Canada's Oil Sands Innovation Alliance (COSIA), NSERC, Alberta Innovates, and Alberta Agriculture and Forestry.

Author Contributions: S.E.N., A.T.F., and C.A.D. conceived and designed the experiments; A.T.F. and C.A.D. collected data; S.E.N. and C.A.D. analyzed the data; C.A.D. and S.E.N. wrote the paper.

Conflicts of Interest: The authors declare no conflict of interest. The funding sponsors had no role in the design of the study; in the collection, analyses, or interpretation of data; in the writing of the manuscript, and in the decision to publish the results.

References

1. Bradbury, S.M. Response of the Post-Fire Bryophyte Community to Salvage Logging in Boreal Mixedwood Forests of Northeastern Alberta, Canada. *For. Ecol. Manag.* **2006**, *234*, 313–322. [CrossRef]
2. Kalamees, R.; Püssa, K.; Tamm, S.; Zobel, K. Adaptation to Boreal Forest Wildfire in Herbs: Responses to Post-Fire Environmental Cues in Two Pulsatilla Species. *Acta Oecol.* **2012**, *38*, 1–7. [CrossRef]
3. Wein, R.W.; MacLean, D.A. *The Role of Fire in Northern Circumpolar Ecosystems*; Published on behalf of the Scientific Committee on Problems of the Environment of the International Council of Scientific Unions; Wiley: Chichester, UK, 1983.
4. Mallik, A.U. Conifer Regeneration Problems in Boreal and Temperate Forests with Ericaceous Understory: Role of Disturbance, Seedbed Limitation, and Keytsone Species Change. *Crit. Rev. Plant Sci.* **2003**, *22*, 341–366. [CrossRef]
5. Beaufait, W. Some Effects of High Temperatures on the Cones and Seeds of Jack Pine. *For. Sci.* **1960**, *6*, 194–198.
6. Radeloff, V.C.; Mladenoff, D.J.; Guries, R.P.; Boyce, M.S. Spatial Patterns of Cone Serotiny in Pinus Banksiana in Relation to Fire Disturbance. *For. Ecol. Manag.* **2004**, *189*, 133–141. [CrossRef]
7. Mallik, A.U. Ecology of a Forest Weed of Newfoundland: Vegetative Regeneration Strategy of *Kalmia angustifolia. Can. J. Bot.* **1993**, *71*, 161–166. [CrossRef]
8. Moola, F.M.; Vasseur, L. The Importance of Clonal Growth to the Recovery of *Gaultheria procumbens* L. (Ericaceae) after Forest Disturbance. *For. Ecol. Recent Adv. Plant Ecol.* **2009**, 319–337. [CrossRef]
9. Ois Hébert, F.; Thiffault, N.; Ruel, J.-C.; Munson, A.D. Ericaceous Shrubs Affect Black Spruce Physiology Independently from Inherent Site Fertility. *For. Ecol. Manag.* **2010**, *260*, 219–228. [CrossRef]
10. Vila, M.; Terradas, J. Effects of Competition and Disturbance on the Resprouting Performance of the Mediterranean Shrub Erica Multiflora L. (Ericaceae). *Am. J. Bot.* **1995**, *82*, 1241. [CrossRef]

11. Neary, D.G.; Klopatek, C.C.; DeBano, L.F.; Ffolliott, P.F. Fire Effects on Belowground Sustainability: A Review and Synthesis. *For. Ecol. Manag.* **1999**, *122*, 51–71. [CrossRef]

12. Pinno, B.D.; Errington, R.C. Burn Severity Dominates Understory Plant Community Response to Fire in Xeric Jack Pine Forests. *Forests* **2016**, *7*. [CrossRef]

13. Alexander, M.E.; Cruz, M.G. Modelling the Effects of Surface and Crown Fire Behaviour on Serotinous Cone Opening in Jack Pine and Lodgepole Pine Forests. *Int. J. Wildland Fire* **2012**, *21*, 709–721. [CrossRef]

14. Ilisson, T.; Chen, H.Y.H. The Direct Regeneration Hypothesis in Northern Forests. *J. Veg. Sci.* **2009**, *20*, 735–744. [CrossRef]

15. Chiang, J.-M.; McEwan, R.W.; Yaussy, D.A.; Brown, K.J. The Effects of Prescribed Fire and Silvicultural Thinning on the Aboveground Carbon Stocks and Net Primary Production of Overstory Trees in an Oak-Hickory Ecosystem in Southern Ohio. *For. Ecol. Manag.* **2008**, *255*, 1584–1594. [CrossRef]

16. Johnstone, J.F.; Chapin, F.S., III; Foote, J.; Kemmett, S.; Price, K.; Viereck, L. Decadal Observations of Tree Regeneration Following Fire in Boreal Forests. *Can. J. For. Res.* **2004**, *34*, 267–273. [CrossRef]

17. Mallik, A.U. Conversion of Temperate Forests into Heaths: Role of Ecosystem Disturbance and Ericaceous Plants. *Environ. Manag.* **1995**, *19*, 675–684. [CrossRef]

18. Feurdean, A.; Florescu, G.; Vannière, B.; Tanțău, I.; O'Hara, R.B.; Pfeiffer, M.; Hutchinson, S.M.; Gałka, M.; Moskal-del Hoyo, M.; Hickler, T. Fire Has Been an Important Driver of Forest Dynamics in the Carpathian Mountains during the Holocene. *For. Ecol. Manag.* **2017**, *389*, 15–26. [CrossRef]

19. Vizcaíno, E.A.D.; Rodríguez, A.I.; Fernández, M. Interannual Variability in Fire-Induced Germination Responses of the Characteristic Ericaceae of the NW Iberian Peninsula. *For. Ecol. Manag.* **2006**, *234*, S179. [CrossRef]

20. Bunnell, F.L. Reproduction of Salal (Gaultheria Shallon) under Forest Canopy. *Can. J. For. Res.* **1990**, *20*, 91–100. [CrossRef]

21. Schimmel, J.; Granström, A. Fire Severity and Vegetation Response in the Boreal Swedish Forest. *Ecology* **1996**, *77*, 1436–1450. [CrossRef]

22. Hall, I.V. Floristic Changes Following the Cutting and Burning of a Woodlot for Blueberry Production. *Can. J. Agric. Sci.* **1955**, *35*, 143–152.

23. Moola, F.M.; Mallik, A.U. Morphological Plasticity and Regeneration Strategies of Velvet Leaf Blueberry (*Vaccinium myrtiltoides* Michx.) Following Canopy Disturbance in Boreal Mixedwood Forests. *For. Ecol. Manag.* **1998**, *111*, 35–50. [CrossRef]

24. Stocks, B.J.; Mason, J.A.; Todd, J.B.; Bosch, E.M.; Wotton, B.M.; Amiro, B.D.; Flannigan, M.D.; Hirsch, K.G.; Logan, K.A.; Martell, D.L.; et al. Large Forest Fires in Canada, 1959–1997. *J. Geophys. Res.* **2002**, *108*, 8149. [CrossRef]

25. Bayne, E.M.; Habib, L.; Boutin, S. Impacts of Chronic Anthropogenic Noise from Energy-Sector Activity on Abundance of Songbirds in the Boreal Forest. *Conserv. Biol.* **2008**, *22*, 1186–1193. [CrossRef] [PubMed]

26. Sheriff, R.E.; Geldart, L.P. Background Mathematics. In *Exploration Seismology*; Cambridge University Press: Cambridge, UK, 1995; pp. 517–568.

27. Canham, C.D.; Denslow, J.S.; Platt, W.J.; Runkle, J.; Spies, T.A.; White, P.S. Light Regimes beneath Closed Canopies and Tree-Fall Gaps in Temperate and Tropical Forests. *Can. J. For. Res.* **1990**, *20*, 620. [CrossRef]

28. Montané, F.; Guixé, D.; Camprodon, J. Canopy Cover and Understory Composition Determine Abundance of *Vaccinium myrtillus* L., a Key Plant for Capercaillie (*Tetrao urogallus*), in Subalpine Forests in the Pyrenees. *Plant Ecol. Divers.* **2016**, *9*, 187–198. [CrossRef]

29. Arienti, M.C.; Cumming, S.G.; Krawchuk, M.A.; Boutin, S. Road Network Density Correlated with Increased Lightning Fire Incidence in the Canadian Western Boreal Forest. *Int. J. Wildland Fire* **2009**, *18*, 970–982. [CrossRef]

30. Lee, P.; Boutin, S. Persistence and Developmental Transition of Wide Seismic Lines in the Western Boreal Plains of Canada. *J. Environ. Manag.* **2006**, *78*, 240–250. [CrossRef] [PubMed]

31. Van Rensen, C.K.; Nielsen, S.E.; White, B.; Vinge, T.; Lieffers, V.J. Natural Regeneration of Forest Vegetation on Legacy Seismic Lines in Boreal Habitats in Alberta's Oil Sands Region. *Biol. Conserv.* **2015**, *184*, 127–135. [CrossRef]

32. Dabros, A.; Hammond, H.E.J.; Pinzon, J.; Pinno, B.; Langor, D. Edge Influence of Low-Impact Seismic Lines for Oil Exploration on Upland Forest Vegetation in Northern Alberta (Canada). *For. Ecol. Manag.* **2017**, *400*, 278–288. [CrossRef]

33. Historical Wildfire Database 1999–2014. AAF—Agriculture and Forestry. Available online: http://wildfire. alberta.ca/resources/historical-data/historical-wildfire-database.aspx (accessed on 14 July 2017).

34. Peet, R.K.; Wentworth, T.R.; White, P.S. A Flexible, Multipurpose Method for Recording Vegetation Composition and Structure. *Castanea* **1998**, *63*, 262–274.

35. Pitelka, L.F.; Stanton, D.S.; Peckenham, M.O. Effects of Light and Density on Resource Allocation in a Forest Herb, Aster Acuminatus (Compositae). *Am. J. Bot.* **1980**, *67*, 942. [CrossRef]

36. Christensen, N.L. Fire and Soil-Plant Nutrient Relations in a Pine-Wiregrass Savanna on the Coastal Plain of North Carolina. *Oecologia* **1977**, *31*, 27–44. [CrossRef] [PubMed]

37. Simms, E.L. The Effect of Nitrogen and Phosphorus Addition on the Growth, Reproduction, and Nutrient Dynamics of Two Ericaceous Shrubs. *Oecologia* **1987**, *71*, 541–547. [CrossRef] [PubMed]

38. Faison, E.K.; Del Tredici, P.; Foster, D.R. To Sprout or Not to Sprout: Multiple Factors Determine the Vigor of Kalmia Latifolia (Ericaceae) in Southwestern Connecticut. *Rhodora* **2014**, *116*, 148–162. [CrossRef]

39. Ballester, A.; Vieitez, A.M.; Vieitez, E. Allelopathic Potential ofErica Vagans, Calluna Vulgaris, and Daboecia Cantabrica. *J. Chem. Ecol.* **1982**, *8*, 851–857. [CrossRef] [PubMed]

40. Watson, M.A.; Casper, B.B. Morphogenetic Constraints on Patterns of Carbon Distribution in Plants. *Annu. Rev. Ecol. Syst.* **1984**, *15*, 233–258. [CrossRef]

41. Reader, R.J. Bog Ericad Flowers: Self-Compatibility and Relative Attractiveness to Bees. *Can. J. Bot.* **1977**, *55*, 2279–2287. [CrossRef]

42. Usui, M.; Kevan, P.G.; Obbard, M. Pollination and Breeding System of Lowbush Blueberries, *Vaccinium angustifolium* Ait. and *V. myrtilloides* Michx. (Ericaceae), in the Boreal Forest. *Can. Field Nat.* **2005**, *119*, 48. [CrossRef]

43. Brodeur, V.; Ouellet, J.-P.; Courtois, R.; Fortin, D. Habitat Selection by Black Bears in an Intensively Logged Boreal Forest. *Can. J. Zool.* **2008**, *86*, 1307–1316. [CrossRef]

44. Tigner, J.; Bayne, E.M.; Boutin, S. Black Bear Use of Seismic Lines in Northern Canada. *J. Wildl. Manag.* **2014**, *78*, 282–292. [CrossRef]

45. Turner, N.J.; Cocksedge, W. Aboriginal Use of Non-Timber Forest Products in Northw Estern North America. *J. Sustain. For.* **2001**, *13*, 31–58. [CrossRef]

46. Gottesfeld, L.M.J. Aboriginal Burning for Vegetation Management in Northwest British Columbia. *Hum. Ecol.* **1994**, *22*, 171–188. [CrossRef]

47. Environment Canada. *Recovery Strategy for the Woodland Caribou (Rangifer tarandus caribou), Boreal Population, in Canada*; Species at Risk Act Recovery Strategy Series; Environment Canada: Ottawa, ON, Canada, 2012.

48. Hebblewhite, M. Billion Dollar Boreal Woodland Caribou and the Biodiversity Impacts of the Global Oil and Gas Industry. *Biol. Conserv.* **2016**, *206*, 102–111. [CrossRef]

49. James, A.R.C.; Stuart-Smith, A.K. Distribution of Caribou and Wolves in Relation to Linear Corridors. *J. Wildl. Manag.* **2000**, *64*, 154. [CrossRef]

50. Edmonds, E.J. Population Status, Distribution, and Movements of Woodland Caribou in West Central Alberta. *Can. J. Zool.* **1988**, *66*, 817–826. [CrossRef]

51. Pinard, V.; Dussault, C.; Ouellet, J.-P.; Fortin, D.; Courtois, R. Calving Rate, Calf Survival Rate, and Habitat Selection of Forest-Dwelling Caribou in a Highly Managed Landscape. *J. Wildl. Manag.* **2012**, *76*, 189–199. [CrossRef]

52. Machtans, C.S. Songbird Response to Seismic Lines in the Western Boreal Forest: A Manipulative Experiment. *Can. J. Zool.* **2006**, *84*, 1421–1430. [CrossRef]

![forests logo] *forests*

MDPI

Article

Fire Effects on Historical Wildfire Refugia in Contemporary Wildfires

Crystal A. Kolden [1,*,†] , Tyler M. Bleeker [2,†], Alistair M. S. Smith [1], Helen M. Poulos [3] and Ann E. Camp [4]

[1]   Department of Forest, Rangeland, and Fire Sciences, University of Idaho, 875 Perimeter Drive MS 1133, Moscow, ID 83844, USA; alistair@uidaho.edu
[2]   Department of Geography, University of Idaho, 875 Perimeter Drive MS 3021, Moscow, ID 83844, USA; tyler.bleeker@gmail.com
[3]   College of the Environment, Wesleyan University, 284 High Street, Middletown, CT 06459, USA; hpoulos@wesleyan.edu
[4]   School of Forestry and Environmental Studies, Yale University, 195 Prospect Street, New Haven, CT 06511, USA; ann.camp@yale.edu
*   Correspondence: ckolden@uidaho.edu; Tel.: +1-208-885-6018
†   Equal contribution.

Received: 2 August 2017; Accepted: 10 October 2017; Published: 20 October 2017

Abstract: Wildfire refugia are forest patches that are minimally-impacted by fire and provide critical habitats for fire-sensitive species and seed sources for post-fire forest regeneration. Wildfire refugia are relatively understudied, particularly concerning the impacts of subsequent fires on existing refugia. We opportunistically re-visited 122 sites classified in 1994 for a prior fire refugia study, which were burned by two wildfires in 2012 in the Cascade mountains of central Washington, USA. We evaluated the fire effects for historically persistent fire refugia and compared them to the surrounding non-refugial forest matrix. Of 122 total refugial (43 plots) and non-refugial (79 plots) sites sampled following the 2012 wildfires, one refugial and five non-refugial plots did not burn in 2012. Refugial sites burned more severely and experienced higher tree mortality than non-refugial plots, potentially due to the greater amount of time since the last fire, producing higher fuel accumulation. Although most sites maintained the pre-fire development stage, 19 percent of sites transitioned to Early development and 31 percent of sites converted from Closed to Open canopy. These structural transitions may contribute to forest restoration in fire-adapted forests where fire has been excluded for over a century, but this requires further analysis.

Keywords: burn severity; forest structure; succession; Cascade Range; restoration; mixed-conifer forest

1. Introduction

Wildfire is an integral natural ecosystem process worldwide [1], including the conifer forests of the inland of northwestern United States [2]. In these fire-adapted ecosystems, not every stand is affected equally by wildfire [3–5] and differential degrees of ecological change can result from variation in burn severity within single fires [6]. Spatially heterogeneous burn severity drives forest structure and composition [7,8] and produces complexity that is critical to supporting biodiversity, ecosystem services, and forest resilience [2,9,10]. Much of the inland Northwest is characterized by a mixed-severity fire regime that includes both frequent, low-severity fire and infrequent, high-severity fire [2]. Less studied, but equally important ecologically, are forest patches that remain unburned or experience a relatively low degree of change from a wildfire event; these patches are often described as 'wildfire refugia' [11–14]. Such patches represent key landscape elements that support the persistence of fire-sensitive flora and fauna both during the fire (as a refuge) and after the fire (as intact habitat), and provide seed sources for the regeneration of adjacent severely burned areas post-fire [14]. One of the key uncertainties about refugia is their persistence, and what characteristics determine persistent versus temporary refugia during successive ecological disturbances [11]. Resolving this uncertainty is necessary for several reasons, including improving our understanding and the predictability of forest succession dynamics, developing conservation management strategies for critical refugia that are vulnerable, and establishing a baseline from which global change impacts can be measured [14].

Refugia have been broadly defined and delineated, depending on the discipline and research query [15]; to-date, there is no single, widely-accepted definition of fire refugia [4,12,14]. Areas within fire perimeters that do not experience any fire effects, e.g., [16,17], or that experience fire effects at a lower severity than surrounding areas, e.g., [11], have both been described as wildfire refugia. In the literature, the scale and characteristics of wildfire refugia vary by organism and ecosystem [14], leading to different definitions of what constitutes a wildfire refugium. The occurrence of wildfire refugia depends upon several environmental factors that vary spatially and temporally, including topography, climate, soils, geomorphology, and ecological disturbances such as meteorological events, insects, pathogens, and fire [18,19]. These factors interact together to create spatial heterogeneity in vegetation and forest structure, which is maintained by fire [20]. For example, vegetation characteristics such as stand age or structure can either increase or decrease the likelihood of fire occurrence [21,22] or can minimize or exacerbate the effects when a fire does occur [23]. Additionally, topographic complexity influences burn severity [23,24] and refugia formation [11,12]. Generally, bottom-up factors such as vegetation and topography exert a greater influence on burn severity than top-down controls such as weather or climate [24–27], but in extreme fire weather events, local weather conditions may override vegetative and landscape effects on burn severity [28,29]. Furthermore, human activities such as stand management have the potential to influence the severity of burn patterns and the formation and persistence of refugial patches [30].

Changes in land cover and land use have greatly altered both fire frequency and intensity in inland northwest forests [31], further confounding our ability to understand the formation of fire refugia. Euro-American settlements and associated timber harvesting and grazing, along with a century of fire exclusion, have impacted the forest stand structure [32,33], as well as the spatial distribution and intensity of wildfires, increasing the risk of stand-replacing fire [5,25]. Inland northwest forests are already experiencing climatic conditions conducive to an increased occurrence and duration of wildfires (e.g., extended heat waves and droughts), and these trends are projected to continue through the 21st century [34–37]. Recent studies suggest that the severity of wildfires may be increasing for some ecosystems [38,39], although the robustness of such trends is questioned by other studies [27,40–42]. If the overall fire severity increases, fire refugia that were historically sheltered from fire or experienced only low severity fires might now burn at a higher severity under the contemporary fire regime. To-date, however, there has been little opportunity to assess the effects of the contemporary, altered fire regime on historically persistent fire refugia.

In 2012, the Wenatchee Complex Fires burned through an area of the central Washington state, USA, that had been previously sampled and classified into refugial and non-refugial patches two decades prior to the two fires [11]. This provided a unique opportunity to assess contemporary fire effects on field-delineated refugial and non-refugial stands to examine the persistence of historic wildfire refugia under changing fire regimes and climatic conditions. To our knowledge, no field-based studies have specifically assessed the persistence of wildfire refugia based on the pre-fire identification of long-term wildfire refugia. Camp and colleagues [11] classified historic wildfire refugia using forest stand structure, tree age, and species composition data within the Wenatchee National Forest, defining historic wildfire refugia as forest patches that had been minimally affected by fire events for at least 140 years (well before the onset of fire exclusion or active fire suppression [43]), while the surrounding forest matrix had experienced greater fire effects in historic fire events. The primary research objective of this study was to determine the effects of the 2012 wildfires on these historical fire refugia, specifically their persistence and changes in forest structure. This question was addressed by (1) quantifying and comparing the fire effects between refugial and non-refugial plots as classified pre-fire by Camp et al. [11] and (2) by investigating changes from inferred pre- to post-fire forest stand structure across stands that had been previously classified as refugial or non-refugial. More broadly, this case study evaluates how the occurrence of wildfire refugia and the mosaic of forest structure may shift in response to changing fire regime characteristics in the contemporary era. While conclusions from a case study are limited in application, such studies are needed to identify threats to critical refugia that are central to conservation plans for key species.

2. Materials and Methods

2.1. Study Area

The study area is located in the 47,794 ha Swauk Late Successional Reserve of the Okanogan-Wenatchee National Forest (Cle Elum Ranger District) of central Washington State, USA (Figure 1). A Late Successional Reserve (LSR) is a management designation for an area created through the Northwest Forest Plan with the objective of protecting and enhancing the condition of late-successional and old-growth forest ecosystems; as such, only limited stand management is permitted in LSR-designated areas [44]. Prior to attaining the LSR status in 1995, this area was subject to more intensive management, including selective timber harvesting, clear-cutting, road building, and mining, particularly in lower drainages [11]. The study area is located at the far eastern edge of the Cascade mountain range, extending into the dry interior Columbia River Plateau to the east. Sampled plots in this study ranged in elevation from 1027 to 1912 m. Vegetative communities in the Swauk LSR form a heterogeneous landscape due to strong responses to the dissected topography, precipitation gradient, and insolation differences [45]. At more xeric sites, lower elevations, and south-facing aspects, open-canopy ponderosa pine (*Pinus ponderosa* Dougl. Ex Forbes) and Douglas-fir (*Psuedotsuga menziesii var. menziesii* (Mirb.) Franco) stands are common. At more mesic sites, higher elevations, and north-facing slopes, stands of subalpine fir (*Abies lasiocarpa* (Hook.) Nutt.) and lodgepole pine (*Pinus contorta* var. *latifolia* Dougl. ex Loud.) are more typical. However, specific site conditions can cause the immediate juxtaposition of disjunct forest stands. Within the surveyed plots, grand fir (*Abies grandis* (Dougl. Ex D. Don.)) accounted for 39% of sampled trees, Douglas-fir accounted for 31%, and subalpine fir a further 11%.

The pre-European settlement fire regime near the Swauk LSR varied spatially by vegetative type with a mean fire return interval of seven to 43 years, and with large fires occurring approximately every 27 years [43]. Pre-European settlement fire severity also varied in the study area, with drier forest types experiencing low-severity fire, while more mesic forest types experienced occasional moderate and high-severity fires. However, fire regimes have been significantly altered since European settlement. Fire frequency declined dramatically around 1900, coinciding with the start of commercial logging [43] and the advent of active fire suppression in the Wenatchee National Forest [46]. Consequently, there is no evidence of fire occurrence in the Swauk LSR from 1900 to 2012 [47].

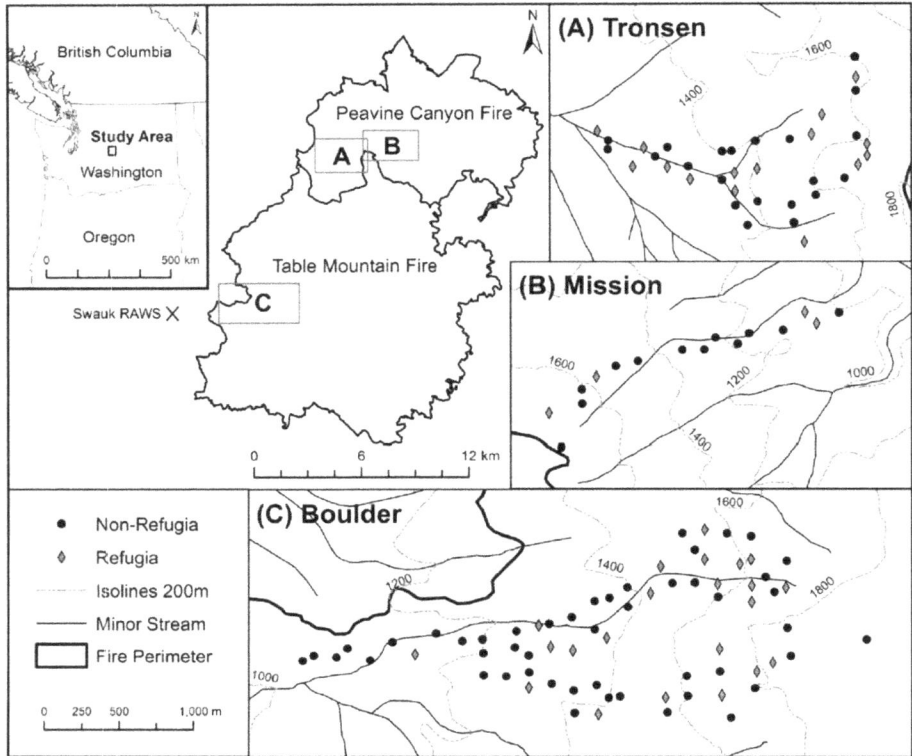

Figure 1. Study area in the east Cascades of central Washington State, USA, with burn perimeters of the 2012 Peavine Canyon and Table Mountain fires and spatial locations of the Camp et al. [11] classified refugial (diamonds) and non-refugial (circles) plots in the three study drainages (Tronsen, Mission, and Boulder).

2.2. The 2012 Fires

The Table Mountain and Peavine Canyon (as part of the Wenatchee Complex) fires burned simultaneously after the fires were ignited from lightning strikes in early September 2012. These fires eventually merged, creating a total burn area of 25,274 ha. These fires burned 226 sample sites of the Camp et al. [11] study in three different drainages, where 43 and 183 plots, respectively, were previously classified as refugial and non-refugial. The fires burned under anomalously dry and warm weather conditions compared to 1985 to 2014 climate data recorded at the Swauk Remote Automatic Weather Station (RAWS) approximately 2 km west of the fire perimeter [48]. The average air temperature for the July through September fire season was higher than normal (83rd percentile), and the average air temperature for September when the fires ignited was much higher than normal (93rd percentile). The average relative humidity for the fire season was lower than normal (35th percentile), and the average relative humidity for September was much lower than normal (3rd percentile). Precipitation was slightly below normal for the fire season (38th percentile), but late-season drought was particularly pertinent as the study area went 52 consecutive days without measurable precipitation prior to fire ignition and 34 days without afterwards [48].

2.3. Field Measurements

Field data were collected in the summer and fall of 2014 by resampling the general locations of the Camp et al. [11] sample sites that burned in 2012 (Figure 1). The original Camp et al. [11] plots were established along multiple transects using a hip chain and sighting compass in the field, and the final plot locations were annotated on topographic maps. To determine the global positioning system (GPS) coordinates of these plots, the annotated topographic maps were first digitized into a geographic information system database and then plot coordinates were navigated to with a handheld GPS unit in the field. Since the prior plots were not permanently monumented, new plots were established as close as possible to the original plots within the same forest stand. GPS locations that fell in barren areas or in unsafe field sites were relocated to the nearest suitable forested site. A subset of plots from the Camp et al. [11] study was sampled across the entire elevational gradient for each drainage. A total of 41 refugial sites and 81 non-refugial sites were sampled across the three study drainages for a total of 122 sampled plots.

At each sample site, a modified replication of the Camp et al. [11] sampling protocol was conducted (Figure 2). Plot centers were monumented at the ascribed GPS coordinates and a 15.2-m diameter circular plot covering 725.8 m^2 was established. Variables collected or derived for the plot level at each site included four topographic, nine vegetative, and four fire effect variables (Table 1). Plot aspect was measured in degrees and then transformed into northness and eastness indices for analysis. We also collected variables at the tree level for each plot, where a sweeping transect from an azimuth of 0° (north) was used to sample the first ten trees in the plot meeting a minimum threshold diameter at breast height (DBH) of 12.6 cm. We tagged and assessed each set of 10 trees for three demographic and five fire effects variables (Table 2). Three additional vegetative variables at the plot level (average DBH, maximum DBH, and pre-fire plot basal area) were derived from the DBH measurements of the 10 sampled trees on each plot. A total of 1220 trees were sampled in this study.

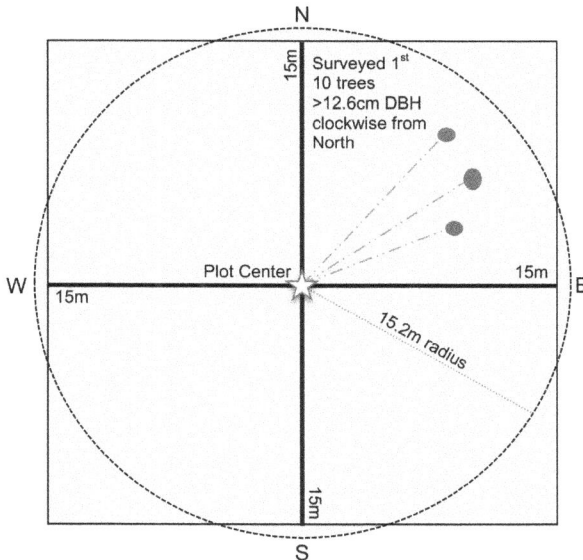

Figure 2. Diagram of the 2014 field sampling protocol.

Table 1. Plot-level topographic, vegetative, and fire effects variables.

Data Type	Variable	Definition
Topographic	Aspect Northness	cos(π/2-aspect); range from -1 (south) to 1 (north)
	Aspect Eastness	sin(π/2-aspect); range from -1 (west) to 1 (east)
	Slope	Degrees; measured by clinometer
	Elevation	Meters; measured by global positioning system
	Topography Type	Ten option categorical classification (from [11])
Vegetative	Max Canopy Height	Meters; measured by Impulse Laser
	Species Present	Dominant tree species
	Canopy Structure	Presence/absence of overstory/subcanopy strata
	All Trees Pre-fire	Count of all trees alive in plot pre-fire
	Overstory Trees Pre-fire	Count of all $\geq$12.6 cm DBH alive in overstory strata pre-fire
	Subcanopy Trees Pre-fire	Count of all trees $\geq$12.6 cm DBH alive in subcanopy strata pre-fire
	Total Canopy Cover Pre-fire	Ocular estimate of pre-fire canopy cover for all tree strata
	Overstory Canopy Cover Pre-fire	Ocular estimate of pre-fire canopy cover for overstory strata
	Subcanopy Canopy Cover Pre-fire	Ocular estimate of pre-fire canopy cover for subcanopy strata
	Average DBH	Average DBH of 10 sampled trees on plot
	Maximum DBH	Maximum DBH of 10 sampled trees on plot
	Pre-fire Plot Basal Area	Average basal area of 10 sampled trees on plot [π(DBH/2)^2] * count of trees on plot. Unit: m^2 basal area/900 m^2 plot area
Fire Effects	Total Tree Mortality	Count of all tree mortality in plot post-fire
	Overstory Tree Mortality	Count of all $\geq$12.6 cm DBH tree mortality in overstory strata post-fire
	Subcanopy Tree Mortality	Count of all $\geq$12.6 cm DBH subcanopy tree mortality in subcanopy strata post-fire
	Total Plot CBI	Composite Burn Index protocol (score from 0 to 3)

DBH—diameter at breast height; CBI—Composite Burn Index [49].

Table 2. Tree-level demographic and fire effect variables.

Data Type	Variable	Definition
Demographic	Species	Field Identification
	Diameter at breast height	Field measure; cm
	Secondary Stress	Presence of: Fire, Freezing, Fungus, Insect, Mechanical, Mistletoe, Rot
Fire Effects	Mortality	Fire-induced tree death
	Percent Bole Char	Maximum percent of basal bole with visible char
	Bole Char Max Height	Maximum height of continuous char on bole (m)
	Percent Foliage Scorch	Ocular estimate of pre-fire living foliage scorched or girdled
	Percent Foliage Torch	Ocular estimate of pre-fire living foliage torched by fire

Burn severity was also assessed at each site using the Composite Burn Index (CBI) protocol [49]. Due to the aggregation of fire effects in the CBI protocol [50], we also assessed the burn severity for each of the five different specific fire effect metrics at the tree level. Per [49], the CBI analysis area was modified to be a 30 by 30-m square to correspond to the size of a Landsat pixel; additionally, the plot was oriented in the cardinal directions to align with the Camp et al. [11] plot azimuths. While CBI is normally conducted one-year post-fire [49], the methodology has been previously utilized to assess fire effects two years post-fire [51]; doing so allows an assessment of longer-term fire effects while also capturing delayed mortality in the tree strata.

2.4. Data Quality Assurance

To ensure that the plots sampled in 2014 were comparable to the plots that Camp et al. [11] originally sampled and classified into potential refugia, we conducted a paired-plot assessment using the original plot data and stand delineations produced by Camp et al. [11]. Stand delineations developed by Camp et al. [11] and aerial imagery were used to determine if the original and resampled plots fell in the same stand. If the original and resampled plot pair was not visually within the same stand, then the topographic and vegetative attributes of each plot were compared for similarities. We were not able to confidently match 13 sampled plots through this qualitative comparison and these were excluded from further paired-plot analysis which resulted in a total of 109 confident paired-plot matches. To confirm that the plots were equivalent to Camp et al. [11], we then used a paired *t*-test with Welch modification for non-normality to test for differences in the topographic setting of both sample sets [52] by calculating the modified Heat Load Index from McCune and Keon [53], using Equation (1) due to the steep slopes at our study site. As we found no significant differences, we contend that these 109 matched plots provide a conservative match to the Camp et al. [11] plot-level data.

2.5. Comparison of Fire Effects between Resampled and Original Plots

We used the 109 plots determined to be confident matches from the data quality assurance step (Section 2.4) and tested for differences in burn occurrence and burn severity between refugial (*n* = 36) and non-refugial (*n* = 73) plots as classified by Camp et al. [11]. Differences in burn occurrence between refugial and non-refugial plots were assessed with a chi-square test of independence.

Fire effects were compared between those plots which Camp et al. [11] classified as refugial versus non-refugial for eight fire effects metrics. Overall plot burn severity was assessed using the Composite Burn Index score and seven individual burn severity metrics (maximum bole char height, percent bole charred, percent foliage scorched, percent foliage torched and percent tree mortality of overstory, and sub-canopy tree strata) were compared between the Camp et al. [11] refugial and non-refugial plots using a Wilcoxon-signed-rank test due to the non-normal distribution of the data (α = 0.1). The Wilcoxon-signed-rank test results in a W-value where lower W-values correspond with lower *p*-values.

2.6. Assessment of Changes in Forest Structure

To investigate fire impacts on forest structure, we assigned successional states to each plot based on the inferred pre-fire and observed post-fire structural characteristics. In order to assign successional forest structure states to our sample plots, we first needed to assign the type of forested ecosystem in which each plot occurred. A nationwide vegetative community classification called the Biophysical Setting (BpS; NatureServe) was developed as part of the LANDFIRE resource management and planning tool [54,55]. BpS and other LANDFIRE vegetation products are widely used in state-and-transition studies because of the ecologically-based successional states and ecological transitions described in each BpS model [56–59]. All 122 plots sampled in 2014 were assigned to one of the three most common Biophysical Settings for the study area through an analysis of the 2014 quantitative plot data, recorded qualitative field observations, and photos for each plot. Since each BpS model covers a broad geographical area and cannot account for more localized variation within a single model, BpS models were refined with locally available information on the habitat types of the Wenatchee National Forest [60]; forest types were cross-walked to the Wenatchee National Forest correlate names of the Douglas-fir (*n* = 17), grand fir (*n* = 92), and subalpine fir series (*n* = 13) (Table 3), to be consistent with and comparable to the three forest series reported by Camp et al. [11] in their original analysis.

Table 3. Cross-walk between the BpS Model and Wenatchee NF Correlate from [60].

Biophysical Setting Model Name	Wenatchee NF Correlate	Plots
Northern Rocky Mountain Dry-Mesic Montane Mixed Conifer Forest	Douglas-fir Series	17
East Cascades Mesic Montane Mixed-Conifer Forest and Woodland	Grand Fir Series	84
Rocky Mountain Subalpine Dry-Mesic Spruce-Fir Forest and Woodland	Subalpine Fir Series	13

Once a BpS model and forest series were assigned to each plot, pre- and post-fire forest structure was classified using the Vegetation Dynamics Development Tool (VDDT) [61] classes for each respective BpS model. The VDDT model for each of the three BpS models uses five distinct successional/structural classes (hereafter, referred to as "successional states"): Early Development, Mid-Development Open Canopy, Mid-Development Closed Canopy, Late Development Open Canopy, and Late Development Closed Canopy (Figure 3). Within each BpS model there are different attribute criteria for what constitutes a particular successional state based on species composition; quantitative criteria used in classification included canopy cover, canopy height, maximum tree size class, and average tree size class, while qualitative criteria included the species present, tree relative canopy position, and fuel model (Table 4). Post-fire successional state was classified through post-fire vegetation observed during the 2014 field season. For the pre-fire successional state, in-field estimates of canopy cover and counts of trees presumed living pre-fire were used as best approximations to infer pre-fire vegetative structure and composition. Field notes and plot photos were used to refine this successional state classification when quantitative data alone proved inconclusive.

Once pre-fire and post-fire successional states were assigned to each plot, the transition from one successional state to another due to fire effects was assessed. There were three distinct transitions a plot could have taken due to fire effects (hereafter referred to as "ecological transitions"): (1) plot was maintained in the current successional state; (2) plot canopy was thinned from a closed to open canopy structure of the same development stage; or (3) plot transitioned to an early development successional state. We conducted the transition analysis for all 122 plots (not just the 109 paired plots) as we were not statistically comparing the refugial and non-refugial plots for this analysis, but rather characterizing changes in forest structure.

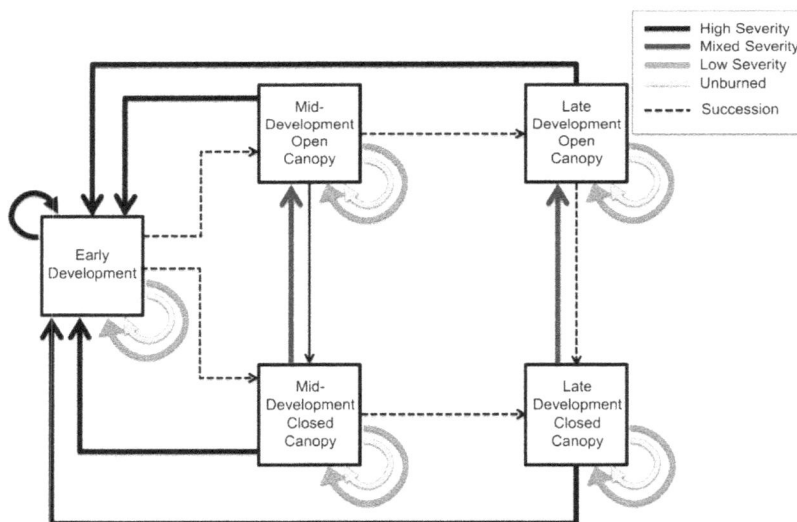

Figure 3. State-and-transition model of the Vegetation Dynamics Development Tool successional states and ecological transitions as affected by differential wildfire severity (modified from [4]).

Table 4. Quantitative criteria and thresholds utilized to classify plots into the five successional states that describe the forest structure for the three Biophysical Setting (BpS) models present in the study area. Both the initial classification of the plots into BpS models and further classification into successional states also utilized qualitative information and species lists in the BpS model descriptions, which can be found on the LANDFIRE website (www.landfire.gov).

Biophysical Setting (BpS)	State	Cover	Height	Tree Size Class
Northern Rocky Mountain Dry-Mesic Montane Mixed Conifer Forest (Douglas Fir Series)	Early	0–20%	Tree 0–5 m	Sapling > 4.5', <5" DBH
	Mid-Open	0–40%	Tree 5.1–25 m	9–21" DBH
	Mid-Closed	41–100%	Tree 5.1–25 m	9–21" DBH
	Late-Open	11–40%	Tree 25.1–50 m	> 33" DBH
	Late-Closed	41–100%	Tree 25.1–50 m	> 33" DBH
East Cascades Mesic Montane Mixed Conifer Forest and Woodland (Grand Fir series)	Early	0–100%	Tree 0–10 m	Sapling >4.5', <5" DBH
	Mid-Open	0–60%	Tree 10.1–25 m	9–21" DBH
	Mid-Closed	61–100%	Tree 10.1–25 m	9–21" DBH
	Late-Open	0–60%	Tree 25.1–>50 m	>33" DBH
	Late-Closed	61–100%	Tree 25.1–>50 m	>33" DBH
Rocky Mountain Subalpine Dry-Mesic Spruce-Fir Forest and Woodland (Subalpine Fir series)	Early	0–40%	Shrub 0–0.5 m	None
	Mid-Open	11–30%	Tree 5.1–10 m	9–21" DBH
	Mid-Closed	31–60%	Tree 5.1–10 m	9–21" DBH
	Late-Open	11–40%	Tree 10.1–25 m	21–33" DBH
	Late-Closed	41–70%	Tree 10.1–25 m	21–33" DBH

2.7. Limitations of Methods

As we did not have pre-fire vegetative data, and the Camp et al. [11] plots were not monumented, we inferred pre-fire vegetation attributes with only the burned post-fire vegetation available to sample, which is standard for the CBI protocol [49]. Vegetative reconstruction estimates can be problematic due to the uncertainties of differentiating the magnitude of observed effects solely due to the fire event from the pre-fire conditions and other ecological changes (such as an insect attack or hydrological flow) that occur between the fire and the post-fire measurements [62–64]. Canopy cover is particularly difficult to measure from ocular estimates [65], especially after the canopy has been partially consumed in a fire, and it is reasonable to assume that the two-year lag between the fire and our field data collection introduced further error. However, we felt that the opportunity to revisit forested stands that were classified for a refugial objective in a prior study in order to describe fire effects was an opportunity that could not be ignored, despite these limitations.

3. Results

3.1. Assessment of Fire Effects in the Camp et al., Refugial and Non-Refugial Plots

Of the 109 paired plots that were established initially by Camp et al. [11] and that we revisited after the 2012 wildfires, only six (5.5%) did not experience any fire effects in 2012 (i.e., they were unburned). Only one of these six unburned plots was classified as a fire refugium by Camp et al. [11] (Table 5), although this difference in burn occurrence between refugial and non-refugial plots was not statistically significant (chi-square test of independence, $\chi^2 = 0.768$, $p = 0.381$) due to the low number of samples. This translates to <1 percent of plots persisting as refugia and 4.5% having no fire effects and becoming refugia under the most conservative definition.

Table 5. Number of plots classified in the original Camp et al. [11] study that burned in the 2012 fires, re-surveyed in the present study, paired to the plots from [11], and found to have no fire effects (i.e., the most conservative definition of refugial) by this study.

Sampling Description	Refugial per [11]	Non-Refugial per [11]	Total
Sampled by [11]	43	183	226
Sampled in 2014, this study	41	81	122
Plots paired to [11]	36	73	109
Plots unburned	1	5	6

A comparison of the fire effects between the 109 matched Camp et al. [11] refugial and non-refugial plots revealed a trend in differences in burn severity, where classified refugial plots generally experienced greater fire effects (Figure 4). Refugial plots burned more severely than non-refugial plots as assessed through the total plot CBI metric (W = 1015.5, p = 0.0582). For the tree-level severity metrics, the percent total tree mortality (W = 921.5, p = 0.0196), percent understory tree mortality (W = 867, p = 0.0116), average bole char height (W = 1042, p = 0.0802), and average foliage scorched (W = 899, p = 0.00754) were all significantly higher for refugial plots (at α = 0.1, used due to non-normal data and a greater number of non-refugial plots). Percent overstory tree mortality and percent bole char observations were also higher for refugial plots, but these differences were not statistically significant. Foliage torched showed no significant difference between refugial and non-refugial plots, but this result is likely attributable to low overall levels of foliar torching, with the exception of a few highly torched plots resulting from crown fires.

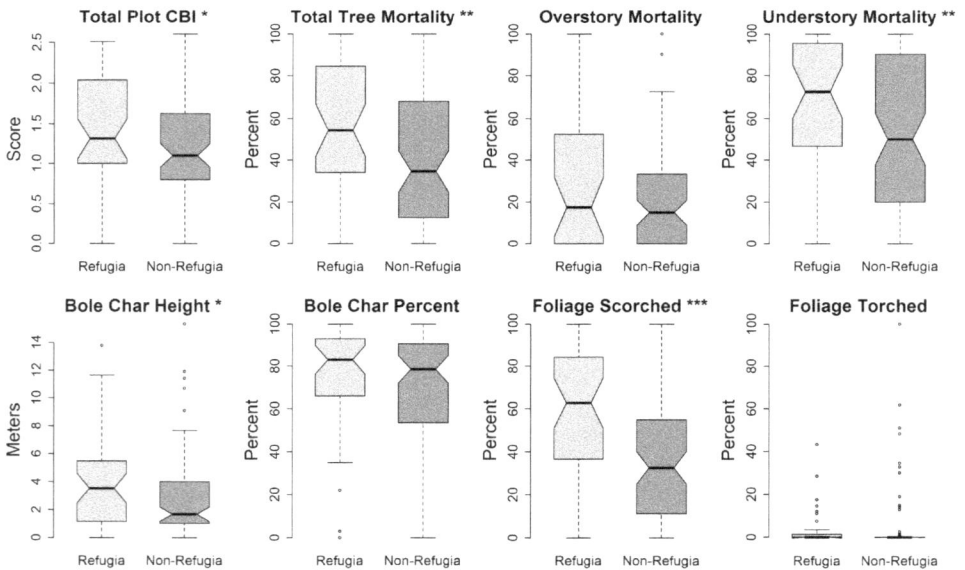

Figure 4. Comparison of fire effects between Camp et al. [11] refugial and non-refugial plots for eight different burn severity metrics. * $p \leq 0.10$, ** $p \leq 0.05$, *** $p \leq 0.01$.

3.2. Changes in Successional State

Based on our reconstruction of forest structure and successional state pre-fire, there were no Early Development successional state plots pre-fire, with most of the plots classified at Mid-Development

(32% Open Canopy and 39% Closed Canopy), and the remaining 29% classified as Late Development (9% Open Canopy and 20% Closed Canopy) (Table 6). Post-fire, 19% of sample plots were classified as an Early Development successional state, 26% were classified as Late Development, and most plots were still Mid-Development (55%). Half (50%) of the sampled plots did not change in terms of the successional state due to fire effects, while nearly one-third (31%) transitioned from a closed to open-canopy structure (for both Mid- and Late Development categories combined). Closed Canopy plots classified as either Mid- or Late Development successional state (59%) were more abundant than Open Canopy plots pre-fire (41%), but Open Canopy plots (65%) were four times more abundant than Closed Canopy plots (16%) post-fire.

Table 6. Distribution of successional states for all 122 sampled plots according to pre-fire and post-fire successional state for all Biophysical Setting models.

Successional State	Pre-Fire State		Post-Fire State	
	Count	Percent	Count	Percent
Early Development	0	0%	23	19%
Mid-Development Open Canopy	39	32%	55	45%
Mid-Development Closed Canopy	47	39%	12	10%
Late Development Open Canopy	11	9%	24	20%
Late Development Closed Canopy	25	20%	8	6%

Assessing structural transitions by whether plots were classified as refugial or non-refugial by Camp et al. [11] reveals that a higher proportion of pre-fire refugia transitioned to Early development successional state (24%) than non-refugial plots (16%) (Table 7). More refugia were also thinned by the fire from Closed to Open Canopy (34%; in both Mid and Late-development states combined) than non-refugia (30%). Accordingly, more non-refugial plots maintained their pre-fire successional state.

Table 7. Primary plot structural transitions by pre-fire refugia classification.

Successional Transition	Refugia		Non-Refugia	
	Count	Percent	Count	Percent
Maintained state	17	42%	44	54%
Thinned from Closed to Open Canopy	14	34%	24	30%
Converted to Early Development state	10	24%	13	16%

4. Discussion

4.1. Comparison of Fire Effects between Sample Years

We found that the plots classified by Camp et al. [11] as refugial experienced more severe fire effects from the 2012 wildfires in comparison to classified non-refugial plots. This finding supports an inference of Camp et al. [11] in their study; they noted that pre-European settlement fire refugia appeared to have higher fire intensities and severities associated with longer fire return intervals. What was particularly surprising in the present study was just how few of the 122 plots sampled were unburned (<6%). Studies quantifying the unburned proportion across entire fires at local to regional scales have found a very broad range of the proportion unburned for individual fires [4,42], with Meddens et al. [13] reporting a regional average of 20% unburned for the inland Northwest. As these plots were specifically located by Camp et al. [11] to capture prospective fire refugia, the low proportion of unburned compared to regional averages further supports their conclusion that refugial plots, when they do eventually burn, do so with a higher severity than the non-refugial surrounding matrix. Because of the longer fire return intervals in refugial patches, tree species with thin bark and minimal self-pruning, such as grand fir, are able to establish and develop as ladder fuels that facilitate the surface fire ignition of crown fuels due to their lower canopy base height [32], a concept that Camp et al. [11] termed 'outgrowing' the refugial status. The additional fuels accumulated in these 'outgrown' refugial patches then lead such sites to burn at a higher severity than the surrounding matrix.

4.2. Distribution of Post-Fire Successional States and Ecological Transitions

We found that less than 30 percent of plots resampled in 2014 were Late Development stage pre-fire, with Mid-Development stage plots being the most common. This successional composition is relatively consistent with the pre-European settlement fire regime [39], although occasional high severity patches of fire (as part of the mixed-fire regime) would have also produced a few Early Development patches scattered across the landscape during that period. Here, however, we encountered no plots that were Early Development successional state pre-fire, which was unsurprising given that prior to 2012, the most recent wildfire in the study area occurred before 1900 due to the effectiveness of US fire suppression policies in the 20th century [46]. Post-fire, 19 percent of sampled plots transitioned to the Early Development state, increasing the structural heterogeneity of the study area [66]. Open Canopy stands were four times more abundant than Closed Canopy stands post-fire, whereas Closed Canopy stands were more abundant pre-fire. These sorts of transitions are consistent with the restoration needs highlighted by Haugo et al. [59] for these forests, and suggest that in this case study, at least, the fire behavior may have played a restorative role by creating both early successional openings and a more open canopy from a previously closed canopy.

The relationship between the more severe fire effects and the forest structural changes becomes evident when stratifying structural changes by whether plots were classified as refugia pre-fire. This fire-induced transition to an increased open canopy structure is consistent with mixed severity fires reducing canopy closure [32,43,67], and occurred in a higher proportion of refugial plots because they burned at a higher severity. Similarly, more refugial plots transitioned to Early Development successional state due to a higher burn severity in refugial plots. One point of interest in this breakdown is the number of plots that maintained their successional state (as there were no Early state plots pre-fire, these were all Mid- and Late-Development sites). Most of these plots (all but six) experienced some fire effects; these effects were of a low enough severity to maintain the forest structure successional state given the definitions in the BpS models.

4.3. Implications for Management

In the face of environmental change, managers can take many actions to increase ecosystem resilience [36,68]. Many plant and animal species are fire-sensitive and require refugial habitats to persist in the landscape [2], including species of high management concern such as the northern spotted owl (*Strix occidentalis caurina*), which was the focal species of the Camp et al. [11] refugia analysis. There is much concern over the persistence of refugia given climate change and changing ecological disturbance regimes [41], but our results demonstrate that even with fire re-entry into sites with over a century of fire exclusion, some refugial plots still maintained pre-fire forest structure and some previously non-refugial plots served as during-fire refugia in the 2012 fires. For managers seeking to conserve refugia for given species, there are likely pathways to doing so that do not require the total exclusion of wildfire, but additional research is needed to improve quantitative landscape composition models and determine when forests have become vulnerable to permanent structural transitions that can eradicate refugia [69].

5. Conclusions

Wildfire refugia in forested landscapes are critical to species survivorship both during and between fires, as well as facilitating forest regeneration in adjacent burned areas. However, refugia research is still nascent, with the definitions and delineation of refugia being highly variable and dependent upon species of interest. This case study provided a unique opportunity to assess the effects of a wildfire on fire refugia in central Washington State, USA, that were classified and delineated by Camp et al. [11]

over two decades ago and subsequently burned in two 2012 wildfires. Our findings that almost all of the plots burned, but half of the plots persisted in their pre-fire forest structure successional state, suggests that definitions of fire refugia focused on the maintenance of forest structure or canopy thresholds may reveal a higher proportion of refugia that are persistent through multiple fires. We also found that nearly one-fifth of the plots converted to Early Development state due to a high burn severity, and almost one-third experienced enough crown loss to transition from Closed to Open Canopy; these types of transitions are critical to forest restoration following over a century of fire exclusion in forested landscapes across the western US. Ultimately, this opportunistic case study highlights that the re-entry of fire into forests where fire has been excluded produces a range of fire effects that include the maintenance of pre-fire forest structure. Additionally, many current fire refugia may be vulnerable to future fire, particularly where fire exclusion has allowed for fuel accumulation and the growth of fire-sensitive species. Refugia more broadly, however, require much additional study to improve our understanding of their persistence, vulnerability, and role in forest structural dynamics.

Acknowledgments: Arjan Meddens, Phil Higuera, and Paul Hessburg contributed insight to the project and helped with data analysis. Additional thanks to Ryan McCarley for help in field data collection and figure editing. Support for this research came from NASA Award NNX10AT77A, NSF Idaho EPSCoR under award EPS-0814387, the USDA Forest Service Western Wildlands Environmental Threat Assessment Center under agreement number 13-JV-11261900-072, and the USGS Northwest Climate Science Center award G14AP00177. Smith was partially supported under NASA Award NNX11AO24G. This work was partially supported by the National Science Foundation under award no. DMS-1520873.

Author Contributions: C.A.K., T.M.B. and H.M.P. conceived and designed the experiments; T.M.B. performed the experiments; H.M.P. and A.E.C. contributed historical data and aerial photos; C.A.K. and T.M.B. analyzed the data; all authors wrote the paper.

Conflicts of Interest: The authors declare no conflict of interest.

References

1. Bowman, D.M.; Balch, J.K.; Artaxo, P.; Bond, W.J.; Carlson, J.M.; Cochrane, M.A.; D'Antonio, C.M.; DeFries, R.S.; Doyle, J.C.; Harrison, S.P.; et al. Fire in the Earth system. *Science* **2009**, *324*, 481–484. [CrossRef] [PubMed]
2. Agee, J.K. *Fire Ecology of Pacific Northwest Forests*; Island Press: Washington, DC, USA, 1996.
3. Lentile, L.B.; Morgan, P.; Hudak, A.T.; Bobbitt, M.J.; Lewis, S.A.; Smith, A.M.S.; Robichaud, P. Post-fire burn severity and vegetation response following eight large wildfires across the western United States. *Fire Ecol.* **2007**, *3*, 91–108. [CrossRef]
4. Kolden, C.A.; Abatzoglou, J.T.; Lutz, J.A.; Cansler, C.A.; Kane, J.T.; Van Wagtendonk, J.W.; Key, C.H. Climate contributors to forest mosaics: Ecological persistence following wildfire. *Northwest Sci.* **2015**, *89*, 219–238. [CrossRef]
5. Reilly, M.J.; Dunn, C.J.; Meigs, G.W.; Spies, T.A.; Kennedy, R.E.; Bailey, J.D.; Briggs, K. Contemporary patterns of fire extent and severity in forests of the Pacific Northwest, USA (1985–2010). *Ecosphere* **2017**, *8*. [CrossRef]
6. Lentile, L.B.; Holden, Z.A.; Smith, A.M.S.; Falkowski, M.J.; Hudak, A.T.; Morgan, P.; Lewis, S.A.; Gessler, P.E.; Benson, N.C. Remote sensing techniques to assess active fire characteristics and post-fire effects. *Int. J. Wildland Fire* **2006**, *15*, 319–345. [CrossRef]
7. Hessburg, P.F.; Salter, R.B.; James, K.M. Re-examining fire severity relations in pre-management era mixed conifer forests: Inferences from landscape patterns of forest structure. *Landsc. Ecol.* **2007**, *22*, 5–24. [CrossRef]
8. Freund, J.A.; Franklin, J.F.; Lutz, J.A. Structure of early old-growth Douglas-fir forests in the Pacific Northwest. *For. Ecol. Manag.* **2015**, *335*, 11–25. [CrossRef]
9. Hutto, R.L. The ecological importance of severe wildfires: Some like it hot. *Ecol. Appl.* **2008**, *18*, 1827–1834. [CrossRef] [PubMed]
10. Vaillant, N.M.; Kolden, C.A.; Smith, A.M.S. Assessing landscape vulnerability to wildfire in the United States. *Curr. For. Rep.* **2016**, *2*, 201–213. [CrossRef]
11. Camp, A.; Oliver, C.; Hessburg, P.; Everett, R. Predicting late-successional fire refugia pre-dating European settlement in the Wenatchee Mountains. *For. Ecol. Manag.* **1997**, *95*, 63–77. [CrossRef]

12. Krawchuk, M.A.; Haire, S.L.; Coop, J.; Parisien, M.A.; Whitman, E.; Chong, G.; Miller, C. Topographic and fire weather controls of fire refugia in forested ecosystems of northwestern North America. *Ecosphere* **2016**, *7*. [CrossRef]

13. Meddens, A.J.H.; Kolden, C.A.; Lutz, J.A. Detecting unburned areas within wildfire perimeters using Landsat and ancillary data across the northwestern United States. *Remote Sens. Environ.* **2016**, *186*, 275–285. [CrossRef]

14. Robinson, N.M.; Leonard, S.W.; Ritchie, E.G.; Bassett, M.; Chia, E.K.; Buckingham, S.; Gibb, H.; Bennett, A.F.; Clarke, M.F. Refuges for fauna in fire-prone landscapes: Their ecological function and importance. *J. Appl. Ecol.* **2013**, *50*, 1321–1329. [CrossRef]

15. Keppel, G.; Van Niel, K.P.; Wardell-Johnson, G.W.; Yates, C.J.; Byrne, M.; Mucina, L.; Schut, A.G.; Hopper, S.D.; Franklin, S.E. Refugia: Identifying and understanding safe havens for biodiversity under climate change. *Glob. Ecol. Biog.* **2012**, *21*, 393–404. [CrossRef]

16. Delong, S.C.; Kessler, D.W. Ecological characteristics of mature forest remnants left by wildfire. *For. Ecol. Manag.* **2000**, *131*, 93–106. [CrossRef]

17. Schwilk, D.W.; Keeley, J.E. The role of wildfire refugia in the distribution of Pinus sabiniana (Pinaceae) in the southern Sierra Nevada. *Madroño* **2006**, *53*, 364–372. [CrossRef]

18. Román-Cuesta, R.M.; Garcia, M.; Retana, J. Factors influencing the formation of unburned forest islands within the perimeter of a large forest fire. *For. Ecol. Manag.* **2009**, *258*, 71–80. [CrossRef]

19. Stine, P.; Hessburg, P.; Spies, T.; Kramer, M.; Fettig, C.J.; Hansen, A.; Lehmkuhl, J.; O'Hara, K.; Polivka, K.; Singleton, P.; et al. *The Ecology and Management of Moist Mixed-Conifer Forests in Eastern Oregon and Washington: A Synthesis of the Relevant Biophysical Science and Implications for Future Land Management*; General Technical Report; PNW-GTR-897; United States Department of Agriculture Forest Service: Portland, OR, USA, 2014.

20. Kane, V.R.; Lutz, J.A.; Roberts, S.L.; Smith, D.F.; McGaughey, R.J.; Povak, N.A.; Brooks, M.L. Landscape-scale effects of fire severity on mixed-conifer and red fir forest structure in Yosemite National Park. *For. Ecol. Manag.* **2013**, *287*, 17–31. [CrossRef]

21. Kushla, J.D.; Ripple, W.J. The role of terrain in a fire mosaic of a temperate coniferous forest. *For. Ecol. Manag.* **1997**, *95*, 97–107. [CrossRef]

22. Alexander, J.D.; Seavy, N.E.; Ralph, C.J.; Hogoboom, B. Vegetation and topographical correlates of fire severity from two fires in the Klamath-Siskiyou region of Oregon and California. *Int. J. Wildland Fire* **2006**, *15*, 237. [CrossRef]

23. Kane, V.R.; Cansler, C.A.; Povak, N.A.; Kane, J.T.; McGaughey, R.J.; Lutz, J.A.; Churchill, D.J.; North, M.P. Mixed severity fire effects within the Rim fire: Relative importance of local climate, fire weather, topography, and forest structure. *For. Ecol. Manag.* **2015**, *358*, 62–79. [CrossRef]

24. Dillon, G.K.; Holden, Z.A.; Morgan, P.; Crimins, M.A.; Heyerdahl, E.K.; Luce, C.H. Both topography and climate affected forest and woodland burn severity in two regions of the western US, 1984–2006. *Ecosphere* **2011**, *2*, 130. [CrossRef]

25. Cansler, C.A.; McKenzie, D. Climate, fire size, and biophysical setting control fire severity and spatial pattern in the northern Cascade Range, USA. *Ecol. Appl.* **2014**, *24*, 1037–1056. [CrossRef] [PubMed]

26. Birch, D.S.; Morgan, P.; Kolden, C.A.; Abatzoglou, J.T.; Dillon, G.K.; Hudak, A.T.; Holden, Z.A.; Smith, A.M.S. Vegetation, topography, and daily weather influenced burn severity in central Idaho and western Montana forests. *Ecosphere* **2015**, *6*, 17. [CrossRef]

27. Abatzoglou, J.T.; Kolden, C.A.; Williams, A.P.; Lutz, J.A.; Smith, A.M.S. Climatic influences on inter-annual variability in regional burn severity across western US forests. *Int. J. Wildland Fire* **2017**, *26*, 269–275. [CrossRef]

28. Bessie, W.C.; Johnson, E.A. Relative importance of fuels and weather on fire behavior in subalpine forests. *Ecology* **1995**, *76*, 747–762. [CrossRef]

29. Birch, D.S.; Morgan, P.; Kolden, C.A.; Hudak, A.T.; Smith, A.M.S. Is proportion burned severely related to daily area burned? *Environ. Res. Lett.* **2014**, *9*, 064011. [CrossRef]

30. Prichard, S.J.; Kennedy, M.C. Fuel treatments and landform modify landscape patterns of burn severity in an extreme fire event. *Ecol. Appl.* **2014**, *24*, 571–590. [CrossRef] [PubMed]

31. Hessburg, P.F.; Agee, J.K. An environmental narrative of Inland Northwest United States forests, 1800–2000. *For. Ecol. Manag.* **2003**, *178*, 23–59. [CrossRef]

32. Hessburg, P.F.; Agee, J.K.; Franklin, J.F. Dry forests and wildland fires of the inland Northwest USA: Contrasting the landscape ecology of the pre-settlement and modern eras. *For. Ecol. Manag.* **2005**, *211*, 117–139. [CrossRef]

33. Hessburg, P.F.; Smith, B.G.; Salter, R.B.; Ottmar, R.D.; Alvarado, E. Recent changes (1930s–1990s) in spatial patterns of interior northwest forests, USA. *For. Ecol. Manag.* **2000**, *136*, 53–83. [CrossRef]

34. Littell, J.S.; McKenzie, D.; Peterson, D.L.; Westerling, A.L. Climate and wildfire area burned in western U.S. ecoprovinces, 1916–2003. *Ecol. Appl.* **2009**, *19*, 1003–1021. [CrossRef] [PubMed]

35. Abatzoglou, J.T.; Kolden, C.A. Relationships between climate and macroscale area burned in the western United States. *Int. J. Wildland Fire* **2013**, *22*, 1003–1020. [CrossRef]

36. Barbero, R.; Abatzoglou, J.T.; Larkin, N.K.; Kolden, C.A.; Stocks, B. Climate change presents increased potential for very large fires in the contiguous United States. *Int. J. Wildland Fire* **2015**, *24*, 892–899. [CrossRef]

37. Abatzoglou, J.T.; Williams, A.P. Impact of anthropogenic climate change on wildfire across western US forests. *Proc. Nat. Acad. Sci. USA* **2016**, *113*, 11770–11775. [CrossRef] [PubMed]

38. Miller, J.D.; Safford, H.D.; Crimmins, M.; Thode, A.E. Quantitative evidence for increasing forest fire severity in the Sierra Nevada and southern Cascade mountains, California and Nevada, USA. *Ecosystems* **2008**, *12*, 16–32. [CrossRef]

39. Miller, J.D.; Safford, H. Trends in wildfire severity: 1984 to 2010 in the Sierra Nevada, Modoc Plateau, and southern Cascades, California, USA. *Fire Ecol.* **2012**, *8*, 41–57. [CrossRef]

40. Picotte, J.J.; Peterson, B.; Meier, G.; Howard, S.M. 1984–2010 trends in fire burn severity and area for the conterminous US. *Int. J. Wildland Fire* **2016**, *25*, 413–420. [CrossRef]

41. Meddens, A.J.H.; Kolden, C.A.; Lutz, J.A.; Abatzoglou, J.T.; Hudak, A.T. Spatiotemporal patterns of unburned areas within fire perimeters in the northwestern United States from 1984 to 2014. *Ecosphere* **2017**, in press.

42. Kolden, C.A.; Lutz, J.A.; Key, C.H.; Kane, J.T.; van Wagtendonk, J.W. Mapped versus actual burned area within wildfire perimeters: Characterizing the unburned. *For. Ecol. Manag.* **2012**, *286*, 38–47. [CrossRef]

43. Wright, C.S.; Agee, J.K. Fire and vegetation history in the eastern Cascade Mountains, Washington. *Ecol. Appl.* **2004**, *14*, 443–459. [CrossRef]

44. United States Department of Agriculture (USDA); United States Department of the Interior (USDOI). *Record of Decision for Amendments to Forest Service and Bureau of Land Management Planning Documents within the Range of the Northern Spotted Owl, Attachment A: Standards and Guidelines for Management of Habitat for Late-Successional and Old-Growth Forest Related Species within the Range of the Northern Spotted Owl*; U.S. Government: Washington, DC, USA, 1994.

45. Williams, C.K.; Smith, B. *Forested Plant Associations of the Wenatchee National Forest*; US Forest Service Pacific Northwest Region, Wenatchee National Forest: Portland, OR, USA, 1991.

46. Holstine, C.E. *An Historical Overview of the Wenatchee National Forest, Washington*; Rep. 100–80; Archaeological and Historical Services, Eastern Washington University: Cheney, WA, USA, 1992.

47. Everett, R.L.; Martin, S.; Bickford, M.; Schellhaas, R.; Forsman, E. *Variability and Dynamics of Spotted Owl Nesting Habitat in Eastern Washington*; General Technical Report; INT-GTR-291; United States Department of Agriculture Forest Service: Ogden, UT, USA, 1991; pp. 35–39.

48. Western Regional Climate Center. Available online: www.wrcc.dri.edu (accessed on 21 March 2015).

49. Key, C.H.; Benson, N.C. *Landscape Assessment (LA) Sampling and Analysis Methods*; US Forest Service: Ogden, UT, USA.

50. Morgan, P.; Keane, R.E.; Dillon, G.K.; Jain, T.B.; Hudak, A.T.; Karau, E.C.; Sikkink, P.G.; Holden, Z.A.; Strand, E.K. Challenges of assessing fire and burn severity using field measures, remote sensing and modelling. *Int. J. Wildland Fire* **2014**, *23*, 1045–1060. [CrossRef]

51. Zhu, Z.; Key, C.H.; Ohlen, D.; Benson, N.C. *Evaluating Sensitivities of Burn Severity-Mapping Algorithms for Different Ecosystems and Fire Histories*; Joint Fire Science Program: Boise, ID, USA, 2006.

52. Welch, B.L. The generalization of "Student's" problem when several different population variances are involved. *Biometrika* **1947**, *34*, 28–35. [CrossRef] [PubMed]

53. McCune, B.; Keon, D. Equations for potential annual direct incident radiation and heat load. *J. Veg. Sci.* **2002**, *13*, 603–606. [CrossRef]

54. LANDFIRE. Available online: www.landfire.gov (accessed on 24 September 2015).

55. Rollins, M.G. LANDFIRE: A nationally consistent vegetation, wildland fire, and fuel assessment. *Int. J. Wildland Fire* **2009**, *18*, 235–249. [CrossRef]

56. Keane, R.E.; Burgan, R.; van Wagtendonk, J.W. Mapping wildland fuels for fire management across multiple scales: Integrating remote sensing, GIS, and biophysical modeling. *Int. J. Wildland Fire* **2001**, *10*, 301–319. [CrossRef]

57. Keane, R.E.; Karau, E.C. Evaluating the ecological benefits of wildfire by integrating fire and ecosystem simulation models. *Ecol. Model.* **2010**, *221*, 1162–1172. [CrossRef]

58. Strand, E.K.; Vierling, L.A.; Bunting, S.C.; Gessler, P.E. Quantifying successional rates in western aspen woodlands: Current conditions, future predictions. *For. Ecol. Manag.* **2009**, *257*, 1705–1715. [CrossRef]

59. Haugo, R.; Zanger, C.; DeMeo, T.; Ringo, C.; Shlisky, A.; Blankenship, K.; Simpson, M.; Mellen-McLean, K.; Kertis, J.; Stern, M. A new approach to evaluate forest structure restoration needs across Oregon and Washington. *For. Ecol. Manag.* **2015**, *335*, 37–50. [CrossRef]

60. Lillybridge, T.R.; Kovalchik, B.L.; Williams, C.K.; Smith, B.G. *Field Guide for Forested Plant Associations of the Wenatchee National Forest*; Gen. Tech. Rep. 359; US Forest Service Pacific Northwest Research Station: Portland, OR, USA, 1995.

61. ESSA Technologies, Ltd. *Vegetation Dynamics Development Tool User Guide, Version 6.0*; ESSA Technologies Ltd.: Vancouver, BC, Canada, 2007.

62. Smith, A.M.S.; Eitel, J.U.H.; Hudak, A.T. Spectral Analysis of Charcoal on Soils: Implications for Wildland Fire Severity Mapping Methods. *Int. J. Wildland Fire* **2010**, *19*, 976–983. [CrossRef]

63. Roy, D.P.; Boschetti, L.; Smith, A.M.S. Satellite remote sensing of fires. In *Fire Phenomena and the Earth System: An Interdisciplinary Guide to Fire Science*; Belcher, C.M., Rein, G., Eds.; John Wiley & Sons Ltd.: Chichester, UK, 2013; ISBN 978-0-470-65748-5.

64. Smith, A.M.S.; Sparks, A.M.; Kolden, C.A.; Abatzoglou, J.T.; Talhelm, A.F.; Johnson, D.M.; Boschetti, L.; Lutz, J.A.; Apostol, K.G.; Yedinak, K.M.; et al. Towards a new paradigm in fire severity research using dose-response experiments. *Int. J. Wildland Fire* **2016**, *25*, 158–166. [CrossRef]

65. Korhonen, L.; Korhonen, K.T.; Rautiainen, M.; Stenberg, P. Estimation of forest canopy cover: A comparison of field measurement techniques. *Silva Fennica* **2006**, *40*, 577–588. [CrossRef]

66. Swanson, M.E.; Franklin, J.F.; Beschta, R.L.; Crisafulli, C.M.; DellaSala, D.A.; Hutto, R.L.; Lindenmayer, D.B.; Swanson, F.J. The forgotten stage of forest succession: Early-successional ecosystems on forest sites. *Front. Ecol. Environ.* **2011**, *9*, 117–125. [CrossRef]

67. Barrett, S. Fire suppression's effects on forest succession within a central Idaho wilderness. *West. J. Appl. For.* **1988**, *3*, 76–80.

68. Hessburg, P.F.; Churchill, D.J.; Larson, A.J.; Haugo, R.D.; Miller, C.; Spies, T.A.; North, M.P.; Povak, N.A.; Belote, R.T.; Singleton, P.H.; et al. Restoring fire-prone Inland Pacific landscapes: Seven core principles. *Landsc. Ecol.* **2015**, *30*, 1805–1835. [CrossRef]

69. Smith, A.M.S.; Kolden, C.A.; Tinkham, W.T.; Talhelm, A.F.; Marshall, J.D.; Hudak, A.T.; Boschetti, L.; Falkowski, M.J.; Greenberg, J.A.; Anderson, J.W.; et al. Remote sensing the vulnerability of vegetation in natural terrestrial ecosystems. *Remote Sens. Environ.* **2014**, *154*, 322–337. [CrossRef]

forests

Article

Mixed-Severity Fire Fosters Heterogeneous Spatial Patterns of Conifer Regeneration in a Dry Conifer Forest

Sparkle L. Malone [1,2,*], Paula J. Fornwalt [1], Mike A. Battaglia [1], Marin E. Chambers [3], Jose M. Iniguez [4] and Carolyn H. Sieg [4]

1   US Department of Agriculture, Forest Service, Rocky Mountain Research Station, 240 W. Prospect Road, Fort Collins, CO 80526, USA; pfornwalt@fs.fed.us (P.J.F.); mbattaglia@fs.fed.us (M.A.B.)
2   Department of Biological Sciences, Florida International University, 11200 S.W. 8th Street, Miami, FL 33199, USA
3   Colorado Forest Restoration Institute, Colorado State University, Department of Forest & Rangeland Stewardship, Mail Delivery 1472, Fort Collins, CO 80523, USA; marin.chambers@colostate.edu
4   US Department of Agriculture, Forest Service, Rocky Mountain Research Station, 2500 S. Pine Knoll Drive, Flagstaff, AZ 86001, USA; jiniguez@fs.fed.us (J.M.I.); csieg@fs.fed.us (C.H.S.)
*   Correspondence: smalone@fiu.edu; Tel.: +1-305-348-1988

Received: 25 November 2017; Accepted: 17 January 2018; Published: 20 January 2018

Abstract: We examined spatial patterns of post-fire regenerating conifers in a Colorado, USA, dry conifer forest 11–12 years following the reintroduction of mixed-severity fire. We mapped and measured all post-fire regenerating conifers, as well as all other post-fire regenerating trees and all residual (i.e., surviving) trees, in three 4-ha plots following the 2002 Hayman Fire. Residual tree density ranged from 167 to 197 trees ha^{-1} (TPH), and these trees were clustered at distances up to 30 m. Post-fire regenerating conifers, which ranged in density from 241 to 1036 TPH, were also clustered at distances up to at least 30 m. Moreover, residual tree locations drove post-fire regenerating conifer locations, with the two showing a pattern of repulsion. Topography and post-fire sprouting tree species locations further drove post-fire conifer regeneration locations. These results provide a foundation for anticipating how the reintroduction of mixed-severity fire may affect long-term forest structure, and also yield insights into how historical mixed-severity fire may have regulated the spatially heterogeneous conditions commonly described for pre-settlement dry conifer forests of Colorado and elsewhere.

Keywords: forest recovery; wildfire effects; stem maps; resilient ecosystems; Pike National Forest; Hayman Fire

1. Introduction

Historically, many dry conifer forests of western North America were regulated by a relatively frequent mixed-severity fire regime [1–3]. Individual fires were typically dominated by low- and moderate-severity effects where many overstory trees survived, but also contained small patches of high-severity effects where most or all overstory trees were killed. These fires also acted as a control on tree recruitment [4,5], thereby resulting in a variety of tree densities and tree size and age distributions [5,6]. The clustered spatial pattern commonly associated with historical dry conifer stands—where well-defined groups of trees and individual trees were interspersed in a matrix of treeless openings is further attributed to the relatively frequent historical fire regime [7–9]. This heterogeneity in historical forest structure is thought to be more resilient to drought and subsequent wildfires [1], yet little is known regarding the process responsible for this spatial structure. Although we know that historically dry conifer forests had a clustered spatial structure, we do not fully understand whether this pattern was a function of fire-caused mortality, post-fire regeneration, or both.

After nearly a century of fire exclusion, wildfire activity in western North American dry conifer forests has increased [9–12]. Many of these recent wildfires contain uncharacteristically large high-severity burn patches, attributed to higher forest density and homogeneity as a result of fire exclusion, as well as livestock grazing [13], logging [14–16], and a warmer and drier climate [9,11]. However, recent wildfires also commonly include areas that burned with mixed-severity fire—that is, areas that are a more heterogeneous combination of low-, moderate-, and high-severity patches [17–19]. In mixed-severity portions of recent wildfires where high-severity patches are small, fires more closely align with the historical fire regime [6,20], and therefore may represent a reintroduction of the dominant historical processes. As a result, recent fires may provide an opportunity to better understand how mixed-severity fires shape tree spatial patterns and subsequent regeneration.

Post-fire forest recovery research in ponderosa pine (*Pinus ponderosa* Douglas ex P. Lawson & C. Lawson) dominated dry conifer forests has been primarily aimed at examining regeneration following recent high-severity fire. In the interiors of large high-severity patches, regeneration tends to be sparse, suggesting that forest recovery may be delayed or may not occur at all [21–26]. Along the edges of high-severity patches, regeneration tends to be concentrated in areas near surviving trees, as most species in dry conifer forests rely on seed production from live trees to regenerate [25–29]. Regeneration is also commonly concentrated in more mesic sites, such as areas at higher elevations or with more northerly aspects, and in the vicinity of nurse structures such as downed logs and other regenerating trees [24–31].

In contrast, little is known about regeneration following recent mixed-severity fire. Some research suggests that recent mixed-severity fire effects are effectively enhancing forest structure, improving drought resilience, and reducing the probability of high severity fire in subsequent fires [32,33]. It is therefore reasonable to expect that the re-introduction of mixed-severity fire is likely to have benefits to post-fire forest structure and function. Moreover, tree spatial patterns are important because they influence forest dynamics including tree establishment, competition, mortality and even fire behavior [7]. Although essential for understanding how fire shapes ponderosa pine forests, the forest structure created by mixed-severity fire and its effect on subsequent regeneration has rarely been explored in a spatial context. It is therefore unclear how the re-introduction of mixed-severity fire will influence the residual forest structure and subsequent forest development in ponderosa pine-dominated forests.

The objective of this research is to evaluate spatial patterns of post-fire conifer regeneration following mixed-severity burning in the 2002 Hayman Fire, Colorado, USA. Specifically, we aim to (1) describe the post-fire forest structure created by the re-introduction of mixed-severity fire; (2) examine the relationship between the residual forest and post-fire conifer regeneration; and (3) determine how abiotic and biotic factors influence post-fire conifer regeneration density (trees m^{-2}). We hypothesized that (1) trees in the residual forest will be clustered, creating openings and opportunities for conifer regeneration; (2) post-fire residual conifers and regeneration will show repulsion, reflecting the preference for high light environments in regenerating conifers; and (3) post-fire regenerating conifer density will not be limited by distance from residual conifers, but by interactions with sprouting trees and by topography. That is, we expect regenerating conifers to be concentrated in more mesic areas, similar to patterns found in high-severity areas [26]. Understanding the characteristics of ponderosa pine-dominated forest and requirements for seed production, germination, and establishment, we expect topography and burn severity to be important drivers of regeneration density [25]. Characterizing these patterns of tree regeneration following wildfire is critical for understanding the drivers of the clustered spatial patterns observed in historical dry conifer forests, and anticipating longer-term stand structure and intertwined ecological properties and processes such as potential fire behavior and understory plant community composition and productivity.

2. Materials and Methods

2.1. Study Area

The Hayman Fire provides an ideal landscape for examining the post-fire forest structure created by mixed-severity fire and subsequent conifer regeneration. This 52,000 ha fire is the largest known wildfire in Colorado (USA) within the last century, making it of considerable interest to land managers, policy makers, researchers, and the public [34]. Located ~60 km southwest of Denver, the Hayman Fire was ignited on 8 June 2002 (Figure 1) in an area that receives 511 mm of precipitation year^{-1} and has a mean annual temperature of 6 °C [35]. The fire predominantly burned through stands of overly dense and homogeneous ponderosa pine-dominated forest from which the historical mixed-severity fire regime had been excluded for about a century [34,36,37]. A large portion of the Hayman Fire burned with high-severity (43%; Monitoring Trends in Burn Severity; [38], much of it in a single day with extreme weather conditions [39]. Low- and moderate-severity burning occurred on 34% and 22% of area, respectively, typically under less extreme weather conditions [39]. Additional information on fire behavior, fire weather, and fire effects can be found in The Hayman Fire Case Study [34].

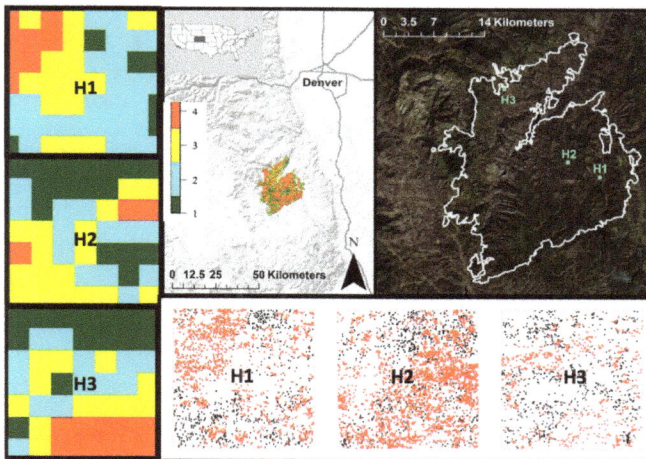

Figure 1. The Hayman Fire and plot (H1, H2, and H3) locations in the Colorado Front Range, USA. Within each plot, residual trees (black points) are shown relative to post-fire regeneration (red points) (bottom row). Esri basemap layers: the World Terrain Reference shows the Hayman Fire in relation to Denver and World Imagery denotes topography and forest cover in relation to plot locations within the fire. Monitoring trends in Burn Severity (MTBS) shows the mixed-severity fire effect in the Hayman Fire. Thematic burn severity classes include unchanged (green), low- (cyan), moderate- (yellow), and high- (red) severity (left side; scale is 4 ha).

The ponderosa pine-dominated forests burned by the Hayman Fire were variable in overstory tree density and composition [37]. Tree density was greater at higher elevations, on northerly slopes, and in draws, where moisture was more available. Douglas-fir (*Pseudotsuga menziesii* Franco) tended to be more abundant in these more moist locales, often becoming codominant with ponderosa pine. Quaking aspen (*Populus tremuloides* Michx.), blue spruce (*Picea pungens* Engelm.) and lodgepole pine (*Pinus contorta* Douglas ex Loudon), also became more common as elevation increased and as aspect became more northerly. Understory plant communities prior to the fire were dominated by graminoids and forbs (e.g., common yarrow (*Achillea millefolium* L.), white sagebrush (*Artemisia ludoviciana* Nutt.), Ross' sedge (*Carex rossii* Boott), and mountain muhly (*Muhlenbergia montana* (Nutt.) Hitchc.) [40].

Relatively short-statured shrubs were also present (e.g., kinnikinnick (*Arctostaphylos uva-ursi* (L.) Spreng.) and alderleaf mountain mahogany (*Cercocarpus montanus* Raf.).

2.2. Study Design

We established 4-ha plots (*n* = 3) within the Hayman Fire perimeter in 2013 and 2014 (Figure 1). Plot locations were determined by first using ArcGIS 10.1 (Esri, Redlands, CA, USA) to identify suitable large-scale (~15 to 35 ha) study sites. Suitable sites were areas that burned with a heterogeneous mosaic of severities (as indicated by the MTBS burn severity product and ArcGIS' aerial imagery basemap), that were on US Forest Service land that were accessible (i.e., within 2–3 km of a road), and not impacted by post-fire logging or planting activities. Once suitable sites were identified, points were randomly generated in ArcGIS, and one point was randomly selected. We then established a plot corner at that point, with the plot oriented to keep the plot within the study site or to avoid undesirable features within the plot (i.e., roads, streams, large rock outcroppings, and areas heavily utilized by the public). All plots contained mixed-severity fire effects, verified by MTBS burn severity, which included elements of low and moderate severity fire, and small portions of high severity that occupied no more than 1 ha within the plot (Figure 1). Post-fire regeneration was measured 12–14 years following wildfire to allow enough time for regeneration to occur.

Within each plot, we mapped all live trees >15 cm tall (Figure 1). A rangefinder (Laser Technologies Inc. TruPulse 360-B) was used in combination with a Trimble global positioning system (GeoXH with Terrasync, accurate to ±20 cm; Trimble Navigation Limited, Sunnyvale, CA, USA) to record the location of every living tree with high accuracy (±38 cm) [30]. In addition to location, we recorded diameter at breast height (DBH), tree height, and species for each overstory tree (i.e., >1.4 m tall), and we recorded tree height, species, and whether germination occurred pre- or post-fire for each regenerating tree (i.e., >15 cm tall and <1.4 m tall). We refer to regeneration that established pre-fire as advanced regeneration and all regeneration that established after the fire as post-fire regeneration. It is important to note that in some portions of the plots we encountered dense clumps of regenerating trees, making it time-consuming to map each one individually. For clumps where regenerating trees were similar in species, height, and age, we mapped the center point of the clump and we recorded the clump radius and the number of regenerating trees in the clump, in addition to recording species, average height, and pre- or post-fire germination status. The radius of the clumps ranged from 0.5 to 4 m, with most being 1 m or less.

2.3. Residual Forest Structure

Forest structure has two principal dimensions: the types, number and sizes of individual structural elements (e.g., individual trees); and their arrangement in space [7,41]. Here, residual forest structure refers to remnant surviving trees and does not include dead standing or downed trees. We described non-spatial aspects of forest structure using tree density (trees per hectare; TPH), diameter at breast height (DBH), tree height, basal area (BA), and quadratic mean diameter (QMD) [42,43]. To estimate canopy cover of the residual forest, we used a crown radius of 3 m for all overstory trees [8]. We measured the nearest neighbor distance for each tree and the distance to the nearest tree for each 1 m pixel within the plot using the R package spatstat [44,45]. We also used spatial quantitative descriptions of forest stand structure that describe three structural characteristics: positioning, mixture, and differentiation [46] (Table 1).

Table 1. Non-spatial and spatial measures of forest structure used to evaluate study objectives and hypotheses.

Objectives	Approach		Inference	Scale	Hypotheses
Describe residual forest structure	Non-spatial Metric	Density, DBH, Height, QMD, Canopy Cover	Non-spatial attributes of the residual forest	Plot	
	Spatial Metric	Nearest Neighbor Distance, Distance to the Nearest Tree	Spatial attributes of trees in the residual forest	Within-plot	
	Spatial Metric (Positioning)	Clark and Evans Index (CEI)	Spatial pattern of the residual forest	Plot	1
	Spatial Analysis (Positioning)	Ripley's K Function	Spatial pattern of trees in the residual forest across distance scales	Within-plot	1
	Spatial Metric (Species Mixture)	Durchmischung Index (DMI)	Residual forest species associations	Plot	
	Spatial Metric (Differentiation)	Differenzierung Index (DZI)	Residual forest horizontal and vertical structural complexity	Plot	
Describe the relationship between the residual forest and post-fire conifer regeneration	Spatial Analysis (Positioning)	Ripley's K Function	Patterning of regenerating trees across distance scales	Within-plot	
	Spatial Analysis (Positioning)	Bivariate K Function	Attraction or repulsion between residual trees and regenerating trees	Within-plot	2
Drivers of post-fire conifer regeneration density	Spatial Analysis (Positioning)	Neyman–Scott Point-Process Model	Drivers of regeneration density	Within-plot	3

Positioning is the spatial distribution of points (i.e., trees) in an area. The Clark and Evans index (CEI) was used to characterize point patterns at the plot level as either random, regular, or clustered [46]. In a completely random pattern (CEI = 1), the position of any one point is independent of the position of all the other points. In a regular pattern (CEI > 1), points are farther away from their nearest neighbors than would be expected for a random pattern. Points are more likely to be found near other points in a clustered pattern (CEI < 1), and the average distance from any arbitrary point to its nearest-neighbor is less than expected in a completely random pattern. We also evaluated positioning with Ripley's K [7,8,47,48] using the R package spatstat [44,45]. The K function estimates spatial dependence between points of the same type (e.g., residual trees) across spatial scales by determining the expected number of points within a distance (r) from any randomly sampled point. For each plot, we evaluate deviations from complete spatial randomness (CSR) by comparing observed data with an inhomogeneous Poisson null model [49–51]. We used an inhomogeneous rather than a homogeneous Poisson process to account for non-constant density gradients in the data [52]. Points were randomly distributed under the inhomogeneous Poisson null model 999 times to test for departure from CSR. It is common to apply a correction for edge effects when calculating the K function. Using the correction = "best" setting in the package spatstat, the best available edge correction, "the isotropic correction" was applied. Observed patterns differ from CSR where the plot of K(r) falls outside the simulated envelope for a random pattern. A clustered point pattern is indicated by a K(r) above the envelope (i.e., more points present than would be expected under CSR) and regular spacing occurs where K(r) falls below the envelope (i.e., fewer points present than would be expected under CSR). Unlike the CEI, the K function describes characteristics of the point processes at many distance scales and uses spatial randomness as a benchmark to define spatial regularity and clustering [7]. Since clustering and regularity are characteristics of a pattern at a specific distance scale, point patterns can exhibit both clustering and regularity at the same time. For example, regeneration can be clustered at longer distances depending on how their parent trees are spaced, but regularly spaced at a small scale because they compete for resources. There is also evidence in some studies of clustering [53,54] and perhaps facilitation at small scales [29].

Mixture refers to spatial interspersion among species within an area and was quantified with the Durchmischung index (DMI) [46]. Ranging between 0 and 1, the DMI describes the degree to which species exist in homogeneous clusters, exhibit repulsion, or show attraction to other species. Strongly represented species or those that exist in homogeneous groups will result in low DMI values, whereas less frequent or regularly positioned species will have high DMI values (indicating that a species nearest neighbor is likely to be a different species) [46]. We calculated DMI on a plot level to understand the average degree of mixing between species in dry conifer forests and use the three nearest neighbors to calculate DMI for each plot [46]. Although we explored the effects of increasing the nearest neighbors up to six trees, results did not change so we report the DMI using the three nearest neighbors to aid comparisons with other studies [46].

Differentiation is a measure of the variation in tree height or DBH among neighboring trees (three nearest neighbors). We calculated the Differenzierung index (DZI) using total tree height (DZI_{TH}) and DBH (DZI_{DBH}) to measure vertical and horizontal structure, respectively. Ranging between 0 and 1, values from 0 to 0.2 represent low differentiation (similar height or DBH between neighboring trees), 0.2 to 0.4 moderate differentiation, 0.4 to 0.6 clear differentiation, 0.6 to 0.8 strong differentiation, and 0.8 to 1 very strong differentiation or heterogeneity between the three nearest neighbors [46]. Measuring horizontal and vertical variation is important for understanding the amount of within-plot variation among neighboring trees.

2.4. Post-Fire Regeneration

Similar to the residual trees, the spatial pattern of post-fire conifer regeneration was analyzed using Ripley's K functions [48], in the R package spatstat [44,45]. This approach was used to test for departure from a spatially random pattern across distance scales to understand at what distance post-fire regeneration exhibits a random, clustered, or regular pattern. The spatial relationship between residual trees and post-fire regeneration was evaluated using a bivariate K-function [50,55]. The bivariate K-function calculates the expected mean number of post-fire regenerating trees within a given radius (r; i.e., distance) of an arbitrary residual tree. This approach evaluates attraction between trees and regeneration within r. The null hypothesis of spatial independence between the two groups is refuted where values of K(r) fall above or below the bivariate inhomogeneous Poisson null model simulation envelope. The inhomogeneous Poisson null model envelope is calculated using random toroidal shifts of one pattern relative to the other during each of the 999 Monte Carlo iterations. A K(r) above the simulation interval is indicative of a positive spatial association (attraction; i.e., there is more regeneration than would be expected within r). Where K(r) falls below the confidence interval, there is a negative spatial association (repulsion; i.e., less regeneration than would be expected within r). This approach explores at what distances trees and regeneration show attraction and repulsion.

Potential drivers of post-fire conifer regeneration density were modeled using a Neyman–Scott point-process model, which is routinely used to describe clumped spatial patterns [50,56]. Least squares techniques were used to optimize the model parameters. Simulation envelopes were generated using the Neyman–Scott distribution, after the parameters had been fit to the data, to test for a departure from the Neyman–Scott process. Simulations of CSR were generated through 999 iterations of a Monte Carlo simulation. If the K(t) plot for the observed data falls within the simulated envelope for the Neyman–Scott process, it suggests that the pattern has the properties of the specified model. Nearest residual tree distance (m) and height (m), burn severity [57,58], topographic wetness index (TWI) [59], topographic position index (TPI) [60,61], elevation (m), and nearest aspen cluster density and distance (m) were explored as drivers of post-fire conifer density. All variables were initially placed in the model. Variables of least significance were removed from the model one at a time until all remaining variables were significant at an α of 0.05. We calculated topographic measurements (TWI, and TPI) using 30 m resolution digital elevation models (Advanced Spaceborne Thermal Emission and Reflection Radiometer (ASTER) Global Digital Elevation Model (GDEM); [62]). We used base tools in ArcGIS 10.3 (ESRI, Redlands, CA, USA) with the Geomorphology and Gradient Metrics toolbox for these calculations.

3. Results

3.1. Residual Forest Structure

The residual forest was dominated by conifers that survived the wildfire. Overstory trees (i.e., >1.4 m tall) accounted for 99% of the residual forest on average (Table 2), and advanced regeneration accounted for the remainder. Ponderosa pine was the dominant species (71% of trees on average) in the residual forest and Douglas-fir was the second most common species (27%). A mean DMI of 0.6 ± 0.02 indicated that ponderosa pine was very dominant and therefore more likely to be surrounded by the same species (Table 2). Residual tree density ranged from 167 to 197 TPH across all plots, while average percent canopy cover was $31.5 \pm 1.8\%$ (Table 2) due to overstory trees occurring in groups with overlapping crowns. The average nearest neighbor distance between overstory trees in the residual forest was 2.9 ± 0.1 m and the average distance to the nearest tree for any point within the plots was 6.1 ± 0.3 m. CEI suggested that trees were clustered, and Ripley's K suggested that this clustering occurred at radii less than 30 m (Table 2; Figure 2). Average indices of horizontal (DZI_{DBH}) and vertical (DZI_{TH}) structure showed moderate vertical differentiation and clear horizontal differentiation (Table 2).

Table 2. Residual forest structure within the Hayman Fire plots (H1, H2, and H3) 11–12 years following burning. This summarization includes all conifers and aspen (POTR) which represents 0.61% of the residual trees. Measures of trees per hectare (TPH), % ponderosa pine (PIPO), % Douglas-fir (PSME), % aspen (POTR), basal area (BA), quadratic mean diameter (QMD), Clark and Evans index (CEI), Durchmischung index (DMI), horizontal differentiation (DZI_{DBH}), and vertical differentiation (DZI_{TH}) are included in this summary.

Plot	H1	H2	H3	Average	Standard Error
Residual Trees (%)	99	99	100	99	0.4
Residual Regeneration (%)	1	1	0	1	0.4
TPH	167	197	173	179	9
Canopy Cover (%)	29	35	30	32	2
% PIPO	74	73	65	71	3
% PMSE	22	25	34	27	4
% POTR	3	1	0.4	1	1
BA (m^2 ha^{-1})	8	11	8	9	1
QMD (cm)	25	27	23	25	1
Nearest Neighbor (m)	3	3	3	3	0.1
Distance from Nearest Tree (m)	6.2	5.6	6.6	6.1	0.3
CEI	−0.3	−0.2	−0.3	−0.3	0.01
DMI	0.6	0.6	0.6	0.6	0.01
DZI_{DBH}	0.5	0.53	0.5	0.5	0.01
DZI_{TH}	0.3	0.3	0.3	0.3	0.01

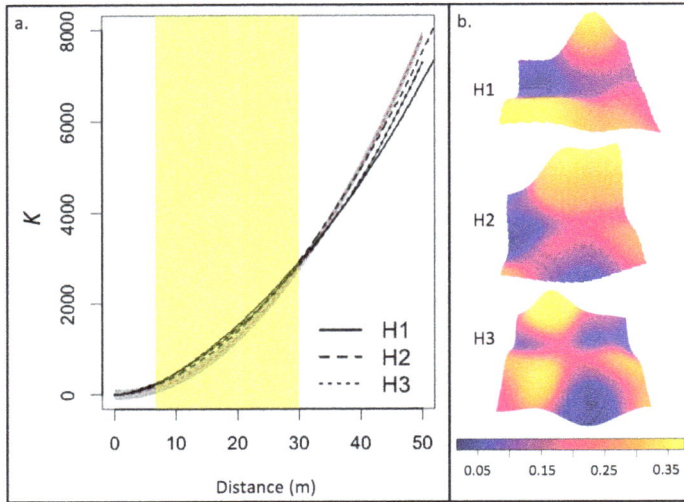

Figure 2. (**a**) Ripley's K for residual trees in the Hayman Fire plots (H1, H2, and H3). A clustered point pattern is indicated by an observed K(r) (black lines) that is above the complete spatial randomness (CSR) simulated envelope (gray shaded region), whereas a regular point pattern is indicated by an observed K(r) that falls below the envelope. Trees exhibited clustering at distances up to ~30 m (yellow shaded region); (**b**) Perspective plots of residual tree density (trees m^{-2}) in the Hayman Fire, with density represented by the height and color of the plot.

3.2. Post-Fire Regeneration

Unlike the sparse advanced regeneration observed in the residual forest, post-fire regeneration was dense, averaging 685 TPH. Ponderosa pine accounted for 73% of post-fire regeneration on average, while Douglas-fir and aspen accounted for 18% and 8%, respectively (Table 3). Focusing on post-fire regenerating conifers, TPH averaged 630, and height averaged 0.33 m. Just 32% of post-fire regenerating conifers were found under residual forest canopy cover, on average. Ripley's K indicated that post-fire conifer regeneration exhibited a clustered pattern across all distances in H1 and H2, and at distances up to 30 m in H3, beyond which patterns were regular (Figure 3). Evaluating bivariate changes in conifer regeneration densities relative to those of residual trees indicated that regeneration and trees were attracted at very short distances (<0.5 m) and repulsed at distances >1 m (Figure 4).

Table 3. Post-fire regeneration in the Hayman Fire plots (H1, H2, and H3) 11–12 years following burning. Measures of stems per hectare (SPH), stem height, and percent of SPH comprised of ponderosa pine (PIPO), Douglas-fir (PSME), and aspen (POTR) are summarized.

Plot	H1	H2	H3	Average	Standard Error
SPH (Conifers)	728 (613)	1071 (1036)	256 (241)	685 (630)	236 (230)
Stem Height (Conifers)	0.31 (0.28)	0.30 (0.29)	0.41 (0.41)	0.34 (0.33)	0.04 (0.04)
% PIPO	67	80	70	73	4
% PMSE	16	16	23	18	2
% POTR	16	3	6	8	4

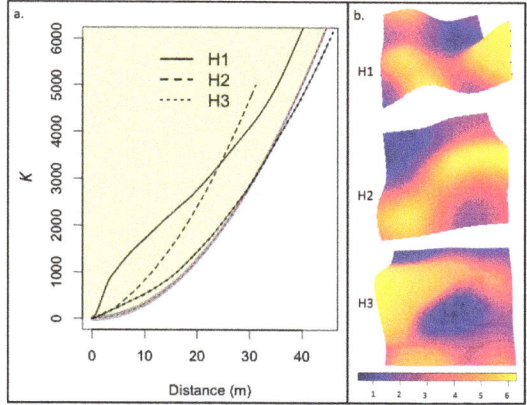

Figure 3. (a) Ripley's *K* for post-fire conifer regeneration in the Hayman Fire plots (H1, H2, and H3). A clustered point pattern is indicated by an observed *K*(r) (black lines) that is above the CSR simulated envelope (gray shaded region) and regular spacing is indicated by an observed *K*(r) that is below the envelope. The clustered region is shown in yellow. Clustering in post-fire conifer regeneration occurred across all distances in H1 and H2 and at distances <30 m in H3. Beyond 30 m regeneration occurred in a regular pattern; **(b)** Perspective plots of post-fire conifer regeneration density (stems m^{-2}) in the Hayman Fire, with density represented by the height and color of the plot.

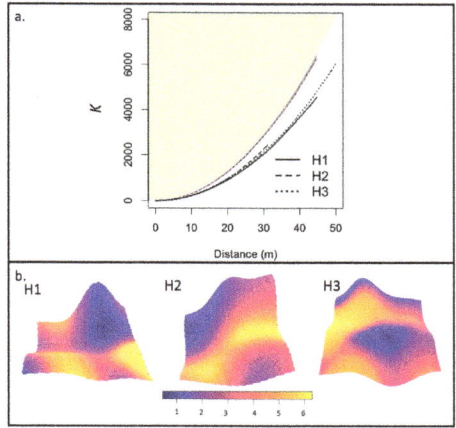

Figure 4. (a) Bivariate *K*-functions of associations between residual trees and post-fire conifer regeneration in the Hayman Fire plots (H1, H2, and H3). Attraction (yellow shaded region) is indicated by an observed *K*(r) that is above the CSR simulated confidence envelope (grey shaded region), while repulsion is indicated by an observed *K*(r) that falls below the envelope. Trees and post-fire conifer regeneration showed repulsion at distances greater than 1 m; **(b)** Perspective plots of residual tree and post-fire conifer regeneration density (trees m^{-2}). The height of the plot represents residual tree density and the color indicates regeneration density. Regeneration densities were higher in areas with lower tree densities.

The Neyman–Scott point-process model adequately described clumped patterns of post-fire conifer regeneration density (stems m^{-2}; Figure 5), while nearest residual tree distance, burn severity, TWI, elevation, and nearest aspen cluster distance were significant drivers of this density (Table 4;

Figure 6). Nearest residual tree distance was negatively correlated with post-fire conifer regeneration, suggesting that post-fire conifer regeneration was more abundant closer to pre-fire residual trees. Unlike the bivariate K-function, the distance to the nearest tree layer considers the location of all trees within a plot to determine a distance value for each location at a 1 m resolution. Burn severity was positively correlated with post-fire regeneration density, indicating that moderate- and high-severity areas were more likely to have higher regeneration densities than low-severity and unburned areas. In plot H1, TWI was negatively correlated with regeneration density and positively correlated with regeneration density in H3. These results suggest that higher moisture levels were associated with greater conifer regeneration except where topographic variation was low. In H1, elevation was negatively correlated with regeneration density and in H3, elevation was positively correlated with regeneration density. The distance to the nearest post-fire aspen cluster was negatively correlated with conifer regeneration in H1 and positively correlated in H3. In the highest elevation plot with more topographic variation (H1), aspen clusters and conifer regeneration showed repulsion, while in the lowest elevation plot with minimal topographic variation (H3), conifer regeneration density was greater closer to aspen clusters.

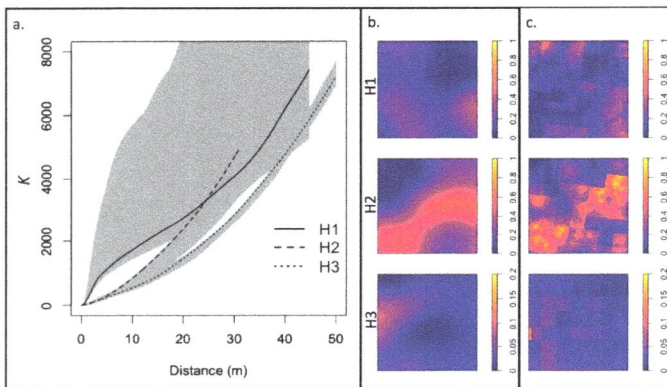

Figure 5. (**a**) Post-fire regeneration density in relation to the Neyman–Scott point process model for plots (H1, H2, and H3) in the Hayman Fire. Observed data (black lines) were within the bounds of the simulated envelope for the Neyman–Scott model (shaded region), confirming that the data were spatially clumped at all distances. Comparing (**b**) observed and (**c**) predicted densities shows that the model is capturing general trends in regeneration density.

Table 4. Estimates of regeneration density (stems m^{-2}) regression coefficients, standard errors (S.E.) and *p*-values for the Hayman Fire plot level point pattern models. The distance to the nearest residual tree, burn severity, topographic wetness index (TWI), elevation, and the distance to the nearest aspen (POTR) cluster were important drivers of post-fire regeneration density.

Plot		Intercept	Nearest Residual Tree Distance (m)	Burn Severity			TWI	Elevation	Nearest POTR Cluster (m)
				Low	Moderate	High			
H1	Estimate	262.627	−0.036	0.357	1.029	0.842	−0.007	−0.112	−0.007
	S.E.	7.530	0.002	0.057	0.063	0.078	−0.014	0.003	−0.014
	p-value	<0.001	<0.001	<0.001	<0.001	<0.001	<0.001	<0.001	<0.001
H2	Estimate	−1.113	−0.042	0.657	1.091	0.939			
	S.E.	0.028	0.002	0.028	0.026	0.042			
	p-value	<0.001	<0.001	<0.001	<0.001	<0.001			
H3	Estimate	−49.284		0.266	0.616	0.301	0.002	0.018	0.002
	S.E.	9.175		0.092	0.087	0.098	0.001	0.004	0.001
	p-value	<0.001		<0.001	<0.001	<0.001	<0.001	<0.001	<0.001

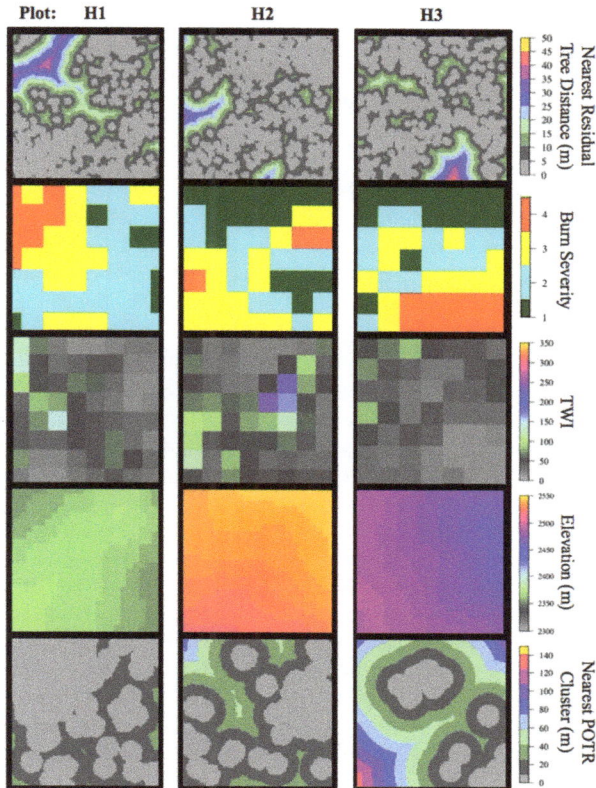

Figure 6. Significant drivers of post-fire regeneration density for the Hayman Fire plots (H1, H2, and H3) include nearest residual tree distance (m), burn severity, topographic wetness index (TWI), elevation, and the nearest aspen (POTR) cluster. Thematic burn severity classes include unburned (green), low- (cyan), moderate- (yellow), and high- (red) severity.

4. Discussion

We examined spatial patterns of post-fire conifer regeneration following the reintroduction of mixed-severity fire in a Colorado ponderosa pine-dominated forest. We hypothesized that (1) trees in the residual forest would be clustered, creating openings and opportunities for regeneration; (2) residual trees and post-fire conifer regeneration would exhibit repulsion; and (3) post-fire regenerating conifer density would not be limited by distance from residual trees. Our results support our hypotheses; moreover, they highlight how other biotic and abiotic factors, such as topography and the location of sprouting species, further complicate spatial patterns in post-fire conifer regeneration.

4.1. Residual Forest Structure

The re-introduction of mixed-severity fire after a century of fire exclusion resulted in a clustered spatial pattern of residual trees, with densities that were still higher than those reported for historical forests elsewhere in the region [8]. However, residual tree densities were lower than those reported for undisturbed forests in the region [31,37,63], as well as for recently restored forests [31,63]. Basal area and canopy cover values following mixed-severity fire were also within the range of those reported for restored forests [31,63]. The lack of advanced regeneration suggests that these trees, if they were present before the fire, were killed. Furthermore, the clustered nature of the residual forest suggests that trees survived in groups. These results indicate that fire-caused mortality is at least partly responsible for creating the historical spatial pattern reported in a number of studies [7,8,64]. Many of these studies have speculated that the spatial clustering was a result of fire, but this is one of the only studies that have documented this pattern immediately post-fire.

4.2. Post-Fire Regeneration

Following recent large fires, one of the limiting factors of natural conifer regeneration is a seed source [65], particularly for ponderosa pine which has a relatively large seed. For species without serotinous cones, tree establishment is largely a function of seed dispersal [66] and competition from other species [23,67]. While the conditions created by high-severity fire can lead to dispersal and recruitment limitations [23,25,29], the residual forest structure following mixed-severity fire contained fire-resistant, seed-bearing conifer trees [68], whose location within the landscape, thick bark, and high crown base heights facilitated their survival and post-fire forest recovery. The residual forest following mixed-severity fire was sufficient to support natural regeneration within 12 years following the fire, in a region where seed masting events are episodic [26,69,70] and occur on average every 3–12 years for dominant conifers (ponderosa pine and Douglas-fir), with limited seed availability in intervening years [71].

The clustered spatial pattern of the residual forest allowed for ample openings, which together had a pronounced effect on conifer regeneration density. Our results suggest that, like the residual forest, post-fire regeneration also had a tendency to aggregate at distances up to 30 m. Similar to patterns observed following high-severity fire, distance from the nearest residual tree was a significant driver of post-fire conifer regeneration density [25,27,29]. However in this case, distance does not represent a limitation of seed dispersal, but the influence of both light and available space. That is, the clustered spatial pattern of post-fire regeneration appears to be a direct inverse of the post-fire residual spatial structure. This interpretation is supported by the significant repulsion between residual trees and post-fire regeneration at distances greater than 1 m. It is also important to note that patterns in regeneration density were very similar to patterns observed in the distribution of nearest residual tree distance across plots. This pattern suggests that seed source was not limiting regeneration patterns, and that high regeneration rates largely followed patterns in available space (Figure 7) with slightly higher regeneration occurring in areas due to variations in elevation that influenced moisture conditions and aspen clusters [27]. This result is important because it suggests that the clustered spatial pattern observed in historical ponderosa pine studies is attained, in part, in the regeneration process [7,8,64].

Moderate increases in the availability of resources such as light and water can be enough to influence regeneration establishment and survival [72,73], and drive spatial patterns in forest structure [74]. Similar attraction patterns have been found in Dahurian larch (*Larix gmelinii* Rupr), where its conical canopy is thought to create a light-facilitating environment for regeneration [74]. Results here indicate that at distances farthest from trees, there was less regeneration than would be expected if there were no benefit to being in close proximity of trees. If light were the dominant driver of regeneration success, then regeneration would occur at higher frequencies, at greater distances from overstory trees [74]. Lower light levels created by overlapping canopies and the distribution of small gaps within a plot would also have influenced the relationship between distance from the nearest tree and regeneration density.

The residual forest structure following the re-introduction of mixed-severity fire, combined with the mix of burn severities and environmental conditions, is perpetuating ponderosa pine-dominated forests. Fire can act as a mechanism of change in plant community composition by altering the composition of species in the residual forest [75], or by producing post-fire conditions that favor the establishment of species that did not dominate the pre-fire forest [75,76]. Moisture requirements, one of the most important characteristics that affects post-fire tree establishment, are likely to change following fire. The moisture preference of trees can have a strong influence on the distribution of species, even at relatively small scales, including after fire, when ambient conditions can be warmer and drier than pre-fire conditions [77]. Unlike high-severity fire, mixed-severity burns had more positive effects on the residual forest structure for conifer species and subsequent conifer regeneration [68]. Similar to patterns observed by Kemp et al. [27], conifer species composition did not vary substantially between the residual trees and post-fire regeneration. Ponderosa pine was the dominant residual tree species and the species composition of regeneration suggests that this pattern will continue.

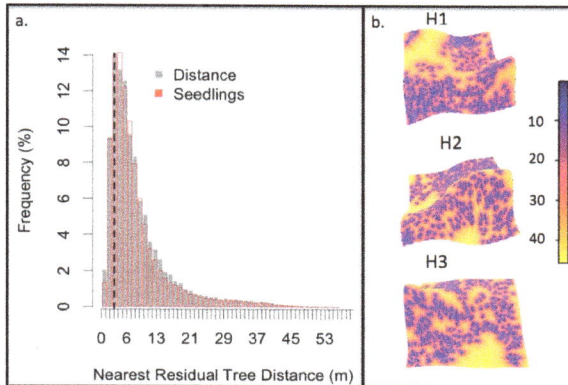

Figure 7. (**a**) The distributions of all post-fire regeneration (red) and nearest residual tree distance (gray) within plots by nearest residual tree distance (m) in the Hayman Fire. The vertical black line marks distances under canopy cover; (**b**) Post-fire perspective plots where the height of the plot represents the density of post-fire regeneration and the color indicates the distance to the nearest residual tree (m) for each 4-ha plot (H1, H2, and H3).

4.3. Study Limitations

Major limitations of this study include the low sample size and the lack of information on the pre-fire forest structure. Although we measured the forest structure following mixed-severity fire, assessing how the forest structure changed as a result of fire was not possible. Additionally, it is important to note that like in many other analysis methods, the driving factors and processes that cause different or even similar spatial patterns (i.e., clustered, random, and regular), including facilitation (positive effect) or competition (negative effect), are open to interpretation [74]. A clustered spatial pattern can be an indication of species having similar ecological requirements [78], facilitation among individuals [79,80], or dispersal limitations at larger scales [81].

5. Conclusions

Forest spatial structure yields important clues to understanding tree interactions with the environment as well as the dynamics of forest communities. The development of spatial patterns resulting from positive and negative associations and environmental factors is an important topic in ecosystem research [74], and has been used here to understand patterns in post-fire regeneration following mixed-severity fire. Predicting plant community responses to changing environmental conditions is a key element of forecasting and mitigating the effects of global change. Disturbance can play an important role in these dynamics, by initiating cycles of secondary succession and generating opportunities for communities of trees to reorganize.

Acknowledgments: This research was funded by the United States Forest Service (USFS) National Fire Plan (NFP-13-16-FWE-43) and the Rocky Mountain Research Station (RMRS). The views expressed in this manuscript are those of the authors alone and do not necessarily reflect the stance of the RMRS or the USFS.

Author Contributions: P.J.F., M.A.B., J.M.I., and C.H.S. conceived and designed the study; M.E.C. and P.J.F. managed the fieldwork; S.L.M. analyzed the data; S.L.M. wrote the paper with input from the other authors.

Conflicts of Interest: The authors declare no conflict of interest.

References

1. Allen, C.D.; Savage, M.; Falk, D.A.; Suckling, K.F.; Swetnam, T.W.; Schulke, T.; Stacey, P.B.; Morgan, P.; Hoffman, M.; Klingel, J.T. Ecological Restoration of Southwestern Ponderosa Pine Ecosystems: A Broad Perspective. *Ecol. Appl.* **2002**, *12*, 1418–1433. [CrossRef]
2. Perry, D.A.; Hessburg, P.F.; Skinner, C.N.; Spies, T.A.; Stephens, S.L.; Taylor, A.H.; Franklin, J.F.; McComb, B.; Riegel, G. The ecology of mixed severity fire regimes in Washington, Oregon, and Northern California. *For. Ecol. Manag.* **2011**, *262*, 703–717. [CrossRef]
3. Stevens, J.T.; Safford, H.D.; North, M.P.; Fried, J.S.; Gray, A.N.; Brown, P.M.; Dolanc, C.R.; Dobrowski, S.Z.; Falk, D.A.; Farris, C.A.; et al. Average stand age from forest inventory plots does not describe historical fire regimes in ponderosa pine and mixed-conifer forests of western North America. *PLoS ONE* **2016**, *11*, e0147688. [CrossRef] [PubMed]
4. Covington, W.W.; Moore, M.M. Southwestern ponderosa pine forest structure: Changes since Euro-American settlement. *J. For.* **1994**, *92*, 39–47.
5. Brown, P.M.; Wu, R. Climate and Disturbance Forcing of Episodic Tree Recruitment in a Southwestern Ponderosa Pine landscape. *Ecology* **2005**, *86*, 3030–3038. [CrossRef]
6. Arno, S.F.; Parsons, D.J.; Keane, R.E. Mixed-severity fire regimes in the northern Rocky Mountains: Consequences of fire exclusion and options for the future. In *Proceedings: Wilderness Science in a Time of Change Conference*; Utah State University: Logan, UT, USA, 2000; Volume RMRS-P-15-5, pp. 225–232.
7. Larson, A.J.; Churchill, D. Tree spatial patterns in fire-frequent forests of western North America, including mechanisms of pattern formation and implications for designing fuel reduction and restoration treatments. *For. Ecol. Manag.* **2012**, *267*, 74–92. [CrossRef]
8. Brown, P.M.; Battaglia, M.A.; Fornwalt, P.J.; Gannon, B.; Huckaby, L.S.; Julian, C.; Cheng, A.S. Historical (1860) forest structure in ponderosa pine forests of the northern Front Range, Colorado. *Can. J. For. Res.* **2015**, *45*, 1462–1473. [CrossRef]
9. Westerling, A.L.; Hidalgo, H.G.; Cayan, D.R.; Swetnam, T.W. Warming and earlier spring increase western U.S. forest wildfire activity. *Science* **2006**, *313*, 940–943. [CrossRef] [PubMed]
10. Miller, J.D.; Safford, H.D.; Crimmins, M.; Thode, A.E. Quantitative Evidence for Increasing Forest Fire Severity in the Sierra Nevada and Southern Cascade Mountains, California and Nevada, USA. *Ecosystems* **2009**, *12*, 16–32. [CrossRef]
11. Dillon, G.K.; Holden, Z.A.; Morgan, P.; Crimmins, M.A.; Heyerdahl, E.K.; Luce, C.H. Both topography and climate affected forest and woodland burn severity in two regions of the western US, 1984 to 2006. *Ecosphere* **2011**, *2*, 1–33. [CrossRef]
12. Miller, J.D.; Safford, H.D. Trends in wildfire severity: 1984 to 2010 in the Sierra Nevada, Modoc Plateau, and southern Cascades, California, USA. *Fire Ecol.* **2012**, *8*, 41–57. [CrossRef]
13. Belsky, A.J.; Blumenthal, D.M. Effects of Livestock Grazing on Stand Dynamics and Soils in Upland Forests of the Interior West. *Conserv. Biol.* **1997**, *11*, 315–327. [CrossRef]
14. O'Connor, C.D.; Falk, D.A.; Lynch, A.M.; Swetnam, T.W. Fire severity, size, and climate associations diverge from historical precedent along an ecological gradient in the Pinaleño Mountains, Arizona, USA. *For. Ecol. Manag.* **2014**, *329*, 264–278. [CrossRef]
15. Guiterman, C.H.; Margolis, E.Q.; Swetnam, T.W. Dendroecological Methods for Reconstructing High-Severity Fire in Pine-Oak Forests. *Tree-Ring Res.* **2015**, *71*, 67–77. [CrossRef]
16. Harris, L.; Taylor, A.H. Topography, fuels, and fire exclusion drive fire severity of the rim fire in an old-growth mixed-conifer forest, Yosemite National Park, USA. *Ecosystems* **2015**, *18*, 1192–1208. [CrossRef]
17. Lentile, L.B.; Smith, F.W.; Shepperd, W.D. Patch structure, fire-scar formation, and tree regeneration in a large mixed-severity fire in the South Dakota Black Hills, USA. *Can. J. For. Res.* **2005**, *35*, 2875–2885. [CrossRef]

18. Haire, S.L.; McGarigal, K. Effects of landscape patterns of fire severity on regenerating ponderosa pine forests (*Pinus ponderosa*) in New Mexico and Arizona, USA. *Landsc. Ecol.* **2010**, *25*, 1055–1069. [CrossRef]
19. Reilly, M.J.; Dunn, C.J.; Meigs, G.W.; Spies, T.A.; Kennedy, R.E.; Bailey, J.D.; Briggs, K. Contemporary patterns of fire extent and severity in forests of the Pacific Northwest, USA (1985–2010). *Ecosphere* **2017**, *8*. [CrossRef]
20. Arno, S.F. Forest Fire History in the Northern Rockies. *J. For.* **1980**, *78*, 460–465.
21. Keyser, T.L.; Lentile, L.B.; Smith, F.W.; Shepperd, W.D. Changes in Forest Structure After a Large, Mixed-Severity Wildfire in Ponderosa Pine Forests of the Black Hills, South Dakota, USA. *For. Sci.* **2008**, *54*, 328–338.
22. Fulé, P.Z.; Crouse, J.E.; Roccaforte, J.P.; Kalies, E.L. Do thinning and/or burning treatments in western USA ponderosa or Jeffrey pine-dominated forests help restore natural fire behavior? *For. Ecol. Manag.* **2012**, *269*, 68–81. [CrossRef]
23. Collins, B.M.; Roller, G.B. Early forest dynamics in stand-replacing fire patches in the northern Sierra Nevada, California, USA. *Landsc. Ecol.* **2013**, *28*, 1801–1813. [CrossRef]
24. Dodson, E.K.; Root, H.T. Conifer regeneration following stand-replacing wildfire varies along an elevation gradient in a ponderosa pine forest, Oregon, USA. *For. Ecol. Manag.* **2013**, *302*, 163–170. [CrossRef]
25. Chambers, M.E.; Fornwalt, P.J.; Malone, S.L.; Battaglia, M.A. Patterns of conifer regeneration following high severity wildfire in ponderosa pine—Dominated forests of the Colorado Front Range. *For. Ecol. Manag.* **2016**, *378*, 57–67. [CrossRef]
26. Rother, M.T.; Veblen, T.T. Limited conifer regeneration following wildfires in dry ponderosa pine forests of the Colorado Front Range. *Ecosphere* **2016**, *7*, e01594. [CrossRef]
27. Kemp, K.B.; Higuera, P.E.; Morgan, P. Fire legacies impact conifer regeneration across environmental gradients in the U.S. northern Rockies. *Landsc. Ecol.* **2016**, *31*, 619–636. [CrossRef]
28. Welch, K.R.; Safford, H.D.; Young, T.P. Predicting conifer establishment post wildfire in mixed conifer forests of the North American Mediterranean-climate zone. *Ecosphere* **2016**, *7*, e01609. [CrossRef]
29. Owen, S.M.; Sieg, C.H.; Sánchez Meador, A.J.; Fulé, P.Z.; Iniguez, J.M.; Baggett, L.S.; Fornwalt, P.J.; Battaglia, M.A. Spatial patterns of ponderosa pine regeneration in high-severity burn patches. *For. Ecol. Manag.* **2017**, *405*, 134–149. [CrossRef]
30. Donato, D.C.; Harvey, B.J.; Turner, M.G. Regeneration of montane forests 24 years after the 1988 Yellowstone fires: A fire-catalyzed shift in lower treelines? *Ecosphere* **2016**, *7*, e01410. [CrossRef]
31. Ziegler, J.P.; Hoffman, C.M.; Fornwalt, P.J.; Sieg, C.H.; Battaglia, M.A.; Chambers, M.E.; Iniguez, J.M. Tree Regeneration Spatial Patterns in Ponderosa Pine Forests Following Stand-Replacing Fire: Influence of Topography and Neighbors. *For. Trees Livelihoods* **2017**, *8*, 391. [CrossRef]
32. Stevens-Rumann, C.S.; Sieg, C.H.; Hunter, M.E. Ten years after wildfires: How does varying tree mortality impact fire hazard and forest resiliency? *For. Ecol. Manag.* **2012**, *267*, 199–208. [CrossRef]
33. Larson, A.J.; Belote, R.T.; Cansler, C.A.; Parks, S.A.; Dietz, M.S. Latent resilience in ponderosa pine forest: Effects of resumed frequent fire. *Ecol. Appl.* **2013**, *23*, 1243–1249. [CrossRef] [PubMed]
34. Graham, R.T. *Hayman Fire Case Study: Summary*; USDA Forest Service: Washington, DC, USA, 2003.
35. Northwest Alliance for Computational Science & Engineering. PRISM Climate Data. Available online: http://prism.nacse.org/normals/ (accessed on 1 May 2014).
36. Brown, P.M.; Kaufmann, M.R.; Shepperd, W.D. Long-term, landscape patterns of past fire events in a montane ponderosa pine forest of central Colorado. *Landsc. Ecol.* **1999**, *14*, 513–532. [CrossRef]
37. Kaufmann, M.R.; Regan, C.M.; Brown, P.M. Heterogeneity in ponderosa pine/Douglas-fir forests: Age and size structure in unlogged and logged landscapes of central Colorado. *Can. J. For. Res.* **2000**, *30*, 698–711. [CrossRef]
38. USGS. Monitoring Trends in Burn Severity. Available online: https://mtbs.gov (accessed on 1 May 2014).
39. Bradshaw, L.; Bartlette, R.; McGinely, J.; Zeller, K. Fire behavior, fuel treatments, and fire suppression on the Hayman Fire—Part 1: Fire weather, meteorology, and climate. In *Hayman Fire Case Study*; Graham, R.T., Ed.; RMRS-GTR-114; Department of Agriculture, Forest Service, Rocky Mountain Research Station: Ogden, UT, USA, 2003; pp. 36–58.
40. Fornwalt, P.J.; Kaufmann, M.R.; Huckaby, L.S.; Stohlgren, T.J. Effects of past logging and grazing on understory plant communities in a montane Colorado forest. *Plant Ecol.* **2009**, *203*, 99–109. [CrossRef]

41. Franklin, J.F.; Spies, T.A.; Van Pelt, R.; Carey, A.B. Disturbances and structural development of natural forest ecosystems with silvicultural implications, using Douglas-fir forests as an example. *For. Ecol. Manag.* **2002**, *155*, 399–423. [CrossRef]

42. Curtis, R.O.; Marshall, D.D. Technical note: Why quadratic mean diameter? *West. J. Appl. For.* **2000**, *15*, 137–139.

43. Reilly, M.J.; Spies, T.A. Regional variation in stand structure and development in forests of Oregon, Washington, and inland Northern California. *Ecosphere* **2015**, *6*, 1–27. [CrossRef]

44. Baddeley, A.; Turner, R. Spatstat: An R Package for Analyzing Spatial Point Patterns. *J. Stat. Softw.* **2005**, *12*, 1–42. [CrossRef]

45. Baddeley, A.; Rubak, E.; Turner, R. *Spatial Point Patterns: Methodology and Applications with R*; Chapman and Hall/CRC Press: London, UK, 2015.

46. Kint, V.; Lust, N.; Ferris, R.; Olsthoorn, A. Quantification of forest stand structure applied to Scots pine (*Pinus sylvestris* L.) forests. *For. Syst.* **2000**, *9*, 147–163.

47. Ripley, B.D. Modelling Spatial Patterns. *J. R. Stat. Soc. Ser. B Stat. Methodol.* **1977**, *39*, 172–212.

48. Moeur, M. Characterizing spatial patterns of trees using stem-mapped data. *For. Sci.* **1993**, *39*, 756–775.

49. Besag, J.; Diggle, P.J. Simple Monte Carlo Tests for Spatial Pattern. *J. R. Stat. Soc. Ser. C Appl. Stat.* **1977**, *26*, 327–333. [CrossRef]

50. Boyden, S.; Binkley, D.; Shepperd, W. Spatial and temporal patterns in structure, regeneration, and mortality of an old-growth ponderosa pine forest in the Colorado Front Range. *For. Ecol. Manag.* **2005**, *219*, 43–55. [CrossRef]

51. Baddeley, A.; Diggle, P.J.; Hardegen, A.; Lawrence, T.; Milne, R.K.; Nair, G. On tests of spatial pattern based on simulation envelopes. *Ecol. Monogr.* **2014**, *84*, 477–489. [CrossRef]

52. Wiegand, T.; Moloney, K.A. *Handbook of Spatial Point-Pattern Analysis in Ecology*; CRC Press: Boca Raton, FL, USA, 2013; ISBN 9781420082555.

53. Fajardo, A.; Goodburn, J.M.; Graham, J. Spatial patterns of regeneration in managed uneven-aged ponderosa pine/Douglas-fir forests of Western Montana, USA. *For. Ecol. Manag.* **2006**, *223*, 255–266. [CrossRef]

54. Fajardo, A.; McIntire, E.J.B. Under strong niche overlap conspecifics do not compete but help each other to survive: Facilitation at the intraspecific level. *J. Ecol.* **2011**, *99*, 642–650. [CrossRef]

55. Lotwick, H.W.; Silverman, B.W. Methods for Analysing Spatial Processes of Several Types of Points. *J. R. Stat. Soc. Ser. B Stat. Methodol.* **1982**, *44*, 406–413.

56. Neyman, J.; Scott, E.L. On a Mathematical Theory of Populations Conceived as Conglomerations of Clusters. *Cold Spring Harb. Symp. Quant. Biol.* **1957**, *22*, 109–120. [CrossRef]

57. Eidenshink, J.; Schwind, B.; Brewer, K.; Zhu, Z.; Quayle, B.; Howard, S. A project for monitoring trends in burn severity. *Fire Ecol.* **2007**, *3*, 3–21. [CrossRef]

58. Finco, M.; Quayle, B.; Zhang, Y.; Lecker, J.; Megown, K.A.; Brewer, C.K. *Monitoring Trends and Burn Severity (MTBS): Monitoring Wildfire Activity for the Past Quarter Century Using Landsat Data*; U.S. Department of Agriculture, Forest Service, Northern Research Station: Newtown Square, PA, USA, 2012.

59. Sörensen, R.; Zinko, U.; Seibert, J. On the calculation of the topographic wetness index: Evaluation of different methods based on field observations. *Hydrol. Earth Syst. Sci. Discuss.* **2006**, *10*, 101–112. [CrossRef]

60. Weiss, A. Topographic position and landforms analysis. Presented at the ESRI User Conference, San Diego, CA, USA, 9–13 July 2001; Volume 200.

61. Jenness, J. Topographic Position Index (tpi_jen. avx) Extension for ArcView 3. x, v. 1.3 a. Jenness Enterprises. Available online: http://www.jennessent.com/arcview/tpi.htm (accessed on 4 July 2017).

62. NASA. Global Data Explorer. Available online: https://gdex.cr.usgs.gov/gdex/ (accessed on 1 May 2014).

63. Briggs, J.S.; Fornwalt, P.J.; Feinstein, J.A. Short-term ecological consequences of collaborative restoration treatments in ponderosa pine forests of Colorado. *For. Ecol. Manag.* **2017**, *395*, 69–80. [CrossRef]

64. Reynolds, R.T.; Meador, A.J.S.; Youtz, J.A.; Nicolet, T.; Matonis, M.S.; Jackson, P.L.; DeLorenzo, D.G.; Graves, A.D. Restoring Composition and Structure in Southwestern Frequent-Fire Forests: A Science-Based Framework for Improving Ecosystem Resiliency. 2013. Available online: https://www.fs.fed.us/rm/pubs/rmrs_gtr310 (accessed on 25 November 2017).

65. Bonnet, V.H.; Schoettle, A.W.; Shepperd, W.D. Postfire environmental conditions influence the spatial pattern of regeneration for *Pinus ponderosa*. *Can. J. For. Res.* **2005**, *35*, 37–47. [CrossRef]

66. Cattelino, P.J.; Noble, I.R.; Slatyer, R.O.; Kessell, S.R. Predicting the multiple pathways of plant succession. *Environ. Manag.* **1979**, *3*, 41–50. [CrossRef]
67. Connell, J.H.; Slatyer, R.O. Mechanisms of Succession in Natural Communities and Their Role in Community Stability and Organization. *Am. Nat.* **1977**, *111*, 1119–1144. [CrossRef]
68. Crotteau, J.S.; Morgan Varner, J.; Ritchie, M.W. Post-fire regeneration across a fire severity gradient in the southern Cascades. *For. Ecol. Manag.* **2013**, *287*, 103–112. [CrossRef]
69. Mooney, K.A.; Linhart, Y.B.; Snyder, M.A. Masting in ponderosa pine: Comparisons of pollen and seed over space and time. *Oecologia* **2011**, *165*, 651–661. [CrossRef] [PubMed]
70. League, K.; Veblen, T. Climatic variability and episodic *Pinus ponderosa* establishment along the forest-grassland ecotones of Colorado. *For. Ecol. Manag.* **2006**, *228*, 98–107. [CrossRef]
71. Shepperd, W.D.; Edminster, C.B.; Mata, S.A. Long-Term Seedfall, Establishment, Survival, and Growth of Natural and Planted Ponderosa Pine in the Colorado Front Range. *West. J. Appl. For.* **2006**, *21*, 19–26.
72. York, R.A.; Battles, J.J.; Heald, R.C. Edge effects in mixed conifer group selection openings: Tree height response to resource gradients. *For. Ecol. Manag.* **2003**, *179*, 107–121. [CrossRef]
73. Moghaddas, J.J.; York, R.A.; Stephens, S.L. Initial response of conifer and California black oak seedlings following fuel reduction activities in a Sierra Nevada mixed conifer forest. *For. Ecol. Manag.* **2008**, *255*, 3141–3150. [CrossRef]
74. Jia, G.; Yu, X.; Fan, D.; Jia, J. Mechanism Underlying the Spatial Pattern Formation of Dominant Tree Species in a Natural Secondary Forest. *PLoS ONE* **2016**, *11*, e0152596. [CrossRef] [PubMed]
75. Johnstone, J.F.; Allen, C.D.; Franklin, J.F.; Frelich, L.E.; Harvey, B.J.; Higuera, P.E.; Mack, M.C.; Meentemeyer, R.K.; Metz, M.R.; Perry, G.L.W.; et al. Changing disturbance regimes, ecological memory, and forest resilience. *Front. Ecol. Environ.* **2016**, *14*, 369–378. [CrossRef]
76. McKenzie, D.A.; Tinker, D.B. Fire-induced shifts in overstory tree species composition and associated understory plant composition in Glacier National Park, Montana. *Plant Ecol.* **2012**, *213*, 207–224. [CrossRef]
77. Denslow, J.S.; Battaglia, L.L. Stand composition and structure across a changing hydrologic gradient: Jean Lafitte National Park, Louisiana, USA. *Wetlands* **2002**, *22*, 738–752. [CrossRef]
78. Rüger, N.; Huth, A.; Hubbell, S.P.; Condit, R. Response of recruitment to light availability across a tropical lowland rain forest community. *J. Ecol.* **2009**, *97*, 1360–1368. [CrossRef]
79. Bever, J.D. Host-specificity of AM fungal population growth rates can generate feedback on plant growth. *Plant Soil* **2002**, *244*, 281–290. [CrossRef]
80. Barker, M.G.; Press, M.C.; Brown, N.D. Photosynthetic characteristics of dipterocarp seedlings in three tropical rain forest light environments: A basis for niche partitioning? *Oecologia* **1997**, *112*, 453–463. [CrossRef] [PubMed]
81. Burslem, D.F.; Garwood, N.C.; Thomas, S.C. Tropical Forest Diversity—The Plot Thickens. *Science* **2001**, *291*, 606–607. [CrossRef] [PubMed]

![forests logo] *forests*

MDPI

Article

Overlapping Bark Beetle Outbreaks, Salvage Logging and Wildfire Restructure a Lodgepole Pine Ecosystem

Charles C. Rhoades [1,*], Kristen A. Pelz [1], Paula J. Fornwalt [1], Brett H. Wolk [2] and Antony S. Cheng [2]

1 USDA Forest Service, Rocky Mountain Research Station, Fort Collins, CO 80526, USA; kpelz@fs.fed.us (K.A.P.); pfornwalt@fs.fed.us (P.J.F.)
2 Colorado Forest Restoration Institute, Colorado State University, Fort Collins, CO 80523, USA; Brett.Wolk@colostate.edu (B.H.W.); Tony.CHENG@colostate.edu (A.S.C.)
* Correspondence: crhoades@fs.fed.us; Tel.: +01-970-498-1250

Received: 14 January 2018; Accepted: 24 February 2018; Published: 27 February 2018

Abstract: The 2010 Church's Park Fire burned beetle-killed lodgepole pine stands in Colorado, including recently salvage-logged areas, creating a fortuitous opportunity to compare the effects of salvage logging, wildfire and the combination of logging followed by wildfire. Here, we examine tree regeneration, surface fuels, understory plants, inorganic soil nitrogen and water infiltration in uncut and logged stands, outside and inside the fire perimeter. Subalpine fir recruitment was abundant in uncut, unburned, beetle-killed stands, whereas lodgepole pine recruitment was abundant in cut stands. Logging roughly doubled woody fuel cover and halved forb and shrub cover. Wildfire consumed all conifer seedlings in uncut and cut stands and did not stimulate new conifer regeneration within four years of the fire. Aspen regeneration, in contrast, was relatively unaffected by logging or burning, alone or combined. Wildfire also drastically reduced cover of soil organic horizons, fine woody fuels, graminoids and shrubs relative to unburned, uncut areas; moreover, the compound effect of logging and wildfire was generally similar to wildfire alone. This case study documents scarce conifer regeneration but ample aspen regeneration after a wildfire that occurred in the later stage of a severe beetle outbreak. Salvage logging had mixed effects on tree regeneration, understory plant and surface cover and soil nitrogen, but neither exacerbated nor ameliorated wildfire effects on those resources.

Keywords: disturbance; forest management; mountain pine beetle; subalpine ecosystem; Colorado; Rocky Mountains

1. Introduction

Lodgepole pine (*Pinus contorta* Dougl. ex. Loud. var. *latifolia*)-dominated ecosystems are adapted to periods of rapid post-disturbance change, as evidenced by the dense, even-aged forests that regenerate after wildfire and timber harvest [1,2]. New cohorts of lodgepole also establish readily after bark beetles (*Dendroctonus ponderosae* Hopkins) kill overstory pine [3,4]. The response of tree regeneration and understory plants following such disturbances determines forest vegetation dynamics and biodiversity [5–9] and has implications for ecosystem productivity and the biogeochemical processes that regulate soil nutrient retention and export [10–12]. For the 13,800 km^2 of forests infested by bark beetles since the early 2000s in Colorado, USA [13], the likelihood of overlapping disturbances increases with time as these forests are salvage logged or affected by wildfire [14]. However, the outcomes of compounding salvage logging and wildfire in beetle-killed lodgepole pine forests remain relatively poorly understood [15,16].

Site conditions and pre-disturbance forest composition and structure influence how individual and compound disturbance events affect forest ecosystem dynamics [17,18]. For example, while lodgepole pine typically regenerates densely after wildfire, seedling densities often vary by orders of magnitude even across a single wildfire [2,19], reflecting spatial patterns of fire behavior, fuel load, slope and other site attributes [15,20,21]. The implications of bark beetle outbreaks on wildfire probability and severity are mixed [22,23]. For example, flammability of canopy fuels increases following beetle infestation at stand scales [24], whereas bark beetle activity is not well related to wildfire severity [23] or extent at regional scales [25]. The severity, specific order and timing of consecutive disturbances determine their ecological outcomes, with greater impacts expected when initial disturbance severity is relatively high and time between disturbances is relatively short [14,16,26]. Wildfires occurring during the initial green-attack stage of beetle outbreaks—when needle flammability is highest—and the red-needle stage—when foliage begins to fall but serotinous cones remain unopened—are likely to have different effects than those occurring in gray-stage forests after needles have fallen and cones have opened [21,24,27–30].

Forest management activities prompted by recent severe beetle outbreaks in lodgepole pine forests of Colorado and elsewhere in the southern Rocky Mountains aim primarily to regenerate forests and to reduce short-term crown fire risk and longer-term risk of severe wildfire effects associated with heavy fuel accumulation after tree fall [31,32]. However, like other types of disturbance, the consequences of post-beetle management vary with forest composition, stand structure and time elapsed since the outbreak [18,33], and such factors have likely consequences for potential fire risk and behavior and other ecosystem attributes. The process of removing the forest canopy during salvage logging, for example, increases the mass of surface fuels and alters their moisture dynamics [3,34], but it also affects light, moisture and soil nutrients that influence plant responses [35]. The initial understory plant response to post-beetle salvage logging can differ between woody and non-woody plants and be affected by logging slash retention [8]. The cohort of trees that regenerate beneath the beetle-killed overstory and following salvage logging can form a new stratum of fuels and a future management concern [3,28,36]. In spite of the continental scale of recent bark beetle outbreaks [37] and the ensuing management response, it is uncertain whether post-beetle logging will aggravate wildfire effects.

In October 2010, the Church's Park Fire burned lodgepole pine forests where bark beetle infestation killed >85% of overstory basal area in the early 2000s. Portions of the burned area were salvage logged one year prior to the fire. The Church's Park Fire provides a fortuitous opportunity to evaluate overlapping effects of salvage logging and wildfire within severely-infested, gray-phase, beetle-killed forests. Our assessment included tree regeneration, surface fuels, understory plants, soil nitrogen and water infiltration under these conditions. All individual and overlapping disturbance events are unique, but in the absence of well-replicated experimental trials, this case study increases understanding of post-fire ecosystem dynamics in gray-stage beetle-impacted forests.

2. Materials and Methods

2.1. Study Area

This research was conducted on the Arapaho-Roosevelt National Forest near Fraser, Colorado, USA, in forests burned by the Church's Park Fire (39°56'25″ N; 105°57'00″ W) and surrounding unburned areas. The study area lies on the western edge of Colorado's Front Range between 2438–3200 m elevation. The area receives ~700 mm of precipitation annually, 75% as snow. Soils are gravelly, sandy-loam Alfisols derived from colluvium and alluvium of granitic gneiss and schist parent material [38].

Forests of the study area are a mix of lodgepole pine, subalpine fir (*Abies lasiocarpa* (Hook.) Nutt.) and Engelmann spruce (*Picea engelmannii* Parry ex. Engelm.) with scattered patches of quaking aspen (*Populus tremuloides* Michx.), and are part of the temperate steppe mountain ecoregion that extends from New Mexico, USA to southwestern Canada [39]. Bark beetles reached epidemic levels around 2000 and their activity peaked around 2006 in this part of Colorado [40,41]. Overstory pine mortality commonly exceeded 70% in mature, pine-dominated stands in this region of Colorado [3,42]. At Church's Park, lodgepole pine comprised 69% of total stand basal area before the outbreak, 89% of which was killed by beetles [43]. Lodgepole pine stands in the area typically contain a mixture of serotinous and non-serotinous cones [29].

Salvage logging occurred in 2009, several years after peak beetle activity. Harvested stands were clear cut and whole-tree yarded to central processing and loading areas (Figure 1) using tracked feller-bunchers and rubber-tired skidders. All harvest areas were on moderate (<35%), south-facing slopes.

Figure 1. Paired photos taken (**a**) one year pre-fire (October 2009) and (**b**) one year post-fire (September 2011) within the Church's Park Fire perimeter, near Fraser, Colorado. The photos are oriented northeast (20–30° azimuth) across an operational-scale Cut + Burn study site centered near 39°56′16.96″ N; 105°56′33.43″ W (See arrow in Figure 2). The log deck visible in photo (**a**) had been removed before the fire.

The Church's Park Fire began on 3 October 2010 and grew rapidly due to a combination of moderate wind speed, unseasonably high temperature, low relative humidity (16–32%), and very low fuel moisture (5%; [44,45]). The following three days were cooler with increasing humidity and the fire was 100% contained on 7 October. A cold front on 8 October effectively terminated the fire.

The fire burned a total of 200 ha of predominantly south-facing, beetle-killed, pine-dominated slopes, interspersed with meadows and aspen (Figure 2). Fire spread was pushed both across and upslope by down-valley winds. Observers noted very active to extreme fire behavior when the fire was burning in beetle-killed lodgepole pine stands, including active crown fire behavior, high rates of spread and flame lengths, and spotting of up to 0.4 km [46]. The crown fires burning through beetle-killed lodgepole pine stands and surface fires burning in salvaged logged units were classified as high- and moderate-severity based on complete or near complete combustion of organic soil layers and tree crowns and attached cones, and 100% mortality of residual live trees (Figure 2) [47]. According to burn-severity maps developed from remotely-sensed imagery and adjusted by on-site visual assessments, these areas comprised roughly half of the Church's Park Fire area (17% high- and 30% moderate-severity) [48]. Low-severity burning occurred primarily within meadow and aspen vegetation. Owing to the small size of and low risk for high intensity rainstorms after this October fire, post-fire mulch treatments were not applied [48].

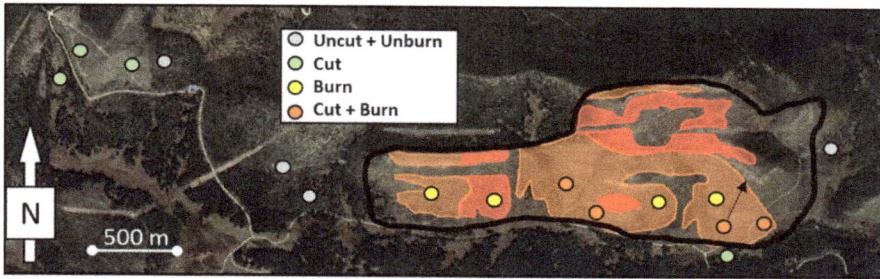

Figure 2. Perimeter of the Church's Park Fire and surroundings, near Fraser, Colorado, denoting areas mapped with moderate- (**orange**) and high- (**red**) burn severity and centers of operational-scale study sites. Unshaded areas within the fire perimeter were burned at low-severity. The dashed arrow is oriented with the general views in Figure 1. Aerial photo image date is 9 October 2015, five years after the fire. Image Citation: Google Earth Pro V7.3.1.4505 (Google LLC, Mountain View, CA, USA); accessed 2 February 2018.

2.2. Sampling and Analysis

We compared tree regeneration, surface fuel and understory plant cover and soil properties among the following ecosystem conditions: (1) Uncut + Unburned (UU); (2) Cut (C); (3) Burned (B) and (4) Cut + Burned (CB) (Figure 2). We established four operational-scale study sites (3–10 ha) for each ecosystem condition with three stand-scale sampling areas in each (~1 ha), then established one randomly-oriented 50-m long transect per sampling area. All study areas were dominated by gray-phase lodgepole pine prior to salvage logging and the fire. Burned study areas were located in high- and moderate-severity patches. All study areas were located on south-facing hillslopes with moderate slope (mean: 34%). Unburned study areas, both cut and uncut, were within 3 km of the fire perimeter.

We examined understory plant and surface cover, tree regeneration, and plant-available soil nitrogen (N) over the course of three years. We measured understory plant and surface cover in August 2012, 2013, 2014 with a gridded point-intercept method in five 1-m^2 quadrats per transect. Common understory plants were identified to genus or species while others were identified to growth form (graminoid, forb, shrub). Surface cover elements included organic horizon (O) soil (litter and duff), mineral soil, 1- to 10-h woody fuels (<2.5 cm diameter), 100-h woody fuels (2.5–7.6 cm diameter), and 1000-h woody fuels (>7.6 cm diameter). Regenerating trees were tallied within the quadrats by species and height classes (1–15 cm, 15–75 cm, $\geq$75 cm but <2.5 cm diameter). We used ion exchange resin (IER) bags to measure plant-available soil N and potential nitrate (NO_3-N) leached in spring snowmelt [49]. We inserted 10 resin bags per transect, 5–10 cm into mineral soil each fall and exchanged them the following spring during 2011/2012, 2012/2013, and 2013/2014. Resin bags consisted of a 1:1 mixture of cation (Sybron Ionic C-249, Type 1 Strong Acid, Na^+ form, Gel Type) and anion (Sybron Ionic ASB-1P Type 1, Strong Base OH^- form, Gel Type) exchange resin beads. After removal from the field, resins were extracted with a 2 M KCl solution, shaken for 60 min, filtered and frozen until analysis. Nitrate (NO_3-N) and ammonium (NH_4-N) concentrations were measured by spectrophotometry using a flow injection analyzer (Lachat Company, Loveland, CO, USA).

As an indicator of post-fire soil hydrologic conditions, in 2012 we also measured soil water infiltration rate with a field infiltrometer designed to assess wildfire effects (Decagon Devices, Pullman, WA, USA). We recorded the volume of water infiltrating into the mineral soil (2 cm depth) during triplicate 60-s subsample periods at five locations per sample transect. We evaluated soil hydrophobicity [50] at a similar sampling intensity by measuring the time that a water drop remained on the soil surface (e.g., water drop penetration resistance) using the following time periods: none (<10 s); weak (10–40 s); moderate (40–180 s); strong (>180 s).

Data were composited within the three stand-scale sampling areas in each of the four study sites (n = 12). Given the close proximity and consistent topographic position, forest composition and degree of beetle-related mortality of the study sites, we assume they are comparable for statistical analysis. The four ecosystem conditions were compared using analysis of variance with Cut, Burn, and a Cut × Burn interaction as fixed effects and stand-scale sampling areas nested within study sites as random effects (SPSS version 22, IBM Co., Chicago, IL, USA). We added a repeated measures term for analysis of tree regeneration, surface and understory plant cover and plant-available soil N. Each water drop penetration measure was placed into a resistance class, plot and transect-scale replicates were averaged then analyzed as a continuous variable. Where fixed effects were significant, we used pairwise, Tukey-adjusted comparisons to identify differences among the four ecosystem conditions. Levene's statistic was used to test assumptions of homogeneity of variance; ion exchange resin data violated this assumption and were log-transformed prior to analyses. Statistical significance is reported where $\alpha \leq 0.05$, unless otherwise stated.

3. Results

Tree seedling density varied among the four ecosystem conditions (Figure 3). Total seedling density in 2014 was highest (~13,000 trees ha^{-1}) in the UU areas, consisting almost entirely of subalpine fir (90% of all seedlings) in the smaller two size classes. Aspen and the other conifer species occurred at much lower densities in UU areas.

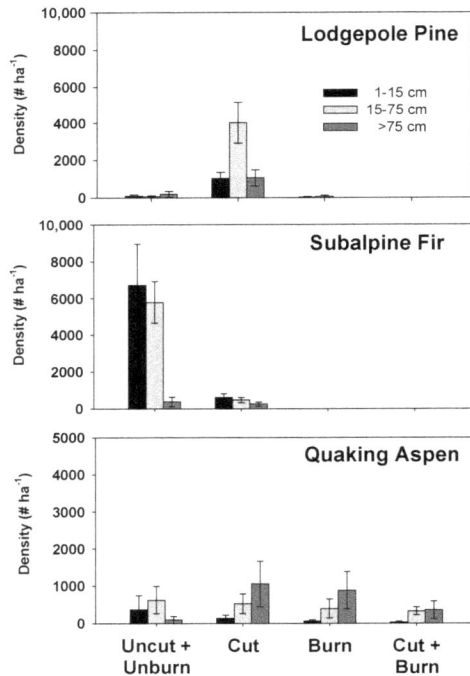

Figure 3. Tree seedling density in August 2014, four growing seasons after the Church's Park Fire and five years after harvesting of the bark beetle-infested lodgepole pine overstory. Data are means with standard error bars for twelve stand-scale sampling areas per ecosystem condition, by seedling height class. Spruce seedlings were absent from both burned conditions (Burn and Cut + Burn) and represented <56 tree ha^{-1} in both unburned conditions (Unburn + Uncut and Cut). Note: The y axis of the bottom panel is half that of the upper two panels.

Fir seedling density was 90% lower in harvested (C) compared to UU areas. Conversely, total lodgepole pine density was 6100 trees ha^{-1} in C areas and 340 trees ha^{-1} in UU areas. Cutting stimulated a 10-fold increase in the density of the tallest class of aspen via sprouting compared to the UU treatment. Burning had a dramatic and lasting effect on conifer seedling density (Figure 3). Pine seedlings were extremely rare (<100 trees ha^{-1} in B; 0 trees ha^{-1} in CB) and there were no fir or spruce [43] tallied in either burned condition (B or CB) during the study. Aspen was the only tree species found both in the B and CB areas. From 2012 to 2014, aspen sprout density increased 6-fold in B and 3-fold in CB areas, but changed little in the other conditions [43]. Conifer density did not increase in UU and C areas during the study.

Soil organic (O) horizon cover was 29% lower in the cut (C) areas compared to UU areas, averaged over all sample years (Table 1). Conversely, fine (1 and 10 h), 100 h, and total woody fuel cover was 1.7-, 4.0 and 1.9 times higher in C compared to UU areas. Burning (B) had greater effects than harvesting on surface cover (Table 1). Soil O horizon extent was 55% lower in B compared to UU areas on average, whereas mineral soil cover was about 30 times higher. Fine woody fuel cover was 73% lower in B areas overall; larger fuel classes and total woody fuel cover did not differ from UU areas. Organic horizon and mineral soil cover for CB treatments were intermediate relative to the UU and B treatments. However, woody fuel cover in the CB combination did not differ from burning (B) alone. Soil and wood cover changed little in UU areas over the course of the study (Table 1). There was no return of O horizon cover in the C, B or CB conditions over the course of the study or decline in mineral soil cover.

Table 1. Surface cover (%) after salvage logging and wildfire in bark beetle-infested lodgepole pine forests. Data are means with standard error for twelve stand-scale sampling areas per ecosystem condition per date. Different letters within columns denote differences within years based on Tukey's pairwise adjusted comparisons.

Year	Condition/Label	Soil Surface Cover		Woody Fuel Cover			
		Organic	Mineral	1 and 10-h	100-h	1000-h	Total Fuel
2012	Uncut + Unburn (UU)	86.9 2.3 a	2.4 0.8 c	15.0 1.9 a	1.5 0.4 b	5.8 1.5 ab	22.5 2.5 b
	Cut (C)	71.4 3.5 ab	12.5 2.4 c	24.8 4.8 a	6.5 0.9 a	11.2 2.8 a	43.9 6.6 a
	Burn (B)	36.6 5.7 c	52.6 5.3 a	3.1 1.6 b	0.9 0.2 b	2.9 1.5 b	8.8 1.7 b
	Cut + Burn (CB)	55.5 5.7 b	31.9 5.4 b	4.1 1.6 b	1.2 0.3 b	5.4 1.4 ab	10.9 2.4 b
2013	Uncut + Unburn (UU)	87.2 2.3 a	1.5 0.3 c	9.5 1.8 ab	0.9 0.3 b	7.3 1.5 a	18.4 1.8 ab
	Cut (C)	59.0 4.2 bc	14.8 3.7 bc	15.1 2.6 a	4.5 0.7 a	10.7 3.2 a	31.2 5.3 a
	Burn (B)	47.8 4.9 c	44.6 5.5 a	3.0 1.2 b	1.2 0.3 b	3.1 1.4 a	9.6 1.5 b
	Cut + Burn (CB)	64.9 4.9 bc	27.1 5.5 ab	1.8 0.4 b	1.8 0.4 b	4.0 1.0 a	10.9 2.1 b
2014	Uncut + Unburn (UU)	85.3 0.9 a	2.0 0.6 c	11.5 1.8 b	2.2 0.5 b	4.2 1.4 ab	20.1 1.9 b
	Cut (C)	53.2 4.8 b	15.3 2.6 c	21.8 3.5 a	5.7 1.2 a	10.8 2.5 a	42.3 5.8 a
	Burn (B)	31.8 3.3 c	60.8 3.8 a	3.4 0.4 c	1.1 0.3 b	2.6 0.7 b	11.4 1.4 b
	Cut + Burn (CB)	44.3 4.9 bc	46.6 5.1 b	3.6 0.6 c	1.4 0.6 b	4.8 1.7 ab	10.7 2.2 b
Effects		F — p	F — p	F — p	F — p	F — p	F — p
Cut		4.9 0.029	1.0 0.317	13.8 <0.001	53.7 <0.001	14.5 <0.001	27.4 <0.001
Burn		123.0 <0.001	270.3 <0.001	101.9 <0.001	47.5 <0.001	19.5 <0.001	102.8 <0.001
Cut * Burn		72.7 <0.001	42.7 <0.001	9.9 0.002	35.1 <0.001	3.3 0.071	23.6 <0.001
Date		8.5 <0.001	6.0 0.004	3.1 0.051	1.01 0.368	0.2 0.800	1.9 0.151

Graminoid cover was similar between C and UU conditions, and forb cover was only marginally lower, but shrub cover was considerably lower in the C conditions (Table 2). Averaged over the three-year study, shrub cover was 66% lower in C than UU areas. Total understory plant cover was 46% lower overall in C compared to UU areas. Graminoid and shrub covers were both 89% lower in B relative to UU areas, though forb cover was similar. Graminoid cover was intermediate for cutting followed by burning (CB) and was from 3 to 11 times higher than B. Understory plant cover was relatively stable over the course of the study in the UU and C areas. In contrast, total plant cover doubled between 2012 and 2014 in B areas. In the B treatment, graminoid, forb and shrub cover increased 5-, 2- and 3-fold during the study. Shrub cover was also 5 times higher in the CB treatment in 2014 compared to 2012.

Table 2. Understory plant cover (%) after salvage logging and wildfire in bark beetle-infested lodgepole pine forests. Data are means with standard error for twelve stand-scale sampling areas per ecosystem condition per year. The sum of plant growth forms may exceed 100% within treatment due to overlapping plant canopy layers. Different letters within columns denote differences within single years based on Tukey's pairwise adjusted comparisons.

Year	Condition	Graminoid			Forb			Shrub			Total		
2012	Uncut + Unburn	43.7	4.7	a	34.8	4.3	a	39.7	3.7	a	118.2	5.9	a
	Cut	33.6	4.2	a	16.8	5.6	a	12.9	2.4	b	63.2	7.4	b
	Burn	1.2	0.3	b	20.0	3.1	a	2.3	0.4	c	23.6	3.1	c
	Cut + Burn	13.9	3.4	b	35.3	7.4	a	0.7	0.3	c	49.8	9.2	bc
2013	Uncut + Unburn	32.8	5.8	a	20.8	3.9	ab	38.2	3.0	a	91.8	8.5	a
	Cut	26.2	3.5	ab	13.1	3.5	b	11.7	2.1	b	51.0	5.5	b
	Burn	3.9	1.2	c	28.9	4.4	a	4.6	0.9	c	37.5	3.9	b
	Cut + Burn	13.9	2.5	bc	30.0	4.3	a	1.8	0.6	c	45.6	4.5	b
2014	Uncut + Unburn	33.3	3.9	a	37.5	5.1	a	39.8	3.2	a	110.5	7.4	a
	Cut	27.1	4.3	ab	16.0	5.0	b	15.7	2.6	b	58.7	6.2	b
	Burn	6.1	1.3	c	39.0	4.7	a	6.3	1.1	c	51.3	5.4	b
	Cut + Burn	18.5	3.4	bc	32.7	3.6	ab	3.3	0.9	c	54.5	5.4	b
Effects		*F*	*p*		*F*	*p*		*F*	*p*		*F*	*p*	
	Cut	1.0	0.331		6.2	0.014		179.0	<0.001		30.2	<0.001	
	Burn	150.4	<0.001		11.2	0.001		466.2	<0.001		111.5	<0.001	
	Cut * Burn	27.2	<0.001		11.1	0.001		120.4	<0.001		70.3	<0.001	
	Date	1.2	0.308		3.4	0.038		2.0	0.145		4.7	0.011	

As of 2014, cover of the most common species in each plant growth form remained low in both C and B treatments (Figure 4). The forb, heartleaf arnica (*Arnica cordifolia* Hook.) was 12% in UU areas and 3% and 1.5% in C and B areas. Fireweed (*Chamerion angustifolium* (L.) Holub.) cover was nearly 2.5 times higher in B relative to UU areas. The dominant shrub, grouse whortleberry (*Vaccinium scoparium* Leiberg ex. Coville), averaged 33% in UU areas compared to 12 and 2% after cutting and burning, respectively. Both arnica and whortleberry were nearly absent where cutting was followed by burning (CB), but the combined treatment more than doubled sedge cover (*Carex* spp., predominately *C. rossii* Boott. and *C. geyeri* Boott.) in areas that were only burned.

Plant available soil N (IER-N) was lowest in UU areas and generally increased with additional disturbance (Figure 5). Cut areas had significantly higher nitrate and total IER-N overall, though C and UU treatments did not differ statistically within individual years. Burned areas had elevated nitrate, ammonium and total IER-N relative to UU stands throughout the study. Averaged across three years, there were 5.4 and 3.5 times more nitrate and total IER-N in B compared to UU areas. The burn effect on IER-ammonium was statistically significant in 2012, when it was 2.2 times higher than in UU areas. Significant cut-by-burn interactions for nitrate and total IER-N indicate an additive effect of burning in salvage-logged areas (CB). Overall, CB areas had 10 and 6 times more nitrate IER-N and total IER-N, respectively, than UU areas. In 2012, CB areas had 14 times more IER-nitrate than UU areas and roughly double that measured in B areas. In the subsequent two years, IER-nitrate was similar in CB and B areas. On average, nitrate represented 76% of total IER-N in BC compared to 48% in UU areas.

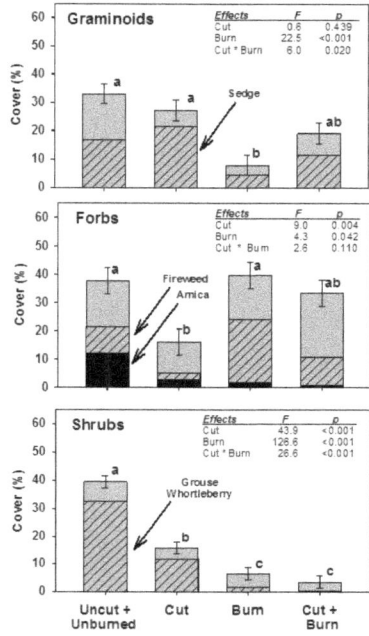

Figure 4. Understory plant cover in bark beetle-infested, cut and burned forest combinations at the Church's Park Fire, Colorado during August 2014. Data are means of total plant cover (gray bar) by plant growth form with standard error bars for twelve stand-scale sampling areas per ecosystem condition. The sum of plant growth forms may exceed 100% within treatment due to overlapping plant canopy layers. Different letters indicate that treatment values differ based on Tukey's pairwise adjusted comparisons. Cover of the most abundant (hatched) or second-most abundant (blackened) species in each growth form are displayed and identified by arrows.

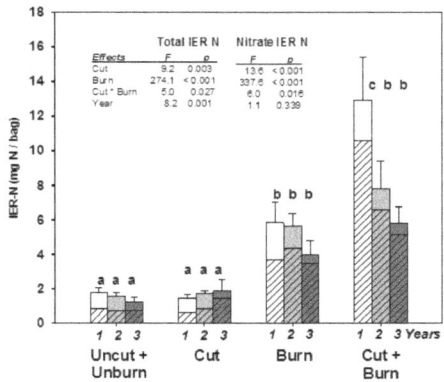

Figure 5. Plant-available soil N in bark beetle-infested, cut and burned forest conditions at the Church's Park Fire, Colorado. Bars are the average total IER-N and nitrate IER-N (hatched) with standard error bars. The three years denote the 2011/2012, 2012/2013 and 2013/2014 overwinter sampling periods. Different letters denote differences within years based on Tukey's pairwise adjusted comparisons of log-transformed total IER-N data.

In 2012, water infiltration in the B and CB areas (1.5 and 1.9 mL min^{-1}) was half the rate measured in UU areas (3.4 mL min^{-1}) (Figure 6a). Wildfire also inhibited water drop penetration, indicating moderate levels of hydrophobicity (Figure 6b) with highest resistance in B areas. Cutting decreased infiltration, though to a marginally lesser extent than burning, and it had no effect on hydrophobicity.

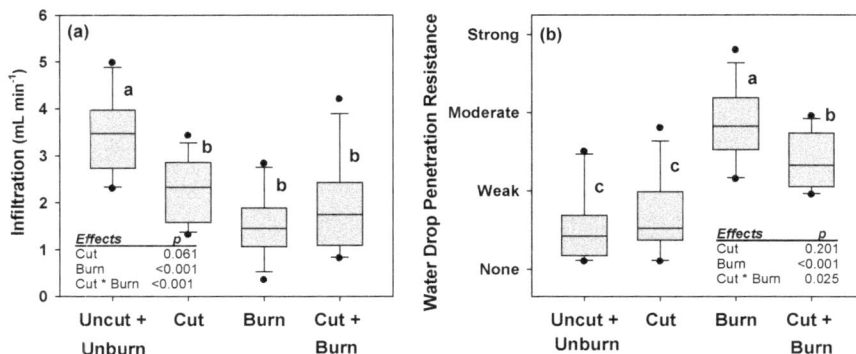

Figure 6. Water infiltration rate (**a**) and water drop penetration resistance (**b**) measured in 2012 after logging and the Church's Park Fire in mountain pine beetle-infested stands. Plots show 25th, 50th and 75th percentiles (box), 10th and 90th percentiles (whiskers) and outliers (filled circles). Different letters indicate that treatment values differ based on Tukey's pairwise adjusted comparisons. Water drop resistance ratings as follows: none <10 s; weak 10–40 s; moderate 40–180 s; strong >180 s.

4. Discussion

4.1. Overlapping Disturbances

The Church's Park Fire concluded a series of disturbances that started with bark beetle infestation and was followed by salvage logging in some stands. The lack of lodgepole pine recruitment for four years after the fire contrasted with dense post-fire seedling establishment that is typical within two to three years of harvesting and burning [2,20,31,51,52]. It also differed from fires in beetle-affected lodgepole stands that occurred during green-attack or red-needle stage [21,53] or gray-stage stands with lower outbreak severity (0–56% beetle-killed basal area) [15]. Cone serotiny is a critical determinant of post-fire and post-bark beetle lodgepole recruitment [15,54–56], and is prevalent in Church's Park area stands. Lodgepole pine seeds remain viable in serotinous cones for over 25 years after trees are infested by bark beetles [29], so there would have been a canopy seed source at the time of the fire. However, the fire scorched and consumed nearly all cones remaining on standing dead pine trees and in logging slash as well as any advance regeneration or seedlings established since the outbreak.

The absence of conifer regeneration within the Church's Park Fire contrasts with surrounding unburned areas as well, where bark beetle mortality alone or salvage logging of beetle-infested stands stimulated conifer recruitment. Observations of stand development 20–30 years after a 1980s-era beetle outbreak [17] confirm projections of stand dynamics based on inventory of seedling establishment after the recent outbreak [3]. Both of these Colorado studies along with those conducted elsewhere [57,58] suggest that (1) uncut beetle-infested stands will develop into well-stocked, conifer-dominated forests with more subalpine fir than prior to the beetle outbreak and that (2) salvage-logged, beetle-infested stands will regenerate into pine-dominated stands, similar to those that existed at the time of the outbreak. In our study, aspen was the only tree species observed regenerating via sprouting in significant numbers after the Church's Park Fire. Aspen density was relatively insensitive to cutting and burning compared to the conifers (Figure 3). Long-term forest development within the Church's Park Fire perimeter is uncertain, but based on our findings it appears likely that aspen will increase and conifers will decline relative to pre-fire conditions and surrounding unburned areas; similar patterns have been reported elsewhere [18].

The outcome of individual disturbances is determined by unique combinations of site conditions, disturbance characteristics and post-disturbance ecological interactions [7]. At Church's Park, overlap of the beetle outbreak and salvage logging or wildfire disturbances are likely to produce even more complexity. The high levels of beetle-induced mortality (>85% of overstory basal area), time elapsed since the outbreak (~8 year post-infestation) and steep slopes were features of the site and wildfire that resulted in crown fire behavior with near-complete crown and cone consumption [16]. Though post-fire regeneration was generally adequate in beetle-infested northwestern Wyoming lodgepole stands, regeneration was nonetheless lowest under conditions such as those we studied, where crown fire in gray-stage beetle kill scorched crowns and consumed cones [15]. Our findings are specific to the site, pre-fire stand structure and fire behavior at Church's Park, but the compositional changes we documented after the fire are likely to be repeated where wildfires burn similar gray-stage, beetle-killed stands with extensive overstory mortality [18,37,59].

4.2. Implications of Post-Bark Beetle Salvage Logging on Wildfire Effects

Widespread overstory mortality associated with severe bark beetle outbreaks increased concerns about fire risk and prompted post-outbreak timber harvesting in Colorado after decades of public opposition [31]. Salvage logging is prescribed to address numerous objectives [60,61] and in response to recent insect outbreaks it has been used to reduce canopy fuels and crown fire potential, capture the value of dead timber, regenerate forests, protect infrastructure and humans from falling trees, and facilitate fire suppression [62]. However, as observed here and elsewhere, logging increases surface fuel loads [34,63], and in the event of a post-harvest wildfire, has the potential to exacerbate fire behavior and effects [64]. Salvage harvesting is controversial where it fails to meet intended objectives [60,65] and at Church's Park there was potential that logging in conjunction with the overlapping beetle and wildfire disturbances would have unintended negative consequences for biodiversity, ecosystem function and delivery of ecosystem services [61]. Regional concerns for management of federal forest lands include regenerating well-stocked forests, retaining native plant diversity and cover, maintaining soil and ecosystem productivity and protecting clean water supply.

At the time of the fire (<two years after harvesting), residual fine fuels likely altered fire spread and large fuels may have increased the duration of combustion in BC areas. Both graminoids and shrubs were negatively impacted by burning. Grouse whortleberry, the shrub that formed >30% cover in UU areas, was reduced to ~2% in B areas and was almost eliminated from BC areas (Figure 4). However, with that exception, other responses we measured suggest that fire effects were no more severe in areas that were logged prior to the fire. Conversely, while salvage logging removed the forest canopy and thus eliminated the risk of crown fire, the surface fire that burned through the harvested areas had similar effects to crown fire in uncut areas.

After the fire, BC areas had less exposed mineral soil and greater O horizon cover than solely burned (B) areas. The higher residual O horizon cover is likely to have contributed to the marginally higher water infiltration (Figure 6) and plant-available N in those areas (Figure 5) relative to B areas. The initial pulse of soil N in BC areas may have resulted from the combustion of accumulated post-harvest fuels; similar to N dynamics after pile burning, it began to recede after one year [66]. After the first year of sampling, soil nitrate was similar between B and BC areas relative to UU and C areas (Figure 5). Both bark beetles and salvage logging are known to increase soil N in unburned lodgepole pine forests [35,67]. In spite of post-bark beetle increases in soil N, and unlike beetle outbreaks in parts of Europe that receive high atmospheric N deposition [68], Colorado beetle outbreaks have not threatened surface water with high N loading [69]. Research in Europe and the US highlights the role of nutrient demand and compensatory growth by recruiting and residual vegetation for intercepting surplus soil nutrients after tree mortality [11,12,68,69]. At Church's Park, post-fire IER-N levels and the risk of nitrate leaching will recede as understory plant cover increases.

The marginally higher understory cover of BC compared to B areas suggests that salvage logging did not exacerbate these concerns at Church's Park.

The US Forest Service is required by the National Forest Management Act of 1976 (United States Public Law 94-588) to monitor and rectify tree regeneration failure associated with management activities. In all stand-level study areas affected by the Church's Park Fire (both B and BC), conifer regeneration fell below US Forest Service density thresholds aimed at ensuring the development of acceptably stocked forests (370 tree ha^{-1}) [70]. Though fire eliminated virtually all conifer regeneration in the B and BC areas (Figure 3), it did not reduce regenerating aspen density relative to unburned areas. Owing to the small spatial extent of the Church's Park Fire and establishment of conifer cohorts in beetle-infested and salvage-logged forests surrounding the burn, scarce regeneration within the fire is not likely to have negative effects on local biodiversity. Limited conifer recruitment into the Church's Park Fire, in fact, should interrupt landscape continuity and thus reduce the spread of future wildfires [19].

Nonetheless, the Church's Park Fire appears to have had a potentially lasting effect on forest species composition relative to pre-outbreak, pre-fire conditions. Aspen was present throughout the Church's Park area prior to the series of disturbances, and sprouts were stimulated or retained within salvage logged (C), burned (B) and combined cut, then burned areas (CB). Post-fire expansion of aspen is common in the Colorado subalpine forest zone and is associated with benefits for floral and faunal biodiversity, fire resistance, and landscape aesthetics [71,72]. Aspen regeneration was abundant across all our study conditions and our findings suggest that the species could play an increasingly important role in similar post-beetle outbreak forests across the Rocky Mountain West.

5. Conclusions

After four years of post-fire recovery, it appears that the overlapping disturbances culminating with the Church's Park Fire will have a long-term effect on forest development. The severe level of bark beetle-related overstory mortality, followed by crown fire in gray-stage stands, virtually eliminated conifer regeneration. In contrast to the conifers, the density of aspen in 2014 was similar inside and outside of the fire (Figure 3) and it has increased more than three-fold in burned areas since the fire. Shrubs were greatly reduced by burning alone and burning in previously-logged areas, though their cover has also begun to increase since the fire. The impacts of the Church's Park Fire on forest regeneration were consistent with patterns documented in northwestern Wyoming where crown fire consumed serotinous cones in gray-stage beetle-killed lodgepole pine [15]. Recent studies suggest that projected increases in drought and associated fire frequency and behavior may detract from the resilience of lodgepole and other forest types of the Rocky Mountain West [73,74]. However, low precipitation did not contribute to the scarce conifer regeneration following the Church's Park Fire [75]. Summer season precipitation in the year of the fire (2010) and the following year were above average. In fact, 2011 received the highest total precipitation during the past 30 years. Complete canopy, cone and seedbank consumption was the probable cause of the scant conifer regeneration following the wildfire. As beetle-killed lodgepole pine forests transition to the gray-stage, the conditions we documented after the Church's Park Fire are likely to become more common, especially throughout portions of Colorado, Wyoming and Montana with concentrated beetle activity and high levels of overstory mortality [41,59]. Future research should take advantage of these expanded possibilities and conduct well-replicated studies to advance understanding and provide critical knowledge for managing and conserving forest processes and biodiversity under changing climatic conditions.

Acknowledgments: Conversations, ideas and technical assistance were generously provided by many colleagues including Greg Aplet, Dan Binkley, Chad Hoffman, Kevin Moriarty, Eric Schroder and Skip Smith. Thanks to Kevin Miller, Peter Pavlowich, Ben Wudtke, Rob Addington and Derek Pierson for careful field and laboratory work. We gratefully acknowledge financial support from the Colorado Forest Restoration Institute, the Arapaho Roosevelt National Forest and USFS R2 Regional Office, and we especially thank Hal Gibbs and Tommy John for their support. Thanks to Susan Miller for editorial comments and Scott Baggett for statistical comments. Comments from three anonymous reviewers greatly improved the accuracy and clarity of the manuscript.

Author Contributions: C.C.R., K.A.P. and P.J.F. located the study areas, designed the sampling, analyzed the data and wrote the manuscript with input and comments from B.H.W. and A.S.C.

Conflicts of Interest: The authors declare no conflict of interest. The funding sponsors had no role in the design of the study; in the collection, analyses, or interpretation of data; in the writing of the manuscript, and in the decision to publish the results.

References

1. Lotan, J.E.; Critchfield, W.B. Pinus contorta Dougl. ex. Loud. lodgepole pine. In *Silvics of North America. Volume 1. Conifers*; Burns, R.M., Honkala, B.H., Eds.; U.S. USDA: Washington, DC, USA, 1990; Volume 1, pp. 302–315.
2. Turner, M.G.; Romme, W.H.; Gardner, R.H.; Hargrove, W.W. Effects of fire size and pattern on early succession in Yellowstone National Park. *Ecol. Monogr.* **1997**, *67*, 411–433. [CrossRef]
3. Collins, B.J.; Rhoades, C.C.; Hubbard, R.M.; Battaglia, M.A. Tree regeneration and future stand development after bark beetle infestation and harvesting in Colorado lodgepole pine stands. *For. Ecol. Manag.* **2011**, *261*, 2168–2175. [CrossRef]
4. Diskin, M.; Rocca, M.E.; Nelson, K.N.; Aoki, C.F.; Romme, W.H. Forest developmental trajectories in mountain pine beetle disturbed forests of Rocky Mountain National Park, Colorado. *Can. J. For. Res.* **2011**, *41*, 782–792. [CrossRef]
5. Stone, W.E.; Wolfe, M.L. Response of understorey vegetation to variable tree mortality following a mountain pine beetle epidemic in lodgepole pine stands in northern Utah. *Vegetatio* **1996**, *122*, 1–12. [CrossRef]
6. Selmants, P.C.; Knight, D.H. Understory plant species composition 30–50 years after clearcutting in southeastern Wyoming coniferous forests. *For. Ecol. Manag.* **2003**, *185*, 275–289. [CrossRef]
7. Turner, M.G. Disturbance and landscape dynamics in a changing world. *Ecology* **2010**, *91*, 2833–2849. [CrossRef] [PubMed]
8. Fornwalt, P.J.; Rhoades, C.C.; Hubbard, R.M.; Harris, R.L.; Faist, A.M.; Bowman, W.D. Short-term understory plant community responses to salvage logging in beetle-affected lodgepole pine forests. *For. Ecol. Manag.* **2018**, *409*, 84–93. [CrossRef]
9. Romme, W.H.; Whitby, T.G.; Tinker, D.B.; Turner, M.G. Deterministic and stochastic processes lead to divergence in plant communities 25 years after the 1988 Yellowstone fires. *Ecol. Monogr.* **2016**, *86*, 327–351. [CrossRef]
10. Parsons, W.F.J.; Knight, D.H.; Miller, S.L. Root gap dynamics in lodgepole pine forest: Nitrogen transformations in gaps of different sizes. *Ecol. Appl.* **1994**, *4*, 354–362. [CrossRef]
11. Griffin, J.M.; Turner, M.G.; Simard, M. Nitrogen cycling following mountain pine beetle disturbance in lodgepole pine forests of Greater Yellowstone. *For. Ecol. Manag.* **2011**, *261*, 1077–1089. [CrossRef]
12. Rhoades, C.C.; McCutchan, J.H.; Cooper, L.A.; Clow, D.W.; Detmer, T.M.; Briggs, J.S.; Stednick, J.D.; Veblen, T.T.; Ertz, R.M.; Likens, G.; et al. Biogeochemistry of beetle kill: Explaining a weak nitrate response. *Proc. Natl. Acad. Sci. USA* **2013**, *110*, 1756–1760. [CrossRef] [PubMed]
13. Colorado State Forest Service (CSFS). *2015 Report on the Health of Colorado's Forests: 15 Years of Change*; Colorado State University: Fort Collins, CO, USA, 2016; Volume 32. Available online: https://csfs.colostate.edu/media/sites/22/2016/02/ForestHealthReport-2015.pdf (accessed on 14 January 2018).
14. Edwards, M.; Krawchuk, M.A.; Burton, P.J. Short-interval disturbance in lodgepole pine forests, British Columbia, Canada: Understory and overstory response to mountain pine beetle and fire. *For. Ecol. Manag.* **2015**, *338*, 163–175. [CrossRef]
15. Harvey, B.J.; Donato, D.C.; Turner, M.G. Recent mountain pine beetle outbreaks, wildfire severity, and postfire tree regeneration in the US Northern Rockies. *Proc. Natl. Acad. Sci. USA* **2016**, *111*, 15120–15125. [CrossRef] [PubMed]
16. Kane, J.M.; Varner, J.M.; Metz, M.R.; van Mantgem, P.J. Characterizing interactions between fire and other disturbances and their impacts on tree mortality in western U.S. Forests. *For. Ecol. Manag.* **2017**, *405*, 188–199. [CrossRef]
17. Pelz, K.A.; Smith, F.W. Thirty year change in lodgepole and lodgepole/mixed conifer forest structure following 1980s mountain pine beetle outbreak in western Colorado, USA. *For. Ecol. Manag.* **2012**, *280*, 93–102. [CrossRef]

18. Kulakowski, D.; Matthews, C.; Jarvis, D.; Veblen, T.T. Compounded disturbances in sub-alpine forests in western Colorado favour future dominance by quaking aspen (*Populus tremuloides*). *J. Veg. Sci.* **2013**, *24*, 168–176. [CrossRef]
19. Seidl, R.; Donato, D.C.; Raffa, K.F.; Turner, M.G. Spatial variability in tree regeneration after wildfire delays and dampens future bark beetle outbreaks. *Proc. Natl. Acad. Sci. USA* **2016**, *113*, 13075–13080. [CrossRef] [PubMed]
20. Turner, M.G.; Gardner, R.H.; Romme, W.H. Prefire heterogeneity, fire severity, and early postfire plant reestablishment in subalpine forests of Yellowstone National Park, Wyoming. *J. Int. Assoc. Wildland Fire* **1999**, *9*, 21–36. [CrossRef]
21. Harvey, B.J.; Donato, D.C.; Romme, W.H.; Turner, M.G. Fire severity and tree regeneration following bark beetle outbreaks: The role of outbreak stage and burning conditions. *Ecol. Appl.* **2014**, *24*, 1608–1625. [CrossRef] [PubMed]
22. Agne, M.C.; Woolley, T.; Fitzgerald, S. Fire severity and cumulative disturbance effects in the post-mountain pine beetle lodgepole pine forests of the Pole Creek Fire. *For. Ecol. Manag.* **2016**, *366*, 73–86.
23. Meigs, G.W.; Zald, H.S.; Campbell, J.L.; Keeton, W.S.; Kennedy, R.E. Do insect outbreaks reduce the severity of subsequent forest fires? *Environ. Res. Lett.* **2016**, *11*, 045008. [CrossRef]
24. Jolly, W.M.; Parsons, R.A.; Hadlow, A.M.; Cohn, G.M.; McAllister, S.S.; Popp, J.B.; Hubbard, R.M.; Negron, J.F. Relationships between moisture, chemistry, and ignition of Pinus contorta needles during the early stages of mountain pine beetle attack. *For. Ecol. Manag.* **2012**, *269*, 52–59. [CrossRef]
25. Hart, S.J.; Schoennagel, T.; Veblen, T.T.; Chapman, T.B. Area burned in the western United States is unaffected by recent mountain pine beetle outbreaks. *Proc. Natl. Acad. Sci. USA* **2015**, *112*, 4375–4380. [CrossRef] [PubMed]
26. Lynch, H.J.; Renkin, R.A.; Crabtree, R.L.; Moorcroft, P.R. The Influence of previous mountain pine beetle (*Dendroctonus ponderosae*) activity on the 1988 Yellowstone Fires. *Ecosystems* **2006**, *9*, 1318–1327. [CrossRef]
27. Hubbard, R.M.; Rhoades, C.C.; Elder, K.; Negron, J.F. Changes in transpiration and foliage growth in lodgepole pine trees following mountain pine beetle attack and mechanical girdling. *For. Ecol. Manag.* **2013**, *289*, 312–317. [CrossRef]
28. Teste, F.P.; Lieffers, V.J.; Landhäusser, S.M. Viability of forest floor and canopy seed banks in *Pinus contorta* var. *latifolia* (Pinaceae) forests after a mountain pine beetle outbreak. *Am. J. Bot.* **2011**, *98*, 630–637. [PubMed]
29. Aoki, C.F.; Romme, W.H.; Rocca, M.E. Lodgepole pine seed germination following tree death from mountain pine beetle attack in Colorado, USA. *Am. Midl. Nat.* **2011**, *165*, 446–451. [CrossRef]
30. Tinker, D.B.; Romme, W.H.; Hargrove, W.W.; Gardner, R.H.; Turner, M.G. Landscape-scale heterogeneity in lodgepole pine serotiny. *Can. J. For. Res.* **1994**, *24*, 897–903. [CrossRef]
31. Collins, B.J.; Rhoades, C.C.; Underhill, J.; Hubbard, R.M. Post-harvest seedling recruitment following mountain pine beetle infestation of Colorado lodgepole pine stands: A comparison using historic survey records. *Can. J. For. Res.* **2010**, *40*, 2452–2456. [CrossRef]
32. Page, W.G.; Alexander, M.E.; Jenkins, M.J. Wildfire's resistance to control in mountain pine beetle-attacked lodgepole pine forests. *For. Chron.* **2013**, *89*, 783–794. [CrossRef]
33. Pelz, K.A.; Rhoades, C.C.; Hubbard, R.M.; Battaglia, M.A.; Smith, F.W. Species composition influences management outcomes following mountain pine beetle in lodgepole pine-dominated forests. *For. Ecol. Manag.* **2015**, *336*, 11–20. [CrossRef]
34. Hood, P.R.; Nelson, K.N.; Rhoades, C.C.; Tinker, D.B. The effect of salvage logging on surface fuel loads and fuel moisture in beetle-infested lodgepole pine forests. *For. Ecol. Manag.* **2017**, *390*, 80–88. [CrossRef]
35. Griffin, J.M.; Simard, M.; Turner, M.G. Salvage harvest effects on advance tree regeneration, soil nitrogen, and fuels following mountain pine beetle outbreak in lodgepole pine. *For. Ecol. Manag.* **2013**, *291*, 228–239. [CrossRef]
36. Donato, D.C.; Simard, M.; Romme, W.H.; Harvey, B.J.; Turner, M.G. Evaluating post-outbreak management effects on future fuel profiles and stand structure in bark beetle-impacted forests of Greater Yellowstone. *For. Ecol. Manag.* **2013**, *303*, 160–174. [CrossRef]
37. Raffa, K.F.; Aukema, B.H.; Bentz, B.J.; Carroll, A.L.; Hicke, J.A.; Turner, M.G.; Romme, W.H. Cross-scale drivers of natural disturbances prone to anthropogenic amplification: The dynamics of bark beetle eruptions. *BioScience* **2008**, *58*, 501–517. [CrossRef]
38. Alstatt, D.; Miles, R.L. *Soil Survey of Grand County Area, Colorado*; USDA: Washington, DC, USA, 1983; p. 174.

39. Bailey, R.G. *Ecoregions: The Ecosystem Geography of the Oceans and the Continents*; Springer: New York, NY, USA, 1998.

40. Tishmack, J.; Mata, S.A.; Schmid, J.M.; Porth, L. *Mountain Pine Beetle Emergence from Lodgepole Pine at Different Elevations Near Fraser, CO*; USDA: Fort Collins, CO, USA, 2004; p. 13.

41. Chapman, T.B.; Veblen, T.T.; Schoennagel, T. Spatiotemporal patterns of mountain pine beetle activity in the southern Rocky Mountains. *Ecology* **2012**, *93*, 2175–2185. [CrossRef] [PubMed]

42. Rhoades, C.C.; Hubbard, R.M.; Elder, K. A Decade of streamwater nitrogen and forest cynamics after a mountain pine beetle outbreak at the Fraser Experimental Forest, Colorado. *Ecosystems* **2016**, *20*, 380–392. [CrossRef]

43. Rhoades, C. *Church's Park Fire Study, Field Data*; US Forest Service, Rocky Mountain Research Station: Fort Collins, CO, USA, 2018.

44. USDA Forest Service (USFS), InciWeb. Church's Park Fire. Fire and Aviation Management, Washington, DC. 2010. Available online: https://www.fs.fed.us/fire/aviation/av_library/FS%20Special%20Mission%20Airworthiness%20Assurance%20Guide_Final_11_5_2010.pdf (accessed on 15 December 2010).

45. Western Regional Climate Center (WRCC). Remote Automatic Weather Station. Western Regional Climate Center, Reno, NV. 2017. Available online: https://www.dri.edu/images/stories/news/media_kits/WRCCFactSheet11_15_3.pdf (accessed on 10 December 2017).

46. Moriarty, K. Firefighter Observations on Mountain Pine Beetle Post-Outbreak Lodgepole Pine Fires: Expectations, Surprises and Decision-Making. Master's Thesis, Colorado State University, Fort Collins, CO, USA, 2014.

47. Parsons, A.; Robichaud, P.R.; Lewis, S.A.; Napper, C.; Clark, J.T. *Field Guide for Mapping Post-Fire Soil Burn Severity*; USDA: Fort Collins, CO, USA, 2010; p. 49.

48. USDA Forest Service (USFS). *Church's Park Fire—Burned Area Emergency Rehabilitation Burn Severity Estimate—Technical Report*; USDA Forest Service (USFS): Denver, CO, USA, 2010.

49. Binkley, D.; Matson, P. Ion exchange resin bag method for assessing forest soil nitrogen availability. *Soil Sci. Soc. Am. J. Abstr.* **1983**, *47*, 1050–1052. [CrossRef]

50. DeBano, L.F. *Water Repellent Soils: A State of the Art*; USDA: Berkeley, CA, USA, 1981; p. 21.

51. Harvey, B.J.; Donato, D.C.; Turner, M.G. High and dry: Post-fire tree seedling establishment in subalpine forests decreases with post-fire drought and large stand-replacing burn patches. *Glob. Ecol. Biogeogr.* **2016**, *25*, 655–669. [CrossRef]

52. Lotan, J.E.; Brown, J.K.; Neuenschwander, L.F. Role of fire in lodgepole pine forests. In *Lodgepole Pine: The Species and Its Management*; Baumgartner, D.M., Krebill, R.G., Arnott, J.T., Weetman, G.F., Eds.; Washington State University: Spokane, WA, USA, 1985; pp. 133–152.

53. Wright, M.; Rocca, M. Do post-fire mulching treatments affect regeneration in serotinous lodgepole pine. *Fire Ecol.* **2017**, *13*, 139–145. [CrossRef]

54. Lotan, J.E. The role of cone serotiny in lodgepole pine forests. In *Management of Lodgepole Pine Ecosystems: Symposium Proceedings*; Washington State University, Cooperative Extension Service, College of Agriculture: Pullman, WA, USA, 1975; pp. 471–495.

55. Schoennagel, T.; Turner, M.G.; Romme, W.H. The influence of fire interval and serotiny on postfire lodgepole pine density in Yellowstone National Park. *Ecology* **2003**, *84*, 2967–2978. [CrossRef]

56. Teste, F.P.; Lieffers, V.J.; Landhäusser, S.M. Seed release in serotinous lodgepole pine forests after mountain pine beetle outbreak. *Ecol. Appl.* **2011**, *21*, 150–162. [CrossRef] [PubMed]

57. Amman, G.D.; Baker, B.H. Mountain pine beetle influence on lodgepole pine stand structure. *J. For.* **1972**, *70*, 204–209.

58. Page, W.G.; Jenkins, M.J. Mountain pine beetle-induced changes to selected lodgepole pine fuel complexes within the intermountain region. *For. Sci.* **2007**, *53*, 507–518.

59. Berner, L.T.; Law, B.E.; Meddens, A.J.H.; Hicke, J.A. Tree mortality from fires, bark beetles, and timber harvest during a hot and dry decade in the western United States (2003–2012). *Environ. Res. Lett.* **2017**, *12*, 065005. [CrossRef]

60. Lindenmayer, D.B.; Foster, D.R.; Franklin, J.F.; Hunter, M.L.; Noss, R.F.; Schmiegelow, F.A.; Perry, D. Salvage harvesting policies after natural disturbance. *Science* **2004**, *80*, 1303. [CrossRef] [PubMed]

61. Leverkus, A.B.; Benayas, J.M.R.; Castro, J.; Boucher, D.; Brewer, S.; Collins, B.M.; Donato, D.; Fraver, S.; Kishchuk, B.E.; Lee, E.-J.; et al. Salvage logging effects on regulating and supporting ecosystem services—A systematic map. *Forests* **2018**, under review.

62. Jenkins, M.J.; Page, W.G.; Hebertson, E.G.; Alexander, M.E. Fuels and fire behavior dynamics in bark beetle-attacked forests in Western North America and implications for fire management. *For. Ecol. Manag.* **2012**, *275*, 23–34. [CrossRef]

63. Collins, B.J.; Rhoades, C.C.; Battaglia, M.A.; Hubbard, R.M. The effects of bark beetle outbreaks on forest development, fuel loads and potential fire behavior in salvage logged and untreated lodgepole pine forests. *For. Ecol. Manag.* **2012**, *284*, 260–268. [CrossRef]

64. Francos, M.; Pereira, P.; Mataix-Solera, J.; Arcenegui, V.; Alcañiz, M.; Úbeda, X. How clear-cutting affects fire severity and soil properties in a Mediterranean ecosystem. *For. Ecol. Manag.* **2018**, *206*, 625–632. [CrossRef] [PubMed]

65. Donato, D.C.; Fontaine, J.B.; Campbell, J.L.; Robinson, W.D.; Kauffman, J.B.; Law, B.E. Post-wildfire logging hinders regeneration and increases fire risk. *Science* **2006**, *311*, 352. [CrossRef] [PubMed]

66. Rhoades, C.C.; Fornwalt, P.J.; Paschke, M.W.; Shanklin, A.; Jonas, J.L. Recovery of small pile burn scars in conifer forests of the Colorado Front Range. *For. Ecol. Manag.* **2015**, *347*, 180–187. [CrossRef]

67. Clow, D.W.; Rhoades, C.; Briggs, J.; Caldwell, M.; Lewis, W.M. Responses of soil and water chemistry to mountain pine beetle induced tree mortality in Grand County, Colorado, USA. *Appl. Geochem.* **2011**, *26*, S174–S178. [CrossRef]

68. Huber, C. Long lasting nitrate leaching after bark beetle attack in the highlands of the Bavarian Forest National Park. *J. Environ. Qual.* **2005**, *34*, 1772–1779. [CrossRef] [PubMed]

69. Rhoades, C.C.; McCutchan, J.H.; Cooper, L.A.; Clow, D.; Detmer, T.M.; Briggs, J.S.; Stednick, J.D.; Veblen, T.T.; Ertz, R.M.; Likens, G.E.; et al. Biogeochemistry of beetle-killed forests: Explaining a weak nitrate response. *Proc. Natl. Acad. Sci. USA* **2013**, *110*, 1756–1760. [CrossRef] [PubMed]

70. USDA Forest Service (USFS). *Revision of the Land Resource Management Plan. Arapaho and Roosevelt National Forests and Pawnee National Grassland, Fort Collins, Colo*; USDA Forest Service (USFS): Denver, CO, USA, 1997.

71. Crouch, G.L. Aspen regeneration after commercial clearcutting in southwestern Colorado. *J. For.* **1983**, *81*, 316–319.

72. Bigler, C.; Kulakowski, D.; Veblen, T. Multiple disturbance interactions and drought influence fire severity in Rocky Mountain subalpine forests. *Ecology* **2005**, *86*, 3018–3029. [CrossRef]

73. Abatzoglou, J.T.; Williams, A.P. Impact of anthropogenic climate change on wildfire across western US forests. *Proc. Natl. Acad. Sci. USA* **2016**, *113*, 11770–11775. [CrossRef] [PubMed]

74. Stevens-Rumann, C.S.; Kemp, K.B.; Higuera, P.E.; Harvey, B.J.; Rother, M.T.; Donato, D.C.; Morgan, P.; Veblen, T.T. Evidence for declining forest resilience to wildfires under climate change. *Ecol. Lett.* **2017**, *21*, 243–252. [CrossRef] [PubMed]

75. Natural Resources Conservation Service (NRCS), National Water and Climate Center, Snowtel Site: CO-335. 2017. Available online: https://wcc.sc.egov.usda.gov/nwcc/site?sitenum=335 (accessed on 6 December 2017).

forests

MDPI

Article

Pine Plantations and Invasion Alter Fuel Structure and Potential Fire Behavior in a Patagonian Forest-Steppe Ecotone

Juan Paritsis [1,*], Jennifer B. Landesmann [1], Thomas Kitzberger [1], Florencia Tiribelli [1], Yamila Sasal [1], Carolina Quintero [1], Romina D. Dimarco [2], María N. Barrios-García [3], Aimé L. Iglesias [1], Juan P. Diez [4], Mauro Sarasola [4] and Martín A. Nuñez [5]

[1] Laboratorio Ecotono, INIBIOMA-Universidad Nacional del Comahue, CONICET, Quintral 1250, 8400 Bariloche, Argentina; jennifer.landesmann@gmail.com (J.B.L.); kitzberger@gmail.com (T.K.); flopitiribelli@gmail.com (F.T.); shamilacony@gmail.com (Y.S.); quintero.carolina@gmail.com (C.Q.); iglesias.aime1987@gmail.com (A.L.I.)

[2] Grupo de Ecología de Poblaciones de Insectos, INTA-CONICET, EEA Bariloche, CC 277, 8400 Bariloche, Argentina; rominadimarco@gmail.com

[3] Consejo Nacional de Investigaciones Científicas y Técnicas (CONICET), CENAC-APN, Fagnano 244, 8400 Bariloche, Argentina; noeliabarrios@gmail.com

[4] Grupo Ecología Forestal, INTA EEA Bariloche, CC 277, 8400 Bariloche, Argentina; diez.juan@inta.gob.ar; (J.P.D.); sarasola.mauro@inta.gob.ar (M.S.)

[5] Grupo de Ecología de Invasiones, INIBIOMA-Universidad Nacional del Comahue, CONICET, Quintral 1250, 8400 Bariloche, Argentina; nunezm@gmail.com

[*] Correspondence: j.paritsis@gmail.com; Tel.: +54-294-443-3040

Received: 28 December 2017; Accepted: 28 February 2018; Published: 3 March 2018

Abstract: Planted and invading non-native plant species can alter fire regimes through changes in fuel loads and in the structure and continuity of fuels, potentially modifying the flammability of native plant communities. Such changes are not easily predicted and deserve system-specific studies. In several regions of the southern hemisphere, exotic pines have been extensively planted in native treeless areas for forestry purposes and have subsequently invaded the native environments. However, studies evaluating alterations in flammability caused by pines in Patagonia are scarce. In the forest-steppe ecotone of northwestern Patagonia, we evaluated fine fuels structure and simulated fire behavior in the native shrubby steppe, pine plantations, pine invasions, and mechanically removed invasions to establish the relative ecological vulnerability of these forestry and invasion scenarios to fire. We found that pine plantations and their subsequent invasion in the Patagonian shrubby steppe produced sharp changes in fine fuel amount and its vertical and horizontal continuity. These changes in fuel properties have the potential to affect fire behavior, increasing fire intensity by almost 30 times. Pruning of basal branches in plantations may substantially reduce fire hazard by lowering the probability of fire crowning, and mechanical removal of invasion seems effective in restoring original fuel structure in the native community. The current expansion of pine plantations and subsequent invasions acting synergistically with climate warming and increased human ignitions warrant a highly vulnerable landscape in the near future for northwestern Patagonia if no management actions are undertaken.

Keywords: fire severity; forestry; fuel build-up; restoration; wildfire

---

1. Introduction

Fire regimes are being altered by climate warming as well as by synergisms between changes in climate and land use [1,2]. Changes in fuel load and vegetation structure, due to a variety of factors including fire exclusion [3], livestock grazing [4] and invasion of non-native plant species [5], have crucial implications on fire activity, possibly altering fire propagation, and the severity and extension of fire events. For instance, anthropogenic fuel build-up can sharply alter fire behavior in fuel-limited systems, increasing the difficulty of fire control, and thus raising the amount of resources needed for suppression. Disruption of fire regimes due to changes in fuel loads and structure can ultimately result in altered or delayed successional trajectories.

Both planted and invading non-native plant species can alter fire regimes through changes in fuel loads and in the structure and continuity of fuels, potentially increasing or decreasing the flammability of native plant communities [6,7]. Well studied cases, such as the invasion of cheatgrass (*Bromus tectorum* Huds.) into the US Great Basin, show that invaded areas burn nearly four times more frequently than native vegetation types due to the higher flammability and faster recovery of cheatgrass compared to native species [8]. Although there is a significant body of knowledge on the relationship between invasive grasses and fire activity, much less is known about the consequences of non-native woody species on fire regimes [7]. Non-native and fire-prone tree species, such as certain eucalypts and pines, are expected to alter fire behavior in sites where they are planted or invading, potentially increasing fire severity [9,10]. On the contrary, other non-native woody species, such as *Hakea sericea* Schrad. and J. C. Wendl and *Acacia saligna* (Labill.) H. L. Wendl. in South Africa can suppress fire occurrence due to understory fuel reduction, higher fuel moisture in their tissues or more densely packed fuel compared to native vegetation [11]. Therefore, the potential response of fire behavior associated to non-native woody plants is complex and unpredictable, and deserves system-specific studies. Careful evaluation of native and non-native fuel attributes is needed in order to understand and predict potential changes in fire regimes in those ecosystems where non-native species are planted or invading.

In several regions of the southern hemisphere, such as New Zealand, South Africa and South America, pines have been planted in native treeless areas for forestry purposes and have subsequently invaded the native environments [12]. In these open environments with naturally low fuel loads (grasses and short shrubs), exotic trees represent a significant alteration in the load and structure of fuels. For example, in northwestern Patagonia, Argentina, pines have been planted along the forest-steppe ecotone since the late 1970s [13] and escaped pines from plantations have rapidly become biological invaders in several formerly treeless areas [14]. In this region, the most commonly planted species is *Pinus ponderosa* Douglas ex C. Lawson and the most widespread invasive pine species is *Pinus contorta* Douglas, both of which experience frequent fire in their native environments and thus have life-history traits that generally make them well adapted to fire [15]. Furthermore, both species have fast growth rates and thus rapid fuel load accumulation, particularly in the introduced range [16,17]. Lastly, frequent observations of fires originating or spreading rapidly in non-native plantations and neighboring invaded areas are generating concerns as to whether pine plantations and invasions increase native vegetation's flammability and may qualitatively alter the historical fire regime [9,18].

Arguments supporting the high flammability of pine stands seem reasonable. Nonetheless, the heterogeneity of fuel loads and structures in pine plantations and invaded areas, as well as the various native plant communities that are replaced, create complex scenarios that prevent generalizations on the relative flammability and the ecological consequences of fire. For instance, well managed plantations where tree thinning treatments were rigorously conducted seem less prone to high severity fires compared to those that were not thinned [19]. Similarly, pine invasions only seem to generate positive fire-invasion feedbacks after certain pine density thresholds are surpassed [16]. Thus, it is necessary to evaluate the relative fire hazard of pine plantations under different scenarios of management and different stages of invasion in comparison to the native vegetation being replaced.

Despite the pervasive presence of pine plantations and invasions in many regions of the southern hemisphere, where currently millions of hectares are occupied by plantations and invasions [10,12], there are remarkably few studies that have evaluated changes in the flammability of the native vegetation caused by pines. Only recently, two studies have explored the changes in fuel loads and structure generated by pine invasions in southern South America [16,20]. However, neither addressed changes in flammability caused by plantations, which are also a main component of these landscapes, nor did they assess the impacts of different management practices on the relative flammability of these scenarios. The objective of this study is to evaluate changes in fuel structure and potential fire behavior due to pine plantations and invasions of two *Pinus* species (*Pinus ponderosa* and *Pinus contorta*) in the northwestern Patagonian forest-steppe ecotone. Although changes in fuel loads may seem obvious (especially above the height of steppe's vegetation), alterations in potential fire behavior originated from fuel build-up are not easily forecasted. We quantified the amount and spatial arrangement of fine fuels in four contrasting scenarios: native steppe, pine plantations, pine invasions and mechanically removed pine invasions. We also compared fuel structure among three levels of invasion with progressively higher basal areas and older ages of establishment and between plantations with pruned basal branches vs. unpruned plantations. We used fuel characterizations and additional quantifications of fuel loads to evaluate the effects of pine-induced fuel changes on fire behavior under two fire danger scenarios of contrasting fuel moisture and wind speed utilizing the modeling software BehavePlus 5.0.5 (US Forest Service and Systems for Environmental Management, Missoula, MT, USA) [21,22]. This will allow the assessment of the relative ecological vulnerability to fire of the most common forestry and invasion scenarios in the region.

2. Materials and Methods

2.1. Study Area

This study was conducted in northwestern Patagonia, Argentina, on the eastern side of the Andes from $39°55'$ S to $41°58'$ S. Low foothills and plains dominate the landscape and the climate in the region is temperate with a Mediterranean precipitation regime where most precipitation occurs during May–September as rain or snow. The elevations of the selected sites ranged from 765 to 950 m above sea level and mean annual precipitation ranges from 700 to 800 mm [23]. Vegetation in the study areas is typically ecotonal with a mosaic of low shrublands and steppes of tussock grasses (e.g., *Festuca pallescens* (St. Yves) Parodi, *Pappostipa speciosa* (Trin. and Rupr.) Romasch.) and scattered low shrubs (e.g., *Mulinum spinosum* Pers., *Acaena splendens* Gillies ex Hook. and Arn., *Berberis buxifolia* Gillies ex Hook. and Arn.). Wildfires and grazing are the two most important broad-scale disturbances in the northwestern Patagonian forest-steppe ecotone, with most fires being anthropogenic in origin [24]. The selected study sites are representative of the typical environments in which pine plantations and invasions are located in the region. Pine plantations in the Andean Patagonia are distributed from ~36° S to ~44° S; typically on steppe areas and less frequently on mixed shrublands or secondary successions of Chilean cypress (*Austrocedrus chilensis* (D. Don) Pic-Serm. and Bizzari) [13]. Most planted species in the ecotone area are *P. ponderosa* followed by *P. contorta* (80% and 7.5% of the planted area respectively) [25,26]. The latter species is an aggressive invader in multiple locations across the southern hemisphere including the steppe in northwestern Patagonia, while *P. ponderosa* is considered less invasive [27].

2.2. Study Design

We conducted the study at five different locations (Table S1) that exhibited various levels of pine invasion from adjacent plantations (from ca. 25 to 40 years old). In each location, we selected a minimum of four 20 × 20 m plots with the following vegetation conditions (hereafter referred to as treatments): (i) native vegetation consisting mostly of steppe with scattered low (<1 m) shrubs, (ii) mature (reproductive) pine plantations dominated by *P. ponderosa* and with variable proportions of *P. contorta*, (iii) invaded native vegetation mainly by *P. contorta* (>95% composition), and (iv) removed (clearcutted) *P. contorta* invasions (Table 1, Figure S1). The group of four plots was replicated two to four times at each site for a total of 14 groups, thus totaling 56 plots. Additionally, we classified the invaded and their paired removed invasion plots into three different categories based on their basal area: (a) low invasion (<5 m²/ha) (and low removed invasion) (5 plots each), (b) intermediate invasion (from 5 to 20 m²/ha) (and intermediate removed invasion) (4 plots each), and (c) high invasion (>20 m²/ha) (and high removed invasion) (5 plots each) (Table 1, Figure S1). The plantations were divided into those that exhibited pruning of basal branches (9 plots, pruned) and those that were not pruned (5 plots, unpruned) (Table 1, Figure S1). Basal branches were pruned up to a height of 2 to 3 m and the removal of forestry residues was variable among plantations. Pine invasion removal was conducted one year before fuel sampling in plots with similar densities and basal areas as their paired invasion plots. Pines were cut at basal height and removed from the plot. Year of invasion initiation on each plot was estimated by counting tree rings on basal discs that were obtained from three to five harvested individuals per plot which presented the largest diameter.

Table 1. Number of sampled plots with each vegetation condition classified according to the level of invasion (invaded plots) or the pruning status (plantations) totaling nine treatments.

Vegetation Condition	Control	Low Invasion	Intermediate Invasion	High Invasion	Pruned	Unpruned	Total
Native steppe	14	-	-	-	-	-	14
Pine plantation	-	-	-	-	9	5	14
Pine invasion	-	5	4	5	-	-	14
Removed pine invasion	-	5	4	5	-	-	14

We measured fuel characteristics in mixed plantations dominated by *P. ponderosa*, and in invasion stands dominated by *P. contorta*. It is important to note that this work is not aimed at describing the specific characteristics of fire behavior in different *Pinus* species stands, which have been quantified in previous studies (e.g., [28,29]), but to describe the most common scenarios in the studied region and evaluate their relative vulnerability to fire. *Pinus ponderosa* plantations are ubiquitous in northwestern Patagonia and also in other countries such as New Zealand [12,30]. On the other hand *P. contorta* invasions are by far the most common pine invasions in the region and in the southern hemisphere [12]. We acknowledge that plantations dominated by *P. ponderosa* may have a different structure to plantations dominated by *P. contorta*, and the same can be said for their invasions. However, neither pure plantations of *P. contorta* nor *P. ponderosa* invasions are common in the area. Therefore, our analysis provides useful information on fuel changes originated by pines for Patagonia and other regions (e.g., New Zealand, Chile) where these scenarios of plantations and invasions are a ubiquitous part of their landscape.

2.3. Fuel Characterization

We used two sampling strategies: one designed to characterize vertical and horizontal fine fuel (i.e., 1 h fuels; <0.6 cm diameter twigs and leaves) structure and the other aimed at creating fuel models to simulate potential fire behavior using BehavePlus 5.0.5 [21]. To quantify fine fuel structure at each treatment, we followed the point-intercept method [4,31]. We limited fine fuel characterization up to a height of 4 m, which includes surface fuels and the lower portion of the canopy of the evaluated stands. We set a grid of 30 points (5 × 6) equally spaced by 2 m intervals (i.e., 10 × 12 m) in the center of each of the 56 plots. At each point, a 4 m long pole with subdivisions every 25 cm (16 height segments) was vertically placed and we recorded the number of segments that were in contact with dead or live fine fuels. Species identity of the intercepted fuel was also recorded. To characterize canopy height and canopy base height, we visually estimated these variables for the closest tree to the pole as the maximum tree height and the height to the lowest portion of the crown, respectively.

To model potential fire behavior, we quantified fuel load variables and constructed custom fuel models in BehavePlus 5.0.5. Because BehavePlus 5.0.5 assumes homogeneous horizontal distribution of fuel loads, fuel models were only created for those treatments with reasonably homogeneous horizontally distributed fuels (including litter) in most height classes: native vegetation, pruned and unpruned pine plantations, high invasion and its paired removed high invasion. Low and intermediate invaded areas exhibited discontinuous horizontal fuel arrangements above 0.5 m, preventing their use in BehavePlus 5.0.5. We selected one representative plot per treatment to collect fuel data for the parameterization of each model. To estimate surface fuel loads on each treatment, we delimited 10 1×1 m micro plots within an area of ~0.25 ha and we harvested and weighted all fuels classified into dead (sub classified into 1 h fuels, 10 h fuels or 0.6 to 2.5 cm diameter twigs, and 100 h fuels or 2.5 to 7.5 cm branches) and live (sub classified into herbaceous and woody). Fuels were weighted in the field with a portable digital field scale and subsamples of dead fuel were oven dried at 60 °C for four days to calculate dry weight. Canopy bulk density (kg/m^3) in *P. contorta* invasions and *P. ponderosa* plantations was estimated using values in Scott and Reinhardt [32]. Live fuel moisture corresponds to values from February (i.e., mid fire season) recorded over three growing seasons for *P. ponderosa* foliage and two of the most representative steppe species (*M. spinosum* and *P. speciosa*) that were subsequently averaged. Live fuel moisture was calculated as follows [(fresh mass–dry mass)/dry mass] $\times$ 100. *Pinus contorta* live fuel moisture was assumed to be similar to that of *P. ponderosa* based on Qi, et al. [33].

2.4. Fire Behavior Modeling

We used the fire behavior modeling software BehavePlus 5.0.5 [21] to estimate differences in potential fire behavior among the five selected treatments. This simulation software is based on Rothermel's mathematical model of wildfire spread [34] and has been widely used to characterize potential fire behavior (e.g., [35,36]). BehavePlus 5.0.5 was used in this study to assess an envelope of possible fire behaviors rather than as a precise predictor of fire characteristics. We focused the comparison of fire behavior on the following variables: (i) surface and crown rate of spread (ROS, m/min), (ii) surface flame length (m), (iii) transition ratio to crown fire (dimensionless), and (iv) heat per unit area (kJ/m^2). Rate of fire spread and flame length are useful indicators of the difficulty of fire control, potential for fire escapes, and equipment required for suppression [37]. The transition ratio to crown fire indicates the probability of a surface fire to transition into a crown fire; and values equal to or above one suggest enough intensity to reach the canopy. Crown fires are inherently more difficult to control and imply a radical change in fire behavior in the native steppe. Heat per unit area is the heat energy released within the flaming front and is closely related to severity, providing an indication of the ecological effects of fire on vegetation and other organisms [38].

Fuel models for the five selected treatments were created by substituting fuel parameters into standard fuel behavior models [39]. Specifically, we used the low load, dry climate grass-shrub model (GS1) for the steppe, the long-needle litter model (TL8) for the pruned and unpruned plantations and the short needle litter model (8) for the high invasion and the removed high invasion sites. These models were selected due to the similarity to fuel conditions within each treatment. Except for the GS1 model used for the steppe, these standard fuel behavior models were originally created for North American conifers and thus are adequate for the fuel types and species found in pine plantations and invasions in Patagonia. We defined two contrasting fire danger scenarios: a high danger scenario with fine fuel moisture content of 5% and a 7-m wind speed of 20 km/h, and an extreme danger scenario with fine fuel moisture of 1% and a 7-m wind speed of 40 km/h. These scenarios were based on observed meteorological conditions during documented fires in the region [40]. The slope was set to 5% for all simulations because most plantations are typically located in flat or low angle slope terrain. Fuel load data for models were collected in October (i.e., austral spring) and not during peak fire season. Nevertheless, because of the virtual absence of live annual fuels within the closed canopy of plantations and high invasion levels, and the perennial condition of dominant species at the steppe, there is limited seasonal variation in fuel loads in these environments. Comparisons of potential fire behavior among the five selected treatments were conducted qualitatively because we only characterized one representative site per treatment.

2.5. Data Analysis

To relate fuel amount and structure to pine plantations and invasion, we computed mean fuel intercepts and mean horizontal fuel continuity using the fuel data collected with the point-intercept method. Mean fuel intercepts (%) for a given height class were calculated as the number of segments of that height class that were intercepted by fine fuels over the amount of points in the grid (30 points). We calculated this value for all the fuel types together, for live and dead fuels separately and for pine and non-pine species fuels separately. Mean horizontal fuel continuity corresponds to the mean distance between adjacent fine fuel intercepts in a given height class transformed to a percentage, where 2 m (minimum possible distance between two adjacent intercepts) is 100% connectivity and 10 m (maximum possible distance between two adjacent intercepts) is 0% connectivity. We calculated this value by pooling all fine fuel intercepts and species together (total fine fuels). Mean fuel intercepts and mean horizontal fuel continuity values were then averaged per height class among plots of the same treatment to plot their vertical distribution.

We used generalized linear mixed-effects models to compare the vertical classes of fuel and horizontal fuel continuity as response variables, and treatments (i.e., steppe, plantation, invasion and removed invasion) as a fixed factor. We also compared height classes between pruned and unpruned plantations, among the three levels of invasion and among the three levels of removed invasion. To simplify statistical analyses, original 0.25 m height classes were grouped into three broader height classes (i.e., 0–0.50 m; 0.51–2.00 m; 2.01–4 m). All models included "sites" as a random effect. Based on graphical analysis (i.e., residual vs. predicted values), all models satisfied the underlying statistical assumptions, including linearity and the expected relation of the variance to the mean given the nature of the dependent variable error distribution. The multiple mean comparison between treatments and conditions was conducted with Tukey tests (a = 0.05). All models were implemented with the statistical software R version 3.4.1 (R Core Team, R Foundation for Statistical Computing, Vienna, Austria) [41] using the function nlm from the package nlme [42] and the function glmer from the package lme4 [43].

3. Results

3.1. Fuel Characterization

Fine fuels in the native steppe were limited to the first 1.5 m and were more abundant and horizontally continuous within the first 0.5 m (Figures 1a and 2a). At the lowest height class (i.e., 0 to 0.5 m), fuel amount and horizontal continuity in the steppe were significantly higher than in the plantations and the removed invasion but similar to the invasion treatment (Figure 3a,b). Conversely, at higher height classes (>0.5 m) fuel amount and horizontal continuity were significantly lower in the steppe compared to the plantations and the invasion treatment, but similar to the removed invasion treatment (Figure 3a,b). For instance, for the 0.51–2.00 height class, mean fuel intercepts were 1% in the steppe and 4% and 17% in plantations and invasions respectively, while mean horizontal fuel continuity for the same height class was 5% in the steppe and 12% and 37% in plantations and invasions respectively (Figure 3a,b). Plantations had the lowest amount of fine fuels and horizontal continuity within the lower 0.5 m (except for litter) compared to all other treatments (Figure 3a,b). Pruned plantations exhibited a significantly lower amount of fuel across height classes compared to the unpruned plantations (Figure 3c). However, while there was no difference in mean horizontal fine fuel continuity between pruned and unpruned plantations in the 0–0.5 m height class, fine fuels were significantly more continuous horizontally for the higher height classes in the unpruned plantations (Figure 3d).

There were no significant differences in the amount of fine fuels across the vertical distribution and the horizontal continuity between low and intermediate invasions (Figure 3e,f). Low and intermediate invasion levels, each with mean (±SE) basal areas of 3.6 ± 0.2 and 12.1 ± 1.7 m^2/ha, and mean (±SE) ages of establishment of 12.4 ± 0.9 and 15.5 ± 1.4 respectively, did not form a closed canopy, generating a similar amount and horizontal continuity of fine fuels in the lower segments as in the uninvaded steppe (Figure 1d,e and Figure 2d,e). On the contrary, high levels of invasion, with a mean ($\pm$SE) basal area and age of establishment of 27.8 ± 2.6 m^2/ha and 20.0 ± 2.7 years respectively, generated a closed canopy with vertically and horizontally continuous fine fuels largely composed of pine needles and twigs but less fine fuels within the first 0.5 m (except for needle litter) than the low and intermediate invasion levels (Figures 1f, 2f and 3e,f). Fuel amount and horizontal continuity of fine fuels at high invasion levels were significantly higher than those of low and intermediate invasions in the height classes above 0.5 m (Figure 3e,f). In the removed invasion treatments, regardless of the original invasion level, vertical and horizontal distribution of fine fuel resembled that of the native steppe but with significantly less fuels and less horizontal continuity (Figure 3a,b). Fuel amount was not significantly different among the three levels of removed invasion (Figure 3g) but horizontal continuity was significantly higher in the removed low invasion (Figure 3h).

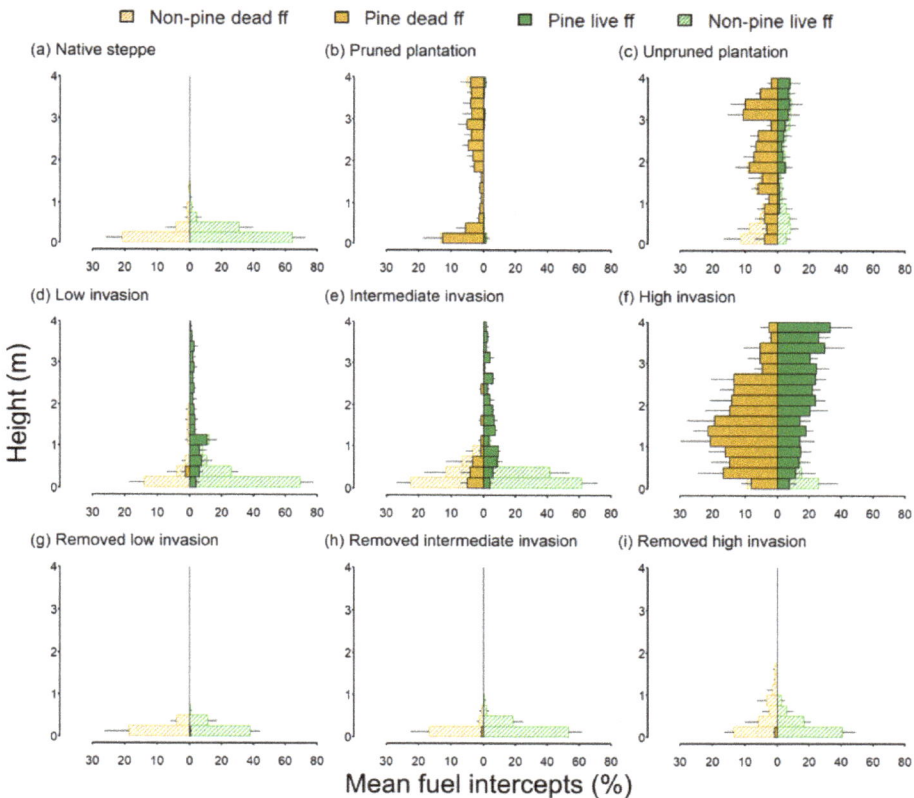

Figure 1. Vertical distribution of the mean fuel intercepts ($\pm$SE) at each of the nine treatments divided into 0.25 m height classes: native steppe (**a**), pruned plantation (**b**), unpruned plantation (**c**), low invasion level (**d**), intermediate invasion level (**e**), high invasion level (**f**), removed low level invasion (**g**), removed intermediate level invasion (**h**), and removed high level invasion (**i**). Bars represent mean dead and live fine fuel (ff) intercepts classified into pine and all fuels together.

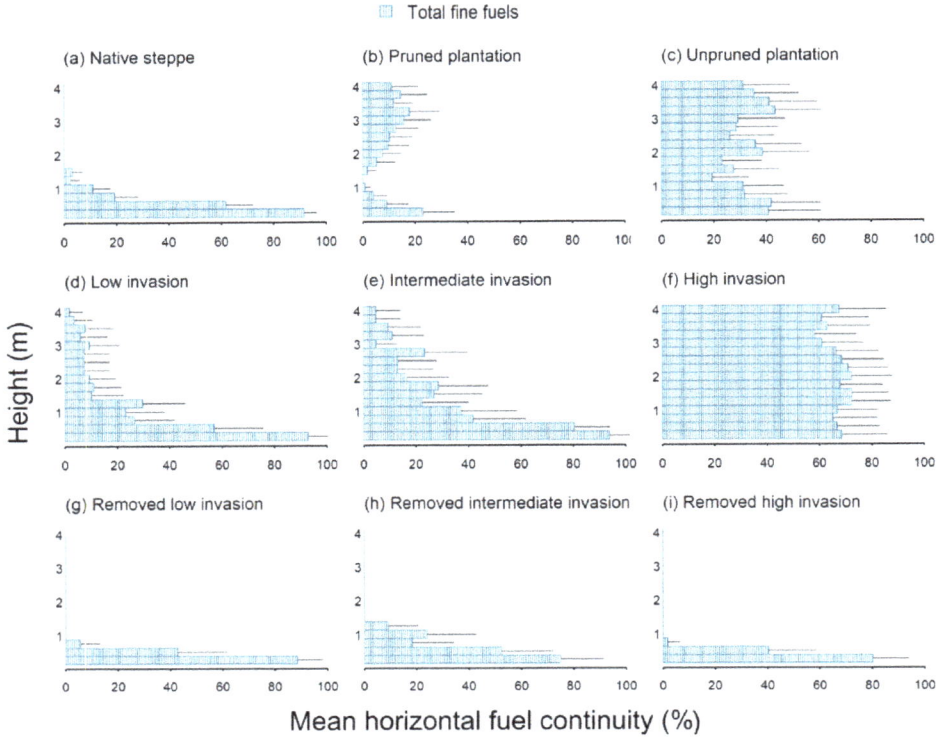

Figure 2. Mean horizontal total (live and dead) fuel continuity (expressed as a percentage) (±SE) for each 0.25 m height class at each of the nine treatments: native steppe (**a**), pruned plantation (**b**), unpruned plantation (**c**), low invasion level (**d**), intermediate invasion level (**e**), high invasion level (**f**), removed low level invasion (**g**), removed intermediate level invasion (**h**), and removed high level invasion (**i**).

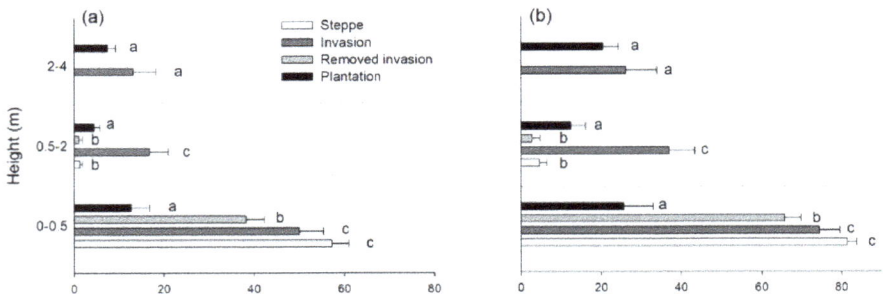

Figure 3. *Cont.*

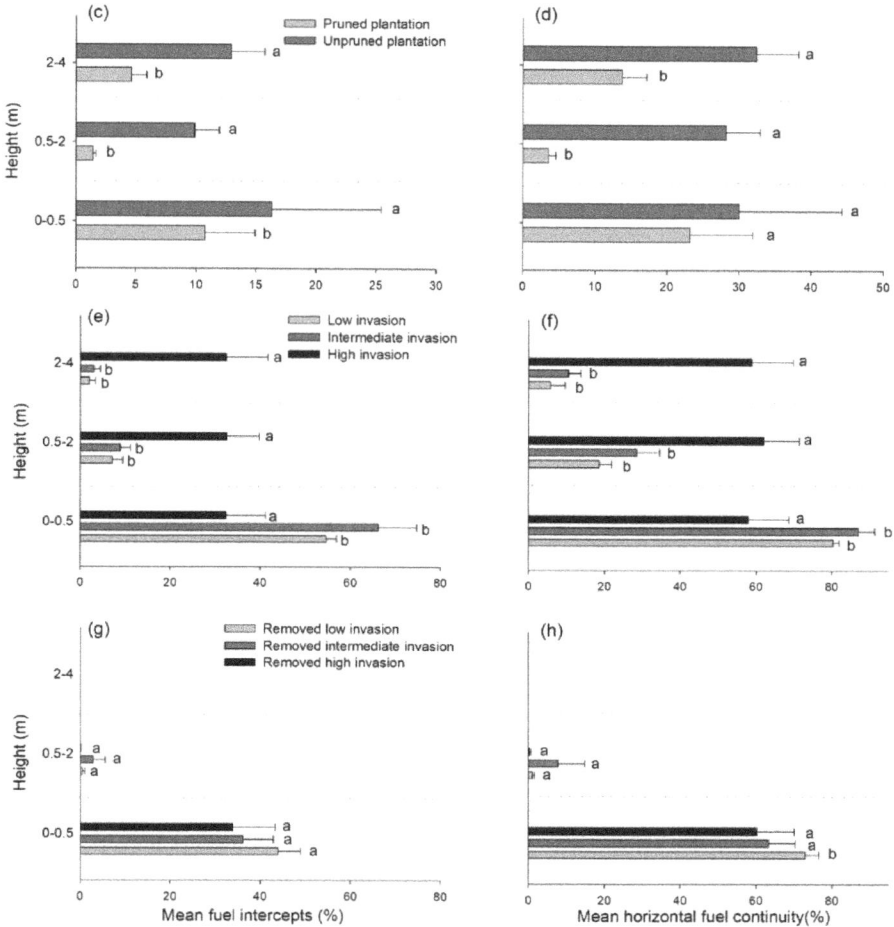

Figure 3. Vertical distribution of the mean fuel intercepts and mean horizontal continuity (±SE) of each treatment grouped into three height classes. Bars represent mean total (live plus dead) fine fuel intercepts. We evaluated the differences in fuel structure among steppe, invasion, removed invasion, and plantation treatments (**a,b**); between pruned and unpruned plantations (**c,d**); among the three levels of invasion (**e,f**); and among the three levels of removed invasion (**g,h**). Different letters indicate significant statistical differences (*p* < 0.05) among treatments within each height class based on the generalized linear mixed-effects models.

3.2. Fire Behavior Modeling

The most noticeable differences between the fuel model for the native steppe site and the fuel models for the pine stands (both invasion and plantations) stem from the absence of trees and the presence of grasses and short shrubs in the former (Table 2). The pruned plantation had a higher canopy base height compared to the unpruned plantation site (Table 2). The high invasion level site had higher tree density but lower basal area and shorter canopy base height compared to the plantation sites. The steppe site exhibited the lowest simulated surface ROS, flame length and heat per unit area compared to the plantations, the high invasion level and the removed high invasion (Table 3). Slightly higher fine fuel loads and fuel bed depth were recorded in the pruned plantation compared to the unpruned plantation site, which originated higher flame length in the former. Despite the slightly higher flame length in the pruned plantation, the transition ratio to crown fire was larger in the unpruned plantation due to its lower canopy base height (Table 3). The high invasion site showed lower surface ROS, flame length and heat per unit area compared to the plantation sites. However, the low canopy base height of the high invasion stand resulted in a transition ratio to crown fire well above 1 in both fire danger scenarios, which is notoriously higher than in the plantations, implying high chances for fire crowning even if conditions are not extreme. Simulated crown fires in the three pine-dominated sites (i.e., high invasion, pruned and unpruned plantations) involved a ~30-fold increase in total heat per unit area (i.e., surface plus canopy) compared to the native steppe and exhibited three times faster crown ROS than the surface ROS in the steppe. The removed high invasion treatment had similar but slightly higher values of ROS, flame length and heat per unit area compared to the steppe (Table 3).

In order to verify if the simulated fire behavior was reasonably realistic for the treatments, we contrasted our results with available information collected during actual fires. Specifically, we compared ROS with reports conducted by firefighters which were gathered and analyzed by Sagarzazu and Defossé [40]. The simulated ROSs for the steppe and pine plantations are reasonable based on the estimated range of ROS of actual fires occurring in the region on similar vegetation types. For instance, estimated surface ROSs for actual fires on shrubby steppe (Río Percey, 1979 and Rinconada, 1998) were 2–3 m/min on a <5% slope with 20 km/h winds and 10–12 m/min on a <5% slope with 50 km/h winds [40]; while simulated surface ROSs in this study for similar conditions were 5.9 and 11.2 m/min respectively. For an actual fire in a pine plantation (Lago Puelo, 1987), estimated crown ROS was 30–46 m/min on 15% to 40% slopes with 40–50 km/h winds [40], while our simulated crown ROS for a 5% slope and similar wind conditions was 32.1 m/min.

Table 2. Description of the five selected treatments sampled for constructing the fuel models to simulate fire behavior. Fuel load values correspond to surface fuels only (<2 m height). nd: no data

Site/fuel variables	Steppe	Pruned Plantation	Unprunned Plantation	High Invasion	Removed High Invasion
Dominant Species	*M. spinosum/ P. speciosa/ A. splendens*	*P. ponderosa*	*P. ponderosa*	*P. contorta*	*P. speciosa*
Tree age (years)	-	~30	~26	~3 to 29	-
Tree density (ind/ha)	-	1220	1330	8400	-
Basal Area (m²/ha)	-	85.4	59.7	25.1	-
1 h fuel load (ton/ha)	0.45	1.61	1.41	1.22	0.96
10 h fuel load (ton/ha)	0	0.31	0.45	0.05	0.09
100 h fuel load (ton/ha)	0	0	0	0	0
Live herbaceous fuel load (ton/ha)	0.32	0	0	0.03	0.08
Live woody fuel load (ton/ha)	0.61	0	0	0.02	0
Fuel bed depth (m)	0.35	0.11	0.09	0.05	0.08
Canopy height (m)	-	13.8	12.3	11.2	-
Canopy base height (m)	-	3.25	1.62	0.27	-
Canopy bulk density (kg/m³)	-	0.16	0.16	0.18	-
Mid-season live woody moisture (%)	80	100	100	100	-
Mid-season foliar/herbaceous moisture (%)	140/62/nd	135	135	135	62

Table 3. Potential fire behavior in two contrasting fire danger scenarios (high and extreme danger) for the five selected treatments. High danger consists of a scenario with fine fuel moisture content of 5% and a 7-m wind speed of 20 km/h, while extreme danger implies fine fuel moisture of 1% and a 7-m wind speed of 40 km/h.

Fire Behavior Variables	Steppe		Pruned Plantation		Unpruned Plantation		High Invasion		Removed High Invasion	
	High	Extreme	High	Extreme	High	Extreme	High	Extreme	High	Extreme
Surface rate of spread (m/min)	5.9	11.2	16.6	38.8	11.4	24	7.4	17.1	7.7	16.5
Surface flame length (m)	0.6	1.0	1.6	2.6	1.3	2.0	0.9	1.6	0.6	1.4
Transition ratio to crown fire	-	-	0.49	1.45	0.89	2.35	6.66	20.19	-	-
Crown rate of spread (m/min)	-	-	9.9	32.1	9.9	32.1	9.9	32.1	-	-
Surface heat per unit area (kJ/m^2)	1033	1331	2549	3264	2306	2946	1870	2461	1478	1906
Canopy heat per unit area (kJ/m^2)	0	0	31435	31435	31822	31822	36638	36638	0	0

4. Discussion

Pine plantations and their subsequent invasion in the Patagonian native steppe produce sharp changes in vegetation structure and fuel loads. As expected, we found that pine plantations and invasion increase the amount and continuity of fine fuels above ~0.5 m height. Most interestingly, these changes in fuel attributes have the potential to affect fire behavior, increasing fire intensity (measured as heat per area) ~30 times compared to the intensity of fires in the native steppe. However, significant changes in fuel structure do not occur until advanced stages of invasion in which fire hazard, measured as transition to crown fire and total heat per unit area, surpasses that found in plantations. Our results also show that pruning of basal branches in plantations can substantially reduce fire hazard by lowering the probability of fire crowning, and that mechanical removal of invasion seems effective in restoring original fuel structure in the steppe, at least in the short term. Overall, pine plantations and invasions significantly alter the amount and structure of fuels in the native steppe, potentially allowing more severe fires. Nevertheless, adequate silvicultural practices and invasion management techniques can contribute to reduce fire hazard in these areas.

Pruning of basal branches was effective in reducing potential fire severity in *P. ponderosa* plantations by reducing the probability of fire transition to the crown. However, the slightly higher surface fuel loads found in the pruned plantation compared to the unpruned plantation originated from longer flames and higher ROS in the pruned vs. the unpruned plantation, most likely because pruned branches were left on the site for some time after their removal, thus increasing the amount of needles on the ground. This implies that pruning treatments without immediate removal of the residuals created by such activity may exacerbate fire hazard rather than ameliorate it [44]. Accordingly, adequate silvicultural management is needed not only to improve the quality of wood and the logistics of timber production but also for fire hazard mitigation [19,45]. Basic methods for fire mitigation in tree stands involve reduction of surface fuels, increasing the height to live crown and decreasing crown density [44]. Equally important for fire hazard mitigation in a plantation is the removal or thinning of adjacent invasions since these provide effective ladder fuels, substantially increasing the chances of a crown fire in the plantation. Thus, to mitigate fire hazard associated with plantations, an integrated approach is needed that not only includes management of the plantation itself but also control of the escaped individuals in the surroundings.

Pine invasion did not significantly alter fuel structure until advanced stages of invasion. A closed pine canopy was only observed at the high invasion level, with an average age of establishment of 20.0 ± 2.7 years. This is only 4.5 years more than the average age of establishment at intermediate invasion levels, which showed extremely low horizontal continuity at height classes above 0.5 m. This is consistent with a recent study conducted in native grasslands in the Chilean Patagonia, where fuel build-up generated by *P. contorta* invasions started growing exponentially between 15 and 20 years after invasion initiation [16]. Although we did not simulate fire behavior in low and intermediate levels of invasion, their fuel characteristics implies a much lower fire hazard compared to high invasion levels [46]. The relatively rapid change in fuel conditions suggests the existence of a threshold in fuel build-up below which there is apparently no significant increase in fire hazard. However, once this proposed threshold is surpassed, fires may radically increase their intensity and severity in pine-invaded areas. The existence of this threshold for *P. contorta* invasions in the Patagonian steppe was previously shown by Taylor et al. [16], where simulated soil heating due to fire drastically increased after the invasion reaches 10 years [16]. In addition to increased fire hazard, high invasion levels significantly raise pine removal costs [47] and magnify the chances of positive fire-vegetation feedbacks [16], thus decreasing the probabilities of successful restoration of the native community. Therefore, it is critical to plan for an early control of pine-invaded areas before fuels reach the critical threshold.

Although the current area planted with pines in northwestern Patagonia is still relatively small (ca. 100 thousand ha), there are strong socioeconomic incentives to continue planting over an extensive area of ca. 800 thousand ha in the Patagonian steppe, near the ecotone at the Andean foothills [25]. Concurrently, pine invasions in northwestern Patagonia are at an incipient state but show strong trends of rapid expansion and densification [10,18]. Most densely invaded areas are still nearby plantations (i.e., a few hundred meters), but long-distance seed dispersed individuals (i.e., a few kilometers) that have already reached reproductive maturity are increasingly common across the landscape (pers. obs.). Furthermore, a large proportion of established plantations have not reached a reproductive stage yet, so more sources of invasion will be available in the near future [18]. These rapid changes in landscape cover will likely surpass landscape flammability thresholds in the future if no actions are taken. At this initial stage of plantations expansion and invasion spread, management practices have a realistic chance of lessening the potential consequences of increased fire hazard at a relative low cost compared to control measures applied at advanced stages.

Despite several native plant species in the forest-steppe ecotone showing post-fire regeneration strategies, elevated temperatures due to pine-promoted high intensity fire may still kill individuals of native species or slow down their rate of recovery [48]. This may promote further invasion and/or conversion from steppe to pine forest stands, as extirpation of native species due to death of belowground resprouting structures eliminates competition and favors pine establishment [49]. Moreover, many of the pine species planted and invading in Patagonia have fire-adapted life-history traits (e.g., serotiny) that enhance their colonization of post-fire native areas [50]. In pine-invaded sites or plantations, there are already documented reductions in the abundance and richness of native plant species. For example, in southern Chile, plantations and invasions filtered out most specialist and endemic plants [51,52], while in the forest-steppe ecotone of NW Patagonia in Argentina richness of herbaceous species has decreased almost 50% and abundance has decreased 70 times within plantations compared to the native steppe [53]. Therefore, the native steppe ecosystem becomes even more vulnerable to biodiversity loss considering the potential higher fire severity of pine stands, which may further eliminate the few surviving individuals of native species inside the planted or invaded stands.

Pine plantations and invasions are not only threatening the forest-steppe ecotone but also adjacent fire-sensitive native ecosystems. Increased connectivity and flammable landscape elements may threaten native forests in several ways. Pines are being planted near fire-sensitive tree species such as *Austrocedrus chilensis* and *Nothofagus pumilio* (Poepp. and Endl.) Krasser which may favor fire spread into these forests. Also in the region, native fire-resistant tree species such as the endangered monkey puzzle tree *Araucaria araucana* (Molina) K. Koch [54], are threatened by changes in fuel structure due to *P. contorta* invasions [20]. Therefore, spatially explicit models of fire spread are necessary to identify critical levels of invasion that may increase landscape level vulnerability of native ecosystems. Furthermore, spatially explicit models of fire spread can be used to test for the accuracy of fire behavior simulations by contrasting actual fire extension (size and shape) against modeled fires. Future research in Patagonia should combine quantification approaches of the effects of stand structure and composition on fire behavior with the spatial characterization of fire spread and severity, especially in the context of climate warming.

Current climatic trends and forecasted warming scenarios for the southern hemisphere are associated with large and more severe fires [55,56]. The upward trend observed in the Southern Annular Mode (SAM), linked to current warming conditions, is tightlty coupled with fire activity in southern South American forests and woodlands, resulting in increased fire synchrony and activity [56]. Clearly, increased warming in Patagonia promotes low fuel moisture levels, favoring extreme fire danger conditions, such as the scenario described in this work, which in turn may allow more intense and severe fires. These changes occur in synergy with land use trends, such as the expansion of the wildland–urban interface and the increase in non-native plantations, implying increased anthropogenic ignitions and greater exposure of societies to wildfire hazards [9,57,58]. Adaptation measures such as fuel reduction in both planted and incipiently invaded sites following adequate silvicultural practices may sensitively diminish ecological and socioeconomic vulnerability to these altered fire regimes. Furthermore, these practices may reduce the probability of transition from native to novel pine-dominated states that could further result in more frequent and severe fire events.

5. Conclusions

Pine plantations and invasions in the Patagonian forest-steppe ecotone originate significant increases in fuel loads and produce changes in fuel structure affecting the potential behavior of fire. Wildfires in pine plantations or areas with high levels of pine invasion can increase the potential fire intensity ~30 times compared to the native steppe. This drastic rise in intensity may have profound impacts on post-fire regeneration due to the likely extirpation of the few native resprouting species in the understory. Pruning of basal branches can contribute to reducing fire hazard within plantations by lowering the probability of fire crowning. Likewise, the mechanical removal of invasion can result in a similar fuel structure to that of the native steppe. As pines are becoming ubiquitous in southern hemisphere landscapes and fire activity is increasing, active management to prevent unnaturally severe fires is urgently needed.

Supplementary Materials: The following are available online at www.mdpi.com/1999-4907/9/3/117/s1, Table S1: Location and elevation of the five study sites. Figure S1: Photographs depicting representative sites of each treatment: (**a**) native steppe, (**b**) *P. ponderosa* pruned plantation, (**c**) *P. ponderosa* unpruned plantation, (**d**) low level *P. contorta* invasion, (**e**) intermediate level *P. contorta* invasion, (**f**) high level *P. contorta* invasion, (**g**) removed low level *P. contorta* invasion, (**h**) removed intermediate level *P. contorta* invasion, and (**i**) removed high level *P. contorta* invasion.

Acknowledgments: We are grateful to K. Heinemann for assistance in the field. Part of this work was funded by BioSilva #005 and PIA #12026 of the MAEyP of Argentina.

Author Contributions: J.P., M.A.N., Y.S., C.Q., R.D.D., M.N.B.-G., T.K., J.B.L., J.P.D. and M.S. conceived and designed the study; J.P., F.T., A.L.I., M.A.N., J.P.D. and M.S. implemented the field design and collected the data; J.P., Y.S., F.T., J.B.L. and T.K. analyzed the data, all the authors wrote the manuscript.

Conflicts of Interest: The authors declare no conflict of interest. The founding sponsors had no role in the design of the study; in the collection, analyses, or interpretation of data; in the writing of the manuscript, and in the decision to publish the results.

References

1. Flannigan, M.D.; Amiro, B.D.; Logan, K.A.; Stocks, B.J.; Wotton, B.M. Forest fires and climate change in the 21ST century. *Mitig. Adapt. Strateg. Glob. Chang.* **2006**, *11*, 847–859. [CrossRef]
2. Cochrane, M.A.; Barber, C.P. Climate change, human land use and future fires in the Amazon. *Glob. Chang. Biol.* **2009**, *15*, 601–612. [CrossRef]
3. Veblen, T.T.; Kitzberger, T.; Donnegan, J. Climatic and human influences on fire regimes in *Ponderosa pine* forests in the Colorado Front Range. *Ecol. Appl.* **2000**, *10*, 1178–1195. [CrossRef]

4.   Blackhall, M.; Raffaele, E.; Paritsis, J.; Tiribelli, F.; Morales, J.M.; Kitzberger, T.; Gowda, J.H.; Veblen, T.T. Effects of biological legacies and herbivory on fuels and flammability traits: A long-term experimental study of alternative stable states. *J. Ecol.* **2017**, *105*, 1309–1322. [CrossRef]

5.   D'Antonio, C.M.; Hughes, R.F.; Tunison, J.T. Long-term impacts of invasive grasses and subsequent fire in seasonally dry Hawaiian woodlands. *Ecol. Appl.* **2011**, *21*, 1617–1628. [CrossRef] [PubMed]

6.   Brooks, M.L.; D'Antonio, C.M.; Richardson, D.M.; Grace, J.B.; Keeley, J.E.; DiTomaso, J.M.; Hobbs, R.J.; Pellant, M.; Pyke, D. Effects of invasive alien plants on fire regimes. *BioScience* **2004**, *54*, 677–688. [CrossRef]

7.   Mandle, L.; Bufford, J.L.; Schmidt, I.B.; Daehler, C.C. Woody exotic plant invasions and fire: Reciprocal impacts and consequences for native ecosystems. *Biol. Invasions* **2011**, *13*, 1815–1827. [CrossRef]

8.   Balch, J.K.; Bradley, B.A.; D'Antonio, C.M.; Gómez-Dans, J. Introduced annual grass increases regional fire activity across the arid western USA (1980–2009). *Glob. Chang. Biol.* **2013**, *19*, 173–183. [CrossRef] [PubMed]

9.   Veblen, T.T.; Holz, A.; Paritsis, J.; Raffaele, E.; Kitzberger, T.; Blackhall, M. Adapting to global environmental change in Patagonia: What role for disturbance ecology? *Austral Ecol.* **2011**, *36*, 891–903. [CrossRef]

10. Simberloff, D.; Nuñez, M.A.; Ledgard, N.J.; Pauchard, A.; Richardson, D.M.; Sarasola, M.; Van Wilgen, B.W.; Zalba, S.M.; Zenni, R.D.; Bustamante, R.; et al. Spread and impact of introduced conifers in South America: Lessons from other southern hemisphere regions. *Austral Ecol.* **2010**, *35*, 489–504. [CrossRef]

11. Van Wilgen, B.W.; Richardson, D.M. The effects of alien shrub Invasions on vegetation structure and fire behaviour in South African fynbos shrublands: A simulation study. *J. Appl. Ecol.* **1985**, *22*, 955–966. [CrossRef]

12. Nuñez, M.A.; Chiuffo, M.C.; Torres, A.; Paul, T.; Dimarco, R.D.; Raal, P.; Policelli, N.; Moyano, J.; García, R.A.; Van Wilgen, B.W.; et al. Ecology and management of invasive Pinaceae around the world: Progress and challenges. *Biol. Invasions* **2017**, *19*, 3099–3120. [CrossRef]

13. Bava, J.O.; Logercio, G.A.; Salvador, G. Por qué plantar en Patagonia? Estado actual y el rol futuro de los bosques plantados. *Ecol. Austral* **2015**, *21*, 101–111.

14. Richardson, D.M.; Van Wilgen, B.W.; Nuñez, M.A. Alien conifer invasions in South America: Short fuse burning? *Biol. Invasions* **2008**, *10*, 573–577. [CrossRef]

15. Keeley, J.E. Ecology and evolution of pine life histories. *Ann. For. Sci.* **2012**, *69*, 445–453. [CrossRef]

16. Taylor, K.T.; Maxwell, B.D.; McWethy, D.B.; Pauchard, A.; Nuñez, M.A.; Whitlock, C. *Pinus contorta* invasions increase wildfire fuel loads and may create a positive feedback with fire. *Ecology* **2017**, *98*, 678–687. [CrossRef] [PubMed]

17. Gyenge, J.E.; Fernández, M.E.; Schlichter, T.M. Are differences in productivity between native and exotic trees in N.W. Patagonia related to differences in hydraulic conductance? *Trees Struct. Funct.* **2008**, *22*, 483–490. [CrossRef]

18. Raffaele, E.; Nuñez, M.A.; Relva, M.A. Plantaciones de coníferas exóticas en Patagonia: Los riesgos de plantar sin un manejo adecuado. *Ecol. Austral* **2015**, *25*, 89–92.

19. Godoy, M.M.; Defossé, G.E.; Bianchi, L.O.; Davel, M.M.; Withington, T.E. Fire-caused tree mortality in thinned Douglas-fir stands in Patagonia, Argentina. *Int. J. Wildland Fire* **2013**, *22*, 810–814. [CrossRef]

20. Cóbar-Carranza, A.J.; García, R.A.; Pauchard, A.; Peña, E. Effect of *Pinus contorta* invasion on forest fuel properties and its potential implications on the fire regime of *Araucaria araucana* and *Nothofagus antarctica* forests. *Biol. Invasions* **2014**, *16*, 2273–2291. [CrossRef]

21. Andrews, P.L. Current status and future needs of the BehavePlus Fire Modeling System. *Int. J. Wildland Fire* **2013**, *23*, 21–33. [CrossRef]

22. Andrews, P.L.; Bevins, C.D.; Seli, R.C. *BehavePlus Fire Modeling System, Version 4.0: User's Guide (General Technical Report)*; USDA Forest Service. Rocky Mountain Research Station: Fort Collins, CO, USA, 2008.

23. Hijmans, R.J.; Cameron, S.E.; Parra, J.L.; Jones, P.G.; Jarvis, A. Very high resolution interpolated climate surfaces for global land areas. *Int. J. Climatol.* **2005**, *25*, 1965–1978. [CrossRef]

24. Ghermandi, L.; Guthmann, N.; Bran, D. Early post-fire succession in northwestern Patagonia grasslands. *J. Veg. Sci.* **2004**, *15*, 67–76. [CrossRef]

25. Defossé, G.E. Conviene seguir fomentando las plantaciones forestales en el norte de la Patagonia? *Ecol. Austral* **2015**, *25*, 93–100.

26. Nuñez, M.A.; Sarasola, M. Invasión de pinos: Una problemática para no descuidarse. *Prod. For.* **2014**, *3*, 15–17.

27. Sarasola, M.M.; Rusch, V.E.; Schlichter, T.M.; Ghersa, C.M. Invasión de coníferas forestales en áreas de estepa y bosques de ciprés de la cordillera en la Región Andino Patagónica. *Ecol. Austral* **2006**, *16*, 143–156.

28. Reiner, A.L.; Vaillant, N.M.; Dailey, S.N. Mastication and prescribed fire influences on tree mortality and predicted fire behavior in *Ponderosa pine*. *West. J. Appl. For.* **2012**, *27*, 36–41.

29. Nelson, K.N.; Turner, M.G.; Romme, W.H.; Tinker, D.B. Simulated fire behaviour in young, postfire lodgepole pine forests. *Int. J. Wildland Fire* **2017**, *26*, 852–865. [CrossRef]

30. Raffaele, E.; Schlichter, T. Efectos de las plantaciones de pino ponderosa sobre la heterogeneidad de micrositios en estepas del noroeste patagónico. *Ecol. Austral* **2000**, *10*, 151–158.

31. Gosper, C.R.; Yates, C.J.; Prober, S.M.; Wiehl, G. Application and validation of visual fuel hazard assessments in dry Mediterranean-climate woodlands. *Int. J. Wildland Fire* **2014**, *23*, 385–393. [CrossRef]

32. Scott, J.H.; Reinhardt, E.D. *Stereo Photo Guide for Estimating Canopy Fuel Characteristics in Conifer Stands*; General Technical Report RMRS-GTR-145; USDA Forest Service, Rocky Mountain Research Station: Ft. Collins, CO, USA, 2005.

33. Qi, Y.; Jolly, W.M.; Dennison, P.E.; Kropp, R.C. Seasonal relationships between foliar moisture content, heat content and biochemistry of lodgepole line and big sagebrush foliage. *Int. J. Wildland Fire* **2016**, *25*, 574–578. [CrossRef]

34. Rothermel, R.C. *A Mathematical Model for Predicting Fire Spread in Wildland Fuels*; INT-115 U.S. Forest Service Research Paper; USDA Forest Service: Ogden, UT, USA, 1972.

35. Yospin, G.I.; Bridgham, S.D.; Kertis, J.; Johnson, B.R. Ecological correlates of fuel dynamics and potential fire behavior in former upland prairie and oak savanna. *For. Ecol. Manag.* **2012**, *266*, 54–65. [CrossRef]

36. Diamond, J.M.; Call, C.A.; Devoe, N. Effects of targeted grazing and prescribed burning on community and seed dynamics of a downy brome (*Bromus tectorum*)–dominated landscape. *Invasive Plant Sci. Manag.* **2012**, *5*, 259–269. [CrossRef]

37. Rothermel, R.C. *How to Predict the Spread and Intensity of Forest and Range Fires*; General Technical Report INT-143; USDA Forest Service: Ogden, UT, USA, 1983.

38. Turner, M.G.; Baker, W.L.; Peterson, C.J.; Peet, R.K. Factors influencing succession: Lessons from large, infrequent natural disturbances. *Ecosystems* **1998**, *1*, 511–523. [CrossRef]

39. Scott, J.H.; Burgan, R.E. *Standard Fire Behavior Fuel Models: A Comprehensive Set for Use with Rothermel's Surface Fire Spread Model*; General Technical Report RMRS-GTR-153; USDA Forest Service, Rocky Mountain Research Station: Fort Collins, CO, USA, 2005.

40. Sagarzazu, M.S.; Defossé, G.E. *Study and Analysis of Large Past Fires in Western Chubut Province, Patagonia, Argentina, Deliverable D8.3-1 of the Integrated project "Fire Paradox"*, Project no. FP6-018505; European Commission: Brussels, Belgium, 2009; p. 82.

41. Team, R.C. *R: A Language and Environment for Statistical Computing, Version 3.4.1*; R Foundation for Statistical Computing: Vienna, Austria, 2017. Available online: http://www.R-project.org (accessed on 4 September 2017).

42. Pinheiro, J.; Bates, D.; DebRoy, S.; Sarkar, D.; Core-Team, R. Nlme: Linear and nonli Near Mixed Effects Models. R Package Version 3.1-128. 2016. Available online: https://cran.r-project.org/web/packages/nlme/ (accessed on 4 September 2017).

43. Bates, D.; Maechler, M.; Bolker, B.; Walker, S. Fitting linear mixed-effects models using lme4. *J. Stat. Softw.* **2015**, *67*, 1–48. [CrossRef]

44. Agee, J.K.; Skinner, C.N. Basic principles of forest fuel reduction treatments. *For. Ecol. Manag.* **2005**, *211*, 83–96. [CrossRef]

45. Fulé, P.Z.; Crouse, J.E.; Roccaforte, J.P.; Kalies, E.L. Do thinning and/or burning treatments in western USA ponderosa or Jeffrey pine-dominated forests help restore natural fire behavior? *For. Ecol. Manag.* **2012**, *269*, 68–81. [CrossRef]

46. Scott, J.H.; Reinhardt, E.D. Estimating canopy fuels in conifer forests. *Fire Manag. Today* **2002**, *62*, 45–50.

47. Sarasola, M.M.; Diez, J.P.; Jaque, N.F.; Nuñez, M.A. Desarrollo de protocolos para la prevención, monitoreo y control de las invasiones de coníferas introducidas en el noroeste patagónico: Análisis de efectividad y costos. In *Investigación Forestal 2011–2915 Los Proyectos de Investigación Aplicada*; UCAR, Ministerio de Agroindustria: Buenos Aires, Argentina, 2016.

48. González, S.; Ghermandi, L.; Pelaez, D. Growth and reproductive post-fire responses of two shrubs in semiarid Patagonian grasslands. *Int. J. Wildland Fire* **2015**, *24*, 809–818. [CrossRef]

49. Nuñez, M.A.; Raffaele, E. Afforestation causes changes in post-fire regeneration in native shrubland communities of northwestern Patagonia, Argentina. *J. Veg. Sci.* **2007**, *18*, 827–834. [CrossRef]

50. Franzese, J.; Raffaele, E. Fire as a driver of pine invasions in the Southern Hemisphere: A review. *Biol. Invasions* **2017**, *19*, 2237–2246. [CrossRef]

51. Bravo-Monasterio, P.; Pauchard, A.; Fajardo, A. *Pinus contorta* invasion into treeless steppe reduces species richness and alters species traits of the local community. *Biol. Invasions* **2016**, *18*, 1883–1894. [CrossRef]

52. Braun, A.C.; Troeger, D.; Garcia, R.; Aguayo, M.; Barra, R.; Vogt, J. Assessing the impact of plantation forestry on plant biodiversity: A comparison of sites in Central Chile and Chilean Patagonia. *Glob. Ecol. Conserv.* **2017**, *10*, 159–172. [CrossRef]

53. Victoria Lantschner, M.; Rusch, V.; Hayes John, P. Influences of pine plantations on small mammal assemblages of the Patagonian forest-steppe ecotone. *Mammalia* **2011**, *75*, 24–255. [CrossRef]

54. Premoli, A.; Quiroga, P.; Gardner, M. Araucaria Araucana. The IUCN Red List of Threatened Species 2013: e.T31355A2805113. 2013. Available online: http://dx.doi.org/10.2305/IUCN.UK.2013-1.RLTS.T31355A2805113.en (accessed on 10 October 2017).

55. Liu, Y.; Stanturf, J.; Goodrick, S. Trends in global wildfire potential in a changing climate. *For. Ecol. Manag.* **2010**, *259*, 685–697. [CrossRef]

56. Holz, A.; Paritsis, J.; Mundo, I.A.; Veblen, T.T.; Kitzberger, T.; Williamson, G.J.; Aráoz, E.; Bustos-Schindler, C.; González, M.E.; Grau, H.R.; et al. Southern Annular Mode drives multicentury wildfire activity in southern South America. *Proc. Natl. Acad. Sci. USA* **2017**, *114*, 9552–9557. [CrossRef] [PubMed]

57. De Torres Curth, M.D.; Biscayart, C.; Ghermandi, L.; Pfister, G. Wildland-urban interface fires and socioeconomic conditions: A case study of a northwestern Patagonia city. *Environ. Manag.* **2012**, *49*, 876–891. [CrossRef] [PubMed]

58. Van Wilgen, B.W.; Forsyth, G.G.; Prins, P. The management of fire-adapted ecosystems in an urban setting: The case of Table Mountain National Park, South Africa. *Ecol. Soc.* **2012**, *17*. [CrossRef]

forests

MDPI

Article

Predicting Potential Fire Severity Using Vegetation, Topography and Surface Moisture Availability in a Eurasian Boreal Forest Landscape

Lei Fang [1], Jian Yang [2,*], Megan White [2] and Zhihua Liu [1]

[1] CAS Key Laboratory of Forest Ecology and Management, Institute of Applied Ecology,
    Chinese Academy of Sciences, Shenyang 110016, China; fanglei@iae.ac.cn (L.F.); liuzh@iae.ac.cn (Z.L.)
[2] Department of Forestry and Natural Resources, TP Cooper Building, University of Kentucky, Lexington,
    KY 40546, USA; 12mgn11@gmail.com
*   Correspondence: jian.yang@uky.edu; Tel.: +1-859-257-5820

Received: 16 January 2018; Accepted: 6 March 2018; Published: 8 March 2018

Abstract: Severity of wildfires is a critical component of the fire regime and plays an important role in determining forest ecosystem response to fire disturbance. Predicting spatial distribution of potential fire severity can be valuable in guiding fire and fuel management planning. Spatial controls on fire severity patterns have attracted growing interest, but few studies have attempted to predict potential fire severity in fire-prone Eurasian boreal forests. Furthermore, the influences of fire weather variation on spatial heterogeneity of fire severity remain poorly understood at fine scales. We assessed the relative importance and influence of pre-fire vegetation, topography, and surface moisture availability (SMA) on fire severity in 21 lightning-ignited fires occurring in two different fire years (3 fires in 2000, 18 fires in 2010) of the Great Xing'an Mountains with an ensemble modeling approach of boosted regression tree (BRT). SMA was derived from 8-day moderate resolution imaging spectroradiometer (MODIS) evapotranspiration products. We predicted the potential distribution of fire severity in two fire years and evaluated the prediction accuracies. BRT modeling revealed that vegetation, topography, and SMA explained more than 70% of variations in fire severity (mean 83.0% for 2000, mean 73.8% for 2010). Our analysis showed that evergreen coniferous forests were more likely to experience higher severity fires than the dominant deciduous larch forests of this region, and deciduous broadleaf forests and shrublands usually burned at a significantly lower fire severity. High-severity fires tended to occur in gentle and well-drained slopes at high altitudes, especially those with north-facing aspects. SMA exhibited notable and consistent negative association with severity. Predicted fire severity from our model exhibited strong agreement with the observed fire severity (mean r^2 = 0.795 for 2000, 0.618 for 2010). Our results verified that spatial variation of fire severity within a burned patch is predictable at the landscape scale, and the prediction of potential fire severity could be improved by incorporating remotely sensed biophysical variables related to weather conditions.

Keywords: fire severity; surface moisture; remote sensing; spatial controls; boreal forest; Great Xing'an Mountains

1. Introduction

Wildfires, ignited by human or natural agents, are crucial disturbances in the boreal forests of Eurasia and North America [1–3]. Wildfires can strongly influence regional land surface processes such as carbon cycling [4,5] and energy and water budgets [6,7]. Severe burns can result in tree mortality and soil erosion, thereby degrading ecosystem functions [8]. Nevertheless, there is a growing consensus that forest wildfires can also provide a unique opportunity for ecosystem restoration [9,10]. In particular, fire severity can exert profound impacts on the successional trajectories of the early post-fire vegetation [11,12], which may ultimately determine the future forest structure and function [13–15]. Many recent studies have indicated that warming and drying climates tend to shift the current fire regimes toward more frequent, large burns with high severities [16–18]. Understanding the causal mechanisms of fire severity patterns is essential for mitigating the adverse effects of fires, and for maintaining beneficial ecosystem functions and services.

Fire severity is generally defined as the magnitude of ecosystem changes caused by fire [19,20]. It is often based on metrics obtained from the field that represent short-term fire effects in the immediate post-fire environment (e.g., tree mortality and soil organic matter loss) [19,21], or based on the validated relationships between remotely sensed spectral indices (e.g., normalized burn ratio (NBR) and differenced normalized burned ratio (dNBR)) and field-measured metrics [19–22]. It has been shown that variations in fire severity patterns are at the heart of how an ecosystem will change in response to fire events [23–25]. Spatially explicit fire severity maps can assist resource managers in evaluating post-fire biomass loss and developing sound strategies for ecological restoration [26,27]. Numerous studies that incorporate field investigation and remotely sensed images have been conducted to map fire severity variability within burned patches [19,20,27,28]. However, a potential fire severity map may be even more useful in allowing managers to anticipate the hotspot areas that are likely to burn severely, and thus to prioritize resources for fuel treatments for those areas [29,30].

Fire behavior is regulated by three major controlling factors (i.e., weather, topography, and fuels) that form a fire triangle at spatial scales ranging from a few hectares to thousands of square kilometers [31]. Fuel composition and fuel loading interact with terrain features and fire weather to influence burning duration and heat flux, which ultimately determine the spatial pattern of fire severity [20,32]. Various methods have been developed to represent the cause-and-effect relationships between environmental factors (drivers), fire behavior (process), and fire effects on an ecosystem (patterns). Those methods generally fall into two categories: physical models (e.g., FIREHARM and FOFEM) that simulate fire spread and effects based on fire spread physics developed from laboratory experiments [29,33–35], and the empirical approaches that draw relationships from the analysis of existing fire severity patterns [36–38]. In the past two decades, empirical relationships between fire severity pattern and its spatial controls have gained considerable attention [37–40], with scientific aims to identify the primary drivers and quantify their explanatory power on spatial heterogeneity of fire severity.

Since fire severity can be evaluated across a range of spatial scales, environmental variables in empirical models need to be consistently described at comparable spatial scales for maximum predictive power. Variables representing fuels and topography are considered to be bottom-up controls because their influences on fire patterns are pronounced largely at fine spatial scales [41]. Those variables can be easily observed and/or resampled at various scales without losing their precision. For example, when fire severity is assessed in the field (site scale), the fuel properties and topography measured in situ are often used to build relationships with fire severity [39,40]. When fire severity is quantified at coarser scales (e.g., fire patch level), vegetation variables derived from remotely sensed forest maps or forest inventory analysis databases, and topographic features developed from digital elevation models (DEMs) are often re-sampled to a coarser spatial resolution (upscaling) [42–44]. Vegetation and topography have been widely recognized as the dominant controls on fire severity in many types of forest ecosystems [37,38,42,43,45].

Fire weather is considered a top-down control of wildfires as its influences on fire behavior are pronounced at broad scales. Although fire weather has been proven an influential driver of fire severity at the fire patch level [37,42,43,46,47], its influences and predictive power on spatial heterogeneity of fire severity remain poorly understood at fine scales. Current meteorological data are usually collected at a coarse spatial resolution, which is inconsistent with the fine scale at which site-level fire severity is assessed and predicted. The viewpoints that emphasize that fire weather is less important than vegetation or topography on regulating fire severity may be problematic as the low-resolution fire weather may not represent the spatial variations of in situ meteorological conditions. Spatially interpolated data from weather stations can mitigate this issue, but in regions where the weather stations are scarce, this technique is inadequate. The lack of availability of credible and timely weather data limits the understanding of spatial controls on fire severity.

Recent advances in remote sensing techniques have provided a new opportunity to examine fine-scale fire weather effects on fire severity. Fire weather conditions are often characterized using the metric of fuel moisture content, which reflects dryness of dead fuels and water deficit of live biomass. Remotely sensed surface evapotranspiration (ET) products, which capture broad relationships between surface moisture availability and fuel dryness and vegetation drought-stress [44,48,49], have seldom been applied in modeling fire severity, even though they possess relatively high temporal and spatial resolutions.

Furthermore, prediction of fire severity patterns based on the empirical understanding of spatial controls has been well studied in North American boreal and western US forest landscapes [50–52], but there is still a lack of comprehensive analysis in Eurasian boreal forests, which are expected to become more fire prone with climate warming and drying [1,53,54]. This study conducts a comprehensive analysis of within-patch fire severity variations in response to pre-fire vegetation, topography, and surface moisture availability at fine spatial scales in a Eurasian boreal forest landscape. The Great Xing'an boreal forest in northeastern China is an important forest ecosystem that stores 1.0–1.5 Pg C and provides 30% of the total timber yield in China [55]. It is located near the southern frontier of Eurasian boreal forests where the fire regime is very sensitive to climate changes [43,54,56]. Since April 2014, commercial logging has been completely forbidden in this region in an effort to restore and protect its valuable ecosystem services. Fire is the primary disturbance in this area, and there is an increasing urgency to understand the driving mechanisms of fire severity patterns to mitigate fire-induced ecological damage. In this paper, our objectives are three-fold. First, we investigate how fire severity varies across the landscape in response to environmental factors characterized at a consistent spatial resolution. Second, we verify the predictive power of remotely sensed pre-fire surface moisture conditions on determining fire severity patterns. Third, we develop an empirically based model to identify the distribution of potential high-severity burns.

2. Materials and Methods

2.1. Study Area

The boreal forest in the Great Xing'an Mountains of China is a fire-prone ecosystem that generally experiences frequent, moderate- to low-severity surface burns, mixed with infrequent high-severity crown fires. The climate is classified as mid-latitude continental cold-temperate type with short, warm, humid summers and long, cold, dry winters [28]. The study area is mainly located in the Huzhong Forestry Bureau (Figure 1), which is situated in the central part of Great Xing'an region and represents a typical boreal forest landscape of northeastern China. It has a mean annual precipitation of approximately 460 mm that mostly occurs between July and September, and a mean annual temperature of approximately −4.7 °C. The topography is mountainous, with elevations ranging between 360 m and 1511 m above sea level. In contrast to the boreal forests dominated by evergreen coniferous tree species in North America and Europe, the forests in this area are dominated primarily by a deciduous coniferous tree species, Dahurian larch (*Larix gmelinii* (Rupr.) *Rupr.*), and mixed with some evergreen coniferous tree/shrub species including Korean spruce (*Picea koraiensis Nakai*), Scotch pine (*Pinus sylvestris* var. *mongolica*), and Siberian dwarf pine (*Pinus pumila* (Pall.) *Regel*), and a few deciduous broadleaf species of birch (*Betula platyphylla*) and aspen (*Populus davidiana* and *Populus suaveolens Fisch.*). The understory species are composed of evergreen shrubs (e.g., *Ledum L.* and *Vaccinium vitis-idaea L.*), deciduous shrubs (e.g., *Betula fruticose Pall.* and *Rhododendron dauricum L.*), and some herbaceous plants (e.g., *Chamaenerion angustifolium (L.)* and *Carex appendiculata* (Trautv.) *Kukenth.*) [28], whose distributions are influenced by the topographic and soil conditions [57].

Based on fire occurrence records published by the Chinese Forestry Science Data Center, there were 146 fires in the Huzhong Forestry Bureau's jurisdiction between 1991 and 2010, of which 111 were lightning-ignited [58]. Most fires occurred in June, July, and August, which suggests that summer is the primary fire season in our study area. Similar fire occurrence tendencies for the entire Great Xing'an Mountains were reported from 1980 to 2005 in Fan et al. (2017) [54]. Here we focused on 21 fires (Table 1) occurring in two fire years: 2000 (3 fires) and 2010 (18 fires). These two fire years exhibited the greatest burned areas of any other years in the past two decades, together accounting for about 82.5% of the total burned area over a 20-year period [28,43]. Another important reason for selecting these fires is that they were all lighting-ignited in mid-to-late June and located within similar biotic and abiotic environments. In addition, field measurements of fire severity and forest regeneration in the area of these fires have been conducted by our research team since 2010 [28,59,60].

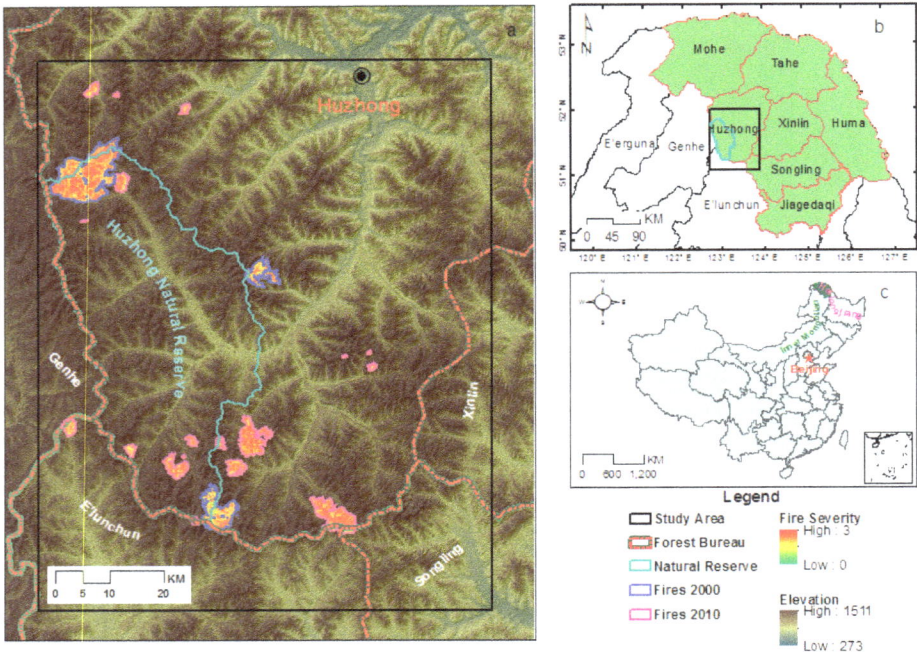

Figure 1. Location of study area (**a**) showing severity of wildfires occurring in 2000 (blue perimeters) and 2010 (pink perimeters). Most of the study area is located in the Huzhong Forestry Bureau (**b**) in the middle of Great Xing'an boreal forests (green patch in b and c) which administratively belongs to Heilongjiang province in northeastern China (**c**). One fire was located in E'lunchun County (**a**), which belongs to the Inner Mongolian part of Great Xing'an boreal forests. Forests within the Huzhong Natural Reserve are primarily natural forests because of a strictly enforced ban on commercial and salvage logging within the reserve since 1958, while forests outside the natural reserve experienced severe cutting since the 1950s.

Table 1. Detailed information of 21 fires included in this study.

Occurrence Date	DOY [†]	Duration (Day)	Longitude	Latitude	Burned Area (ha)
17 June 2000	168	7	122.830	51.891	8518.5
17 June 2000	168	5	123.175	51.314	2918.3
18 June 2000	169	3	123.294	51.724	1443.6
12 June 2010	163	1	123.092	52.003	207.4
12 June 2010	163	1	122.947	51.420	320.1
13 June 2010	164	1	122.844	52.036	394.8
15 June 2010	166	1	122.821	51.813	26.3
15 June 2010	166	1	123.579	51.583	29.0
15 June 2010	166	1	123.587	51.559	104.5
20 June 2010	171	1	122.908	52.027	47.0
25 June 2010	176	1	123.513	51.577	17.4
26 June 2010	177	5	123.486	51.305	2891.5
26 June 2010	177	5	123.252	51.472	1926.1
27 June 2010	178	1	123.116	51.300	102.4
27 June 2010	178	3	123.182	51.431	255.1
27 June 2010	178	3	123.224	51.390	734.6
27 June 2010	178	3	123.108	51.435	258.8
28 June 2010	179	1	123.302	51.450	260.4
28 June 2010	179	1	122.784	51.459	536.0
28 June 2010	179	3	123.065	51.391	984.3
29 June 2010	180	1	122.922	51.879	670.8

[†] DOY: day of year corresponding to fire occurrence date.

2.2. Remote Sensing Imagery Processing

We obtained four L-1 terrain-corrected Landsat TM and ETM+ images (path-row 121/24) with very good quality from 1999, 2000, 2007 and 2010 from the USGS website. To minimize spectral bias caused by phenology differences, the four Landsat images selected for study were all acquired in September. The raw digital number (DN) images of each spectral band were first calibrated into at-satellite radiance using the sensor-specific parameters cited in Chander et al. [61]. A consistent radiometric response between multitemporal Landsat images is critically important for regional fire severity assessment over long time scales. Atmospherically-corrected Landsat surface reflectance products recently have been provided by the USGS, but to our knowledge, a further radiometric normalization can still be useful for consistent monitoring of forest changes. Thus, before using these Landsat images, we applied the 6S atmospheric correction method [62] and an absolute radiometric normalization approach, the iteratively reweighted multivariate alteration detection (IR-MAD) [63], to eliminate atmospheric effects and improve radiometric consistency between the Landsat time-series datasets. We selected a 2802 × 3483-pixel subset from each image (Figure 2), which covers all 21 fires, to carry out the normalization procedure in ENVI/IDL 4.7 (ITT Industries Inc., White Plains, NY, USA, 2009). We selected the 2007 image as the common reference image due to the least cloud coverage and used the bandwise regression parameters generated by IR-MAD to normalize the other three reflectance images. All the images were normalized to a consistent radiometric standard of the 2007 reference image. The dNBR is a well-known spectral index and has been proven to correlate well with fire severity in many types of forest ecosystems, including the study area of this research [28]. We calculated the NBR and dNBR indices using Landsat bands 4 and 7 as proposed in Key and Benson [19]:

$$NBR = \frac{TM_4 - TM_7}{TM_4 + TM_7} \tag{1}$$

$$dNBR = NBR_{pre} - NBR_{post} \tag{2}$$

The two images acquired in 1999 and 2007 were used to calculate pre-fie NBR for the years of 2000 and 2010, respectively, while the images acquired in 2000 and 2010 were used to calculate post-fire NBR.

The 8-day moderate resolution imaging spectroradiometer (MODIS) ET product (MOD16A2) includes actual ET (AET), latent heat flux, potential ET (PET), and potential latent heat flux data. It is composed of daily canopy evaporation, plant transpiration, and soil evaporation and is calculated as the average value of cloud-free ET during an 8-day period [64]. MOD16A2 products have been evaluated based on flux tower measurements in many ecosystems and exhibit agreement with the ground-measured ET in eastern Asian forests [64,65]. In this study, we assumed that the MODIS-derived SMA index captures the broad relationship between remote sensing spectral signals and fuel moisture content. Although we did not develop an empirical relationship model to retrieve the actual pre-fire fuel moisture content, many recent publications have tested this hypothesis and indicated that MODIS data could be operationally integrated into fire danger systems [48,49,66]. Currently, there are two versions of MOD16A2 products available. The latest (version 6) can provide 500 m ET observations, but it is not available for year 2000 as its collection began in 2001. Version 5 is available for both fire years, but its spatial resolution is coarser (1 km) (Figure A1). In addition, because these two versions used different model input datasets (e.g., land cover product, leaf area index) and algorithms [67], the final ET outputs differ between version 5 and version 6 (Figure A2). We obtained both version 5 and version 6 of MOD16A2 product for our study area and assessed the performance of both versions in our modeling. In our study, we mainly used the version 5 product for a consistent comparison of the models for 2000 and 2010. We also compared model performance and the ability of versions 5 and 6 in explaining spatial variation of fire severity for the year 2010.

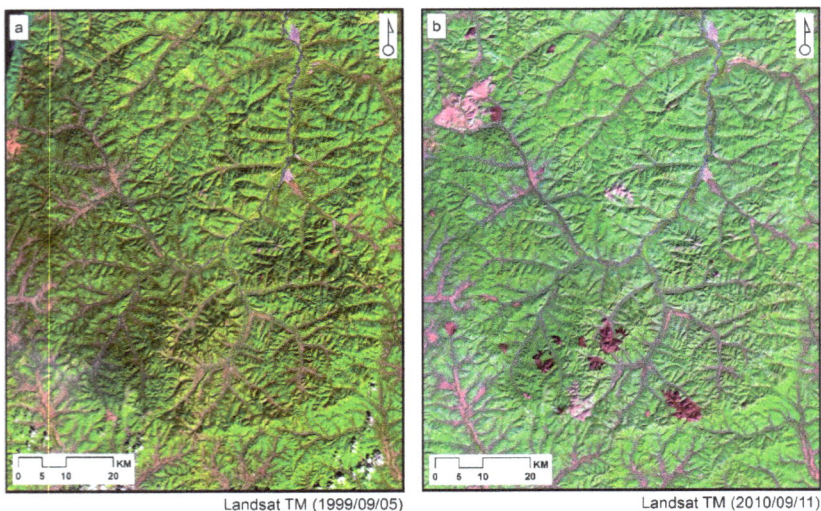

Landsat TM (1999/09/05)   Landsat TM (2010/09/11)

Figure 2. Pre-fire (**a**) and post-fire (**b**) false color Landsat TM images (R-TM5, G-TM4, and B-TM3) of the study area. The light pink patches in (**b**) indicate fires occurred in 2000, while dark red patches indicate fires occurred in 2010.

Based on occurrence dates of 21 fires (Table 1), we obtained five pre-fire MOD16A2 ET data (tile: H25V03, day of year from 129 to 168) for 2000 and six 1 km pre-fire MOD16A2 ET data (tile: H25V03, day of year from 129 to 176) for 2010 from the University of Montana's Numerical Terra-dynamic Simulation group. The six 500 m MOD16A2 ET data (tile: H25V03, day of year from 129 to 176) were obtained from the USGS Land Processes Distributed Active Archive Center. With the assistance of MOD16A2 Quality Control (QC) data, we only selected high-quality pixels (QC equal 0) from each 8-day ET dataset and used them to generate two integrated pre-fire ET datasets for 2000 and 2010. For a given burned pixel, the maximum of six observations with high quality were ranked by the time since fire occurrence date. We gave preference to the observation whose acquisition time was closest to the fire occurrence date to ensure the selection reflects the latest pre-fire surface ET conditions.

2.3. Fire Severity Mapping

Burned pixels (30 m) were extracted based on a thresholding process for the forest disturbance index following the protocol of the Landsat Ecosystem Disturbance Adaptive Processing System (LEDAPS) [68]. The detailed description of this approach can be found in Fang et al. (2015) [43]. Twenty-one fires burned about 22,650 ha in total; the three fire events in 2000 burned over 12,880 ha (about 251, 670 Landsat pixels), and the 18 fires in 2010 burned over 9760 ha (about 108, 450 Landsat pixels). These burned pixels exhibited significantly different spectral features compared to the unburned pixels in the Landsat images, as shown in Figure 2. Our previous study established a quadratic polynomial relationship between the dNBR and field measured composite burn index (CBI):

$$CBI = -0.0425 + 2.753 \times dNBR - 0.8142 \times dNBR^2 \tag{3}$$

which was confirmed to explain 84.6% variance in 74 CBI plots of 2010 fires and to produce accurate severity maps of the largest fire in 2000 (Kappa = 0.72) [28]. The CBI values of 1.1 and 2.0 were applied as boundaries of three severity levels (i.e., low severity ($0.1 \leq CBI \leq 1.1$), moderate severity ($1.1 < CBI \leq 2$), and high severity ($2 < CBI \leq 3$)) as they were corresponding to about 10% and 80% of canopy mortality, respectively, in our ecosystem, where the live mature trees were very important for vegetation restoration [28]. The two dNBR thresholds associated with CBI values of 1.1 and 2.0 were 0.484 and 1.099, respectively; these thresholds were applied to classify burned pixels of 2000 and 2010 into three severity levels.

2.4. Environmental Metrics

We used a suite of explanatory variables to describe various aspects of fuel, topography, and surface moisture conditions, and modeled their relationships with fire severity (Table 2). Using Landsat imagery, we developed two vegetation cover maps for two different years (1999 and 2007) and reconstructed the pre-fire vegetation conditions. Using a stratified decision tree classification method, we combined the conspicuous differences of phenology and spectral characteristics among vegetation types, and used Landsat surface reflectance, spectral indices, and components of Tasseled Cap transformation to produce vegetation cover maps at 30 m resolution [69]. The whole area was classified into six vegetated categories consisting of deciduous coniferous forest (DCF, i.e., larch forest), deciduous broadleaf forest (DBF), evergreen coniferous forest (ECF), mixed forest (MF), grassland (GRS), and shrublands (SRB), as well as five non-vegetation categories (water bodies, bare rock, bare soil, urban land, and shade from mountains and clouds). Accuracy assessments were carried out in Fang et al. (2015) [43], which reported that the overall accuracy and the Kappa coefficient of the vegetated areas were 64% and 57% respectively. In detail, the mapping accuracy of DCF is 72%, DBF is 40%, ECF is 81%, MF is 49%, GRS is 82%, and SRB is 70%, indicating suitability for subsequent analysis.

We used a 30 m ASTER global digital elevation model (GDEM) product published by the National Aeronautics and Space Administration (NASA) to derive a number of commonly used topographic indices, including elevation (ELV), slope (SLP), aspect, and topographic wetness index (TWI). Aspect was further converted to better represent the potential influence of solar radiation (PSR) on site moisture conditions, following the equation:

$$PSR = \cos((\theta - 225) \times \pi/180) \tag{4}$$

where θ is the aspect in degrees [70]. The higher *PSR* values represent higher potential solar insolation. The TWI was designed based on the assumption that water movement is controlled by the topography of the slopes [71]. TWI is defined as:

$$\text{TWI} = \ln(\alpha / \tan \beta) \tag{5}$$

where α is the local upslope area draining through a certain point per contour length and $\tan \beta$ is the slope angle at the point [72]. *TWI* is often used to describe the spatial distribution of soil moisture and surface saturation, and has shown good correlation with soil moisture and depth to ground water in Eurasian boreal forests [72,73]. *TWI* values typically range from 3 to 30, where a higher value indicates higher soil moisture potential.

Fuel moisture content is a good surrogate for representing the indirect impacts of weather on fire behavior [74] because it influences many fire processes, such as ignition, combustion, and smoldering [75]. Fuel moisture is also controlled by weather conditions such as precipitation and ET and is closely associated with soil moisture [48,75]. Here, we calculated surface moisture availability (SMA) as the indicator of live fuel moisture content using the equation:

$$\text{SMA} = \text{AET/PET} \tag{6}$$

where AET is actual ET and PET is the potential ET [76]. The SMA is a critical parameter in governing the partition between sensible and latent heat flux at the surface, which is the key determinant of soil moisture, and also determines the water stress and water content of live plants [49,76]. In general, when AET is equal to PET, the surface will reach moisture-saturated conditions and the SMA will be equal to 1. However, when the SMA is below certain threshold values, the vegetation may begin to suffer from drought stress and the soil may approach a limiting dryness [76].

Table 2. Abbreviations and descriptions of explanatory variables in this study.

Category	Variable	Description	Mean ± SD (2000 & 2010)
Vegetation [†]	ECF	Percentage of Landsat pixels classified into evergreen coniferous trees within 240 m burned pixels, were primarily *Pinus pumila* shrublands and *Larix gmelinii-Pinus pumila* forest.	0.158 ± 0.261 0.169 ± 0.278
	DCF	Percentage of larch forest. The three dominant larch forests are *Larix gmelinii-Ledum palustre* L., *Larix gmelinii-grass*, and *Larix gmelinii-Rhododendron dahurica* L.	0.615 ± 0.324 0.350 ± 0.302
	DBF	Percentage of broad leaf forest. The white birch and aspen are dominant broad leaf species.	0.017 ± 0.057 0.177 ± 0.218
	MF	Percentage of mixed forest. Composited by broad leaf trees and coniferous trees.	0.086 ± 0.145 0.150 ± 0.218
	GRS	Percentage of grassland.	0.047 ± 0.135 0.086 ± 0.166
	SRB	Percentage of shrublands, typically distributed in open land along the river and disturbed areas.	0.035 ± 0.097 0.041 ± 0.134
Topography	ELV	Elevation (meters) derived from the aggregated ASTER GDEM at 240 m spatial resolution.	1089 ± 92.862 1017 ± 107.316
	PSR	Potential solar radiation. It ranges from −1 to 1, where high values represent xeric exposures.	−0.024 ± 0.715 −0.173 ± 0.683
	SLP	Slope (degree) computed from aggregated DEM.	8.840 ± 4.090 8.958 ± 3.992
	TWI	Topographic wetness index (unitless) is computed from the slope and the upslope contributing area per unit contour length.	13.953 ± 1.313 14.024 ± 1.528
Surface moisture	SMA	SMA is calculated from MOD16A2 and represents land surface moisture availability. The higher SMA value indicates the wetter land surface.	0.494 ± 0.099 0.411 ± 0.153

[†] Statistics of five non-vegetation land cover types were not listed.

2.5. Spatial Data Processing

To mitigate the issue of inconsistent spatial resolutions between various spatial datasets, we conservatively selected a 240 m spatial resolution as the overall standard. This resolution (240 m) is exactly eightfold the resolution (30 m) of the Landsat-derived vegetation maps and the ASTER GDEM data. This resolution is particularly suitable for aggregating the fine-resolution categorical map to the coarse-resolution continuous imagery. In addition, it is close to 250 m, which is the finest spatial resolution of MODIS products, and many remote sensing studies use 250 m to conduct downscaling for 1 km MODIS ET products [77]. To downscale the SMA data, we used a nearest neighbor interpolation approach to resample the data from 1 km to 240 m.

Three different spatial aggregation methods were employed to upscale the vegetation, topographic, and fire severity data. Based on the Landsat-derived vegetation cover maps, we calculated the fractional coverage of each vegetation type by calculating the proportion of occurrence within an 8 × 8 pixel window (8-PW). Thus, the 30 m categorical vegetation maps were converted into 240 m continuous vegetation coverage images. For the numerical ASTER GDEM data, we calculated an average value for each 8-PW and subsequently used the aggregated DEM data to calculate the abovementioned topographic metrics.

To generate a fire severity value that can be consistently interpreted at different spatial scales, we used an area-weighted average method to aggregate the 30 m categorical fire severity map to 240 m. We first analyzed the proportions of unburned, low, moderate, and high severity pixels within each 8-PW; pixel proportions in the 8-PW serve as the area-weight values in the aggregation function. For instance, if 32 pixels within a given 8-PW (64 pixels) were classified as high severity, the weight value of high severity would be 50% (i.e., 32/64). If no pixel was classified as high severity in the 8-PW, the weight value of high severity would be 0. Then, a set of fire severity rating integers from 0 to 3 was used to indicate the fire severity gradient from unburned to high severity, respectively:

$$Severity = 0 \times \omega_{unburned} + 1 \times \omega_{low} + 2 \times \omega_{moderate} + 3 \times \omega_{high} \qquad (7)$$

where ω is the weight value of four severity gradients. The area-weighted average value of fire severity ranged from 0 to 3, with higher values representing higher fire severity. The area-weighted averaging process considers the relative importance of each severity level and provides a more balanced interpretation of the data.

2.6. Statistical Modeling

A boosted regression tree (BRT) model was applied to explain the relationships between remote sensed fire severity and selected environmental variables. The BRT model is a useful machine learning approach that uses recursive binary splits and a boosting technique to combine a large number of sequential trees to improve the fit and predictive performance [78,79]. It has the capacity to handle complex relationships among numerical and categorical variables, quantify interactions between explanatory variables and overcome inaccuracies associated with regression and classification methods [78]. We applied the "gbm" package (version 2.1.3) in *R* 3.4.1 (*R* Development Core Team 2017, Boston, MA, USA) to run the BRT analysis. We ran two sets of BRT models for the two different fire years, 2000 and 2010. We used the same parameter settings and amounts of training data to ensure that the outcomes of the two models were comparable. The random subsampling and bagging procedures in a BRT model may introduce stochasticity into the model outcomes. To mitigate the stochastic errors and create stable model outputs, we carried out 50 BRT modeling trials independently and calculated an average as the final result.

Before running the BRT models, we carried out a data exploration procedure to assess whether there was any conspicuous spatial autocorrelation or collinearity. Using functions in the "ape" package [80], we calculated the Moran's I index to examine spatial autocorrelation in fire severity and explanatory variables. Moran's I is similar to a correlation coefficient and represents the similarity of an observation to its nearby observations [81]. It ranges between -1 and 1. Higher positive values indicate greater similarity, which can be interpreted as a spatially clustered distribution, while lower negative values indicate stronger dissimilarity (i.e., a more dispersed spatial pattern), and the zero value indicates a random spatial pattern (i.e., perfect randomness). We set a 300 m sample spacing distance (>1 pixel) when computing the Moran's I index. Although *p*-values were less than the 0.05 level, the Moran's I values of both response and explanatory variables were small (mostly less than 0.20, see Table A1), suggesting weak spatial autocorrelations among sampling pixels. Similar or higher Moran's I thresholds for determining suitable sampling spaces have been used in wildfire-related literature [36,45,50]. We also calculated the pairwise Pearson correlation coefficient (r) using the "Hmisc" package to evaluate potential collinearity among predictor variables. All variables selected for modeling had low pairwise Pearson correlations ($|r| < 0.60$), suggesting relatively low levels of collinearity. We used a subset of 200 random sampling pixels to build the BRT model, while the remaining 100 pixels were used for validation.

The optimization of a BRT model is jointly controlled by the learning rate, tree complexity, bagging fraction, and number of trees. In this study, we set these parameters at 0.01, 5 and 0.5, respectively, following the recommended model inputs of Elith et al. (2008) [79]. The number of trees in each BRT trial was automatically selected based on a 5-fold cross-validation procedure to avoid over-fitting problem. The relative importance of each explanatory variable to fire severity was measured by averaging the frequency it was selected for splitting among all trees, and the importance was weighted by the squared improvement to the model as a result of each split [79]. The relative importance values of the predictors were scaled as a percentage, the sum of which equals 100%. Higher importance values represented stronger impacts on controlling fire severity. In addition, we explored the dependency relationships between several important variables and fire severity by plotting the effect of a specific explanatory variable on the response variable after averaging the effects of remaining explanatory variables in the same model.

The coefficient of determination (R^2) reported by the BRT model was used to evaluate how well the model fits the training data. We used the 100 samples that were independent of the training data to assess the predictive power of our BRT models. We examined the goodness of fit between simulated fire severity values and observed fire severity values by applying a linear regression model and calculating the squared multiple correlation coefficients (denoted r^2). To provide a standalone evaluation of model performance, we used the error matrix method to evaluate the predicted fire severity maps. We selected an optimal simulation image for each fire year based on the r^2 value and classified the burned pixels into low (fire severity $\leq$ 1), moderate (1 < fire severity $\leq$ 2), and high (2 < fire severity $\leq$ 3) severity levels. The two aggregated fire severity images were also converted into categorical severity maps using these same thresholds and were then used as reference maps. We used all burned pixels from each fire severity level to ensure unbiased evaluation for both fire years.

3. Results

3.1. Evaluation of Model Performance

We found that the two sets of BRT models fit the training data very well, as they explained 83.0% ($R^2_{max} = 85.0\%$, SD = 0.008) of the variation in fire severity in 2000 (Figure 3a) and 73.8% ($R^2_{max} = 81.3\%$, SD = 0.035) of the variation in fire severity in 2010 (Figure 3b). Similarly, the 50 validation models of the two years showed goodness of fits of 0.795 ($r^2_{max} = 0.820$, SD = 0.040) (Figure 3c) and 0.618 ($r^2_{max} = 0.656$, SD = 0.012) (Figure 3d), respectively, suggesting that the predicted fire severity results were well correlated with the reference values.

The confusion matrices indicated that the optimal BRT model in 2000 achieved higher prediction accuracy than the optimal model in 2010 (Table 3). The predicted severity map of 2000 produced fewer commission errors in moderate (44.3% vs. 71.6%) and high (11.2% vs. 24.0%) severity pixels than 2010, but it generated a high commission error for low severity (85.8%) pixels, indicating more pixels were erroneously predicted as low severity than were actually observed in the severity results. The predicted severity map of 2010 generated high commission error (71.6%) for moderate severity level and high omission error (64.6%) for low severity level, indicating overestimated moderate severity area and underestimated low and high severity area (Figure A3). Overall, we found the prediction map of 2000 to represent very good agreement with the observed severity map, while the prediction map of 2010 underestimated the severity of most of the 18 fires (Figure 4).

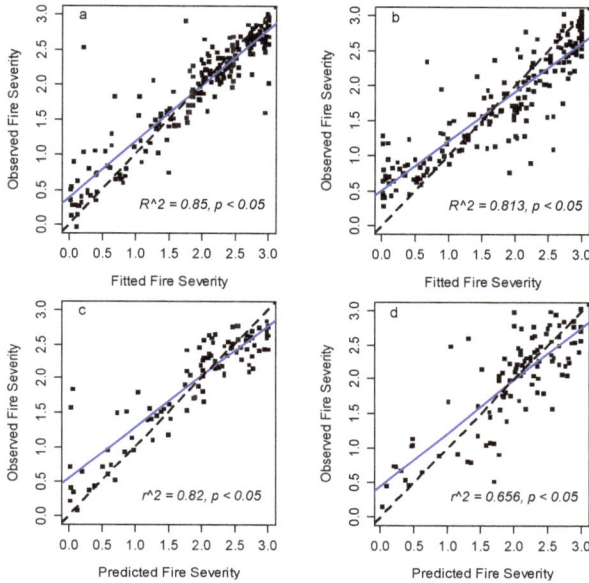

Figure 3. Two examples of linear relationships used for validating model performance of 2000 (**a,c**) and 2010 (**b,d**) based on 200 training samples (**a,b**) and 100 verification samples (**c,d**). Coefficients of determination (R^2 for training samples), and the squared multiple correlation coefficients (r^2 for verification samples) are also plotted. Blue solid lines show predicted linear regression fit. Black dashed lines represent 1:1 line.

Table 3. Accuracy assessment of predicted fire severity classification for fire year 2000 and 2010, respectively.

Severity Class	Predicted Severity of 2000				Predicted Severity of 2010			
	Low	Moderate	High	Producer's Accuracy	Low	Moderate	High	Producer's Accuracy
Low	50	26	7	60.2%	196	263	94	35.4%
Moderate	261	496	142	55.2%	95	276	115	56.8%
High	42	369	1184	74.2%	33	431	662	58.8%
User's Accuracy	14.2%	55.7%	88.8%	-	60.5%	28.5%	76.0%	-
Overall Accuracy			67.1%				52.4%	

3.2. Relative Importance of Environmental Variables

The relative importance of individual environmental variables varied substantially in the different fire years (Figure 5). For the fires in 2000, the six most important predictors of fire severity in decreasing order were DBF, SRB, MF, TWI, SLP and ELV. These predictors contributed to a total of 76.8% of the relative importance. Three vegetation variables together contributed over 55% relative importance, indicating the spatial pattern of fire severity in 2000 was largely driven by the distribution of these three vegetation types. For the 2010 burns, the six most important predictors were ECF, SMA, SLP, PSR, ELV and TWI, which together contributed to 75.1% of the relative importance. The ECF and SMA variables independently contributed as much as 23.3% and 12.8% relative importance respectively. The 500 m version 6 SMA was found to have a slightly lower (11.0%) relative importance than the 1 km version 5 SMA product when modeling spatial variability of fire severity in 2010 (Figure A4). An overall ranking of 11 variables was calculated based on a weighted average approach (see *y*-axis of Figure 5), which identified four vegetation variables (ECF, DBF, SRB, and MF) as the primary controls of fire severity, followed by SMA and topographic variables.

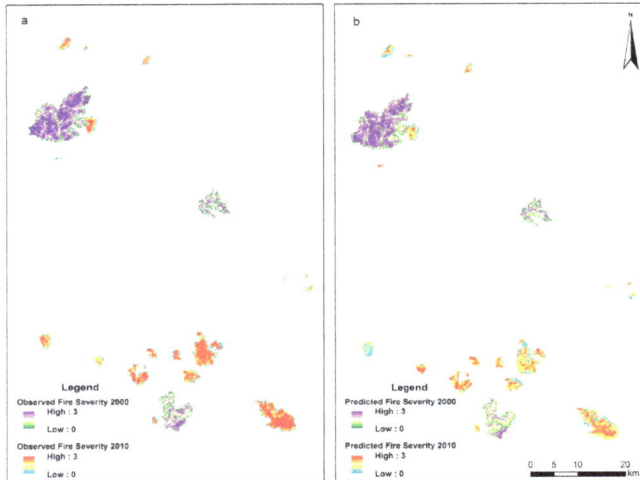

Figure 4. Observed fire severity (**a**) aggregated from Landsat observations versus the modeled fire severity (**b**) for 21 fires based on boosted regression tree models. All fire severity images were plotted at 240 m spatial resolution.

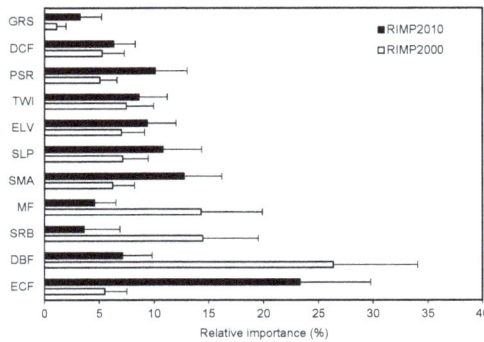

Figure 5. Relative importance proportion (RIMP) of explanatory variables for 50 boosted regression tree models (Mean + SD) of fire severity in 2000 and 2010. The *y*-axis (ECF to GRS) shows an overall rank of 11 variables in descending order according to a weighted average of RIMP, which is calculated by multiplying two mean R^2 (i.e., 0.830 for RIMPs of 2000, 0.738 for RIMPs 2010) with two mean RIMP values (i.e., RIMP2000 and RIMP2010) of each variable. The ECF has the highest overall relative importance while the GRS has the lowest value. See Table 2 for definition of variable abbreviations.

When the 11 variables were grouped into three broad control types of vegetation, topography, or SMA, the relative importance of these three control types became quite similar between the two fire years. Vegetation consistently played a strong role in determining fire severity in the two fire years because it contributed about 40–80% of total relative importance (Figure 6a). Together with 20–50% of the total relative importance contributed by topographic variables, the results revealed that potential fire severity in our study area is mostly controlled by vegetation and topography. At the same time, we found SMA has similar relative importance to the maximum importance value of topographic variables, indicating that SMA also has considerable predictive power over fire severity (Figure 6b).

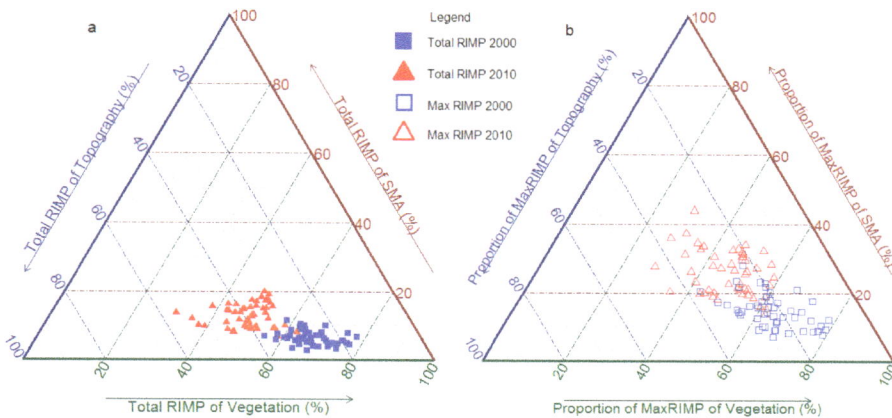

Figure 6. Ternary plot (**a**) showing the total relative importance proportion (RIMP) of three groups of explanatory variables (i.e., Vegetation, Topography and Surface Moisture Availability (SMA)) in controlling fire severity of 2000 (blue) and 2010 (red). In considering the difference of variable numbers among three groups, a variable which has the maximum RIMP (MaxRIMP) of each group is selected and compared (**b**). Axis in ternary plot (**b**) representing relative proportions of MaxRIMP for three types of variables.

3.3. Relationships between Environmental Variables and Fire Severity

Partial dependence plots were helpful for visualizing the response of fire severity to explanatory variables. We found that the relationships between fire severity and environmental variables were mostly nonlinear and varied little in different years (Figure 7). In general, high severity fires were more likely to occur in dense evergreen forests (Figure 7a), especially on well-drained, gentle slopes at high altitudes (Figure 7f–h). On the other hand, the partial dependence of fire severity generally decreased with increasing DBF and SRB (Figure 7b,c), suggesting that the fire-resistant traits of these two vegetation types may mitigate the adverse effects of severe burning. We found that increased coverage of mixed forests generally had a negative impact on fire severity (Figure 7d), especially in the year 2000, in which MF was recognized as an important variable. Fire severity in deciduous coniferous forests was significantly lower than in evergreen coniferous forests (Figure 7j vs. Figure 7a) but higher than in the other forest types. DCF did not exhibit a strong relationship with fire severity in 2010, but did exhibit a positive relationship with fire severity in 2000. The GRS variable was found to have little relative importance to fire severity and represented a weak negative impact on fire severity in 2010 (Figure 7k). The south facing slopes and areas with high TWI values typically experienced moderate severity fires (Figure 7h,i). Our results also indicated that higher probabilities of moderate-to high-severity fires were related to canopy or surface dryness (Figure 7e), especially when the 1 km SMA ranged from 0.13 to 0.30 in the summer. In contrast, we found a strong positive association between fire severity and 500 m SMA (Figure A5), as a result of inconsistent SMA values.

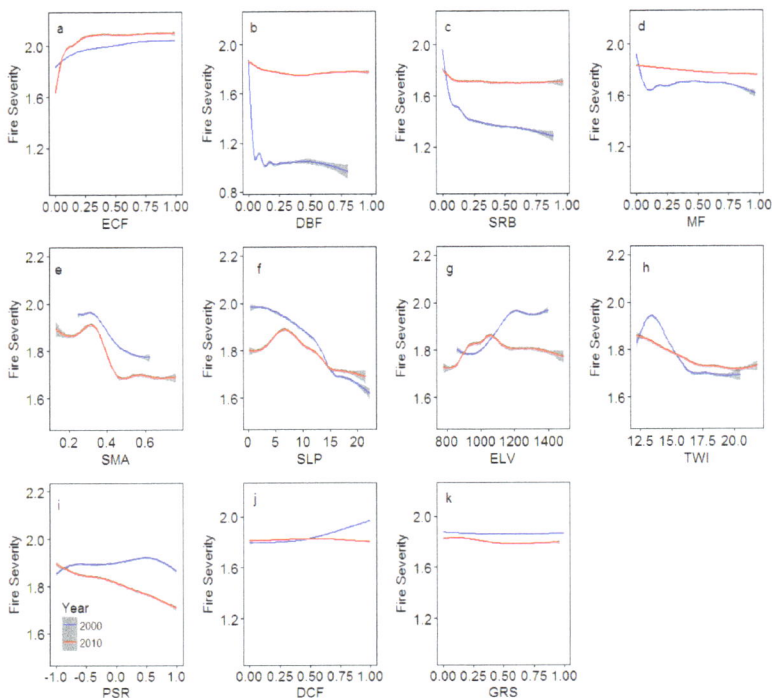

Figure 7. Partial dependence plots of explanatory variables on regulating fire severity in different fire years (2000 in red and 2010 in cyan). This grouping of variables was following the rank as shown in *y*-axis of Figure 5. The curves demonstrate the smoothed means (solid line) and 95% confidence intervals (gray zone) of an ensemble of 50 models. Variable abbreviations were described in Table 2.

4. Discussion

4.1. Environmental Influences on Fire Severity

We found that the BRT method was effective in investigating the relationships between fire severity and environmental gradients, such as vegetation composition, terrain, and surface moisture status. Fire severity is a complex function of these environmental gradients and such relationships may vary in different years and locations. With thousands of 240 m sampling data from representative historical fires, we found that the spatial distribution of fire severity could be predicted with adequate precision in a Great Xing'an boreal forest landscape. In our exploration of the relative importance of these spatial controls on fire severity, we found that fuel conditions are the most influential predictor in determining the magnitude of fire severity. This finding is in accordance with our previous study [43], which was conducted at burned patch level for the entire Great Xing'an boreal forest, and studies in similar ecosystems, such as Canadian boreal forests [82,83] and subalpine forests [37,84,85].

Coniferous forests/shrubs experience higher severity fires than broadleaf forests and shrublands in our ecosystem. Such result is generally consistent with observations in North America boreal forests where deciduous tree dominated stands are found to be fire break and reduce landscape flammability owing to higher foliage moisture and less surface fuels [86,87]. *Larix gmelinii* is the dominant coniferous tree species in the boreal forests of Northeastern China (Figure 8a). Unlike the dominant evergreen coniferous species (e.g., black spruce) in Northern American boreal forests or subalpine forests (e.g., spruce-fir, ponderosa pine) in Western US, larch is a deciduous coniferous species; it is considered fire-tolerant because its self-pruning trait can reduce ladder fuels, and the open crown can reduce the bulk density and connectivity of canopy fuels [88]. However, we found that larch forests usually experience high mortality in flat areas and valley bottoms due to heat-induced root damage (Figure 8c). To adapt to shallow soil and permafrost, *Larix gmelinii* develops lateral roots in the thick, flammable moss and organic soil layers to improve adhesion and nutrient supply, but this also means it can be easily injured or killed by surface fires [43]. Larch forests are also characterized by abundant understory plants due to their open-canopy environment, which can contribute higher fire severity than the forests without an abundant understory component. Another important coniferous species in this region is *Pinus pumila* (Figure 8b), which is evergreen and usually mixed with *Larix gmelinii* in open forests at altitudes of 800–1200 m or grows densely on rocky ridges at altitudes higher than 1200 m [89]. *Pinus pumila* is highly flammable because it contains abundant volatile organic compounds in its needles, twigs, and seeds [90]. Furthermore, windy and dry conditions can accelerate the spreading of fires on the ridges of mountains. Therefore, as demonstrated in our analysis, the increases in *Pinus pumila* coverage may considerably increase the probability of high severity fires (Figure 8d). Such finding is in consistent with Estes et al. (2017) [38], who reported that shrub species that are favored by fires can generate higher fire severity than mixed hardwood/coniferous forests and hardwood forests in northern California.

Figure 8. Photographs of two unburned stands dominated by the deciduous coniferous tree *Larix gmelinii* (**a**) and the evergreen coniferous shrub *Pinus pumila* (**b**), and two 1-year post-fire stands previously dominated by *Larix gmelinii* (**c**, surface fire) and *Pinus pumila* (**d**, canopy fire) in Huzhong Natural Reserve, China.

Our results indicated that topographic factors had a considerable influence on fire severity. It is well known that topography can influence fire behavior by impacting fuel moisture, local wind patterns, fire spread direction, and vegetation composition [32], but quantitative demonstrations of these relationships are still needed for optimal mitigation of adverse fire effects. Our results showed that slope and elevation are the two most important topographic variables, followed by TWI and PSR. This suggests that the primary pathway by which topography regulates fire severity in this Siberian boreal ecosystem is by governing fuel moisture and by strongly interacting with vegetation conditions. For example, we found that severe fires were more likely to occur in high altitude regions.

A similar pattern was found in dry ponderosa pine forests of the western US and boreal forests in Europe [42,51,91]. Surface fuels on the upper slopes could dry quickly due to efficient drainage and greater degrees of solar exposure, which may increase the flammability of fuels and facilitate more severe burns. The preheating effects of upslope fires on the adjacent fuels can also increase fire severity [32]. In addition, it cannot be ignored that the ECF are generally densely distributed at high altitudes in this region. In general, north-facing slopes possess higher fuel moisture and lower surface temperature than south-facing slopes. However, in our ecosystem, we found that the north-facing slopes burned more severely due to higher biomass coverage [60]. Similar findings were also reported in studies conducted in boreal and subalpine ecosystems [42,43,84].

Fuel moisture interacts with many complex ecological and physical processes, making it important yet difficult to represent in a modeling framework used to study its spatiotemporal dynamics and influences on fire regime [75]. Topographic gradients could provide a partial explanation for the spatial variation of fuel moisture, especially in light of our finding that TWI is inversely associated with fire severity, as expected. Previous studies proved that TWI is closely associated with soil moisture in European boreal forests [72,92]. Thus, we believe the drainage condition largely determines the moisture gradient of a forest stand and further influences fuel moisture dynamics. Dead fuel moisture dynamics are driven by three mechanisms—capillary forces, infiltration, and diffusion—among which, infiltration and diffusion are the primary driving mechanisms and are both influenced by moisture gradients [75]. Although live fuel moisture is driven by different mechanisms than dead fuel moisture dynamics, soil water dynamics are still an important part of those mechanisms and can directly influence plant transpiration.

The MODIS-derived 1 km SMA exerted considerable influences on model performance, and its relationship with fire severity was negative, as expected. This finding aligns with the work of van Mantgem et al. (2013) [18], which suggests that a pre-fire water deficit can increase fire severity (tree mortality) because the drought-stressed trees are vulnerable to fire-induced injury. Similarly, Xiao and Zhuang [93] found that drought directly affected fire activity in Canadian and Alaskan boreal forests by enhancing fuel flammability and increasing ignitions. However, it should be noted that the pre-fire SMA applied in this study can only represent the short-term temporal variability of the surface moisture conditions, which may not necessarily reflect the long-term effects of drought stress on fire severity. It has been reported that plant communities within a forest stand, especially understory vegetation layers, may be influenced by the long-term drought stress that is regulated by topographic and climatic factors [94].

Although the ranking of relative importance of vegetation, topography, and SMA was similar between the two fire years (Figure 6), the relative importance of individual explanatory variables differed between the two fire years (Figure 5). Such differences may be attributable to their different pre-fire vegetation composition, structure, and disturbance history. By comparing the pre-fire vegetation compositions (Figure 9), we found that fires in 2000 had lower proportions of DBF and MF but higher proportions of DCF than the 2010 fires. Fires in 2000 burned more areas in the Huzhong Natural Reserve where it is dominated by mature (>100 years) larch forest (DCF) due to the strictly enforced cutting ban, and most DBF, MF, and shrubs are located in recently burned/disturbed areas that carry significantly less fuels. Consequently, the proportion of DBF, MF, and shrubs were considered more important than proportion of DCF in modeling severity of 2000 fires. In contrast, fires in 2010 burned more areas in the Huzhong Forestry Bureau jurisdiction that had been disturbed by clear cutting since 1950s, and as recent as 2000s, leading to greater abundance of young stands irrespective of forest type [95]. This could partly explain why fire severity of 2010 in the areas with high proportions of DCF was similar to the areas with high proportion of DF or MF (Figure 7j vs. Figure 7b,d). In contrast, because the highly flammable evergreen coniferous shrub species *Pinus pumila* is not an economically viable species to cut, its fuel loading is generally higher than young stands of other forest types. Consequently, the fire severity in areas with high proportion of ECF was high (Figure 7a), and ECF was considered a more important vegetation variable in driving overall fire severity variability of 2010 fires.

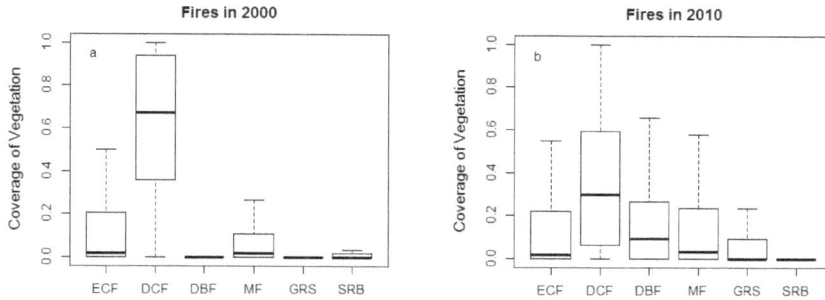

Figure 9. An overview of the pre-fire vegetation composition within fires of 2000 (**a**) and 2010 (**b**). Variable abbreviations were described in Table 2.

4.2. Prediction of Fire Severity

Our results revealed that landscape-level fire severity is primarily determined by fuel and terrain features, suggesting that fire severity is predictable at this scale. The prediction accuracy is determined by the quality of the relevant predictors or proxies. Fuel conditions are usually complicated and difficult to characterize, not only because of the endogenous variety of plant communities throughout the landscape but also because of exogenous factors, such as disturbance regime, climate, and anthropogenic activities [75]. Although various remote sensing methods were proposed to improve fuel mapping, there are still challenges in accurately quantifying the critical fuel parameters that can regulate fire effects, such as fuel loading and canopy bulk density [96,97]. Remotely sensed spectral information is usually applied as ancillary data or proxies when modeling fuel parameters, as it is strongly correlated with many biophysical vegetation parameters, such as biomass, leaf area index, and productivity [26,98,99].

However, remotely sensed imagery cannot detect surface fuels obscured by the forest canopy and insufficiently distinguishes fine fuels from dead biomass pools due to the inconsistency between particle size and the spatial resolution of the image [100]. Compared with the spatial complexity and high variability of fuel parameters at the landscape scale, vegetation cover type is relatively identifiable due to the unique spectral features of plant communities. The Landsat-derived vegetation cover data can reflect the general variability of vegetation composition but may not necessarily depict parameters related to fuel type or fuel particle size. Our results indicated that vegetation coverage could reliably explain the variability of fire severity at a 240 m spatial scale. The classification accuracy of vegetation mapping could also affect the predictability of fire severity models. The spatial aggregation process may confound the accuracy of the 240 m vegetation coverage data, thus increasing the uncertainty of the modeling results.

The large-scale terrain features of wildlands are usually invariant over long periods. In addition, terrain features can be accurately characterized using various traditional or modern survey technologies. Together with their close relationship with fire behavior, topographic variables are commonly used as predictors of fire severity. Digital elevation models provide essential topographic information from which aspect, slope, and other terrain features can be derived. However, it should be noted that some topographic conditions are inherently scale-dependent. Thus, the upscaling process used for the ASTER GDEM data may filter out some detailed terrain features that would be reflected at the 30 m spatial resolution. Because we focused on the 240 m spatial resolution, we did not examine the sensitivity of the relationships between fire severity and topographic variables to the scaling process. Many studies have reported scale dependency in the relationships between topographic characteristics and fire attributes, such as severity, frequency, burned area, and burn probability [101–104]. Despite the good performance of our BRT models, we conclude that 240 m may not be the optimal spatial resolution for predicting fire severity. Enlarging spatial scales (e.g., from 30 to 240 m) can be beneficial in refining the relationships between fire severity and environmental gradients, but it may decrease the visual effect of the prediction maps due to the coarse spatial resolution. Therefore, we suggest that future applications should weigh the model performance against the practical applicability of the prediction maps.

4.3. Limitation and Uncertainty

Despite efforts to improve the prediction of fire severity by incorporating sound explanatory variables, some knowledge gaps should be noted as they may influence the interpretation of the modeling results. First, each MOD16A2 pixel contains the best possible daily ET estimation during the 8-day period, which was selected based on the imaging conditions and observation coverage. The arbitrary application of SMA as a proxy for fuel moisture may increase the alternative quantification of land surface status for modeling fire severity, but its real relationship with actual fuel moisture still needs to be validated in our ecosystem. Due to the lack of daily fire progression maps, and to very sparse weather station coverage in our study area, we cannot address how day-to-day weather impacts fire severity in this study. With the advantage of high-frequency MODIS observations, many recent studies have begun to incorporate spatial interpolation approaches using MODIS data to characterize daily progressions of large fires [105,106]. We believe such efforts can greatly improve the understanding of spatial controls on fire severity. For example, based on Landsat-derived fire progression maps and fire weather observations at 4-km spatial resolution, Birch et al. (2015) [37] investigated the influences of vegetation, topography, and daily fire weather on severity patterns of wildfires in the Western US and reported that vegetation cover had the greatest influence on fire severity; this is quite similar to our findings in this study, as well as in our previous patch-scale analysis [43]. They also acknowledged that the coarse weather conditions may not fully reflect the influences of microscale meteorological conditions on severity patterns. The inconsistent temporal and spatial resolution among daily weather observations, vegetation, and topography can obscure the real effects of weather on fire severity.

Fire severity is a result of accumulated fire effects on forest ecosystems because the thick and moist organic layers in boreal regions can prolong the fire duration. Although we tried to balance the spatial resolution among explanatory variables, it is somewhat challenging to reflect environmental gradients and fire activities at both fine spatial and temporal scales using the data sources available to us. At the same time, because MODIS ET product updates led to considerable changes in SMA values, we found that 500 m SMA has different relative importance and influences on fire severity. Our intentions were not to arbitrarily justify which kind of ET product is more reliable for fire severity prediction, especially without sufficient validation of 500 m ET products with site-based flux observation, but any improvement in spatial resolution is valuable and further efforts are encouraged to verify the suitability of these products for specific ecosystems.

Although our study area, the Huzhong Forestry Bureau jurisdiction, is a representative forest landscape of Great Xing'an Mountains and shares a similar fire regime as other nearby areas [54,56], we could not conclude that fire severity patterns of the entire region are following the same causal mechanism at finer scales. The purpose of this study is not to establish a global prediction model for fire severity that can be generalized for all Chinese boreal ecosystems. There were fires occurring in meadow and wetland ecosystems, as well as in forests dominated by broadleaf trees in the southern part of the Great Xing'an region. We believe our results may not be suitable for predicting the severity of those kinds of wildfires. In addition, sampling data were extracted from fires occurring in summer; although the vegetation coverage and topographic variables adopted in this study are insensitive to intra-annual variability, the relationship between SMA and fire severity may change in different seasons and should be further investigated.

5. Conclusions

Although the parallel comparison of the two models did not show strictly consistent modeling relationships, the models generally demonstrated that fire severity was strongly controlled by the coverage of certain vegetation types that have high flammability or fire resistance. The topographic conditions can help determine the distribution of flammable plant types and communities. Topography can also directly influence fuel moisture and create firebreaks through the drainage systems. Remotely sensed fuel moisture proxies (such as MODIS ET products) were also proven to play important roles in modeling fire severity. These findings reveal that fire severity is predictable at the landscape scale in our study area, and its prediction can be improved by incorporating spatial variables related to fire behavior. Our study provides an overview of the hotspot areas within the landscape where severe fires are most likely distributed. Such mapping capabilities can allow managers to optimize fuel treatment strategies by considering the vegetation, topography, and spatial patterns of land surface moisture. The modeling framework employed in our study can readily incorporate new observations and simulated spatial datasets, promoting the more reliable prediction of fire severity in the future.

Acknowledgments: This work is funded by the National Key R&D Program of China (2017YFA0604403 and 2016YFA0600804), the National Natural Science Foundation of China (Project No. 31500387, 31270511, and 31470517) and the CAS Pioneer Hundred Talents Program. We thank three anonymous reviewers and academic editor for comments that improved this manuscript.

Author Contributions: L.F. and J.Y. conceived and designed the study. L.F. and M.W. analyzed data. L.F., J.Y. and M.W. wrote and revised the manuscript. Z.L. contributed to collecting data and discussing the results.

Conflicts of Interest: The authors declare no conflict of interest.

Appendix A

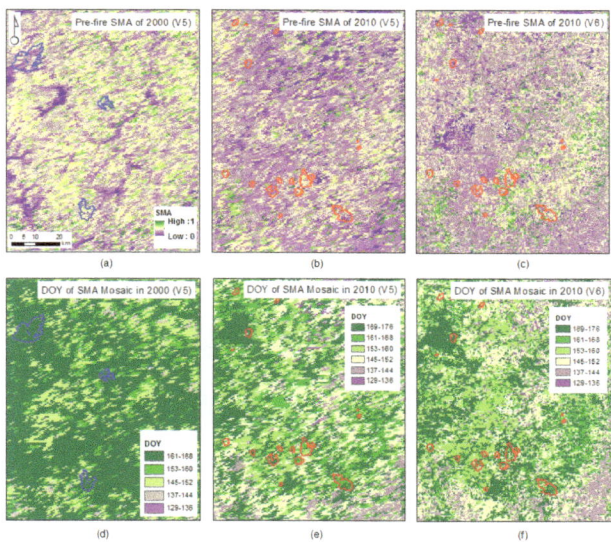

Figure A1. The integrated pre-fire surface moisture availability (SMA) of 2000 (**a**) and 2010 (**b,c**) derived from 8-day MODIS MOD16A2 Version 5 (**b**) and Version 6 (**c**) product. The good quality pixels selected from five 8-day MOD16A2 datasets (**d–f**) were composited. Fire patches of 2000 (in blue, **a,d**) and 2010 (in red, **b,d–f**) were also plotted.

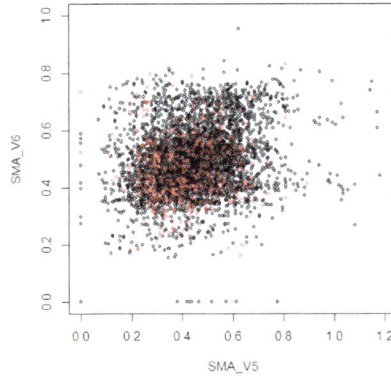

Figure A2. The scatter plot representing inconsistent surface moisture availability (SMA) values of the whole study area (black dot) and burned pixels of 2010 fires (red dot) derived from MOD16A2 Verison 5 (V5) and Version 6 (V6) with good quality.

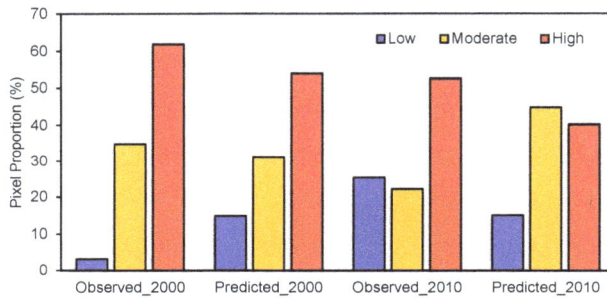

Figure A3. Comparison of pixel proportions of low, moderate and high severity levels between observed fire severity maps and predicted severity maps.

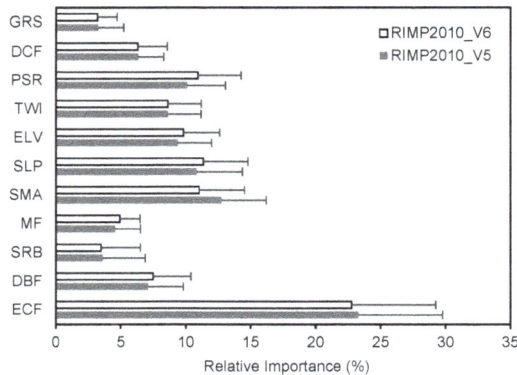

Figure A4. The relative importance (RIMP) of variables generated by 50 boosting regression tree (BRT) models using two different versions (V5 and V6) of surface moisture availability (SMA) derived from MODIS ET products. Variable abbreviations were described in Table 2.

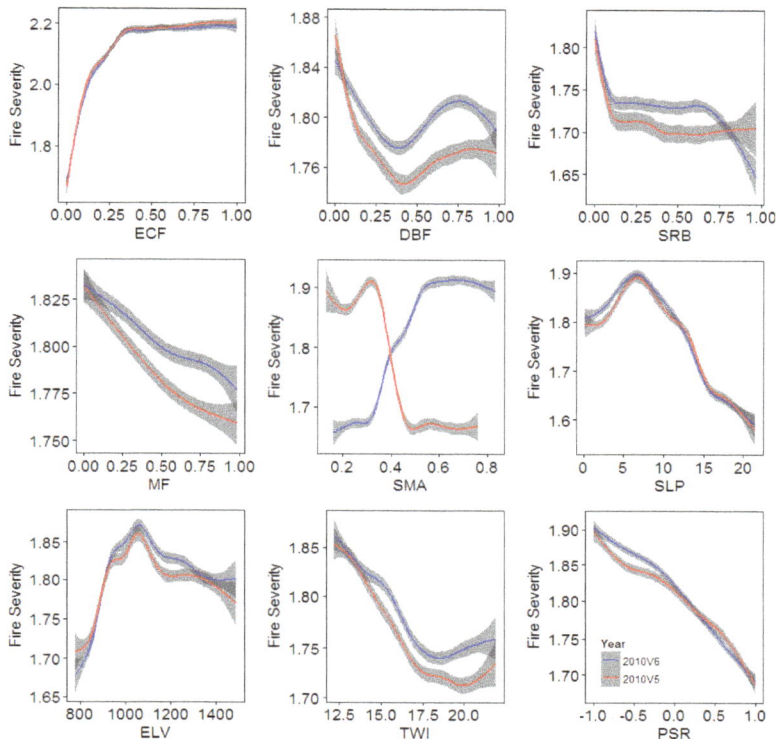

Figure A5. Partial dependence plots of nine variables on regulating fire severity in fire year 2010 using different versions of MODIS derived surface moisture availability (SMA) (Version 5 in cyan and Version 6 in red). This grouping of variables was selected as the top 9 variables following the rank as shown in *y*-axis of Figure 5. The curves demonstrate the smoothed means (solid line) and 95% confidence intervals (gray zone) of an ensemble of 50 models. Variable abbreviations were described in Table 2.

Appendix B

Table A1. The Moran's I for examining spatial autocorrelation of fire severity and explanatory variables. Variable abbreviations were described in Table 2.

Variables	Moran's I of Fires 2000	Moran's I of Fires 2010
Fire Severity	0.132	0.078
SMA_V5	0.224	0.124
ELV	0.192	0.236
PSR	0.061	0.066
SLP	0.120	0.075
TWI	0.054	0.06
ECF	0.078	0.092
DCF	0.178	0.107
DBF	0.041	0.19
MF	0.134	0.07
GRS	0.104	0.056
SRB	0.035	0.156

References

1. De Groot, W.J.; Flannigan, M.D.; Cantin, A.S. Climate change impacts on future boreal fire regimes. *For. Ecol. Manag.* **2013**, *294*, 35–44. [CrossRef]
2. Flannigan, M.; Stocks, B.; Turetsky, M.; Wotton, M. Impacts of climate change on fire activity and fire management in the circumboreal forest. *Glob. Chang. Biol.* **2009**, *15*, 549–560. [CrossRef]
3. Stephens, S.L.; Burrows, N.; Buyantuyev, A.; Gray, R.W.; Keane, R.E.; Kubian, R.; Liu, S.; Seijo, F.; Shu, L.; Tolhurst, K.G.; et al. Temperate and boreal forest mega-fires: Characteristics and challenges. *Front. Ecol. Environ.* **2014**, *12*, 115–122. [CrossRef]
4. Conard, S.G.; Solomon, A.M. Effects of wildland fire on regional and global carbon stocks in a changing environment. *Dev. Environ. Sci.* **2008**, *8*, 109–138.
5. Harden, J.; Trumbore, S.; Stocks, B.; Hirsch, A.; Gower, S.; O'neill, K.; Kasischke, E. The role of fire in the boreal carbon budget. *Glob. Chang. Biol.* **2000**, *6*, 174–184. [CrossRef]
6. Bond-Lamberty, B.; Peckham, S.D.; Gower, S.T.; Ewers, B.E. Effects of fire on regional evapotranspiration in the central Canadian boreal forest. *Glob. Chang. Biol.* **2009**, *15*, 1242–1254. [CrossRef]
7. Jin, Y.; Roy, D. Fire-induced albedo change and its radiative forcing at the surface in northern Australia. *Geophys. Res. Lett.* **2005**, *32*. [CrossRef]
8. Keane, R.E.; Agee, J.K.; Fulé, P.; Keeley, J.E.; Key, C.; Kitchen, S.G.; Miller, R.; Schulte, L.A. Ecological effects of large fires on US landscapes: Benefit or catastrophe? *Int. J. Wildland Fire* **2008**, *17*, 696–712. [CrossRef]
9. Anderson-Teixeira, K.J.; Miller, A.D.; Mohan, J.E.; Hudiburg, T.W.; Duval, B.D.; Delucia, E.H. Altered dynamics of forest recovery under a changing climate. *Glob. Chang. Biol.* **2013**, *19*, 2001–2021. [CrossRef] [PubMed]
10. Bond, W.J.; Keeley, J.E. Fire as a global 'herbivore': The ecology and evolution of flammable ecosystems. *Trends Ecol. Evol.* **2005**, *20*, 387–394. [CrossRef] [PubMed]
11. Johnstone, J.F.; Hollingsworth, T.N.; Chapin, F.S.; Mack, M.C. Changes in fire regime break the legacy lock on successional trajectories in Alaskan boreal forest. *Glob. Chang. Biol.* **2010**, *16*, 1281–1295. [CrossRef]
12. Turner, M.G.; Hargrove, W.W.; Gardner, R.H.; Romme, W.H. Effects of fire on landscape heterogeneity in Yellowstone National Park, Wyoming. *J. Veg. Sci.* **1994**, *5*, 731–742. [CrossRef]
13. Amiro, B.; Orchansky, A.; Barr, A.; Black, T.; Chambers, S.; Chapin, F., III; Goulden, M.; Litvak, M.; Liu, H.; McCaughey, J. The effect of post-fire stand age on the boreal forest energy balance. *Agric. For. Meteorol.* **2006**, *140*, 41–50. [CrossRef]
14. Beck, P.S.; Goetz, S.J.; Mack, M.C.; Alexander, H.D.; Jin, Y.; Randerson, J.T.; Loranty, M. The impacts and implications of an intensifying fire regime on Alaskan boreal forest composition and albedo. *Glob. Chang. Biol.* **2011**, *17*, 2853–2866. [CrossRef]
15. Rogers, B.M.; Soja, A.J.; Goulden, M.L.; Randerson, J.T. Influence of tree species on continental differences in boreal fires and climate feedbacks. *Nat. Geosci.* **2015**, *8*, 228–234. [CrossRef]
16. De Groot, W.J.; Cantin, A.S.; Flannigan, M.D.; Soja, A.J.; Gowman, L.M.; Newbery, A. A comparison of Canadian and Russian boreal forest fire regimes. *For. Ecol. Manag.* **2013**, *294*, 23–34. [CrossRef]
17. Running, S.W. Is global warming causing more, larger wildfires? *Science* **2006**, *313*, 927–928. [CrossRef] [PubMed]
18. Van Mantgem, P.J.; Nesmith, J.C.; Keifer, M.; Knapp, E.E.; Flint, A.; Flint, L. Climatic stress increases forest fire severity across the western United States. *Ecol. Lett.* **2013**, *16*, 1151–1156. [CrossRef] [PubMed]
19. Key, C.H.; Benson, N.C. *Landscape Assessment (LA). Firemon: Fire Effects Monitoring and Inventory System*; U.S. Department of Agriculture, Forest Service, Rocky Mountain Research Station: Fort Collins, CO, USA, 2006.
20. Lentile, L.B.; Holden, Z.A.; Smith, A.M.S.; Falkowski, M.J.; Hudak, A.T.; Morgan, P.; Lewis, S.A.; Gessler, P.E.; Benson, N.C. Remote sensing techniques to assess active fire characteristics and post-fire effects. *Int. J. Wildland Fire* **2006**, *15*, 319–345. [CrossRef]
21. De Santis, A.; Chuvieco, E. Geocbi: A modified version of the composite burn index for the initial assessment of the short-term burn severity from remotely sensed data. *Remote Sens. Environ.* **2009**, *113*, 554–562. [CrossRef]

22. French, N.H.F.; Kasischke, E.S.; Hall, R.J.; Murphy, K.A.; Verbyla, D.L.; Hoy, E.E.; Allen, J.L. Using landsat data to assess fire and burn severity in the north American boreal forest region: An overview and summary of results. *Int. J. Wildland Fire* **2008**, *17*, 443–462. [CrossRef]

23. Hollingsworth, T.N.; Johnstone, J.F.; Bernhardt, E.L.; Chapin, F.S. Fire severity filters regeneration traits to shape community assembly in Alaska's boreal forest. *PLoS ONE* **2013**, *8*, e56033. [CrossRef] [PubMed]

24. Liu, Z.; Yang, J. Quantifying ecological drivers of ecosystem productivity of the early-successional boreal *Larix gmelinii* forest. *Ecosphere* **2014**, *5*, 1–16. [CrossRef]

25. Turner, M.G. Disturbance and landscape dynamics in a changing world. *Ecology* **2010**, *91*, 2833–2849. [CrossRef] [PubMed]

26. Rollins, M.G.; Keane, R.E.; Parsons, R.A. Mapping fuels and fire regimes using remote sensing, ecosystem simulation, and gradient modeling. *Ecol. Appl.* **2004**, *14*, 75–95. [CrossRef]

27. Morgan, P.; Keane, R.E.; Dillon, G.K.; Jain, T.B.; Hudak, A.T.; Karau, E.C.; Sikkink, P.G.; Holden, Z.A.; Strand, E.K. Challenges of assessing fire and burn severity using field measures, remote sensing and modelling. *Int. J. Wildland Fire* **2014**, *23*, 1045–1460. [CrossRef]

28. Fang, L.; Yang, J. Atmospheric effects on the performance and threshold extrapolation of multi-temporal landsat derived dNBR for burn severity assessment. *Int. J. Appl. Earth Obs. Geoinf.* **2014**, *33*, 10–20. [CrossRef]

29. Keane, R.E.; Drury, S.A.; Karau, E.C.; Hessburg, P.F.; Reynolds, K.M. A method for mapping fire hazard and risk across multiple scales and its application in fire management. *Ecol. Model.* **2010**, *221*, 2–18. [CrossRef]

30. Stephens, S.L.; Moghaddas, J.J.; Edminster, C.; Fiedler, C.E.; Haase, S.; Harrington, M.; Keeley, J.E.; Knapp, E.E.; McIver, J.D.; Metlen, K.; et al. Fire treatment effects on vegetation structure, fuels, and potential fire severity in western U.S. Forests. *Ecol. Appl.* **2009**, *19*, 305–320. [CrossRef] [PubMed]

31. Whitlock, C.; Higuera, P.E.; McWethy, D.B.; Briles, C.E. Paleoecological perspectives on fire ecology: Revisiting the fire-regime concept. *Open Ecol. J.* **2010**, *3*, 6–23. [CrossRef]

32. Whelan, R.J. *The Ecology of Fire*; Cambridge University Press: Cambridge, UK, 1995.

33. Cary, G.J.; Keane, R.E.; Gardner, R.H.; Lavorel, S.; Flannigan, M.D.; Davies, I.D.; Li, C.; Lenihan, J.M.; Rupp, T.S.; Mouillot, F. Comparison of the sensitivity of landscape-fire-succession models to variation in terrain, fuel pattern, climate and weather. *Landsc. Ecol.* **2006**, *21*, 121–137. [CrossRef]

34. Karau, E.C.; Keane, R.E. Burn severity mapping using simulation modelling and satellite imagery. *Int. J. Wildland Fire* **2010**, *19*, 710–724. [CrossRef]

35. Keane, R.E.; Cary, G.J.; Davies, I.D.; Flannigan, M.D.; Gardner, R.H.; Lavorel, S.; Lenihan, J.M.; Li, C.; Rupp, T.S. A classification of landscape fire succession models: Spatial simulations of fire and vegetation dynamics. *Ecol. Model.* **2004**, *179*, 3–27. [CrossRef]

36. Kane, V.R.; Cansler, C.A.; Povak, N.A.; Kane, J.T.; McGaughey, R.J.; Lutz, J.A.; Churchill, D.J.; North, M.P. Mixed severity fire effects within the rim fire: Relative importance of local climate, fire weather, topography, and forest structure. *For. Ecol. Manag.* **2015**, *358*, 62–79. [CrossRef]

37. Birch, D.S.; Morgan, P.; Kolden, C.A.; Abatzoglou, J.T.; Dillon, G.K.; Hudak, A.T.; Smith, A.M.S. Vegetation, topography and daily weather influenced burn severity in central Idaho and western Montana forests. *Ecosphere* **2015**, *6*, 1–23. [CrossRef]

38. Estes, B.L.; Knapp, E.E.; Skinner, C.N.; Miller, J.D.; Preisler, H.K. Factors influencing fire severity under moderate burning conditions in the Klamath Mountains, northern California, USA. *Ecosphere* **2017**, *8*. [CrossRef]

39. Alexander, J.D.; Seavy, N.E.; Ralph, C.J.; Hogoboom, B. Vegetation and topographical correlates of fire severity from two fires in the Klamath-Siskiyou region of Oregon and California. *Int. J. Wildland Fire* **2006**, *15*, 237–245. [CrossRef]

40. Lentile, L.B.; Smith, F.W.; Shepperd, W.D. Influence of topography and forest structure on patterns of mixed severity fire in ponderosa pine forests of the South Dakota Black Hills, USA. *Int. J. Wildland Fire* **2006**, *15*, 557–566. [CrossRef]

41. Peters, D.P.; Pielke, R.A., Sr.; Bestelmeyer, B.T.; Allen, C.D.; Munson-McGee, S.; Havstad, K.M. Cross-scale interactions, nonlinearities, and forecasting catastrophic events. *Proc. Natl. Acad. Sci. USA* **2004**, *101*, 15130–15135. [CrossRef] [PubMed]

42. Dillon, G.K.; Holden, Z.A.; Morgan, P.; Crimmins, M.A.; Heyerdahl, E.K.; Luce, C.H. Both topography and climate affected forest and woodland burn severity in two regions of the western US, 1984 to 2006. *Ecosphere* **2011**, *2*, 1–33. [CrossRef]

43. Fang, L.; Yang, J.; Zu, J.; Li, G.; Zhang, J. Quantifying influences and relative importance of fire weather, topography, and vegetation on fire size and fire severity in a Chinese boreal forest landscape. *For. Ecol. Manag.* **2015**, *356*, 2–12. [CrossRef]

44. Parks, S.A.; Parisien, M.A.; Miller, C.; Dobrowski, S.Z. Fire activity and severity in the western US vary along proxy gradients representing fuel amount and fuel moisture. *PLoS ONE* **2014**, *9*, e99699. [CrossRef] [PubMed]

45. Viedma, O.; Quesada, J.; Torres, I.; De Santis, A.; Moreno, J.M. Fire severity in a large fire in a pinus pinaster forest is highly predictable from burning conditions, stand structure, and topography. *Ecosystems* **2014**, *18*, 237–250. [CrossRef]

46. Cansler, C.A.; McKenzie, D. Climate, fire size, and biophysical setting control fire severity and spatial pattern in the northern Cascade Range, USA. *Ecol. Appl.* **2014**, *24*, 1037–1056. [CrossRef] [PubMed]

47. Clarke, P.J.; Knox, K.J.E.; Bradstock, R.A.; Munoz-Robles, C.; Kumar, L.; Ward, D. Vegetation, terrain and fire history shape the impact of extreme weather on fire severity and ecosystem response. *J. Veg. Sci.* **2014**, *25*, 1033–1044. [CrossRef]

48. Qi, Y.; Dennison, P.E.; Spencer, J.; Riano, D. Monitoring live fuel moisture using soil moisture and remote sensing proxies. *Fire Ecol.* **2012**, *8*, 71–87.

49. Yebra, M.; Dennison, P.E.; Chuvieco, E.; Riaño, D.; Zylstra, P.; Hunt, E.R.; Danson, F.M.; Qi, Y.; Jurdao, S. A global review of remote sensing of live fuel moisture content for fire danger assessment: Moving towards operational products. *Remote Sens. Environ.* **2013**, *136*, 455–468. [CrossRef]

50. Barrett, K.; Kasischke, E.; McGuire, A.; Turetsky, M.; Kane, E. Modeling fire severity in black spruce stands in the Alaskan boreal forest using spectral and non-spectral geospatial data. *Remote Sens. Environ.* **2010**, *114*, 1494–1503. [CrossRef]

51. Holden, Z.A.; Morgan, P.; Evans, J.S. A predictive model of burn severity based on 20-year satellite-inferred burn severity data in a large southwestern US wilderness area. *For. Ecol. Manag.* **2009**, *258*, 2399–2406.

52. Kane, V.R.; Lutz, J.A.; Alina Cansler, C.; Povak, N.A.; Churchill, D.J.; Smith, D.F.; Kane, J.T.; North, M.P. Water balance and topography predict fire and forest structure patterns. *For. Ecol. Manag.* **2015**, *338*, 1–13. [CrossRef]

53. Ponomarev, E.; Kharuk, V.; Ranson, K. Wildfires dynamics in siberian larch forests. *Forests* **2016**, *7*, 125. [CrossRef]

54. Fan, Q.; Wang, C.; Zhang, D.; Zang, S. Environmental influences on forest fire regime in the Greater Hinggan Mountains, northeast China. *Forests* **2017**, *8*, 372. [CrossRef]

55. Fang, J.; Chen, A.; Peng, C.; Zhao, S.; Ci, L. Changes in forest biomass carbon storage in China between 1949 and 1998. *Science* **2001**, *292*, 2320–2322. [CrossRef] [PubMed]

56. Liu, Z.; Yang, J.; Chang, Y.; Weisberg, P.J.; He, H.S. Spatial patterns and drivers of fire occurrence and its future trend under climate change in a boreal forest of northeast China. *Glob. Chang. Biol.* **2012**, *18*, 2041–2056. [CrossRef]

57. Wang, C.; Gower, S.T.; Wang, Y.; Zhao, H.; Yan, P.; Lamberty, B.P. The influence of fire on carbon distribution and net primary production of boreal larix gmelinii forests in north-eastern China. *Glob. Chang. Biol.* **2001**, *7*, 719–730. [CrossRef]

58. Liu, Z.; Yang, J.; He, H.S. Identifying the threshold of dominant controls on fire spread in a boreal forest landscape of northeast China. *PLoS ONE* **2013**, *8*, e55618. [CrossRef] [PubMed]

59. Cai, W.; Yang, J.; Liu, Z.; Hu, Y.; Weisberg, P.J. Post-fire tree recruitment of a boreal larch forest in northeast China. *For. Ecol. Manag.* **2013**, *307*, 20–29. [CrossRef]

60. Kong, J.; Yang, J.; Chu, H.; Xiang, X. Effects of wildfire and topography on soil nitrogen availability in a boreal larch forest of northeastern China. *Int. J. Wildland Fire* **2015**, *24*, 433–442. [CrossRef]

61. Chander, G.; Markham, B.L.; Helder, D.L. Summary of current radiometric calibration coefficients for landsat MSS, TM, ETM+, and EO-1 ALI sensors. *Remote Sens. Environ.* **2009**, *113*, 893–903. [CrossRef]

62. Vermote, E.F.; Tanre, D.; Deuze, J.L.; Herman, M.; Morcette, J.J. Second simulation of the satellite signal in the solar spectrum, 6S: An overview. *IEEE Trans. Geosci. Remote Sens.* **1997**, *35*, 675–686. [CrossRef]

63. Canty, M.J.; Nielsen, A.A. Automatic radiometric normalization of multitemporal satellite imagery with the iteratively re-weighted mad transformation. *Remote Sens. Environ.* **2008**, *112*, 1025–1036. [CrossRef]

64. Mu, Q.; Zhao, M.; Running, S.W. Improvements to a modis global terrestrial evapotranspiration algorithm. *Remote Sens. Environ.* **2011**, *115*, 1781–1800. [CrossRef]

65. Kim, H.W.; Hwang, K.; Mu, Q.; Lee, S.O.; Choi, M. Validation of modis 16 global terrestrial evapotranspiration products in various climates and land cover types in Asia. *KSCE J. Civ. Eng.* **2012**, *16*, 229–238. [CrossRef]
66. Wang, L.; Hunt, E.R.; Qu, J.J.; Hao, X.; Daughtry, C.S.T. Remote sensing of fuel moisture content from ratios of narrow-band vegetation water and dry-matter indices. *Remote Sens. Environ.* **2013**, *129*, 103–110. [CrossRef]
67. Running, S.W.; Mu, Q.; Zhao, M.; Moreno, A. *Modis Global Terrestrial Evapotranspiration (ET) Product (NASA MOD16A2/A3) NASA Earth Observing System Modis Land Algorithm*; NASA: Washington, DC, USA, 2017.
68. Masek, J.G.; Huang, C.; Wolfe, R.; Cohen, W.; Hall, F.; Kutler, J.; Nelson, P. North American forest disturbance mapped from a decadal landsat record. *Remote Sens. Environ.* **2008**, *112*, 2914–2926. [CrossRef]
69. Friedl, M.A.; Brodley, C.E. Decision tree classification of land cover from remotely sensed data. *Remote Sens. Environ.* **1997**, *61*, 399–409. [CrossRef]
70. Yang, J.; He, H.S.; Shifley, S.R.; Gustafson, E.J. Spatial patterns of modern period human-caused fire occurrence in the Missouri Ozark Highlands. *For. Sci.* **2007**, *53*, 1–15.
71. Quinn, P.; Beven, K. Spatial and temporal predictions of soil moisture dynamics, runoff, variable source areas and evapotranspiration for Plynlimon, Mid-Wales. *Hydrol. Process.* **1993**, *7*, 425–448. [CrossRef]
72. Sörensen, R.; Zinko, U.; Seibert, J. On the calculation of the topographic wetness index: Evaluation of different methods based on field observations. *Hydrol. Earth Syst. Sci. Discuss.* **2006**, *10*, 101–112. [CrossRef]
73. Qin, C.; Zhu, A.; Yang, L.; Li, B.; Pei, T. Topographic wetness index computed using multiple flow direction algorithm and local maximum downslope gradient. In Proceedings of the 7th International Workshop of Geographical Information System, Beijing, China, 12–14 September 2007.
74. Wotton, B.M. Interpreting and using outputs from the Canadian forest fire danger rating system in research applications. *Environ. Ecol. Stat.* **2009**, *16*, 107–131. [CrossRef]
75. Keane, R.E. *Wildland Fuel Fundamentals and Applications*; Springer: Berlin, Germany, 2015.
76. Carlson, T.N. Regional-scale estimates of surface moisture availability and thermal inertia using remote thermal measurements. *Remote Sens. Rev.* **1986**, *1*, 197–247. [CrossRef]
77. Ha, W.; Gowda, P.H.; Howell, T.A. A review of downscaling methods for remote sensing-based irrigation management: Part I. *Irrig. Sci.* **2012**, *31*, 831–850. [CrossRef]
78. De'ath, G. Boosted trees for ecological modeling and prediction. *Ecology* **2007**, *88*, 243–251. [CrossRef]
79. Elith, J.; Leathwick, J.R.; Hastie, T. A working guide to boosted regression trees. *J. Anim. Ecol.* **2008**, *77*, 802–813. [CrossRef] [PubMed]
80. Paradis, E.; Claude, J.; Strimmer, K. Ape: Analyses of phylogenetics and evolution in R language. *Bioinformatics* **2004**, *20*, 289–290. [CrossRef] [PubMed]
81. Legendre, P. Spatial autocorrelation: Trouble or new paradigm? *Ecology* **1993**, *74*, 1659–1673. [CrossRef]
82. Ferster, C.; Eskelson, B.; Andison, D.; LeMay, V. Vegetation Mortality within Natural Wildfire Events in the Western Canadian Boreal Forest: What Burns and Why? *Forests* **2016**, *7*, 187. [CrossRef]
83. Whitman, E.; Parisien, M.-A.; Thompson, D.; Hall, R.; Skakun, R.; Flannigan, M. Variability and drivers of burn severity in the northwestern Canadian boreal forest. *Ecosphere* **2018**, *9*, e02128. [CrossRef]
84. Bigler, C.; Kulakowski, D.; Veblen, T.T. Multiple disturbance interactions and drought influence fire severity in Rocky Mountain subalpine forests. *Ecology* **2005**, *86*, 3018–3029. [CrossRef]
85. Schoennagel, T.; Veblen, T.T.; Romme, W.H. The interaction of fire, fuels, and climate across Rocky Mountain forests. *BioScience* **2004**, *54*, 661–676. [CrossRef]
86. Rupp, T.S.; Starfield, A.M.; Chapin, F.S.; Duffy, P. Modeling the impact of black spruce on the fire regime of Alaskan boreal forest. *Clim. Chang.* **2002**, *55*, 213–233. [CrossRef]
87. Johnstone, J.F.; Rupp, T.S.; Olson, M.; Verbyla, D. Modeling impacts of fire severity on successional trajectories and future fire behavior in Alaskan boreal forests. *Landsc. Ecol.* **2011**, *26*, 487–500. [CrossRef]
88. Kobayashi, M.; Yury, P.N.; Olga, A.Z.; Takuya, K.; Yojiro, M.; Toshiya, Y.; Fuyuki, S.; Kaichiro, S.; Takayoshi, K. Regeneration after forest fires in mixed conifer broad-leaved forests of the amur region in far eastern Russia: The relationship between species specific traits against fire and recent fire regimes. *Eurasian J. For. Res.* **2007**, *10*, 51–58.
89. Zhao, F.; Shu, L.; Wang, M.; Liu, B.; Yang, L. Influencing factors on early vegetation restoration in burned area of Pinus Pumila—Larch forest. *Acta Ecol. Sin.* **2012**, *32*, 57–61. [CrossRef]
90. Zhao, F.; Shu, L.; Wang, Q.; Wang, M.; Tian, X. Emissions of volatile organic compounds from heated needles and twigs of Pinus Pumila. *J. For. Res.* **2011**, *22*, 243–248. [CrossRef]

91. Ryan, K.C. Dynamic interactions between forest structure and fire behavior in boreal ecosystems. *Silva Fenn.* **2002**, *36*, 13–39. [CrossRef]

92. Kopecký, M.; Čížková, Š. Using topographic wetness index in vegetation ecology: Does the algorithm matter? *Appl. Veg. Sci.* **2010**, *13*, 450–459. [CrossRef]

93. Xiao, J.; Zhuang, Q. Drought effects on large fire activity in Canadian and Alaskan forests. *Environ. Res. Lett.* **2007**, *2*, 044003. [CrossRef]

94. Parks, S.A.; Miller, C.; Nelson, C.R.; Holden, Z.A. Previous fires moderate burn severity of subsequent wildland fires in two large western US wilderness areas. *Ecosystems* **2014**, *17*, 29–42. [CrossRef]

95. Li, X.; He, H.S.; Wu, Z.; Liang, Y.; Schneiderman, J.E. Comparing effects of climate warming, fire, and timber harvesting on a boreal forest landscape in northeastern China. *PLoS ONE* **2013**, *8*, e59747. [CrossRef] [PubMed]

96. Arroyo, L.A.; Pascual, C.; Manzanera, J.A. Fire models and methods to map fuel types: The role of remote sensing. *For. Ecol. Manag.* **2008**, *256*, 1239–1252. [CrossRef]

97. Tian, X.; Mcrae, D.J.; Shu, L.; Wang, M.; Li, H. Satellite remote-sensing technologies used in forest fire management. *J. For. Res.* **2005**, *16*, 73–78.

98. Pierce, A.D.; Farris, C.A.; Taylor, A.H. Use of random forests for modeling and mapping forest canopy fuels for fire behavior analysis in Lassen Volcanic National Park, California, USA. *For. Ecol. Manag.* **2012**, *279*, 77–89. [CrossRef]

99. Song, C. Optical remote sensing of forest leaf area index and biomass. *Prog. Phys. Geogr.* **2012**, *37*, 98–113. [CrossRef]

100. Keane, R.E.; Burgan, R.; van Wagtendonk, J. Mapping wildland fuels for fire management across multiple scales: Integrating remote sensing, GIS, and biophysical modeling. *Int. J. Wildland Fire* **2001**, *10*, 301–319. [CrossRef]

101. Cyr, D.; Gauthier, S.; Bergeron, Y. Scale-dependent determinants of heterogeneity in fire frequency in a coniferous boreal forest of eastern Canada. *Landsc. Ecol.* **2007**, *22*, 1325–1339. [CrossRef]

102. Parisien, M.A.; Parks, S.A.; Krawchuk, M.A.; Flannigan, M.D.; Bowman, L.M.; Moritz, M.A. Scale-dependent controls on the area burned in the boreal forest of Canada, 1980–2005. *Ecol. Appl.* **2011**, *21*, 789–805. [CrossRef] [PubMed]

103. Parks, S.A.; Parisien, M.-A.; Miller, C. Multi-scale evaluation of the environmental controls on burn probability in a southern Sierra Nevada landscape. *Int. J. Wildland Fire* **2011**, *20*, 815–828. [CrossRef]

104. Wu, Z.; He, H.S.; Bobryk, C.W.; Liang, Y. Scale effects of vegetation and topography on burn severity under prevailing fire weather conditions in boreal forest landscapes of northeastern China. *Scand. J. For. Res.* **2014**, *29*, 60–70. [CrossRef]

105. Veraverbeke, S.; Sedano, F.; Hook, S.J.; Randerson, J.T.; Jin, Y.; Rogers, B.M. Mapping the daily progression of large wildland fires using modis active fire data. *Int. J. Wildland Fire* **2014**, *23*, 655–667. [CrossRef]

106. Parks, S.A. Mapping day-of-burning with coarse-resolution satellite fire-detection data. *Int. J. Wildland Fire* **2014**, *23*, 215–223. [CrossRef]

MDPI

Article

Ecoregional Patterns of Spruce Budworm—Wildfire Interactions in Central Canada's Forests

Jean-Noël Candau *, Richard A. Fleming and Xianli Wang

Natural Resources Canada, Canadian Forest Service, Great Lakes Forestry Centre,
Sault Ste Marie, ON P6A 2E5, Canada; Rich.Fleming@canada.ca (R.A.F.); Xianli.Wang@canada.ca (X.W.)
* Correspondence: Jean-Noel.Candau@Canada.ca; Tel.: +1-705-541-5759

Received: 15 February 2018; Accepted: 13 March 2018; Published: 14 March 2018

Abstract: Wildfires and outbreaks of the spruce budworm, *Choristoneura fumiferana* (Clem.), are the two dominant natural disturbances in Canada's boreal forest. While both disturbances have specific impacts on forest ecosystems, it is increasingly recognized that their interactions also have the potential for non-linear behavior and long-lasting legacies on forest ecosystems' structures and functions. Previously, we showed that, in central Canada, fires occurred with a disproportionately higher frequency during a 'window of opportunity' following spruce budworm defoliation. In this study, we use Ontario's spatial databases for large fires and spruce budworm defoliation to locate where these two disturbances likely interacted. Classification tree and Random Forest procedures were then applied to find how spruce budworm defoliation history, climate, and forest conditions best predict the location of such budworm–fire interactions. Results indicate that such interactions likely occurred in areas geographically bound by hardwood content in the south, the prevalence of the three major spruce budworm host species (balsam fir, white spruce and black spruce) in the north, and climate moisture in the west. The occurrence of a spruce budworm–fire interaction inside these boundaries is related to the frequency of spruce budworm defoliation. These patterns provide a means of distinguishing regions where spruce budworm attacks are likely to increase fire risk.

Keywords: spruce budworm defoliation; forest fire; disturbance interactions; forest composition; weather

1. Introduction

Historically, two main types of natural disturbances have dominated Canada's boreal forest: wildfire and outbreaks of spruce budworm (SBW), *Choristoneura fumiferana* (Clem.) [1,2]. While each type has specific impacts on forest composition and dynamics, biogeochemical cycling and numerous ecological processes, there is an increasing recognition that the interaction of these types of disturbance can also have dramatic long-term effects on the ecosystem's structure and functioning [3,4].

As climate change is expected to affect both types of disturbance regimes (e.g., [5,6]), understanding their interactions will also be critical for appropriate risk-assessment and management planning in the future. In the simplest and most direct form of these interactions, a warmer, drier climate is expected to increase the tendency of SBW-killed stands to burn [7]. This effect would likely be magnified by the fact that the spatial extent of SBW outbreaks, and thus the availability of SBW-attacked stands, already much greater than the extent of fires [2], may increase with climate change [5].

When considered together, these factors suggest that in a drier, warmer climate, the boreal forest may experience accelerated carbon releases due to the interaction of SBW and wildfire disturbance regimes. Indeed, recent carbon budget studies have shown that climate change-induced modifications of disturbance regimes have critical impacts on the net atmospheric carbon exchange [8,9].

The idea that SBW-damaged stands represent an increased risk of wildfire has long been based on anecdotal observations of severe forest fires occurring shortly after SBW outbreaks (e.g., [10–13]).

The first attempt to examine empirical evidence of an interaction between SBW defoliation and fire was carried out through a series of experimental burns in northern Ontario from 1976 to 1982 [14,15]. Although the number of experimental plots successfully burnt was low ($n = 5$) and no control plots were established, the results suggested that the abundance of 'ladder fuels' could make SBW-killed stands an extreme fire risk [14,15]. 'Ladder fuels' are dead and broken treetops and branches that became snagged and entangled by other branches before falling completely to the ground. These 'ladder fuels' present a vertical structure that increases the risk of conducting relatively harmless surface fires up into the crown where the fires can become much more dangerous. In practice, wildfire risk assessment uses a separate class of fuel types for SBW-killed conifers (i.e., M3 and M4: Dead Balsam Fir Mixedwood, leafless and Green respectively [7]). The presence of this class testifies to the importance of the relation between SBW damage and the risk of wildfire for fire managers, but the calculation of risk is still based almost entirely on the results from Stocks' single experiment.

Following Stocks' experiment, Péch [16] conducted a long-term monitoring of fuel distribution in plots affected by extensive spruce budworm defoliation in Cape Breton (Nova Scotia). The study showed that, despite heavy mortality, there was no accumulation of fine fuels, and fuel loadings were decreasing after the outbreak except for the larger, less flammable, size class. The risk of fire was further decreased by the proliferation of new growth quickly after the stand was opened. Differences between Stocks [14,15] and Péch's [16] results were attributed to cooler, wetter weather in Cape Breton compared to Ontario that would accelerate the decomposition of spruce budworm-related fuels and decrease the overall fire risk. The comparison of these studies highlighted the importance of local conditions in mediating the influence of insect damage on subsequent fire risk.

Later, a statistical analysis of the spatio-temporal patterns of spruce budworm defoliation and large (i.e., crown) wildfires in Ontario [17] revealed that, within the area defoliated at least once by spruce budworm since 1941, (1) areas that suffered moderate frequencies of defoliation (9–11 years) were the most likely to be burnt; (2) large fires (>2 km^2) rarely occur shortly before defoliation (presumably because spruce budworm populations cannot reach outbreak level in burnt stands); (3) large fires tended to occur disproportionately more often during a 'window of opportunity' of 3–9 years after a spruce budworm outbreak. Fleming et al. [17] hypothesized that this 'window of opportunity' was related to the accumulation of 'ladder fuel' from the breakage of SBW-killed top trees and windthrow of SBW-killed trees. To test this hypothesis, Watt [18] investigated differences in the vertical fuel structure of boreal mixedwood stands that suffered varying durations of SBW defoliation. The results of this investigation show that vertical fuel continuity (i.e., "ladder fuel") increases with the duration of continuous defoliation. Using his estimates of stand fuel characteristics in a crown fire model, Watt was then able to demonstrate how the potential for surface fires to reach the canopy and become crown fires increases with the duration of SBW defoliation. More recently, in a landscape-scale analysis of SBW–fire interaction in Central Canada, James et al. [19] found that lagged cumulative defoliation increased the risk of fire ignition, thus supporting further Fleming et al.'s [17] conceptual model.

The notion of "window of opportunity" is a key component of Fleming et al.'s [17] model. In their analysis, they found that the timing and the duration of this time window both vary geographically, presumably because regional biogeographical factors affect fuel dynamics after defoliation. James et al. [19] reached the same conclusion. More specifically, regional differences in the lag between the end of the defoliation and the start of the "window of opportunity" are thought to be related to the varying speeds at which SBW-killed trees break down depending on weather and forest composition while the end of the "window of opportunity" (i.e., a reduction of the fire risk to each pre-defoliation level) might be more related to decomposition of the accumulated fine fuel and the 'greening up' of the understory as herbaceous plants and suppressed trees fill in the opening created in the stand by SBW defoliation.

In this paper, we expand on the analyses of Fleming et al. [17] and James et al. [19] by examining how the timing and duration of the "window of opportunity" vary over the landscape as a function of factors related to insect damage history, climate, and fire.

2. Data and Methods

Our study area covers a latitudinal belt that spans across the province of Ontario between the 45th and 52nd parallels. The area corresponds to an updated version of what Candau et al. [20] referred to as the 'defoliation belt', i.e., the area within which moderate to severe SBW defoliation occurred at least once between 1941 and 2005. Large-scale, spatially explicit data of historical SBW defoliation, large fires (>2 km²), forest composition and climate were compiled for the entire study area. Since 2005, SBW defoliation in Central Canada has been limited to small areas and sporadic but historical patterns suggest that a new outbreak is to be expected in the next few years [20]. All the data were entered into a Geographic Information System (GIS) and transformed into a 10-km grid before analysis. Once areas covering large lakes and those with missing data were removed, the remaining study area covered 386×10^3 km².

2.1. Spruce Budworm Defoliation Data

The Forest Insect and Disease Survey (FIDS) of the Canadian Forest Service conducted aerial reconnaissance of large-scale defoliation events throughout Ontario's productive, exploitable forest from 1941 to 2005. Each year, survey flights are organized as soon as the current season's defoliation is completed, usually in mid- to late-July. In the aircraft, areas within which defoliation has occurred are sketched on 1:125,000 or 1:250,000 maps [21]. For each area, the level of defoliation is recorded as light, moderate or severe, based on the percentage of new foliage lost (0–25%, 26–75%, 76–100%, respectively). All the maps collected one year were later compiled and transferred to smaller scale maps (e.g., 1:600,000). In the early 1990s, annual maps of defoliation since 1941 were digitized and stored into a spatial database. Since then, areas sketched on 1:125,000 or 1:250,000 maps are directly digitized and stored in the database. Records of light defoliation are often considered relatively unreliable [21,22], so only records of moderate and severe defoliation were included in the present analyses.

The map of the frequency of defoliation by spruce budworm converted to a 10-km grid (Figure 1A) shows the patterns reported in Candau et al. [20]: areas defoliated at least one year during the period 1941–2005 extend over a continuous east–west 'defoliation belt' divided into three zones centered around 'hot spots' of frequent defoliation which are separated longitudinally by two corridors where defoliation is less frequent. In a previous study, Candau and Fleming [23] showed that areas of high defoliation frequency were associated with dry Junes and cool springs. Conversely, low frequencies were associated with cold winters in the north and a low abundance of host species in the south.

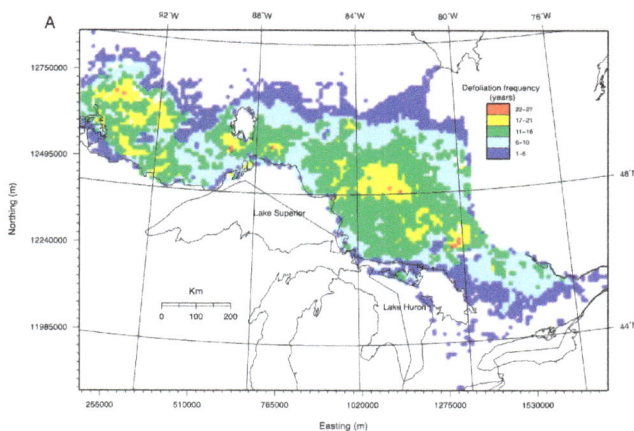

Figure 1. *Cont.*

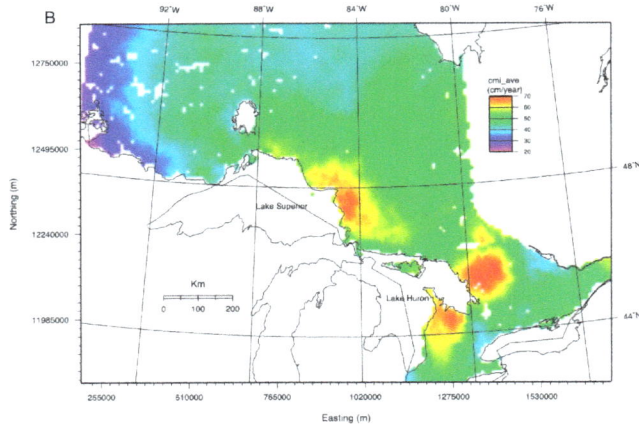

Figure 1. (**A**) Frequency (number of years) of moderate to severe defoliation by spruce budworm between 1941 and 2005 on a 10-km grid; (**B**) Average climate moisture index (cm/year) between 1 November and 31 October on a 10-km grid from 1941 to 2003. (See text for calculating algorithm).

2.2. Fire Data

Fire data were extracted from the same spatial database used in Fleming et al. [17] updated to 2005. Two different sources were used to compile fire data. B.J. Stocks (Canadian Forest Service) provided records extracted from microfiche compiled by the Ontario Ministry of Natural Resources for the period 1941–1979. The rest of the records (i.e., for fires from 1980 to 2005) were extracted from an updated version of the Canadian Large Fire Database [24]. Although the database contains less than 5% of fires reported in Canada, these large fires account for more than 97% of the area burned and thus represent the vast majority of the fire impacts [24].

Independently of their origin, fire polygons included in the final database are fire perimeters mapped from aerial photography, satellite imagery, and aircraft observation (more recently using global positioning system units). Indeed, although most large fires leave unburned islands [25], only a small percentage of the polygons have this information [26], as only the outside perimeter was mapped for most of the fires. Fire data accuracy has likely improved through time, as recent technological developments facilitate mapping and increase accuracy. The area where fires were recorded has probably varied considerably between 1941–2005 with new areas being monitored, particularly in the north of the province. However, most of these areas are located north of the spruce budworm 'defoliation belt' so they were excluded in our analyses.

Previously, Fleming et al. [17] showed that, inside the spruce budworm defoliation belt, large fires occurred disproportionately more often during a period of a few years after a spruce budworm outbreak. This period of time, hereafter called the spruce budworm–fire interaction period (or SFIP), during which fire probability increases in areas previously defoliated does not occur immediately after the defoliation ends but with a delay of a few years. Therefore, SFIPs can be characterized by their duration and by the delay between the end of the outbreak (defined as the last year of moderate–severe defoliation there) and the onset of the SFIP. Both the duration and the delay of the SFIP, varied geographically. In the eastern part of the defoliation belt, the SFIP occurs between 3 and 6 years after an outbreak; in the western part of the defoliation belt, it occurs between 6 and 16 years after an outbreak; in the central part, it occurs between 4 and 9 years.

In this paper, we use both spatial and temporal conditions to define a 'likely interaction' between spruce budworm defoliation and a large fire. First, there must be geographical overlap between the fire and defoliation. Second, the fire must have occurred within the SFIP for the region of concern.

The spruce budworm defoliation and fire spatial databases were merged so every fire could be assessed against these two conditions. Fires that met both conditions were classified as 'likely interaction'. Clearly, this classification system has shortcomings. A fire falling within the SFIP may have occurred regardless of any previous defoliation. On the other hand, a fire occurring after the SFIP finished may have been promoted by longer-term effects of defoliation than were recognized by Fleming et al. [17]. A fire starting inside (or outside) a defoliated area may have spread widely in a non-defoliated (or defoliated) area. Perhaps, without starting where it did, no fire would have occurred at all. Perhaps there were special conditions (e.g., previous defoliation) that allowed the fire to spread as well as it did. The problem is that even with 'boots on the ground' it is often extremely difficult to distinguish between these possibilities. Consequently, we view our approach to defining a 'likely interaction' as a practical compromise, which is not without difficulties. The opposite interaction (of fire affecting the likelihood of subsequent spruce budworm defoliation) has received more attention [17,27–29].

Several hypotheses have been proposed to explain the spatial variation in the timing and duration of SFIPs. In particular, forest composition and climate may play an important role by affecting the decomposition rate of surface fuel and the duration of vertical continuity of 'ladder fuel' after a spruce budworm outbreak.

2.3. Forest Data

Forest composition has been found to affect the distribution of spruce budworm defoliation [23] and fire hazard [30] and was thus considered as a potential factor in explaining the location of spruce budworm–fire interactions. The northern part of the study area is part of the Boreal forest region, dominated by conifer (mainly spruce, jack pine and fir), while the southern part is in the Great Lakes—St Lawrence region, dominated by hardwoods (mainly tolerant hardwoods, birch, poplar). Forest data were extracted from the forest resource inventory (FRI) conducted by the Ontario Ministry of Natural Resources [31]. In the FRI, forest characteristics are determined at the stand level with a combination of aerial photo interpretation and ground surveys. We used a large-scale version of the inventory summarized over grid cells varying in size between 5×5 km and 20×20 km. For each grid cell, the data include the percentage of the total basal area of balsam fir and white spruce (i.e., 'FbSw'), and of balsam fir, white spruce and black spruce (i.e., 'FbSwSb'), and hardwood (i.e., 'hw'). FbSw accounts for the tree species (balsam fir and white spruce) on which spruce budworm feeds primarily, while FbSwSb covers all major hosts in Ontario. Hardwood content was included as it may affect fire behavior [30]. Forest resource inventory data are not available in the northernmost part of the province, but aerially visible defoliation is quite rare there (G. Howse, personal communication). For this study, we used the earliest large-scale Ontario forest inventory available on GIS. This inventory was compiled in 1996 [31] from data acquired between 1988 and 1992. Although, at the stand level, forest composition has likely changed during the 65 years of the study, we assumed that at the large scale, low resolution of our study, the relative proportions of each broad forest type remained relatively stable.

2.4. Climate Data

The historical climate data are spatial interpolations of monthly minimums and maximums for temperature (°C) and precipitation (mm) from 471 meteorological stations across Ontario, eastern Manitoba, and western Quebec, over the period 1901–2000 [32] updated with data from 2001 to 2003.

These data were used to calculate a climate moisture index (cmi) according to the algorithm published by Hogg [33] modified to replace Tdew = Tmin − 2.5 °C with Tdew = Tmin (Tdew is the mean dew point temperature and Tmin the mean daily minimum temperature) to take into account moister conditions in Ontario than in Alberta (E. Hogg, personal communication).

Annual CMIs were calculated from 1941 to 2003 for years starting 1 November and ending 31 October [i.e., CMI(year) = ∑CMI(m), m = Nov(year − 1) − Oct(year)]. This division of months into years follows Girardin et al. [34] and highlights how annual moisture fluctuations relate to the seasonal fire cycle.

The average of the annual CMI (cmi_ave) was then calculated over the period 1941–2003 for each cell of a 10-km grid. Based on this variable, the province is clearly partitioned into a dryer (cmi_ave < 40 cm/year) zone in the West along the Manitoba border, wet (cmi_ave > 60 cm/year) areas along the eastern shores of Lakes Huron and Superior and average areas in the north and northeast (Figure 1B).

2.5. Classification Tree (CART) Analysis

We used classification trees (CART, [35]) to model how climate, forest composition and spruce budworm defoliation history affect wildfire potential. Such models belong to the classification and regression tree family of analysis methods. Compared to classical methods for predictive modelling (e.g., Generalized Linear Models), CART models do not require the restrictive assumptions of (a) Gaussian relations between response and predictor variables; (b) uniform effects of predictors and their interactions on the response over their range of values; and (c) constant interactions among predictors over their range of values. Classification trees also have several advantages over linear discriminant and multiple regression analyses. They can capture non-linear and non-additive behavior, as well as general interactions among predictors, such as when relationships between a response variable and certain predictors are conditional on the values of other predictors. Classification trees can also accommodate both continuous and categorical predictor variables without transformation.

Classification trees can be unstable in the variables retained, in their branching patterns, and in the values of their split points. In this sense, a particular classification tree is but one realization of an ensemble of possible trees and the issue then centers on how well this particular classification tree represents the ensemble. To address this question, we assessed the robustness of this classification tree against each possible source of variation and verified that it was representative of the general relation between the predictor and the response variables.

The first source of instability stems from the fact that the pruning procedure used to reduce the size of a classification tree is based on a 10-fold cross-validation. The cross-validation algorithm separates the original data set into 10 mutually exclusive random subsets and then uses each subset once to independently calculate a cross-validation relative error for the subtrees grown on the 9 remaining subsets. The algorithm uses the cross-validation error to determine at which level (i.e., split) the pruning is performed. Different random samples taken during the cross-validation procedure could produce classification trees of different sizes (but with the same split variables and values up the point of pruning) because the procedure is based on samples drawn randomly from the dataset. We performed 50 independent pruning procedures on the classification tree described above to test the stability of classification tree size after pruning.

The second source of instability, i.e., multicollinearity in explanatory variables, often produces misleading coefficients in linear or nonlinear regressions [36]. One common approach to dealing with multicollinearity is to drop collinear explanatory variables from the analysis but in CART this approach reduces efficiency in finding the best explanatory variable at each split. For this reason, we did not drop collinear explanatory variables. Instead, we verified how representative the variables included in the model were by testing their importance with the randomForest procedure [37,38]. In this procedure, classification trees are constructed using different random subsamples of the originally selected pixels used to build the original tree. At each split, the randomForest procedure finds the best split possible that can be found among the randomly chosen subset of explanatory variables that are available. We constructed 500 classification trees, with two explanatory variables randomly chosen at each split. The importance of each explanatory variable was estimated as the mean decrease in accuracy in the test sample (the 10% of the data held back for testing the classification tree) when data for only that variable are randomized. The random selection of explanatory variables in each classification tree of the randomForest alleviates the instability related to multicollinearity.

Using a different random sample of cells from the dataset might produce different classification trees. To test the stability of our results over variation in the cells sampled, we used a resampling process which involved drawing 50 random samples from the original grids (Figure 1) and then fitting a classification tree and a randomForest to each of these samples. This allowed us to build distributions of the various classification tree 'characteristics' (e.g., misclassification error rates, number of terminal leaves after pruning, the explanatory variables used in the splits) and variable importance rankings as measured by the mean decrease in accuracy in the randomForests procedure.

We also assessed the spatial variability among the areas of likely spruce budworm–fire interaction that were predicted by the classification trees built on the 50 random samples. We used each of these 50 classification trees to predict areas of likely spruce budworm–fire interaction. The probability of spruce budworm–fire interaction was calculated based on the 50 predictions for each 10-km grid cell.

3. Results

3.1. Locating Where Spruce Budworm Defoliation Contributed to Fire Potential

We began by distinguishing the bioclimatic conditions in the areas of the spruce budworm belt where a 'likely interaction' (as defined above) between spruce budworm defoliation and a large fire occurred, from those areas of the belt where there was no 'likely interaction'. These latter areas may have never been burnt, or large fires may have occurred there but not during the SFIP. 'Likely interactions' occurred in 450 of the 3865 cells used to map the spruce budworm belt on a 10-km resolution grid (Figure 2). Unbalanced samples can affect the performance of CART models, particularly in the prediction of the minority class, which is of particular importance in this analysis. For this reason, we re-balanced the sample by keeping all the observations of the minority class (i.e., 450 cells of 'likely interactions') and randomly sampling (without replacement) an equivalent number of cells with no 'likely interactions'.

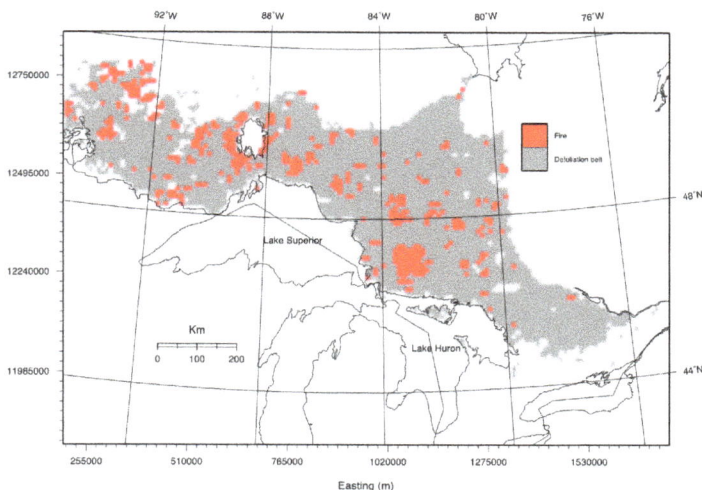

Figure 2. 'Likely interactions' (in red) of spruce budworm defoliation and large (>2 km^2) fires in the spruce budworm belt (gray) mapped on a 10-km grid for 1941–2005.

The classification tree used to distinguish areas of the defoliation belt with 'likely interactions' from those areas without is shown in Figure 3. Four bioclimatic variables were retained: the percentage of the total basal area contributed by hardwood species (hw), and by balsam fir, white spruce and black spruce combined (fbswsb), the average climate moisture index (cmi_ave), and the frequency of defoliation (sbwfreq).

The classification tree has six leaves (colored circles numbered 1–6 from left to right) and an overall rate of correct classification of 69.2%. The number '1' directly under the circles for leaves 5 (red) and 6 (brown) indicates the presence of 'likely interactions' between spruce budworm defoliation and large fires. Working down from the top of the classification tree (Figure 3) toward these leaves reveals the predicted conditions for these interactions. Areas where such interaction likely occurred are characterized by a mix of tree species according to the top two splits in Figure 3 (total basal area was less than 54.4% hardwood (highest split) but also less than 77.6% of the spruce budworm's host species in Ontario (balsam fir, white spruce and black spruce)). The third split (cmi_ave > 33.2) eliminates roughly 4% of the driest remaining areas from further consideration as locations where 'likely interaction' occurred. The fourth split, leads to leaf 6 (brown) suggesting that one condition for 'likely interaction' in the spruce budworm belt was moderate climate moisture (33.2 < cmi_ave < 45.1). The fifth split leads to leaf 5 (red) which suggests that likely spruce budworm–fire interaction also tended to occur in moist areas (cmi_ave > 45.1) which had experienced at least 8 years of defoliation from 1941 to 2003.

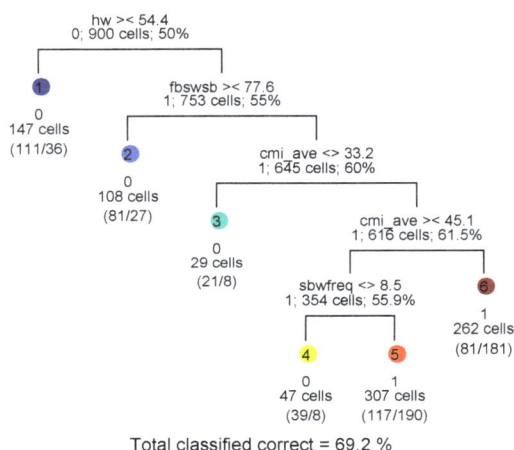

Figure 3. Classification tree of the presence (1) or absence (0) of 'likely interaction' between large fires and spruce budworm defoliation in Ontario's spruce budworm belt (Figure 1A) from 1941 to 2005. Five splits (horizontal bars) and six leaves (circles numbered 1–6 from left to right) are shown. A variable is shown above each bar followed by the two inequality signs, '>' and '<'. The inequality sign on the left (right) applies to the left (right) end of the bar. The variables are the number of years of moderate–severe defoliation (sbwfreq), the average climate moisture index (cmi_ave), the percentage of the total basal area that is hardwood (hw) or balsam fir, white spruce and black spruce combined (fbswsb). The '0' or '1' directly below the circle at each leaf and leading the second line above each bar indicate whether the previous split classified this group of cells as having conditions conducive to the presence (1) or absence (0) of interaction. The '0' or '1' is followed by the group size (# cells) and then, for the bars, the correct classification %, and for leaves, parentheses enclosing the numbers of cells where absence/presence [of a "likely interaction"] is predicted. See text for further explanation.

Misclassification error rates can be calculated for each leaf using the two numbers there (Figure 3) in parentheses (number of cells where absence/presence (of a 'likely interaction') is predicted). The misclassification error inside the terminal leaves is generally higher in the leaves predicting a 'likely interaction' between spruce budworm defoliation and fire (38.1% in leaf 5 and 30.9% in leaf 6) than in the leaves predicting no interaction (from 17.0% in leaf 4 to 25.5% in leaf 3).

Figure 4 maps the unique area defined by each leaf of the classification tree (Figure 3). Leaves 1 and 2 respectively demark areas with high content of either hardwood or Ontario's spruce budworm host trees (balsam fir, white spruce and black spruce). These areas largely define the southern and northern boundaries of spruce budworm defoliation, respectively, particularly in the eastern part of the province. The areas defined by these two splits are also characterized by a low frequency of defoliation (Figure 1A). The third leaf identifies the dry western edge of the defoliation belt where the climate moisture index is lowest. The boundary between this western edge and the rest of the defoliation belt is closely associated with a gradient in the climate moisture index (Figure 1B). Leaf 4 defines a scattered group of moist areas that experienced relatively little defoliation from 1941 to 2005. Leaf 5 accounts for the largest area in Figure 4. It is moist and has at least 9 years of moderate–severe defoliation in its 1941–2005 history. Spatially, leaf 6 identifies a largely contiguous area of moderate climate moisture in the western part of the defoliation belt. Defoliation history does not factor in delimiting leaf 6. The classification tree suggests that the areas defined by leaves 5 and 6 were conducive to spruce budworm–fire interaction.

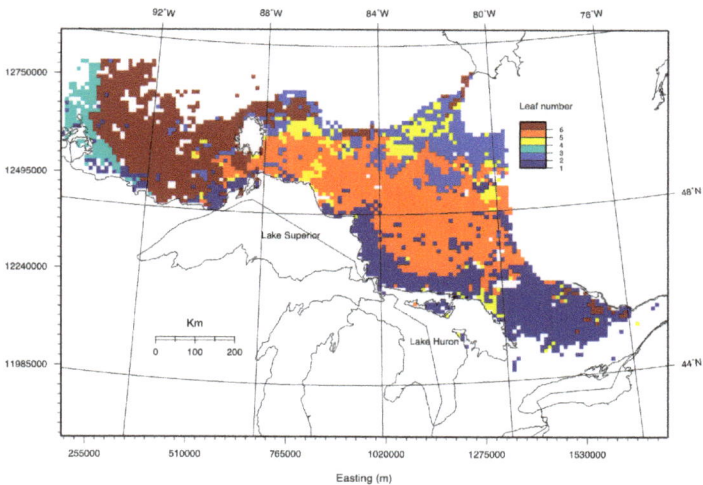

Figure 4. Areas uniquely associated with each leaf of the classification tree in Figure 3.

3.2. Error Analysis

The representiveness of the classification tree in Figure 3 was assessed against three sources of instability. To test for variations in tree size after pruning, 50 independent pruning procedures were performed. The final 50 classification trees ranged in size from 6 to 14 leaves with 6 being the most common (i.e., 48% of the time). Multicollinearity in explanatory variables (Table 1) was addressed by running a randomForest procedure which produces a measure of variable importance based on the mean decrease in accuracy of the model when the data for a variable is randomized. According to this criterion, the climate moisture index and the frequency of defoliation were the most important explanatory variables (Figure 5). When a classification tree was run using only these two explanatory variables (Figure 6), the cross-validated misclassification rate was 100% − 69.4% = 30.6%. This is only 0.2% lower than the misclassification rate of the original classification tree (Figure 3).

The misclassification error rates of the 50-fold random sample ranged from 25–35%. The 30.8% rate produced by our classification tree (Figure 3) is near the middle of this range. The distribution of the number of leaves after pruning ranged from three to 24 leaves, with most of the classification trees having either three or, as in the initial tree (Figure 3), six leaves. Both of these results suggest that our classification tree (Figure 3) is representative of other potential trees for these characteristics.

Table 1. Cross-correlations between explanatory variables. These variables are the average climate moisture index (cmi_ave); the percentages of the total basal area, that is, hardwood (hw), balsam fir and white spruce combined (FbSw), or balsam fir, white spruce and black spruce combined (FbSwSb); and the number of years of moderate–severe defoliation (sbwfreq).

	cmi_ave	FbSw	FbSwSb	hw	sbwfreq
cmi_ave	1	−0.02	−0.09	0.33	−0.09
FbSw	−0.02	1	0.22	0.01	0.18
FbSwSb	−0.09	0.22	1	−0.77	−0.03
Hw	0.33	0.01	−0.77	1	−0.07
sbwfreq	−0.09	0.18	−0.03	−0.07	1

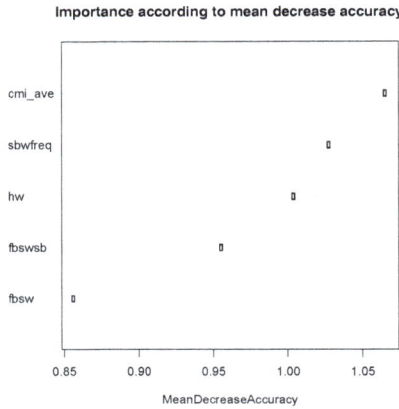

Figure 5. Importance of the explanatory variables used to construct the classification tree (Figure 3), as assessed by the randomForest procedure. These variables are the average climate moisture index (cmi_ave); the number of years of moderate–severe defoliation (sbwfreq); and the percentages of the total basal area, that is, hardwood (hw), balsam fir and white spruce combined (fbsw), or balsam fir, white spruce and black spruce combined (FbSwSb).

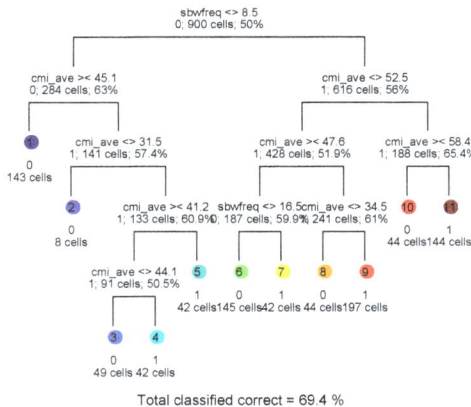

Figure 6. Classification tree of the presence (1) or absence (0) of the interaction between fire and spruce budworm defoliation as a function of just the two most important explanatory variables (Figure 5) from the data used to grow the tree shown in Figure 3. These two variables are the frequency of defoliation by spruce budworm (sbwfreq) and the annual average climate moisture index (cmi_ave). See the caption to Figure 3 for additional detail.

On the other hand, the identity and importance of the explanatory variables retained in the final 50 classification trees after pruning vary slightly from the corresponding results for our original classification tree (Figure 3). In the classification trees based on the 50 random samples, climate moisture index (cmi_ave) and hardwood content (hw) were the explanatory variables most often retained. The proportion of balsam fir, white spruce and black spruce (FbSwSb) and the proportion of balsam fir and white spruce (FbSw) were next. Defoliation frequency (sbwfreq) was the least often selected. However, selection frequency is only one of many possible measures of importance. As shown above, an explanatory variable selected near the bottom of a classification tree (e.g., defoliation frequency in Figure 3), and consequently susceptible to pruning, can have a considerable importance for the accuracy of the classification tree (Figure 5).

For each of these 50 classification trees, the explanatory variables were ranked according to their importance. Figure 7 shows the distribution of these ranks. The importance of average climate moisture index (cmi_ave) and hardwood content (hw) is confirmed as they tend to rank first or second. In contrast, the proportion of balsam fir, white spruce and black spruce combined (FbSwSb), and defoliation frequency (sbwfreq) tend to rank third or fourth. The proportion of balsam fir and white spruce combined (FbSw) almost always ranks fifth (last) in importance in this analysis.

Figure 8 maps the location-specific probabilities of likely spruce budworm–fire interaction based on predictions from 50 random samples. Areas with a high probability of interaction (0.8–1) are mostly located in large patches in (1) the northwest of the province; (2) on the southwestern side of Lake Nipigon and (3) in a fairly narrow band running NW to SE from the southeastern side of Lake Nipigon to the Mississagi river watershed. In between these areas, the probability of predicting an area of interaction is still high (80%) or medium (60%).

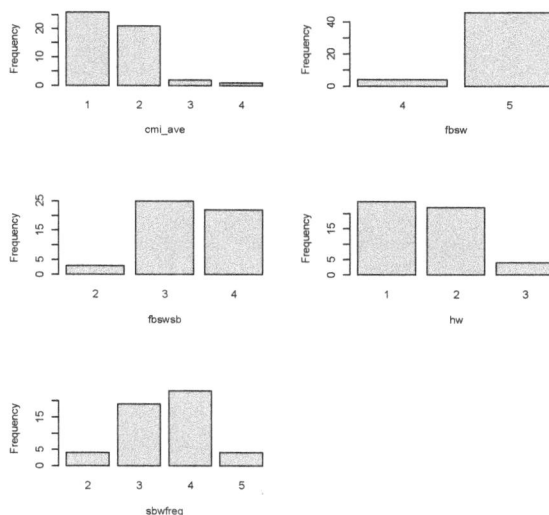

Figure 7. Distributions of the importance rankings of the explanatory variables in 50 classification trees calculated on random samples of the original dataset. Importance was measured by the mean decrease accuracy in the randomForests procedure.

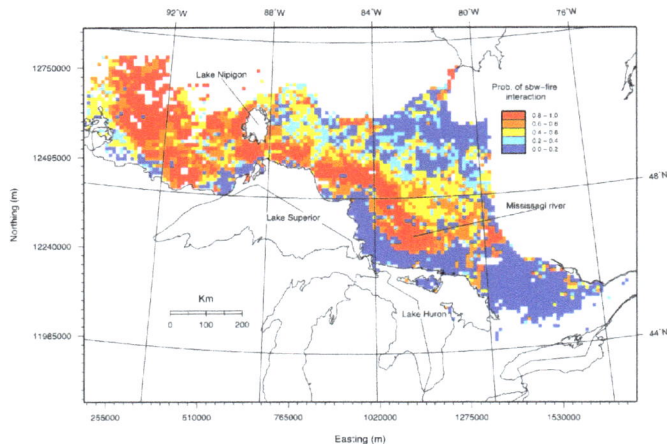

Figure 8. Probability of interaction between spruce budworm and fire based on predictions from 50 classification trees calculated on random samples of the original dataset.

4. Discussion

In our analysis, hardwood content, closely followed by climate moisture (Figure 1B), were the two dominant explanatory variables for predicting where spruce budworm defoliation most likely promoted subsequent large fires in Ontario (Figure 7). The third most important explanatory variable was the prevalence of the spruce budworm's host species (i.e., balsam fir and white and black spruce, combined) and fourth, by the propensity (number of years recorded) for spruce budworm defoliation to occur at that location (Figure 1A). Of least importance was the content of just the two principal spruce budworm host species (i.e., balsam fir and white spruce, combined).

The relative importance of these explanatory variables is evident in (a) the maps (Figures 4 and 8) showing the estimated likelihoods that spruce budworm defoliation will promote subsequent large fires; and (b), the classification trees on which these maps are based (e.g., Figure 3). The areas where spruce budworm defoliation most likely promoted subsequent large fires are best defined by a geographical limit related to hardwood content in the south, balsam fir, white spruce and black spruce together in the north, and moisture in the west. Inside these limits, areas of spruce budworm–fire interaction are related to spruce budworm defoliation frequency. There is little evidence that spruce budworm defoliation promotes subsequent large (>2 km^2) fires in the southeast where hardwood content was high and SBW defoliation rare, in the northeast where there was also little history of defoliation, and in the dry western and southwestern regions. Within the area defined by these boundaries and towards the northern limit of the spruce budworm belt in the west, spruce budworm–fire interaction seems likely.

The steepness of the probability gradients in Figure 8 indicates the relative spatial certainty in locating the borders that separate regions where spruce budworm–fire interaction is likely from those where it is not. For instance, the abrupt shift from a probability of 1 to a probability of 0 in the southeast clearly defines the border's location there. The location of the border is not as easily located in the northeast where the shift from probability 1 to 0 is gradual, nor in the west where few cells have a probability less than 0.4. In the northwest, areas of likely spruce budworm–fire interaction reach the northern limit of the data, suggesting that they could extend further north, beyond the data, if SBW defoliation occurs there.

These results appear robust against several sources of variation: over the ensemble of regression trees examined, hardwood content is consistently one of the most important variables in explaining the areas of spruce budworm–fire interaction (Figure 7). In southeastern Ontario, the latitudinal gradient of hardwood content matches a gradient of increasing urbanization and, as a result, of increasing fire protection that might also explain the absence of fire in these areas [17]. At the northeastern limit of the spruce budworm belt, the high proportion of balsam fir, white spruce and black spruce (FbSwSb > 77.6%) is mainly due to the high proportion of black spruce, a tree that is less supportive of large spruce budworm populations than balsam fir or white spruce [39], and a tree which can proliferate on the wet soils there. In addition, these wet soils coupled with weather patterns bringing cold, moist air from Hudson's Bay limit the occurrence of fire (Figure 2 in [17]). Consequently, it is unlikely that spruce budworm defoliation will promote fire here.

After hardwood content, the climate moisture index is the explanatory variable most consistently high in importance over the ensemble of 50 regression trees constructed (Figure 7). Climate moisture affects the risk of fire directly, but also indirectly through its influence on the rate of decomposition of dead trees and branches and other fuels following spruce budworm defoliation. For instance, compared to western Ontario, Fleming et al. [17] suggest that it is increased rates of decomposition in the wetter climates of eastern Ontario (Figure 1B) which shorten the time-window (their Figure 6) following spruce budworm defoliation during which fire potential remains high. In dry climates (low climate moisture index) such as in the western reaches of the province (Figure 1B), large fires are relatively common (their Figure 2) and seem to burn independently of spruce budworm defoliation (Figure 8). In wetter climates such as in the red zone in Figure 1B east of Lake Superior, large fires are rare (their Figure 2), presumably often prevented by this climate despite the prevalence of spruce budworm defoliation there (Figure 1A). It is in the areas of moderate climate moisture that the presence of spruce budworm defoliation is most likely to elevate the subsequent risk of large fires.

The prevalence (number of years recorded) of spruce budworm defoliation at a given location ranks fourth in importance over the ensemble of 50 regression trees constructed (Figure 7). Hardwood content, climate moisture, and the prevalence of spruce budworm host species all rank higher. The relatively poor explanatory power of defoliation prevalence is partly explained by its curvilinear relationship with fire (Figure 4 in [17]). In this relationship, areas within the spruce budworm belt that experienced moderate frequencies of defoliation were the most likely burned. After a large fire, the forest needs time to recover before it is again suitable for spruce budworm defoliation, so fires tend to be relatively rare in areas with high frequencies of defoliation. Defoliation prevalence is also low in the northeast (due to black spruce prevalence, as explained above), and in the southeast of the defoliation belt where farms and large pockets of dense deciduous forest interrupt the continuity of host trees species that is otherwise found further north. Large fire is rare in these areas due to climate (northeast, see above) and aggressive fire response in the relatively urbanized southeast.

In this analysis, we searched for broad tendencies in the patterns of budworm–fire interaction over decades of historical records at very large spatial scales. Local, instantaneous conditions such as fire weather, topography, and fuel condition at the ignition point are important factors that directly affect fire ignition and spread in particular sites at specific times [40,41], but over the large spatio-temporal scales of this study, variations in weather, topography and fuel condition become so 'smoothed out' that they are no longer useful predictors. Hence such variables were not included in our analysis. However, other ecological and climate factors may affect ecoregional patterns of spruce budworm and fire interactions. The nature of understory vegetation (composition, age, and structure) is likely one of these factors as it could affect the inter- and intra-annual dynamics of fuel moisture. Early on [15,17], it was hypothesized that crown breakage following sustained defoliation would release the understory by opening the canopy. The proliferation of the understory would then increase surface fuel moisture, thus decreasing the risk of surface fire and the risk for surface fires to reach the canopy. As such, the nature of the understory vegetation (composition, age, and structure) would likely affect its post-release dynamics and its effect on fire intra-annually in the timing of leaf-out in the spring (fire risk is generally thought to be higher prior to leaf-out), and inter-annually in the time that it will require to grow enough in size and complexity to achieve a reduction in fire risk. Understory composition, age and structure could not be included in this analysis for lack of data over the study area. Although the three classes of overstory vegetation we used in this analysis may somewhat correspond to broad classes of understory vegetation [42], a better characterization of the understory would certainly be desirable. Wind is another factor that was not included in our analysis for lack of data although it likely increases the risk of crown breakage, which may accelerate the accumulation of "ladder fuel". Long-term mean wind speed in the spring and summer is spatially homogeneous over our study area and has low inter-annual variability [43].

The elevation of fire risk by spruce budworm defoliation may seem to be a relatively small problem. For instance, Fleming et al. [17] reported that of the 417,000 km^2 defoliated by spruce budworm in Ontario at least once between 1941 and 1996, only about 5% experienced large fires. In this paper, we have shown that from 1941 to 2005, spruce budworm defoliated 418,000 km^2 at least once, of which 21% constitutes areas where the probability of spruce budworm–fire interaction is greater than 0.8 (Figure 8). This percentage is larger than the former because, with only 1.5 outbreaks in our data, many of these areas with elevated fire risk have yet to realize their spruce budworm-related fires.

According to our results, climate change could potentially affect spruce budworm–fire interaction through changes in the bioclimatic variables that were retained in the model. Global circulation models predict temperature increases in southern Ontario of 3–5 °C in summer and 4–6 °C in winter before the end of the century. The corresponding predictions in northern Ontario are for seasonal temperature increases of 3–6°C and 4–10 °C, respectively. Precipitation is predicted (with less certainty than temperature) to decrease by 20% in the summer and 10–20% in the winter (20% and 20–30% decrease in northern Ontario, respectively) [44]. An increase in temperature combined with a decrease (or even no change) in precipitation can be expected to decrease climate moisture. As a result, some areas of moderate climate moisture might experience a drier climate under which spruce budworm defoliation has less influence on the subsequent risk of large fires. Climate change can also potentially affect the distribution of the frequency of spruce budworm defoliation. The application of climate projections for 2011–2040 to a bioclimatic model of spruce budworm defoliation in Ontario suggests (1) a northern extension of the area of defoliation combined with a persistence of the southern limit, effectively increasing the total area of defoliation by more than 20% compared to the area observed in the last outbreak (1967–1998); and (2) a decrease of the frequency of defoliation in the center of the historical defoliation belt [5]. A northward extension of the area of defoliation could create more opportunity for interactions with fire, especially because historically, more area has been burnt by large fires north of the defoliation belt (particularly in the northwest) than in it [45].

Changes in temperature and precipitation regimes are also expected to affect forest composition and distribution through their effects on the physiology and ecology of tree species. For instance, white and black spruce respond negatively to temperature increases [46], and Lenihan and Neilson [47] predict that future climate warming could potentially reduce their area of dominance by 20–30%. Balsam fir has a wide distribution that could be displaced by the combination of the northward expansion of the temperate conifer and hardwood species of the Great Lakes—St Lawrence Forest Region south of the Boreal zone, and a northward shift of its climatically optimal habitat. While there is certainty that changes in forest composition and distribution will occur, the rate, magnitude, and location of such changes are all highly uncertain. In the boreal forest, changes in natural disturbance regimes are expected to exert a stronger effect than changes in the climate itself. There is potential for positive (or negative) feedback: disturbances may accelerate changes in forest composition and distribution 'imposed' by a different climate which, in turn, may create new conditions which favor (or hinder) more disturbances and even further forest changes.

While climate change adds another level of complexity to the interactions between spruce budworm defoliation and fire, and the forests in which they occur, the likelihood that it will affect these interactions and the potential impacts of spruce budworm-caused fires as described above, point to the need for further research in this area.

5. Conclusions

The existence of an interaction between spruce budworm defoliation and wildfire in central Canada's boreal forests is supported by an increasing body of experimental [14,15,18] and statistical [17,19] results. The driving factor behind this interaction is the accumulation of "ladder fuel" (i.e., highly flammable tree tops and branches arranged vertically) that increases the probability for surface fires to reach the canopy, thus increasing the risk of severe fires.

In this study, we integrate and extend previous work on the influence of spruce budworm outbreaks on the subsequent potential for wildfire. We show that factors such as climate, defoliation history, and forest condition all help explain characteristics of this influence and its spatial variation across the region. We use this new information to distinguish, at the landscape scale, those areas of Central Canada's boreal forest where spruce budworm defoliation is likely to increase subsequent fire risk from those areas where it is not.

In the short term, these results may help fire managers in geographically allocating resources among areas that were previously considered as having similar fire risk. In the long term, further research is required to better understand how the increase in fire risk and changes in spruce budworm defoliation patterns predicted under climate change will affect the interaction between these two disturbances.

Acknowledgments: We thank Ron Fournier and Tim Burns for support with data analysis. We thank CFS and OMNR for access to data. We thank CCAF, PERD, and OMNR CRA (SPA 90170) for financial support. We thank Paul Gray, OMNR, for his support and encouragement.

Author Contributions: J.N.C. and R.A.F. conceived and designed the study; J.N.C. and X.W. analyzed the data; J.N.C., R.A.F. and X.W. wrote the paper.

Conflicts of Interest: The authors declare no conflict of interest.

References

1. Weber, M.G.; Flannigan, M.D. Canadian Boreal Forest Ecosystem Structure and Function in a Changing Climate: Impact on Fire Regimes. *Environ. Rev.* **1997**, *5*, 145–166. [CrossRef]
2. Fleming, R.A. Climate Change and Insect Disturbance Regimes in Canada's Boreal Forests. *World Resour. Rev.* **2000**, *12*, 520–554.
3. Jasinski, J.P.P.; Payette, S. The creation of alternative stable states in the southern boreal forest, Québec, Canada. *Ecol. Monogr.* **2005**, *75*, 561–583. [CrossRef]
4. Girard, F.; Payette, S.; Gagnon, R. Rapid expansion of lichen woodlands within the closed-crown boreal forest zone over the last 50 years caused by stand disturbances in eastern Canada. *J. Biogeogr.* **2008**, *35*, 529–537. [CrossRef]
5. Candau, J.-N.; Fleming, R.A. Forecasting the response of spruce budworm defoliation to climate change in Ontario. *Can. J. For. Res.* **2001**, *41*, 1948–1960. [CrossRef]
6. Wang, X.; Thompson, D.; Marshall, G.A.; Tymstra, C.; Carr, R.; Flannigan, M.D. Increasing frequency of extreme fire weather in Canada with climate change. *Clim. Chang.* **2015**, *130*, 573–586. [CrossRef]
7. Forestry Canada Fire Danger Group. *Development and Structure of the Canadian Forest Fire Behavior Prediction System*; Forestry Canada Science and Sustainable Development Directorate, Information Report ST-X-3 (Ottawa, ON); Forestry Canada Fire Danger Group: Ottawa, ON, Canada, 1992; p. 64.
8. Kurz, W.A.; Stinson, G.; Rampley, G.J.; Dymond, C.C.; Neilson, E.T. Risk of natural disturbances makes future contribution of Canada's forests to the global carbon cycle highly uncertain. *Proc. Natl. Acad. Sci. USA* **2008**, *105*, 1551–1555. [CrossRef] [PubMed]
9. Dymond, C.C.; Neilson, E.T.; Stinson, G.; Porter, K.; MacLean, D.A.; Gray, D.R.; Campagna, M.; Kurz, W.A. Future Spruce Budworm Outbreak May Create a Carbon Source in Eastern Canadian Forests. *Ecosystems* **2010**, *13*, 917–931. [CrossRef]
10. Graham, S.A. *The Dying Balsam Fir and Spruce in Minnesota*; University of Minnesota Agricultural Experiment Station, Special Bulletin 68, St. Paul, MN; University of Minnesota Agricultural Experiment Station: Minneapolis, MN, USA, 1923.

11. Swaine, J.M.; Craighead, F.C.; Bailey, I.W. *Studies on the Spruce Budworm (Cacoecia fumiferana Clem.)*; Department of Agriculture: Ottawa, ON, Canada, 1924.

12. Prebble, M.L. The Battle of the Budworm. *Pulp Pap. Can.* **1950**, *51*, 150–155.

13. Baskerville, G.L. Spruce Budworm: Super Silviculturist. *For. Chron.* **1975**, *61*, 138–140.

14. Stocks, B.J. Forest Fire Behavior in Spruce Budworm-killed Balsam Fir. In *Advances in Spruce Budworms Research, Proceedings of the CANUSA Spruce Budworms Research Symposium Recent, Bangor, ME, USA, 16–20 September 1984*; Canadian Forest Service: Ottawa, ON, Canada, 1985; pp. 198–209.

15. Stocks, B.J. Fire Potential in the Spruce Budworm-damaged Forests of Ontario. *For. Chron.* **1987**, *63*, 8–14. [CrossRef]

16. Péch, G. Fire hazard in budworm-killed balsam fir stands on Cape Breton Highlands. *For. Chron.* **1993**, *69*, 178–186. [CrossRef]

17. Fleming, R.A.; Candau, J.-N.; McAlpine, R.S. Landscape analysis of interactions between insect defoliation and forest fire in central Canada. *Clim. Chang.* **2002**, *55*, 251–272. [CrossRef]

18. Watt, G.A. Enhanced Vertical Fuel Continuity in Forests Defoliated by Spruce Budworm (*Choristoneura fumiferana* Clem.) Promotes the Transition of a Surface Fire into a Crown Fire. Ph.D. Thesis, Faculty of Forestry, University of Toronto, Toronto, ON, Canada, 2014.

19. James, P.M.; Robert, L.E.; Wotton, B.M.; Martell, D.L.; Fleming, R.A. Lagged cumulative spruce budworm defoliation affects the risk of fire ignition in Ontario, Canada. *Ecol. Appl.* **2017**, *27*, 532–544. [CrossRef] [PubMed]

20. Candau, J.-N.; Fleming, R.; Hopkin, A. Spatiotemporal patterns of large-scale defoliation caused by the spruce budworm in Ontario since 1941. *Can. J. For. Res.* **1998**, *28*, 1733–1741. [CrossRef]

21. Sippell, W.L. A review of the spruce budworm and its outbreak history. In Proceedings of the Spruce Budworm Problem in Ontario—Real or Imaginary? Timmins, ON, Canada, 14–16 September 1983; Sanders, C.J., Carrow, J.R., Eds.; Canadian Forest Service: Sault Ste. Marie, ON, Canada, 1983; pp. 17–25.

22. Hardy, Y.; Mainville, M.; Schmitt, D.M. *An Atlas of Spruce Budworm Defoliation in Eastern North America, 1938–1980*; USDA Forest Service, Miscellaneous Publication Number 1449; USDA Forest Service: Beltsville, MD, USA, 1986.

23. Candau, J.-N.; Fleming, R.A. Landscape scale spatial distribution of spruce budworm defoliation in relation to bioclimatic conditions. *Can. J. For. Res.* **2005**, *35*, 2218–2232. [CrossRef]

24. Stocks, B.J.; Mason, J.A.; Todd, J.B.; Bosch, E.M.; Wotton, B.M.; Amiro, B.D.; Flannigan, M.D.; Hirsch, K.G.; Logan, K.A.; Martell, D.L.; et al. Large forest fires in Canada, 1959–1997. *J. Geophys. Res.* **2002**, *107*. [CrossRef]

25. Bergeron, Y.; Leduc, A.; Harvey, B.D.; Gauthier, S. Natural fire regime: A guide for sustainable management of the Canadian boreal forest. *Silva Fenn.* **2002**, *36*, 81–95. [CrossRef]

26. Parisien, M.-A.; Peters, V.S.; Wang, Y.; Little, J.M.; Bosch, E.M.; Stocks, B.J. Spatial patterns of forest fires in Canada, 1980–1999. *Int. J. Wildland Fire* **2006**, *15*, 361–374. [CrossRef]

27. Bergeron, Y.; Leduc, A. Relationships between change in fire frequency and mortality due to spruce budworm outbreak in the southeastern Canadian boreal forest. *J. Veg. Sci.* **1998**, *9*, 493–500. [CrossRef]

28. Bouchard, M.; Kneeshaw, D.; Bergeron, Y. Forest dynamics after successive spruce budworm outbreaks in mixedwood forests. *Ecology* **2006**, *87*, 2319–2329. [CrossRef]

29. Simard, I.; Morin, H.; Lavoie, C. A millennial-scale reconstruction of spruce budworm abundance in Saguenay, Quebec, Canada. *Holocene* **2006**, *16*, 31–37. [CrossRef]

30. Hély, C.; Bergeron, Y.; Flannigan, M.D. Effects of stand composition on fire hazard in mixed-wood Canadian boreal forest. *J. Veg. Sci.* **2000**, *11*, 813–824. [CrossRef]

31. Ontario Ministry of Natural Resources. *Forest Resources of Ontario 1996*; Ontario Ministry of Natural Resources: Peterborough, ON, USA, 1996.

32. McKenney, D.W.; Pedlar, J.H.; Papadopol, P.; Hutchinson, M.F. The development of 1901–2000 historical monthly climate models for Canada and the United States. *Agric. For. Meteorol.* **2006**, *138*, 69–81. [CrossRef]

33. Hogg, E.H. Temporal scaling of moisture and the forest–grassland boundary in Western Canada. *Agric. For. Meteorol.* **1997**, *84*, 115–122. [CrossRef]

34. Girardin, M.P.; Tardif, J.; Flannigan, M.D.; Wotton, B.M.; Bergeron, Y. Trends and periodicities in the Canadian Drought Code and their relationships with atmospheric circulation for the southern Canadian boreal forest. *Can. J. For. Res.* **2004**, *34*, 103–119. [CrossRef]

35. De'ath, G.; Fabricius, K.E. Classification and regression trees: A powerful yet simple technique for ecological data analysis. *Ecology* **2000**, *81*, 3178–3192. [CrossRef]
36. Draper, N.R.; Smith, H. *Applied Regression Analysis*; Wiley: New York, NY, USA, 1981.
37. Breiman, L. Random forests. *Mach. Learn.* **2001**, *45*, 5–32. [CrossRef]
38. Liaw, A.; Wiener, M. Classification and regression by random Forest. *R News* **2002**, *2*, 18–22.
39. Nealis, V.G.; Régnière, J. Insect–host relationships influencing disturbance by the spruce budworm in a boreal mixedwood forest. *Can. J. For. Res.* **2004**, *34*, 1870–1882. [CrossRef]
40. Taylor, S.W.; Woolford, D.G.; Dean, C.B.; Martell, D.L. Wildfire prediction to inform fire management: Statistical science challenges. *Stat. Sci.* **2013**, *28*, 586–615. [CrossRef]
41. Wang, X.; Parisien, M.-A.; Flannigan, M.D.; Parks, S.A.; Anderson, K.R.; Little, J.M.; Taylor, S.W. The potential and realized spread of wildfires across Canada. *Glob. Chang. Biol.* **2014**, *20*, 2518–2530. [CrossRef] [PubMed]
42. Légaré, S.; Bergeron, Y.; Paré, D. Influence of forest composition on understory cover in Boreal mixedwood forests of western Quebec. *Silva Fenn.* **2002**, *36*, 353–366. [CrossRef]
43. Wan, H.; Wang, X.L.; Swail, V.R. Homogenization and trend analysis of Canadian near-surface wind speeds. *J. Clim.* **2010**, *23*, 1209–1225. [CrossRef]
44. Colombo, S.J.; McKenney, D.W.; Lawrence, K.M.; Gray, P.A. *Climate Change Projections for Ontario: Practical Information for Policymakers and Planners*; Ontario Ministry of Natural Resources, Applied Research and Development Branch, Climate Change Research Report CCRR-05; Ontario Ministry of Natural Resources: Sault Ste. Marie, ON, Canada, 2007; p. 38.
45. Li, C. Fire regimes and their simulation with reference to Ontario. In *Ecology of a Managed Terrestrial Landscape: Patterns and Processes of Forest Landscapes in Ontario*; Perera, A.H., Euler, D.L., Thompson, I.D., Eds.; UBC Press: Vancouver, BC, Canada, 2000; pp. 115–140.
46. Brooks, J.R.; Flanagan, L.B.; Ehleringer, J.R. Responses of boreal conifers to climate fluctuations: Indications from tree-ring widths and carbon isotope analyses. *Can. J. For. Res.* **1998**, *28*, 524–533. [CrossRef]
47. Lenihan, J.M.; Neilson, R.P. Canadian vegetation sensitivity to projected climatic change at three organizational levels. *Clim. Chang.* **1995**, *30*, 27–56. [CrossRef]

Article

Topoedaphic and Forest Controls on Post-Fire Vegetation Assemblies Are Modified by Fire History and Burn Severity in the Northwestern Canadian Boreal Forest

Ellen Whitman [1,2,*], Marc-André Parisien [2], Dan K. Thompson [2] and Mike D. Flannigan [1]

[1]   Department of Renewable Resources, University of Alberta, 751 General Services Building, Edmonton, AB T6G 2H1, Canada; mike.flannigan@ualberta.ca
[2]   Northern Forestry Centre, Canadian Forest Service, Natural Resources Canada, 5320–122nd St., Edmonton, AB T6H 3S5, Canada; marc-andre.parisien@canada.ca (M.-A.P.); daniel.thompson@canada.ca (D.K.T.)
*   Correspondence: ewhitman@ualberta.ca

Received: 15 February 2018; Accepted: 13 March 2018; Published: 17 March 2018

Abstract: Wildfires, which constitute the most extensive natural disturbance of the boreal biome, produce a broad range of ecological impacts to vegetation and soils that may influence post-fire vegetation assemblies and seedling recruitment. We inventoried post-fire understory vascular plant communities and tree seedling recruitment in the northwestern Canadian boreal forest and characterized the relative importance of fire effects and fire history, as well as non-fire drivers (i.e., the topoedaphic context and climate), to post-fire vegetation assemblies. Topoedaphic context, pre-fire forest structure and composition, and climate primarily controlled the understory plant communities and shifts in the ranked dominance of tree species (***8% and **13% of variance explained, respectively); however, fire and fire-affected soils were significant secondary drivers of post-fire vegetation. Wildfire had a significant indirect effect on understory vegetation communities through post-fire soil properties (**5%), and fire history and burn severity explained the dominance shifts of tree species (*7%). Fire-related variables were important explanatory variables in classification and regression tree models explaining the dominance shifts of four tree species (R^2 = 0.43–0.65). The dominance of jack pine (*Pinus banksiana* Lamb.) and trembling aspen (*Populus tremuloides* Michx.) increased following fires, whereas that of black spruce (*Picea mariana* (Mill.) BSP.) and white spruce (*Picea glauca* (Moench) Voss) declined. The overriding importance of site and climate to post-fire vegetation assemblies may confer some resilience to disturbed forests; however, if projected increases in fire activity in the northwestern boreal forest are borne out, secondary pathways of burn severity, fire frequency, and fire effects on soils are likely to accelerate ongoing climate-driven shifts in species compositions.

Keywords: boreal forest; burn severity; disturbance; fire effects; fire history; forest fire; regeneration; species richness

---

1. Introduction

Wildfires are the most extensive stand-initiating disturbance in the northwestern Canadian boreal forest, typically recurring every 50–100 years [1,2]. When wildfires occur, they burn with varying intensities (energy release) in response to fire weather, topography, and fuel type, producing a range of burn severities. Burn severity is defined as the ecological impacts of fire on vegetation and soils [3,4]. Many boreal forest plants have adapted to repeated wildfires though traits such as resprouting or suckering, seed banking, or, in the case of some tree species, serotiny. Serotinous and semi-serotinous conifer tree species have cones that may open in response to and survive some heating, and retain some viable seeds in the canopy following wildfires. Through this mechanism, serotinous species can produce extensive seed rains from aerial seedbanks immediately following fire [1,5]. Wildfire burn severity has important implications for post-fire understory vegetation communities and recruitment of seedlings. Heating and combustion from wildfires kill some trees and may reduce the viability of seeds in aerial seedbanks (including those of serotinous species) beyond a threshold of fire intensity or if the duration of heating is extensive [6,7]. Variable combustion of organic soils provides diverse seedbeds for plants and trees, ranging from thick remnant organic layers to exposed mineral soils, and alters the composition and exposure of post-fire soil seed banks [8,9]. Some burning of organic soils promotes vegetative regeneration, but deep burning may damage roots and rhizomes, negatively affecting the capacity of resprouting species to regenerate following fires [5].

In many ecosystems, burn severity is a dominant and enduring control on post-fire understory vegetation assemblies [10–12] and seedling recruitment [11,13,14], influencing the resulting structure and composition of forests. Although burned sites in the boreal forest generally return to a mature forested stand structure within 100 years [15], researchers using remote sensing to examine the post-fire recovery of vegetation following wildfires have found different rates of revegetation amongst burn severity classes. Severely burned sites demonstrated the highest decline in vegetation immediately post-fire [16,17]. In the years following a wildfire, severely burned sites subsequently experienced the largest increases in vegetation, indicating either forest recovery or colonization of these sites by disturbance-favouring plants and trees [16,17]. In North American boreal forests, post-fire understory vegetation communities in black spruce (*Picea mariana* (Mill.) BSP.) [18], jack pine (*Pinus banksiana* Lamb.) [19], and mixed broadleaf and coniferous stands [20] are influenced by surface burn severity and depth of burn, in conjunction with the availability of seed sources and vegetative propagules. In these studies, colonizing species such as graminoids and annual forbs established themselves broadly in severely burned areas, whereas slow-growing lichens, evergreen shrubs, and higher species richness were more prominent in low severity and scorched areas [18–20]. Lower densities of recruitment of coniferous trees have been observed when sites burned severely and at short intervals [21,22], and increased proportions of early-successional tree species, such as jack pine and trembling aspen (*Populus tremuloides* Michx.) are associated with high severity burning [23,24]. The relative dominance of different species of trees and the density of post-fire forests are lasting legacies of boreal wildfire severity [21,23,25].

When burn severity is studied at a broader landscape scale, that is, across multiple forest types and wildfires, the effects of burn severity on post-fire vegetation communities and recruitment may be challenging to detect. Burn severity is correlated to pre-fire forest type and stand structure [8,26–28], potentially obscuring or explaining observed effects of burn severity on post-fire plants and trees. Studies of burn severity that encompass multiple forest types have identified topoedaphic and pre-fire forest conditions as the primary post-fire drivers of understory plant communities and site suitability for tree species [18,29–32], leading some researchers to characterize burn severity as a secondary "filtering" effect beneath the dominant landscape and climatological controls.

Ranges of burn severity and the relatively infrequent occurrence of large wildfires ($\geq$200 ha) produce a mosaic of stand ages and patterns on the landscape, in regions with mixed- and high-severity fire regimes [33–35]. Wildfires interact with past burns, as previous fires and burn severity determine current fuels. Abnormally short fire frequencies are implicated in the dominance shifts of tree species [36], low stocking in post-fire forests [22], and even near-deforestation [37], with implications for forest resilience [38]. Furthermore, burn severity interacts with fire frequency, potentially reinforcing vegetation type conversions [39]. Wildfires are a weather and, therefore, climate-driven disturbance. Fires are expected to increase in size, frequency, and intensity (and therefore in severity) [40–42] in North America as the climate warms and severe fire weather increases [43]. The forests of the Canadian Northwest Territories provide an interesting opportunity to study the effects of extensive free-burning wildfires in an ecosystem with multiple dominant coniferous and broadleaf tree species, across a moisture gradient ranging from hydric to xeric. Given the ecologically important role and actively changing patterns of fire in the boreal forest, studies characterizing the relative importance of fire effects and fire history, and non-fire and climate drivers in determining post-fire vegetation assemblies and species composition shifts will provide insights into the trajectories of future forests.

This study describes post-fire vegetation communities and seedling recruitment across a broad range of topoedaphic vegetation classes and levels of burn severity, to identify direct and indirect drivers of these assemblies in the northwestern Canadian boreal forest. In support of this goal, our objectives were: 1. To characterize post-fire vegetation assemblies and recruitment of seedlings across burn severity and topoedaphic gradients; 2. To assess the relative importance of climate and pre-fire forests, burn severity and fire history, and post-fire soils to understory vegetation communities and shifts in the dominance of tree species; and 3. To identify direct and indirect effects of fires on post-fire vegetation, as well as drivers of shifts in the dominance of tree species in the post-fire cohort.

2. Methods

2.1. Study Area

Field sites were established in six, large, lightning-caused wildfires (14,000 to 700,000 ha) that burned in 2014 (Figure 1). The year of 2014 was an extreme fire season in the northwestern Canadian boreal forest region, with drought-driven wildfires burning a total area > 3 million ha [44]. The sampled fires burned in the Northwest Territories and Wood Buffalo National Park. The fire regime of this area is one of infrequent stand-initiating wildfires [45,46]. In the Canadian boreal forest, these large wildfires comprise a small fraction of the total number of fires, but they are responsible for the vast majority of the area that was burned [35].

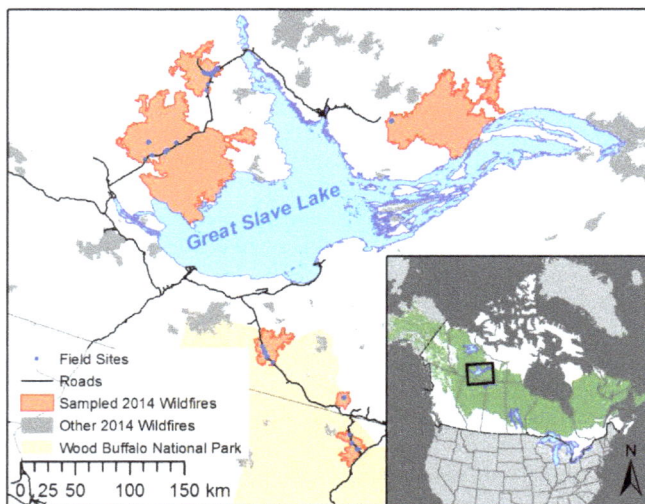

Figure 1. The sampled 2014 wildfires and field site locations in the Northwest Territories and Wood Buffalo National Park. The study area is indicated in black on the inset map, within the context of the North American boreal forest (shown in green) [47].

The study area experiences long cold winters and short hot summers. Mean annual temperatures at the field sites ranged from −4.3 °C at the furthest north site to −1.8 °C at the furthest south [48,49]. Topography of the study area is minimal, consisting of level terrain in the southwestern part of the study area, on the boreal plain, and rolling granitic hills on the boreal shield in the northeast [41]. The forests of the study area are dominated by jack pine, black spruce, white spruce (*Picea glauca* (Moench) Voss), and trembling aspen. Important secondary tree species include eastern larch (*Larix laricina* (Du Roi) K. Koch), paper birch (*Betula papyrifera* Marsh.), and balsam poplar (*Populus balsamifera* L.) [48]. There is also a substantial wetland (chiefly peatlands) component to the region. Peat-forming wetlands may form extensive complexes and cover approximately a third of the total area [50]. Although the study area falls within the discontinuous permafrost zone of Canada [51], no field sites had frozen active layers in the top metre of soil.

2.2. Field Methods

We sampled 51 field sites one year post-fire and resampled 30 sites three years post-fire. The sites were selected using a stratified random sample that was evenly distributed across high-, moderate-, and low-burn severity classes. The mapped burn severity was produced using an initial assessment of a differenced normalized burn ratio (dNBR) image [52], classified with thresholds developed by Hall et al. [53]. Field sites were >100 m and ≤2 km from roads. More isolated sites were also opportunistically accessed by helicopter. The field sites accessed by helicopter were located in order to capture the locally available range of burn severity and topoedaphic vegetation communities (ecosites), ensuring that each sampled site offered a distinct combination of severity and vegetation type. Field sites were positioned in an area of homogenous burn severity, topoedaphic setting (upland or wetland), and dominant vegetation that extended ≥60 m in any direction. The site moisture (from hydric to xeric) and ecosite categories were classified according to Beckingham and Archibald [54]. Ecosites were generalized into the dominant topoedaphic vegetation classes of open wetland, treed wetland, upland spruce, upland mixedwood, and upland jack pine (from wettest to driest). All the sampled wetlands were peat-forming wetlands (peatlands). Plot centres were recorded with a differential GPS unit. The mean distance between the plot centres of all the field sites was 170 km, with a minimum distance of 103 m.

When sampling one year after the fire, the sample plots were 30 × 30 m, with two 30-m transects oriented in the cardinal directions, crossing at the plot centre. A detailed figure of the plot layouts used for field sampling is included in Appendix A: Figure A1. Compositions of tree species, percent overstory mortality due to fire, stem density (stems ha^{-1}), and basal area (m^2 ha^{-1}) of mature trees in the pre-fire stand were measured at this time for 32 trees ≥ 3 cm diameter at breast height (DBH) using the point-centered quarter method [55,56] at eight evenly-spaced points along the two transects. In very low stem-density areas (i.e., open wetlands), a variable-radius circle plot with a minimum radius of 15 m was used to sample overstory trees. Pre-fire understory stem densities of seedlings and saplings (stems ha^{-1}) were measured using 3-m radius plots at the endpoints of each transect. The number of understory density plots sampled ranged from one to four, depending on the density and evenness of the seedling and saplings.

We collected basal sections from fire-scarred trees to determine the time since the stand origin (TSO) and time since the last fire (TSLF) at each plot. If no scarred trees were identified nearby, a section of a mature dominant tree was sampled. Some open wetlands (fens) had no trees. Samples were sanded and digitally scanned, and annual growth rings and fire scars were dated in CooRecorder [57].

Burn severity was measured in 10 × 10 m subplots at the four corners of each plot. Surface burn severity was measured using the surface Burn Severity Index (BSI) [58]. BSI values range from zero (unburned) to four (ash, mineral soil exposed) using classes defined by Dyrness and Norum [59]. Overstory burn severity was measured using the Canopy Fire Severity Index (CFSI) [60]. CFSI classes range from zero (no tree mortality) to six (no primary branches remaining, pole charring occurred). The percent cover of each BSI and CFSI severity class was estimated within the four subplots, and final values of the two severity metrics were calculated using area-weighted means of each class value, and then averaged for each field site.

We measured the post-fire organic soil depth (cm; up to a maximum of 10 cm) at the inner corners of the same subplots used for estimates of severity and seedling density. The soil cores (13.5 cm in depth, 5.5 cm in diameter) were taken one year post-fire at the plot centre and inner corners of the southwest and northeast subplots, as well as at a complementary set of neighbouring unburned control sites (n = 12) representing unburned examples of all sampled vegetation communities. Cores were inserted to a minimum depth of 8.5 cm and the soil samples were separated into organic and mineral horizons; the three samples from each site were pooled by the horizon. If mineral soil was not present in the top 13.5 cm of the soil profile, it was not collected. Soils were oven-dried and the physicochemical properties of both organic and mineral samples were measured in the lab. These properties were: pH, electrical conductivity (EC; mS cm^{-1}), percent total nitrogen (N), percent total carbon (C) measured by loss on ignition, calcium (Ca; mg kg^{-1}), potassium (K; mg kg^{-1}), magnesium (Mg; mg kg^{-1}), and sodium (Na; mg kg^{-1}). The percentages of sand, silt, and clay in mineral soils were also measured. Measurements from the two pooled horizons from each site were combined using sums weighted by the mean proportion of the core occupied by each horizon.

Estimates of percentage cover of understory vascular plant species were made one year post-fire in five 1 × 1 m plots per field site. Vegetation plots were located at the plot centre and at the inner corners of subplots. Species were identified according to Moss [61] and Cody [62], and the estimated percentage cover for each species was summed across the five plots and scaled to sum to 100%. *Carex* spp. and *Salix* spp. were distinguished for counts of species richness but were not identified beyond genus for ordination or indicator species analyses (vegetation analysis explained in detail in Section 2.3).

The density of seedling recruitment was measured one year post-fire (2015), and subsequently re-measured three years post-fire (2017) in 30 forested sites (excluding open wetlands). Initial measures of seedling density were made in the 10 × 10 m subplots in 2015. In 2017, seedling density was re-measured using a 2-m wide 35-m long belt transect that was oriented north-south, crossing the original plot centre at 17.5 m. Belt-transect length varied by seedling and sapling size classes. Seedlings that were 0–10 cm were counted for the first 10 m of the transect (area 20 m^2) and seedlings that were 10–50 cm were counted for the first 20 m (area 40 m^2). Seedlings > 50 cm and saplings (live trees > 1.33 m with a DBH < 3 cm) were counted for the entire transect length. In cases of very uneven seedling density, transects of all size classes were extended to better represent the actual composition and density. This set of resampled sites excluded non-forested open wetlands (n = 11) and inaccessible sites (no helicopter or road access, n = 7). A further three sites were abandoned due to subsequent disturbances. The two datasets were combined and the latest available seedling density measurement for each site was used.

We calculated site climatic variables that described the average heat load and moisture stress from 30-year normals (1981–2010) of PRISM climate data [63] downscaled to local elevation [64] using bilinear interpolation and elevation adjustment in ClimateWNA [49]. The climatic moisture deficit (CMD; mm) was calculated as the sum of the monthly difference between Hargrave's atmospheric evaporative demand and monthly precipitation. Annual heat-to-moisture index (AHM) was calculated as the scaled ratio of mean annual temperature and mean annual precipitation [49].

2.3. Analysis

All statistical analyses were conducted in R [65]. The variance of burn severity explained by the topoedaphic vegetation classes was assessed using a linear mixed-effects model with a random term of the fire name, fitted in the lme4 [66] and lmerTest [67] packages. We examined all model residuals and found them to be normally distributed. The statistical significance of fixed effects was estimated using an analysis of variance (ANOVA) with Type II sums of squares and a Satterthwaite approximation of degrees of freedom [68]. We conducted post-hoc comparisons of least-squares means with a Tukey test for multiple comparisons in the lsmeans package [69]. Species richness and Shannon diversity index of understory vascular plant communities of each site were calculated in the vegan package [70]. Bray-Curtis dissimilarities between understory vegetation communities were ordinated using non-metric multidimensional scaling (NMDS). We fitted vectors of environmental variables to the NMDS axes and assessed the goodness of fit (R^2) of these relationships also using vegan

(Table 1). Indicator species for each topoedaphic vegetation class were identified from understory vascular plant assemblies using 1000 permutations of a multi-level pattern analysis in the indicspecies package [71]. We assessed the influence of burn severity on soil properties when controlling for topoedaphic vegetation class (as a proxy for pre-fire site conditions) using an ANOVA of multivariable linear mixed-effects models with a random term of the fire name. Once again, model residuals were examined for normality. We employed Type II sums of squares where interactions between independent variables were not significant. We applied a Type III ANOVA if there were significant interactions between independent variables. The same approach was used to assess the influence of burn severity, TSLF, and topoedaphic vegetation class (pre-fire conditions) on the Shannon diversity index and seedling density. We used comparisons of least-squares means with a Tukey test for multiple comparisons to assess significant differences in species dominance shifts and seedling density between different topoedaphic vegetation classes.

Table 1. Significant (* $p \leq 0.05$) explanatory environmental variables fitted to nonmetric multidimensional scaling (NMDS) of the understory vegetation community data (Figure 2). Unitless variables are identified with a hyphen in the Units column.

Environmental Variable	Abbreviation	Units	Mean	Range
Basal Area	BA	$m^2\,ha^{-1}$	10.8	0.00–53.39
Burn Severity Index	BSI	-	2.39	0.54–4.00
Electrical conductivity of soil	EC	$mS\,cm^{-1}$	0.73	0.05–3.53
Organic soil depth	OSD	cm	4.7	0–10
Percentage sand in mineral soil	% Sand	%	43.9	0–95
pH	pH	-	6.29	3.21–8.12
Potassium	K	$mg\,kg^{-1}$	411.1	74.1–1148.4
Site moisture	Moisture	-	-	Xeric–Hydric
Sodium	Na	$mg\,kg^{-1}$	137.8	38.52–494.48
Time since last fire	TSLF	$year^{-1}$	58	9–151
Total carbon	Total C	% mass	21.9	0.61–52.8
Total nitrogen	Total N	% mass	0.81	0.18–2.66
Total stem density of overstory and understory trees	Density	$stems\,ha^{-1}$	5822	0–29,012

Differences in the pre-fire and post-fire cohorts of trees were examined using compositional data. Overstory basal areas and total (understory and overstory) stem densities of each dominant tree species were converted to proportions relative to the absolute basal area and stem density for each site. Seedling counts were also converted to proportions by species, and these proportions, or compositions, were transformed with a centred log-ratio using the compositions package [72]. We then used paired *t* tests to identify statistical differences in the pre-fire and post-fire composition of trees, by species. We compared the natural logarithm ($\log_e$) of seedling density in sites that experienced very short fire return intervals to that of sites experiencing more typical fire return intervals with a Wilcoxon signed-rank test. Significant differences in $\log_e$ seedling density between topoedaphic vegetation classes were also tested using a Tukey test of least-squares means.

To examine shifts in the relative importance or dominance of tree species we calculated fractional ranks of pre-fire overstory tree species proportions of jack pine, white spruce, black spruce, trembling aspen, and all other tree species combined, by basal area. The most prevalent species received a rank of 1 and the least dominant (or absent) species received a rank of 5. In the case of ties, ranks were split between species, so that total rank values always summed to 15. We chose to use the pre-fire basal area rather than the number of stems as a measure of dominance as we felt that stem density did not adequately capture the potential fecundity and relative importance of less-common but large trees in mixedwood stands (e.g., white spruce). Because the basal area of trees established after fire represents only a small fraction of the pre-fire measure, post-fire tree species proportions were assigned fractional ranks by seedling stem density. The pre- and post-fire fractional rank scores of each species at each site were differenced to characterize shifts in tree species dominance in the post-fire cohort, producing a matrix of shifts in ranked dominance for each species by site. Rank shifts of near-zero indicated minimal change in the species' prevalence in the post-fire cohort, whereas negative values indicated a decrease and positive values indicated an increase in ranked dominance. Rank shift data had a theoretical range of −4 to 4. Analyses using the shift in rank dominance data were only performed for the four dominant tree species, excluding the combined "other" category. This application of fractional rank shifts characterizes the proportional change of tree species dominance relative to all tree species present in the community, rather than considering a single species at a time (as is the case with ratio data), and offers a normally distributed variable for analysis of dominance changes.

Subsequently, we assessed the relative importance of three groups of variables in the categories "Soils", "Site", and "Fire" to understory vegetation community dissimilarities and shifts in the dominance of tree species using variance partitioning. Soils were represented by post-fire soil properties, whereas Site category variables were pre-fire forests, topoedaphic context, and recent spatial climate averages. The Fire category included burn severity and fire history variables (Table 2). All measured and downscaled environmental variables were considered for inclusion in variance partitioning models. If the variables were highly correlated (Spearman's $|\rho| \geq 0.7$) one explanatory variable of the pair was selected for inclusion in the model. Several highly correlated soil properties were decomposed using a principal components analysis (PCA; Table 2). Sites with incomplete data were removed ($n = 5$), and explanatory variables were standardized before variance partitioning. The significance ($\alpha = 0.05$) of the unique variation explained by each group of environmental drivers (Soils, Site, and Fire) was tested using distance-based redundancy analysis, also in the vegan package.

Table 2. Environmental variables incorporated in the explanatory variance partitioning of understory vegetation community dissimilarities and shifts in tree species dominance. Correlated soil properties collapsed with a principal components analysis for inclusion in variance partitioning are indicated with a [†]. Unitless variables are identified with a hyphen in the Units column.

Environmental Variable	Units	Mean	Range	Variance Partitioning Category
Calcium [†]	mg kg^{-1}	13,648.6	217.1–60,815.4	Soils
Electrical conductivity [†]	mS cm^{-1}	0.73	0.05–3.53	Soils
Magnesium [†]	mg kg^{-1}	1461.9	22.0–5191.3	Soils
Percentage sand in mineral soil	% mass	44	0–95	Soils
Percentage silt in mineral soil	% mass	14	0–51	Soils
pH	-	6.29	3.21–8.12	Soils
Potassium [†]	mg kg^{-1}	411.1	74.1–1148.4	Soils
Sodium [†]	mg kg^{-1}	137.8	38.5–494.5	Soils
Total carbon [†]	% mass	21.9	0.61–52.8	Soils
Total nitrogen [†]	% mass	0.81	0.02–2.66	Soils
Absolute stem density of overstory and understory trees	stems ha^{-1}	5822	0–29,012	Site
Annual Heat-Moisture Index	-	20.32	16.5–23.6	Site
Black spruce basal area	m^2 ha^{-1}	2.37	0–29.07	Site
Climatic Moisture Deficit	mm	191	171–214	Site
Jack pine basal area	m^2 ha^{-1}	5.61	0–51.38	Site
Site moisture	-		Xeric–Hydric	Site
Trembling aspen basal area	m^2 ha^{-1}	0.87	0–6.22	Site
Total overstory basal area	m^2 ha^{-1}	10.8	0–53.4	Site
White spruce basal area	m^2 ha^{-1}	1.55	0–36.53	Site
Burn Severity Index	-	2.38	0.54–4	Fire
Percentage overstory mortality	%	89	6–100	Fire
Post-fire organic soil depth	cm	4.7	0–10	Fire
Time since last fire	year^{-1}	58	9–151	Fire
Time since stand origin	year^{-1}	104	9–237	Fire

Finally, we fit explanatory classification and regression trees (CARTs) to shifts in the ranked dominance of each tree species derived from ranked proportions of the pre-fire basal area and post-fire stem density, using the tree package [73]. The regression trees were constrained by requiring a minimum of five field sites per node, and a minimum within-node deviance of 0.05. We intentionally excluded pre-fire basal area and stem density of any tree species as regression tree predictor variables in order to learn about secondary climatic, soil, and burn severity effects on the dominance shifts of tree species. The same suite of environmental, burn severity and fire history, and climate variables were included as potential predictors of shifts in tree species dominance for each species' CART model (Table 3).

Table 3. Environmental, climate, burn severity, and fire history variables included in the classification and regression tree models of the dominance shifts of tree species. Unitless variables are identified with a hyphen in the Units column.

Variable	Units	Mean	Range
Annual Heat-Moisture Index	-	20.3	16.5–23.6
Burn Severity Index	-	2.38	0.54–4
Canopy Fire Severity Index	-	2.5	0–6
Climatic Moisture Deficit	mm	191	171–214
Electrical conductivity	$mS\,cm^{-1}$	0.73	0.05–3.53
Percentage overstory mortality	%	89	6–100
Percentage sand in mineral soil	%	44	0–95
pH	-	6.29	3.21–8.12
Post-fire organic soil depth	cm	4.7	0–10
Total nitrogen	%	0.81	0.02–2.66
Time since last fire	$year^{-1}$	58	9–151
Time since stand origin	$year^{-1}$	104	9–237
Wetland	-	-	Upland or Wetland

3. Results

A broad range of burn severity was represented in the field sites (Appendix A: Figure A2). The BSI values of field sites ranged from 0.5 to 4, the CFSI values from 0 to 6, and percent overstory mortality ranged from 6.25% to 100% [28]. The surface (BSI) and overstory (CFSI) burn severity were statistically related to topoedaphic vegetation classes (ANOVA, *** $p < 0.001$ and * $p = 0.02$, respectively; Table 4), but overstory mortality was not. Post-hoc comparisons of least-squares means with a Tukey test confirmed some statistical differences in burn severity amongst topoedaphic vegetation classes for BSI (* $p \leq 0.05$; Appendix A: Figure A2). Surface burn severity was lowest in open wetlands and highest in jack pine uplands. All other topoedaphic vegetation classes had BSI values that were similar to one of these two groups. The differences in least-squares means of CFSI between topoedaphic vegetation classes were not significant at $\alpha = 0.05$ (Appendix A: Figure A2).

The richness of the understory vascular plant species sampled at the field sites ranged from three to 20. Both understory vegetation community diversity and seedling density were statistically related to topoedaphic vegetation classes (Table 4; Appendix A: Figure A2). Interactions between TSLF and topoedaphic vegetation classes, and BSI and topoedaphic vegetation classes significantly explained the variability in understory plant diversity (Type III ANOVA, * $p \leq 0.02$; Table 4). The density of seedlings was significantly explained by both topoedaphic vegetation classes and BSI (Type II ANOVA, $p < 0.04$), but not by TSLF or CFSI, or by the interactions between topoedaphic vegetation classes and these two metrics (Table 4; Appendix A: Figure A2). All sites with zero seedling establishment were open wetlands ($n = 7$). Of those sites that experienced some regeneration, seedlings ha^{-1} ranged from 25 to >75,000. The natural logarithm of the density of seedlings was significantly lower in open wetlands, treed wetlands, and upland spruce, and higher in upland jack pine and upland mixedwood topoedaphic vegetation classes (comparison of least-squares means, with a Tukey test, $\alpha = 0.05$). The post-fire seedling density was statistically greater in sites that experienced >17 years between fires (Wilcoxon signed-rank test, * $p = 0.02$).

Table 4. Multivariable linear mixed-effects models describing surface Burn Severity Index (BSI), Canopy Fire Severity Index (CFSI), time since last fire (TSLF), topoedaphic vegetation classes (TVC), and ecological outcomes of seedling density and Shannon Diversity Index of understory vascular plant communities. The effect size (*F*) and significance (*p*) of terms are tested with Type II sums of squares where there are no significant interactions, and Type III sums of squares in the presence of a significant interaction. Significance of intendent variables is signified as follows: *** $p \leq 0.001$, ** $p \leq 0.01$, * $p \leq 0.05$.

Multivariable Linear Mixed-Effects Model	ANOVA Sums of Squares	Degrees of Freedom	Independent Variable	Sums of Squares	F	p
BSI = TVC + (1 | Fire Name)	II	4	TVC	20.39	14.94	*** < 0.001
CFSI = TVC + (1 | Fire Name)	II	4	TVC	33.40	3.36	* 0.02
Diversity = TVC × BSI + (1 | Fire Name)	III	4	TVC	2.51	3.09	* 0.03
		1	BSI	0.15	0.72	0.40
		4	TVC × BSI	2.64	3.26	* 0.02
Diversity = TVC × TSLF + (1 | Fire Name)	III	4	TVC	2.08	2.52	0.06
		1	TSLF	0.00	0.01	0.91
		4	TVC × TSLF	2.89	3.51	* 0.02
Density = TVC + BSI + (1 | Fire Name)	II	4	TVC	18.01	3.49	* 0.02
		1	BSI	5.62	4.35	* 0.04
Density = TVC + TSLF + (1 | Fire Name)	II	4	TVC	43.94	7.84	*** < 0.001
		1	TSLF	0.59	1.14	0.29

The two-dimensional NMDS of understory vascular plant communities had a stress of 0.20 (Figure 2). Similarity of understory species communities was primarily related to the physicochemical properties of the soil; however, pre-fire forest structural characteristics of basal area and absolute stem density of overstory and understory trees were also influential (Figure 2). BSI was also statistically related to understory species community dissimilarity, and TSLF was nearly significant ($p = 0.052$, 999 permutations). Although soil properties were explained by topoedaphic vegetation classes; the organic soil depth, total nitrogen, total carbon, potassium, calcium, and magnesium were also statistically (Type II ANOVA; $\alpha = 0.05$) related to BSI when controlling for the effect of topoedaphic vegetation class. Therefore, some soil properties were affected by fire (Appendix A: Table A1). Topoedaphic vegetation classes tended to occupy characteristic areas of ordination space, but there was some overlap between the normal confidence ellipses of classes. Upland mixedwood and upland jack pine groups were especially intermingled (Figure 2a), and mixedwood communities occurred in a sub-region of the broader environmental space occupied by jack pine. Similar patterns are identifiable in the post-fire understory indicator species of each topoedaphic vegetation class (Table 5). All topoedaphic vegetation classes had unique significant indicator species, with the exception of jack pine uplands, which shared all significant indicator species with the upland mixedwood group, and some with the upland spruce group (Table 5). *Potentilla palustris* (L.) Scop., *Betula glandulosa* Michx., *Epilobium palustre* L., and *Myrica gale* L. were unique indicator species of open wetlands. Treed wetlands had unique indicator species of *Rubus chamaemorus* L., *Vaccinium caespitosum* Michx., and *Vaccinium oxycoccos* L. *Viburnum edule* (Michx.) Raf. was a unique indicator species of upland mixedwood sites. Furthermore, upland mixedwood sites shared significant indicator species of *Cornus canadensis* L., *Geranium bicknellii* Britt., *Rosa acicularis* Lindl., *Linnaea borealis* L., and *Elymus innovatus* Beal with upland jack pine sites. *Vaccinium uliginosum* L. and *Geocaulon lividum* (Richards.) Fern. were unique indicator species in upland spruce communities.

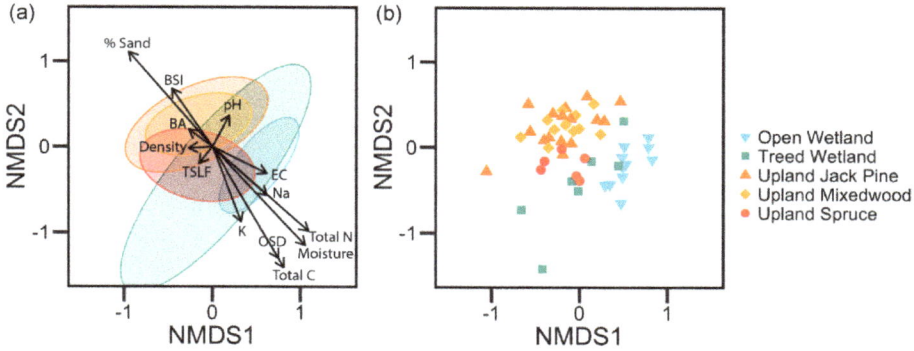

Figure 2. Nonmetric multidimensional scaling (NMDS) of post-fire understory vegetation community dissimilarities. Plot (**a**) shows normal confidence ellipses for topoedaphic vegetation classes (identified by colour) and environmental vectors derived from correlations between environmental variables and the NDMS axes, within the ordination space. Abbreviations of environmental variables are reported in Table 1. The strength of the relationship between an environmental vector and the NMDS (R^2) is indicated by the arrow length. Plot (**b**) shows the individual sites within the ordination space, with topoedaphic vegetation classes identified by point colour and shape.

Table 5. Significant (* $p \leq 0.05$) indicator species identified using multi-level pattern analysis within six topoedaphic vegetation classes. The indicator species uniquely associated with one group are indicated with a [†].

		Vegetation Group		
Open Wetland	**Treed Wetland**	**Upland Mixedwood**	**Upland Jack Pine**	**Upland Spruce**
Potentilla palustris (L.) Scop. [†]	*Rubus* chamaemorus L. [†]	*Viburnum edule* (Michx.) Raf. [†]	*Cornus canadensis* L.	*Vaccinium uliginosum* L. [†]
Betula glandulosa Michx. [†]	*Vaccinium caespitosum* Michx. [†]	*Cornus canadensis* L.	*Geranium bicknellii* Britt.	*Geocaulon lividum* (Richards.) Fern. [†]
Epilobium palustre L. [†]	*Vaccinium oxycoccos* L. [†]	*Geranium bicknellii* Britt.	*Rosa acicularis* Lindl.	*Ledum groenlandicum* Oeder
Myrica gale L. [†]	*Rubus arcticus* L.	*Rosa acicularis* Lindl.	*Linnaea borealis* L.	*Equisetum scirpoides* Michx.
Rubus arcticus L.	*Ledum groenlandicum* Oeder	*Linnaea borealis* L.	*Elymus innovatus* Beal	*Arctostaphylos rubra* (Rehder & Wils.)
Carex L. spp.	*Equisetum scirpoides* Michx.	*Elymus innovatus* Beal		*Rosa acicularis* Lindl.
Salix L. spp	*Arctostaphylos rubra* (Rehder & Wils.) Fern.			*Linnaea borealis* L.
	Carex L. spp.			*Elymus innovatus* Beal
	Salix L. spp.			*Carex* L. spp.
				Salix L. spp.

The dominance of pre-fire and post-fire tree species (represented by log-ratios of compositions of basal area and stem density) were significantly different for jack pine, black spruce, and white spruce (Paired *t* tests, Bonferroni-corrected * $p \leq 0.04$). When pre-fire dominance was characterized using total stem density (Appendix A: Figure A3), post-fire tree species compositions were significantly different for black spruce and trembling aspen (Paired *t* tests, Bonferroni-corrected * $p \leq 0.02$). Having confirmed significant differences between pre-fire and post-fire tree species compositions, we examined the rank shifts in dominance of tree species in order to capture directionality of species-specific changes. Jack pine both increased and decreased in dominance in the post-fire cohort, but the slim majority of sites were neutral (−0.5 to 0.5 shift in rank; 39% of sites). Furthermore, not all plots burned with completely stand-initiating lethal wildfires (Figure 3). Of the sites that experienced declines in the ranked dominance of jack pine, 41% had some surviving jack pine basal area post-fire (Figure 3). Aspen dominance increased in the post-fire cohort, with 54% of sites gaining 1 or more ranks of dominance post-fire, and no sites declining by <−0.5 of a rank (Figure 3). Both varieties of spruce primarily demonstrated no change or declines, in the post-fire cohort (57% of sites were neutral and 37% showed a decrease for black spruce; 69% of sites were neutral and 29% showed a decrease for white spruce). Of those sites with declines in the dominance of black spruce, only 9% (*n* = 2) had incomplete mortality of black spruce trees (Figure 3). No sites demonstrating declines in trembling aspen or white spruce had live individuals of these species post-fire (Figure 3). Declines and increases in tree species dominance were significantly related to topoedaphic vegetation classes. Increases in jack pine dominance were especially associated with wetlands and spruce uplands, whereas jack pine declines occurred in upland communities where jack pine was already established, especially mixedwood stands where the suckering of trembling aspen was prevalent (Appendix A: Figure A4; Tukey test of least-squares means, * $p \leq 0.05$). Although increases in aspen dominance were more pronounced in uplands (Wilcoxon signed-rank test, ** $p = 0.009$), there were no statistical differences in aspen dominance shifts between topoedaphic vegetation classes (Tukey test of least-squares means; Figure A4). Decreases in black spruce dominance were the most pronounced in upland spruce sites, whereas black spruce dominance was largely stable in treed wetlands and other vegetation classes (Tukey test of least-squares means, * $p \leq 0.05$). There were no significant differences in post-fire changes in white spruce dominance between topoedaphic vegetation classes (Figure A4).

When representing the variance in understory vegetation and tree species dominance shifts explained by soils, we found that many soil properties were highly correlated. To address this, we decomposed the correlated soil physicochemical properties using PCA, and included only the first principal component (PC1) as an explanatory variable in the variance partitioning (Table 2). We chose to retain percent sand and PC1, and organic soil depth and site moisture despite high correlations ($\rho = 0.8$) between these two pairs, as these variables characterized important elements of the three environmental driver groups. Post-fire soils (Soils); pre-fire forests, topoedaphic context, and climate (Site); and burn severity and fire history (Fire) together explained 28% of the variance in understory vegetation communities, and 33% of the variance in the dominance shifts of tree species (environmental variables included in model reported in Table 2; Figure 4). There was a substantial shared variance explained between Soils, Site, and Fire. Overall, Site explained the largest portion of the variance in post-fire vegetation communities (8%) and tree species dominance shifts (13%; Figure 4). Soils significantly explained 5% of the variance in understory vegetation but did not significantly explain tree species dominance shifts. Conversely, Fire was of substantial importance to tree species dominance shifts (7% of variance) but did not significantly explain post-fire vegetation communities ($p = 0.08$; Figure 4).

Classification and regression trees of the dominance shifts of tree species had R^2 values ranging from 0.65 to 0.43. Jack pine dominance increased in the post-fire cohort where the total soil N was $\geq 0.48\%$ and decreased in the post-fire cohort in stands that experienced partial mortality (Figure 5). These sites were typically mixedwood stands, and often had some remaining live basal area of jack pine trees, suggesting that these declines in the dominance of the post-fire cohort do not necessarily indicate persistent shifts away from jack pine dominance, although aspen suckering outpaced the establishment of pine seedlings (Appendix A: Figure A3). Trembling aspen increased in dominance in nearly all plots, but increases were somewhat limited in lightly burned plots and plots with higher N availability, both of which tend to be characteristic of wetlands (Figure 5). Changes in black spruce dominance were neutral in young stands where black spruce was essentially absent pre-fire and in wetlands (TSO < 80.5). Declines in black spruce dominance were augmented in moderate aged (TSO < 103.5) uplands and stands experiencing severe canopy burning (CFSI ≥ 4; Figure 5). White spruce dominance increased slightly in sites with low N availability (Total N < 0.11%) and lower moisture deficits (CMD < 191.5). White spruce dominance declined in historically drier sites, especially in those sites that experienced some canopy involvement in the fire (CFSI ≥ 2.65; Figure 5).

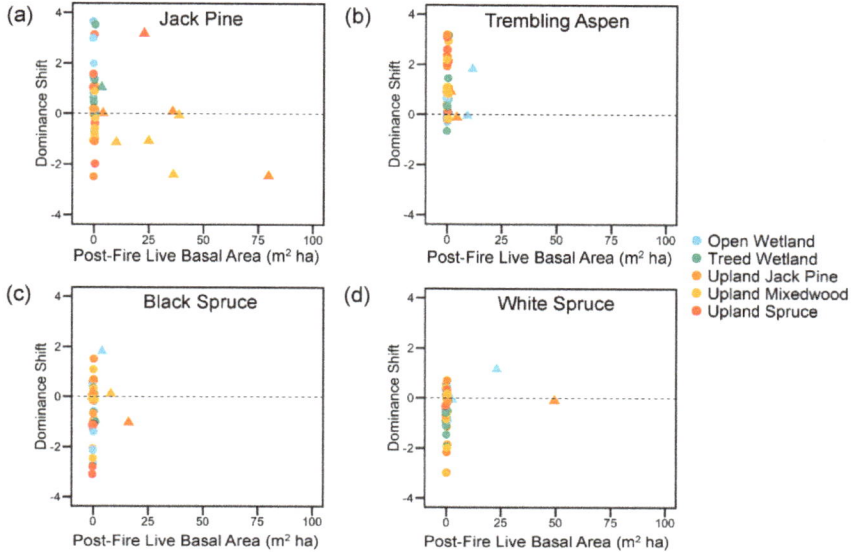

Figure 3. Increases and decreases in the post-fire dominance of (**a**) jack pine; (**b**) trembling aspen; (**c**) black spruce; and (**d**) white spruce, plotted against the post-fire live basal area of the same species. Points are coloured by topoedaphic vegetation classes. Circles indicate sites where the species experienced complete mortality or was absent pre-fire. Triangles indicate sites that had live residual basal areas of the species of interest following a wildfire. Points are offset ("jittered") to reduce overlap. Dashed horizontal lines indicate no change in species dominance post-fire. Points above this line increased in dominance post-fire, and points below are those sites that experienced a decline of the species of interest in the post-fire cohort.

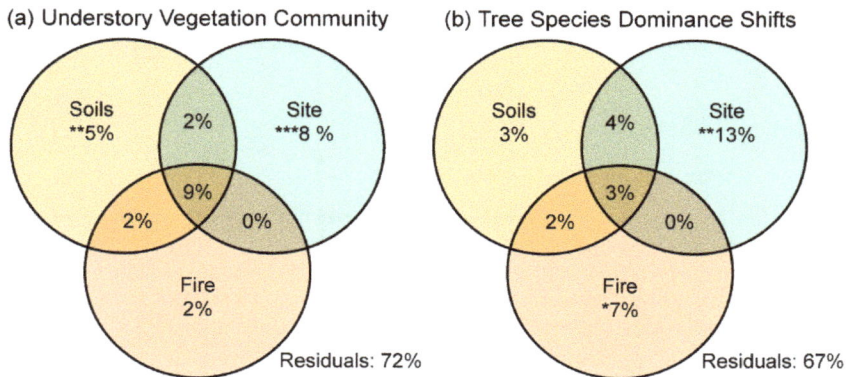

Figure 4. Venn diagrams showing the partitioning of variation in (**a**) post-fire understory vegetation community dissimilarities and (**b**) shifts in tree species dominance, between post-fire soils (Soils); pre-fire forests, site moisture, and climate (Site); burn severity and fire history (Fire); and unexplained residual variance. The significance of unique portions of variance explained is indicated by asterisks (* $p = 0.05$, ** $p = 0.01$, *** $p = 0.001$). The measured and interpolated environmental variables in each explanatory partition are reported in Table 2.

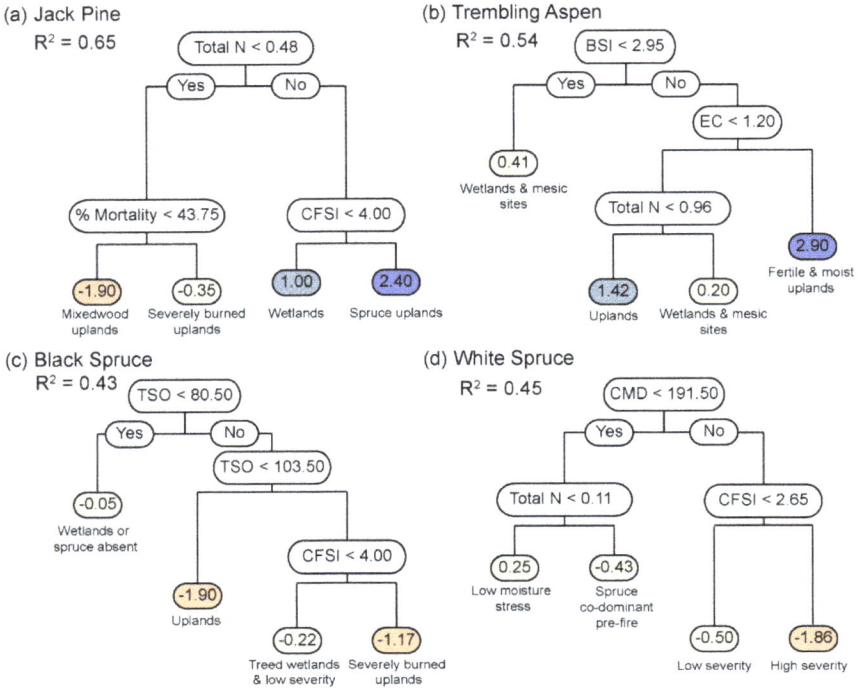

Figure 5. Regression trees of post-fire cohort dominance shifts for (**a**) jack pine; (**b**) trembling aspen, (**c**) black spruce; and (**d**) white spruce. Regression trees were fitted requiring a minimum of five sites per node, and a minimum within-node deviance of 0.05. Light blue terminal nodes indicate increases in dominance post-fire, and dark blue nodes indicate substantial increases in rank dominance (≥ 2). Red terminal nodes indicate decreases in dominance. Yellow nodes may be slightly negative or positive, but do not represent strong shifts (description of dominance shift metric in Section 2.3 Analysis). Descriptions of characteristic sites and drivers appear below each node. The environmental variables included in the regression tree models are reported in Table 3.

4. Discussion

4.1. Post-Fire Vegetation Communities of Vascular Plants

Post-fire understory vegetation communities were primarily explained by site conditions, but burn severity and fire history had significant secondary effects. Topoedaphic vegetation classes occupied distinct areas of the environmental space and had characteristic indicator species, with the exception of jack pine uplands and mixedwood stands. These indicator species were identifiable for topoedaphic vegetation classes despite recent disturbances, underscoring the importance of non-fire drivers to post-fire vascular vegetation. Understory vegetation communities of mixedwood forests appear to predominantly occur in a subset of environmental conditions that are also characterized by jack pine forests. Site moisture, climate, and pre-fire forest structure and composition, all of which were dominant drivers of understory vegetation assemblies, are independent of fire effects. Therefore, these communities may be somewhat robust to disturbance from fire, as they are primarily controlled by non-fire drivers.

Although the climate of the boreal forest is changing, changes to topography through background rates of uplift (1.3 cm year^{-1}) [74] and erosion (~0.005–0.0005 cm year^{-1}) [75] reported in parts of the biome are slow, relative to the velocity of climate change (~1 km/year^{-1} in the boreal forest) [76]. Hydrological feedbacks may also reinforce the persistence of some features, such as peatlands, in the face of drying and warming [77,78]. The persistence of these topoedaphic drivers over time, in the face of ongoing climate change, should encourage the re-establishment of understory vegetation communities following fires. As understory vegetation communities have a substantial influence on below-canopy light availability and nutrient cycling, this vegetation layer's persistence may, in turn, reinforce the similarity of post-fire communities to pre-fire conditions, with implications for seedling establishment [79–81]. Conversely, regions of the boreal forest experiencing rapid topoedaphic changes due to permafrost thaw and thermokarst formation may be more susceptible to shifts in vegetation communities as these dominant drivers undergo substantial short-term changes [82].

Fire affects post-fire vegetation assemblies, both directly through surface burn severity and time since last fire, and indirectly via fire-mediated changes to soil properties. Despite the overarching importance of fire-independent site characteristics, burn severity and fire history were associated with understory vegetation community dissimilarity. Additionally, these variables also had significant effects on diversity, when controlling for topoedaphic vegetation class. Furthermore, post-fire soils alone explained a substantial portion of the variance in understory vegetation communities and several soil properties that were affected by wildfire, as expected in boreal soils [83]. Despite the importance of non-fire site drivers to understory vegetation communities, changes in burn severity or fire return intervals will likely translate to shifts in understory vegetation communities through these secondary pathways. Increases in burn severity, in particular, may lead to lasting, directional compositional changes in understory species assemblies [31]. Where understory vegetation communities exist at the boundaries of their ideal environmental space (or in areas of overlap with other topoedaphic vegetation classes), the effect of burn severity and fire history may be more apparent, and potentially override fire-independent controls.

4.2. Post-Fire Shifts in Tree Species Dominance

Topoedaphic vegetation classes significantly explained total seedling density, but surface burn severity significantly interacted with these conditions, likely indicating an effect of seedbed availability on recruitment. Climate, pre-fire forest composition and structure, and site moisture were important variables in explaining post-fire shifts in the dominance of tree species, and in which sites such shifts occurred. Post-fire seedling density was significantly lower in wetlands and spruce-dominated sites and highest in jack pine and mixedwood uplands. Although the post-fire recruitment was lowest in open wetlands, some forested sites that reburned with very short intervals between stand-initiating fires ($\leq$16 years) also experienced near-failures in the recruitment of all tree species, including those that increased in overall dominance post-fire, and had significantly lower seedling densities than all other sites. Although we observed declines in the post-fire dominance of both spruce species, black spruce maintained its dominance in treed wetlands (peatlands) and white spruce dominance was stable in sites with a lower climatic moisture stress. Furthermore, increases in the dominance of aspen in the post-fire cohort were least prevalent in wetlands. The drivers of species persistence and types of sites where spruce species retained dominance, despite the broader neutral or declining trend across sites, reflect the importance of topoedaphic and climatological drivers to post-fire tree species shifts. Local variability in site moisture may offer refugia from climate change for both of these species, in a landscape with limited topography [78,84].

Despite the importance of climate and pre-fire forests to seedling recruitment, burn severity and fire history had detectable and important effects on post-fire shifts in the dominance of tree species. Canopy fire severity was implicated in both positive and negative shifts in tree species dominance for all conifer species. Black spruce dominance decreased in uplands where stand-initiating wildfires occurred at a frequency of fewer than ~100 years between fires. The dominance of both spruce species was reduced in sites that experienced high-severity crown fire. Surface burn severity was the primary driver of post-fire increases in dominance of trembling aspen, and likely had further indirect effects on the post-fire dominance of white spruce, aspen, and jack pine through nutrient availability due to organic soil combustion and heating. Increases in burn severity and combustion in sites that tended to protect tree species that were susceptible to declines in dominance (i.e., peatlands) or in sites where species declines were particularly pronounced (e.g., upland spruce sites) may have important implications for future tree species compositions [85].

Jack pine and trembling aspen made substantial gains in dominance in the post-fire cohort. These two species are shade-intolerant and require canopy openings from disturbances such as fire to regenerate and are, therefore, successful post-fire species. Spruce species establish shortly after a fire, but appeared to be atypically uncommon in sampled fires from this severe drought-driven fire season, compared to previous studies of the mixedwood boreal zone [86,87] or the northern boreal forest [23,24]. Although spruce trees can persist as suppressed individuals, if seedlings fail to establish following fire they are unlikely to go on to become stand dominants through succession, as the cohort of seedlings established immediately post-fire (1–20 year^{-1}) in boreal forests goes on to make up the future forest [25]. In light of this, jack pine and trembling aspen appear to have gained, at the cost of longer-lived, "late-successional" spruce tree species [88,89]. Additionally, some jack pine stems regularly remained alive post-fire, whereas this was less common for the other three tree species—wildfires killed almost all individuals in burned patches. The successional pathways identified here suggest that increases in burn severity and fire frequency would continue to promote a growing component of jack pine and trembling aspen in northwestern boreal forests, despite topoedaphic, climate, and forest structure controls on post-fire dominance shifts.

In boreal forests, tree species adaptations to wildfire tend to promote "direct regeneration", where post-fire stands return to pre-disturbance compositions over time. Black spruce is a semi-serotinous species that has demonstrated stand self-replacement following fires in the northern boreal forest [90,91]. The post-fire decreases in the ranked proportional dominance of black spruce that we observed may suggest that increasing fire frequencies and severity may surpass the capacity of this species to re-establish following fires at the proportions previously expected, especially in drier uplands, if there is substantial combustion in both the overstory and understory [24,38,85]. Additionally, in the topoedaphic vegetation classes where the dominance of black spruce was stable (wetlands), and where spruce was previously dominant (spruce uplands), the seedling density was significantly lower than that measured in jack pine and mixedwood uplands. Declines in black spruce dominance relative to early-successional tree species, or through deforestation following severe fires, were observed in Alaska [92,93], the Yukon Territory [37], and in the eastern Canadian boreal forest [24], and this research, provides additional evidence for the potential occurrence of this phenomenon in northwestern Canadian forests.

4.3. Implications for Northwestern Boreal Forests

Wildfire is the stand-initiating disturbance with the largest extent in the northwestern Canadian boreal forest [94]. Therefore, drivers of post-fire vegetation assemblies are an important determinant of future forest composition in this region. The post-fire understory vegetation communities, seedling density, and shifts in the dominance of tree species were primarily attributable to pre-fire forests, climate, and topoedaphic context, suggesting that there is substantial capacity for forests and understory vegetation communities to regenerate post-fire. Although some variability in post-fire communities was attributed to burn severity, burn severity in this region is also associated with pre-fire forest structure and composition [28], further reinforcing the importance of pre-fire drivers to observed vegetation assemblies.

Despite this resilience, long-term shifts in tree species compositions are ongoing in parts of the western Canadian boreal forest, with proportions of early-seral shade-intolerant species such as jack pine and trembling aspen demonstrating increasing prevalence, driven by climate change [88].

Simultaneously, droughts appear to have caused decreases in forest productivity, altered seedling establishment and caused large-scale die-offs of mature trees in northern forests [95–97]. Although strong non-fire controls on understory vegetation and seedling establishment offer some resilience to change, the secondary direct and indirect effects of fire will likely serve to accelerate these ongoing changes if fire size, frequency, and severity increase as projected [40,41,98].

An increasing broadleaf component in northwestern boreal forests, such as that observed in this study, may reduce fire severity and flammability of boreal forests [28,99,100] and raise the surface albedo [101], potentially offering a negative feedback to shifts driven by climate change and impeding increases in fire activity [16,102]. This effect would be transient if increases of the proportion of trembling aspen in boreal forests do not persist. Increases in the frequency and severity of droughts may lead to a subsequent decline in this drought-sensitive species [97,103]. Drought stress would also likely further exacerbate black spruce declines and potentially favour more drought-tolerant upland conifers such as jack pine [104–106]. Furthermore, droughts increase the susceptibility of fuel-limited young forests to reburning [33,107], which could yet again reinforce reductions in black spruce dominance through the reduced availability of viable seeds. Observed post-fire seedling density was highly variable, substantially different from pre-fire species compositions, and several sites experienced near regeneration-failures when severely burned at short fire frequencies. This research contributes to the growing body of literature indicating that changes to forests of this region are ongoing, despite the overarching resistance to such shifts conferred by regeneration mechanisms and topoedaphic controls [83,92,93].

4.4. Limitations and Future Research

Due to the opportunistic nature of this study's sampling design, we were unable to measure changes in the composition of the understory vegetation assemblies from pre-fire to post-fire communities. Although we partitioned the variance in post-fire vegetation communities to identify some role of wildfire on their determination, studies where prescribed burns are planned or existing plots are burned over in natural fires are better positioned to measure shifts in the dominance of understory vegetation species from pre-fire to post-fire conditions. Soils are important to post-fire understory vegetation communities and they are also relevant to seedling recruitment through the provision of seedbeds [108–110]. Such studies would allow researchers to measure the changes in soils as a result of fires, including changes in the organic layer depth.

We conducted our field sampling one year and three years post-fire. Studies spanning a longer time period can provide additional insights into post-fire vegetation recovery for both understory plants and trees e.g., [18,26], but this was beyond the scope of this work. An assessment of whether a forest has recovered to a state similar to pre-fire conditions would require an extensive period of time, reflecting the growing conditions at high latitudes and local disturbance regimes (e.g., stand ages at the time of burning ranged from 9 to 237 years in this study). Although the post-fire recruitment pulse for some tree species may not be complete three years post-fire, there is ample evidence that the recruitment occurring within the first few years post-fire largely determines the future species composition and structure of the stand in boreal forests [25,108]. Just over one-third of our plots did not have repeated measures of seedling recruitment data sampled three years post-fire; however, the majority of these sites that were not revisited were non-forested wetlands (fens), with no trees pre-fire. This data gap affected closer to a quarter of the forested sites. Spruce trees are slower to establish, and this may have biased our results; however, we did observe some spruce seedling recruitment in most plots with a pre-fire spruce presence. By converting our measurements of post-fire seedling density to compositional log-ratio data, and calculating shifts in the ranked dominance of species, we captured changes in post-fire tree species composition and normalized the highly skewed seedling density data. This method does not permit us to assess structural changes that may have occurred; for example, whether post-fire forest density increased or decreased and whether these outcomes vary by species. Although we did characterize some variability in seedling density by topoedaphic vegetation communities and fire frequencies, future research could combine ranked dominance shift data with seedling and pre-fire stem densities to directly capture regeneration failures and structural changes, in addition to the shifts in proportional dominance measured here.

At the time of burning there was an ongoing multi-year drought in the study area, which continued into 2015, and may have affected the post-fire recruitment of seedlings, as well as their growth [95,111]. The drought conditions may also have affected burn severity of the fires, as fire weather is significantly related to overstory and understory combustion in boreal forests [28,112]. The identified impacts of fire on understory vegetation and seedling recruitment may have been influenced by these pre- and post-fire environmental conditions and therefore, the observed vegetation assemblies and ecological outcomes may be most representative of severe fire years. Sampling in wildfires that occurred in different years would capture a wider range of pre- and post-fire climates.

5. Conclusions

In this study, the primary determinants of post-fire outcomes for boreal forest vegetation communities and shifts in tree species dominance were pre-fire forests, topoedaphic context and climate. Burn severity, fire history, and post-fire soils were significant secondary drivers. Burn severity and fire history did not significantly explain the variability in understory vegetation communities; however, post-fire soils were related to understory vegetation community dissimilarities. Furthermore, burn severity was significantly related to understory vascular plant diversity. Severely burned vegetation communities tended to have lower understory species richness and diversity, as did very wet sites, which typically burned at low levels of severity. Post-fire shifts in tree species dominance, as characterized by differences in ranked proportional compositions, were significantly related to fire history and burn severity, but this effect was less important than pre-fire and climatological conditions. The overriding control of fire-independent drivers on post-fire vegetation may provide some resilience to forests in the face of climate change, as they are less susceptible to fire-mediated type conversions due to site moisture and pre-fire forest drivers. Despite this potential for resilience, changes to forest vegetation community compositions due to altered climates are occurring, and burn severity and fire history were important explanatory variables in our models of shifts in tree species dominance. In a forest with potentially increasing frequency, size, and severity of fires, the long-term resistance to change conferred by topoedaphic and forest controls may be overwhelmed by the direct and indirect effects of wildfires, which offer pathways to change. Burned sites will also experience altered post-fire climates, with potential increases in moisture stress and droughts which would exert additional pressures on initial post-fire vegetation. Ongoing shifts in the dominance of tree species are the result of both climate and fire. If these disturbances continue to increase, the observed shifts towards early-seral species such as jack pine and trembling aspen could produce large-scale changes in vegetation dominance that may lead to substantial—and perhaps unanticipated—ecological changes.

Acknowledgments: This research was funded by the Natural Sciences and Engineering Research Council of Canada (Funding Reference Number: CGSD3-471480-2015) and the Government of the Northwest Territories. Parks Canada Agency and Jean Morin provided in-kind support. We thank Xinli Cai, Matt Coyle, G. Matt Davies, Kathleen Groenewegen, Derek Hall, Koreen Millard, Sean A. Parks, and Doug Stiff for assistance in the field.

Author Contributions: E.W., M.-A.P., D.K.T., and M.D.F. conceived of the study; E.W., M.-A.P., and D.K.T. designed the experimental methods and collected the data; E.W. analyzed the data and wrote the paper, with contributions from M.-A.P., D.K.T., and M.D.F.

Conflicts of Interest: The authors declare no conflict of interest.

Appendix A. Additional Analyses and Figures

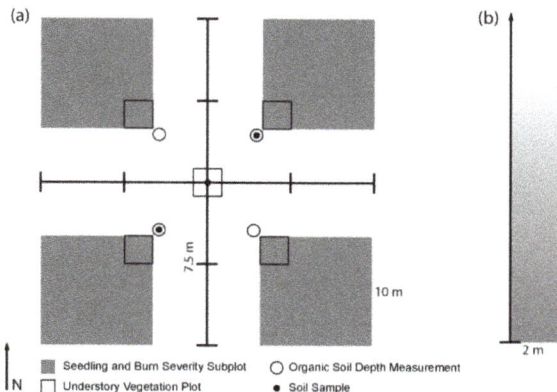

Figure A1. Plot layouts used for field sampling: (**a**) 30 × 30 m square plot one year post-fire; and (**b**) 35 × 2 m belt transect for re-measurement of seedlings three years post-fire. Plot layouts and symbols are not to scale.

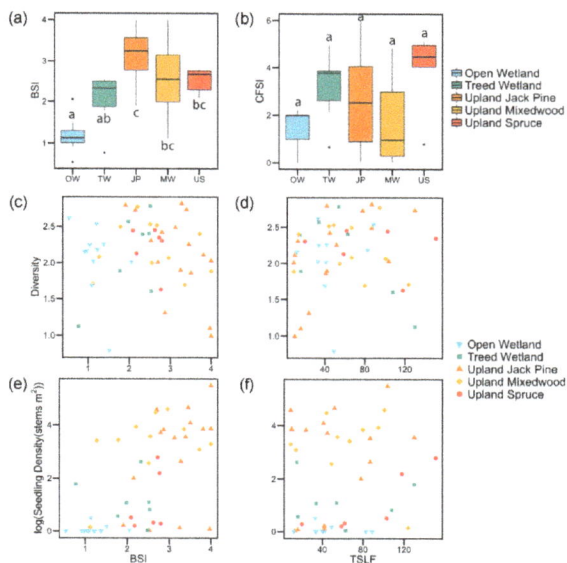

Figure A2. Descriptive plots of observed burn severity, fire history, and topoedaphic vegetation classes in relation to Shannon diversity index and seedling density within sampled plots. Patterns of burn severity within topoedaphic vegetation classes of open wetland (OW), treed wetland (TW), upland mixedwood (MW), upland jack pine (JP), and upland black or white spruce (US) are presented in (**a**) boxplots of surface burn severity index (BSI), and (**b**) boxplots of canopy fire severity index (CFSI). Letters associated with boxplots indicate significant differences ($\alpha = 0.05$) in a Tukey test of least-squares means. Shannon index is shown as a function of BSI (**c**) and time since last fire (TSLF); (**d**), and the natural logarithm of stem density of seedlings is also shown as a function of BSI (**e**) and TSLF (**f**).

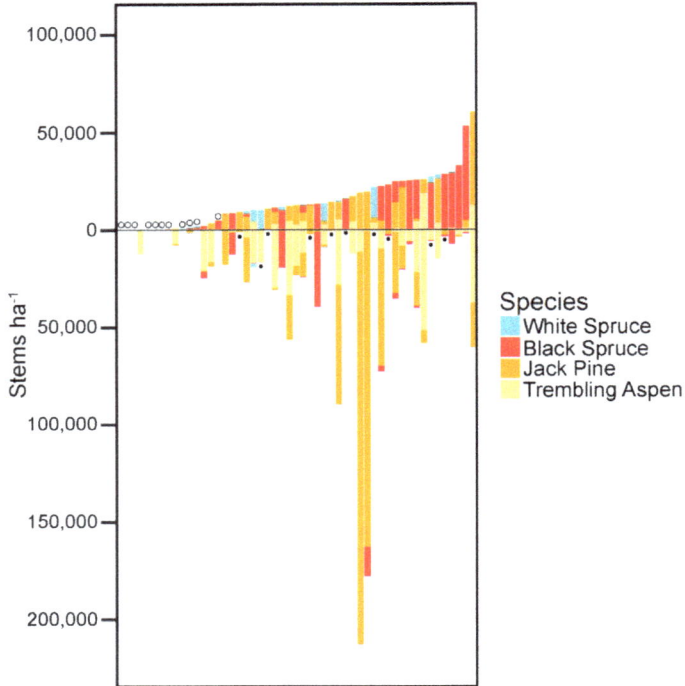

Figure A3. Pre-fire overstory (top) and post-fire (bottom) densities of trees (stems per ha^{-1}) by species at 51 sampled field sites. Sites are ordered from left to right by increasing pre-fire stem density. Non-forested open wetland sites are indicated with a letter O above the pre-fire stem density. Open wetlands and all sites indicated with a black dot below the post-fire density were sampled one year post-fire only. All other sites represent seedling data from three years post-fire.

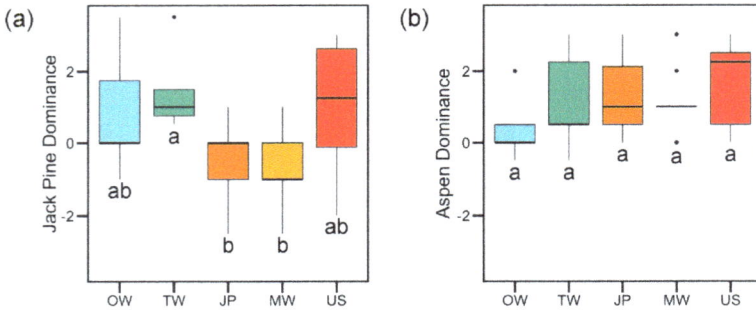

Figure A4. *Cont.*

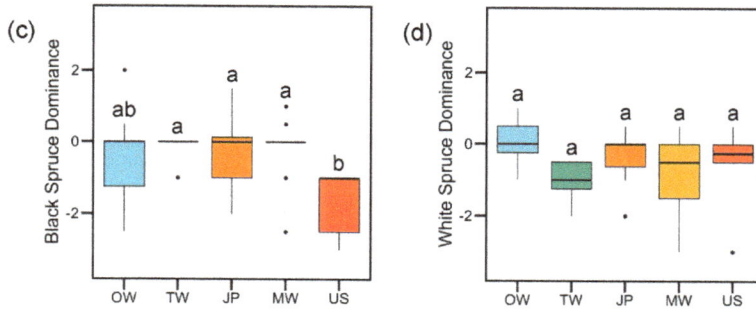

Figure A4. Increases and decreases in post-fire dominance of (**a**) jack pine, (**b**) trembling aspen, (**c**) black spruce, and (**d**) white spruce, within topoedaphic vegetation classes of open wetland (OW; blue), treed wetland (TW; green), upland mixedwood (MW; yellow), upland jack pine (JP; orange), and upland black or white spruce (US; red). Values of 0 indicate no change in dominance, whereas values greater than 0 indicate sites with an increase in dominance post-fire, and those sites that experienced a decline of the species of interest in the post-fire cohort have negative values. Letters above or below boxplots indicate significant differences in least-squares means ($\alpha = 0.05$), with a post-hoc Tukey test for multiple comparisons.

Table A1. Descriptive multivariable linear mixed-effects models explaining post-fire organic soil depth and soil chemical properties as a function of topoedaphic vegetation classes (TVC) and observed surface burn severity, represented by the Burn Severity Index (BSI). The statistical significance of independent predictor variables in explaining soil properties was determined using an ANOVA considering Type II sums of squares. Significance of independent variables to soil properties is signified as follows: *** $p \le 0.001$, ** $p \le 0.01$, * $p \le 0.05$. Some interactions are nearly significant ($\alpha = 0.1$).

Soil Property	Multivariable Linear Mixed-Effects Model	Independent Variable	Degrees of Freedom	Sums of Squares	F	p
Organic Soil Depth (cm)	OSD = TVC + BSI + (TVC × BSI)	Fire Name	TVC	4	91.50	10.10	*** < 0.001
		BSI	1	30.68	13.54	*** < 0.001
		TVC × BSI	4	8.80	0.97	0.43
Total Nitrogen (%)	TN = TVC + BSI + (TVC × BSI)	Fire Name	TVC	4	2.34	8.50	*** < 0.001
		BSI	1	0.81	11.77	** 0.001
		TVC × BSI	4	0.26	0.96	0.44
pH	pH = TVC + BSI + (TVC × BSI)	Fire Name	TVC	4	0.21	0.06	0.99
		BSI	1	002	0.02	0.88
		TVC × BSI	4	0.85	0.25	0.91
Electrical Conductivity	EC = TVC + BSI + (TVC × BSI)	Fire Name	TVC	4	3.16	2.08	0.10
		BSI	1	0.32	0.84	0.36
		TVC × BSI	4	0.59	0.39	0.82
Total Carbon (%)	TC = TVC + BSI + (TVC × BSI)	Fire Name	TVC	4	1697.20	13.20	*** < 0.001
		BSI	1	382.75	11.90	** 0.001
		TVC × BSI	4	91.91	0.71	0.59
Sodium	Na = TVC + BSI + (TVC × BSI)	Fire Name	TVC	4	121,344	6.20	*** < 0.001
		BSI	1	26,453	5.41	* 0.02
		TVC × BSI	4	41,222	2.11	0.09
Potassium	K = TVC + BSI + (TVC × BSI)	Fire Name	TVC	4	59,853	0.60	0.67
		BSI	1	35,436	1.41	0.24
		TVC × BSI	4	226,446	2.26	0.08
Calcium [†]	Ca = TVC + BSI + (TVC × BSI)	Fire Name	TVC	4	2,000,913,013	4.00	** 0.007
		BSI	1	672,020,248	5.37	* 0.02
		TVC × BSI	4	656,747,510	1.31	0.28
Magnesium	Mg = TVC + BSI + (TVC × BSI)	Fire Name	TVC	4	23,579,801	5.54	*** 0.001
		BSI	1	5,452,454	5.13	* 0.03
		TVC × BSI	4	6,042,409	1.42	0.25

[†] Calculated using Kenward–Roger approximation of degrees of freedom, due to the mathematical failure of Satterthwaite's approximation.

References

1. Weber, M.G.; Stocks, B.J. Forest fires in the boreal forests of Canada. In *Large Forest Fires*; Moreno, J.M., Ed.; Backbuys Publishers: Leiden, The Netherlands, 1998; pp. 215–233.
2. Heinselman, M.L. Fire and Succession in the Conifer Forests of Northern North America. In *Forest Succession*; West, D.C., Shugart, H.H., Botkin, D.B., Eds.; Springer: New York, NY, USA, 1981; pp. 374–405.
3. Morgan, P.; Keane, R.E.; Dillon, G.K.; Jain, T.B.; Hudak, A.T.; Karau, E.C.; Sikkink, P.G.; Holden, Z.A.; Strand, E.K. Challenges of assessing fire and burn severity using field measures, remote sensing and modelling. *Int. J. Wildland Fire* **2014**, *23*, 1045–1060. [CrossRef]
4. Keeley, J.E. Fire intensity, fire severity and burn severity: A brief review and suggested usage. *Int. J. Wildland Fire* **2009**, *18*, 116–126. [CrossRef]
5. Whittle, C.A.; Duchesne, L.C.; Needham, T. The importance of buried seeds and vegetative propagation in the development of postfire plant communities. *Environ. Rev.* **1997**, *5*, 79–87. [CrossRef]
6. Alexander, M.E.; Cruz, M.G. Modelling the impacts of surface and crown fire behaviour on serotinous cone opening in jack pine and lodgepole pine forests. *Int. J. Wildland Fire* **2012**, *21*, 709–721. [CrossRef]
7. Knapp, A.K.; Anderson, J.E. Effect of Heat on Germination of Seeds from Serotinous Lodgepole Pine Cones. *Am. Midl. Nat.* **1980**, *104*, 370–372. [CrossRef]
8. Greene, D.F.; Macdonald, S.E.; Cumming, S.; Swift, L. Seedbed variation from the interior through the edge of a large wildfire in Alberta. *Can. J. For. Res.* **2005**, *35*, 1640–1647. [CrossRef]
9. Lee, P. The impact of burn intensity from wildfires on seed and vegetative banks, and emergent understory in aspen-dominated boreal forests. *Can. J. Bot.* **2004**, *82*, 1468–1480. [CrossRef]
10. Schimmel, J.; Granström, A. Fire Severity and Vegetation Response in the Boreal Swedish Forest. *Ecology* **1996**, *77*, 1436–1450. [CrossRef]
11. Turner, M.G.; Romme, W.H.; Gardner, R.H. Prefire heterogeneity, fire severity, and early postfire plant reestablishment in subalpine forests of Yellowstone National Park, Wyoming. *Int. J. Wildland Fire* **1999**, *9*, 21–36. [CrossRef]
12. Lentile, L.B.; Morgan, P.; Hudak, A.T.; Bobbitt, M.J.; Lewis, S.A.; Smith, A.M.S.; Robichaud, P.R. Post-Fire Burn Severity and Vegetation Response Following Eight Large Wildfires Across the Western United States. *Fire Ecol.* **2007**, *3*, 91–108. [CrossRef]
13. Chambers, M.E.; Fornwalt, P.J.; Malone, S.L.; Battaglia, M.A. Patterns of conifer regeneration following high severity wildfire in ponderosa pine–dominated forests of the Colorado Front Range. *For. Ecol. Manag.* **2016**, *378*, 57–67. [CrossRef]
14. Shenoy, A.; Johnstone, J.F.; Kasischke, E.S.; Kielland, K. Persistent effects of fire severity on early successional forests in interior Alaska. *For. Ecol. Manag.* **2011**, *261*, 381–390. [CrossRef]
15. Bartels, S.F.; Chen, H.Y.H.; Wulder, M.A.; White, J.C. Trends in post-disturbance recovery rates of Canada's forests following wildfire and harvest. *For. Ecol. Manag.* **2016**, *361*, 194–207. [CrossRef]
16. Jin, Y.; Randerson, J.T.; Goetz, S.J.; Beck, P.S.A.; Loranty, M.M.; Goulden, M.L. The influence of burn severity on postfire vegetation recovery and albedo change during early succession in North American boreal forests. *J. Geophys. Res. Biogeosci.* **2012**, *117*, 1–15. [CrossRef]
17. Epting, J.; Verbyla, D. Landscape-level interactions of prefire vegetation, burn severity, and postfire vegetation over a 16-year period in interior Alaska. *Can. J. For. Res.* **2005**, *35*, 1367–1377. [CrossRef]
18. Gibson, C.M.; Turetsky, M.R.; Cottenie, K.; Kane, E.S.; Houle, G.; Kasischke, E.S. Variation in plant community composition and vegetation carbon pools a decade following a severe fire season in interior Alaska. *J. Veg. Sci.* **2016**, *27*, 1187–1197. [CrossRef]
19. Pinno, B.D.; Errington, R.C. Burn severity dominates understory plant community response to fire in xeric jack pine forests. *Forests* **2016**, *7*. [CrossRef]
20. Wang, G.G.; Kemball, K.J. Effects of fire severity on early development of understory vegetation. *Can. J. For. Res.* **2005**, *35*, 254–262. [CrossRef]
21. Arseneault, D. Impact of fire behavior on postfire forest development in a homogeneous boreal landscape. *Can. J. For. Res.* **2001**, *31*, 1367–1374. [CrossRef]
22. Pinno, B.D.; Errington, R.C.; Thompson, D.K. Young jack pine and high severity fire combine to create potentially expansive areas of understocked forest. *For. Ecol. Manag.* **2013**, *310*, 517–522. [CrossRef]

23. Johnstone, J.F.; Kasischke, E.S. Stand-level effects of soil burn severity on postfire regeneration in a recently burned black spruce forest. *Can. J. For. Res.* **2005**, *35*, 2151–2163. [CrossRef]

24. Lavoie, L.; Sirois, L. Vegetation changes caused by recent fires in the northern boreal forest of eastern Canada. *J. Veg. Sci.* **1998**, *9*, 483–492. [CrossRef]

25. Johnstone, J.F.; Chapin, F.S., III; Foote, J.; Kemmett, S.; Price, K.; Viereck, L. Decadal observations of tree regeneration following fire in boreal forests. *Can. J. For. Res.* **2004**, *34*, 267–273. [CrossRef]

26. Collins, B.M.; Stephens, S.L. Stand-replacing patches within a "mixed severity" fire regime: Quantitative characterization using recent fires in a long-established natural fire area. *Landsc. Ecol.* **2010**, *25*, 927–939. [CrossRef]

27. Lydersen, J.M.; Collins, B.M.; Brooks, M.L.; Matchett, J.R.; Shive, K.L.; Povak, N.A.; Kane, V.R.; Smith, D.F. Evidence of fuels management and fire weather influencing fire severity in an extreme fire event. *Ecol. Appl.* **2017**, *27*, 1–18. [CrossRef] [PubMed]

28. Whitman, E.; Parisien, M.-A.; Thompson, D.K.; Hall, R.J.; Skakun, R.J.; Flannigan, M.D. Variability and drivers of burn severity in the northwestern Canadian boreal forest. *Ecosphere* **2018**, *9*, e02128. [CrossRef]

29. Fourrier, A.; Bouchard, M.; Pothier, D. Effects of canopy composition and disturbance type on understorey plant assembly in boreal forests. *J. Veg. Sci.* **2015**, *26*, 1225–1237. [CrossRef]

30. Turner, M.G.; Romme, W.H.; Gardner, R.H.; Hargrove, W.W. Effects of Fire Size and Pattern on Early Succession in Yellowstone National Park. *Ecol. Monogr.* **1997**, *67*, 411–433. [CrossRef]

31. Day, N.J.; Carrière, S.; Baltzer, J.L. Annual dynamics and resilience in post-fire boreal understory vascular plant communities. *For. Ecol. Manag.* **2017**, *401*, 264–272. [CrossRef]

32. Boiffin, J.; Aubin, I.; Munson, A.D. Ecological controls on post-fire vegetation assembly at multiple spatial scales in eastern North American boreal forests. *J. Veg. Sci.* **2015**, *26*, 360–372. [CrossRef]

33. Erni, S.; Arseneault, D.; Parisien, M.-A.; Bégin, Y. Spatial and temporal dimensions of fire activity in the fire-prone eastern Canadian taiga. *Glob. Chang. Biol.* **2016**, 1–15. [CrossRef] [PubMed]

34. Turner, M.G.; Romme, W.H. Landscape dynamics in crown fire ecosystems. *Landsc. Ecol.* **1994**, *9*, 59–77. [CrossRef]

35. Stocks, B.J.; Mason, J.A.; Todd, J.B.; Bosch, E.M.; Wotton, B.M.; Amiro, B.D.; Flannigan, M.D.; Hirsch, K.G.; Logan, K.A.; Martell, D.L.; et al. Large forest fires in Canada, 1959–1997. *J. Geophys. Res.* **2002**, *108*, FFR 5-1–FFR 5-12. [CrossRef]

36. Johnstone, J.F.; Chapin, F.S., III. Fire interval effects on successional trajectory in boreal forests of northwest Canada. *Ecosystems* **2006**, *9*, 268–277. [CrossRef]

37. Brown, C.D.; Johnstone, J.F. Once burned, twice shy: Repeat fires reduce seed availability and alter substrate constraints on Picea mariana regeneration. *For. Ecol. Manag.* **2012**, *266*, 34–41. [CrossRef]

38. Johnstone, J.F.; Allen, C.D.; Franklin, J.F.; Frelich, L.E.; Harvey, B.J.; Higuera, P.E.; Mack, M.C.; Meentemeyer, R.K.; Metz, M.R.; Perry, G.L.W.; et al. Changing disturbance regimes, ecological memory, and forest resilience. *Front. Ecol. Environ.* **2016**, in press. [CrossRef]

39. Coop, J.D.; Parks, S.A.; McClernan, S.R.; Holsinger, L.M. Influences of Prior Wildfires on Vegetation Response to Subsequent Fire in a Reburned Southwestern Landscape. *Ecol. Appl.* **2016**, *26*, 346–354. [CrossRef] [PubMed]

40. Wotton, B.M.; Flannigan, M.D.; Marshall, G.A. Potential climate change impacts on fire intensity and key wildfire suppression thresholds in Canada. *Environ. Res. Lett.* **2017**, *12*, 95003. [CrossRef]

41. Wotton, B.M.; Nock, C.A.; Flannigan, M.D. Forest fire occurrence and climate change in Canada. *Int. J. Wildland Fire* **2010**, *19*, 253–271. [CrossRef]

42. Wang, X.; Parisien, M.-A.; Taylor, S.W.; Candau, J.N.; Stralberg, D.; Marshall, G.A.; Little, J.M.; Flannigan, M.D. Projected changes in daily fire spread across Canada over the next century. *Environ. Res. Lett.* **2017**, *12*. [CrossRef]

43. Wang, X.; Thompson, D.K.; Marshall, G.A.; Tymstra, C.; Carr, R.; Flannigan, M.D. Increasing frequency of extreme fire weather in Canada with climate change. *Clim. Chang.* **2015**, *130*, 573–586. [CrossRef]

44. Northwest Territories Environment and Natural Resources. *2014 NWT Fire Season: Review Report*; Government of the Northwest Territories: Yellowknife, NT, Canada, 2015.

45. Boulanger, Y.; Gauthier, S.; Burton, P.J.; Vaillancourt, M.A. An alternative fire regime zonation for Canada. *Int. J. Wildland Fire* **2012**, *21*, 1052–1064. [CrossRef]

46. Johnson, E.A. *Fire and Vegetation Dynamics: Studies from the North American Boreal Forest*; Cambridge University Press: Cambridge, UK, 1992; ISBN 0521341515.

47. Brandt, J.P. The extent of the North American boreal zone. *Environ. Rev.* **2009**, *17*, 101–161. [CrossRef]

48. Ecological Stratification Working Group (ESWG). *A National Ecological Framework for Canada*; Agriculture and Agri-Food Canada and Environment Canada: Ottawa, ON, Canada, 1995; ISBN 066224107X.

49. Wang, T.; Hamann, A.; Spittlehouse, D.L.; Murdock, T.Q. ClimateWNA—High-resolution spatial climate data for western North America. *J. Appl. Meteorol. Climatol.* **2012**, *51*, 16–29. [CrossRef]

50. Tarnocai, C.; Kettles, I.M.; Lacelle, B. *Peatlands of Canada*; Open File 6561; Geological Survey of Canada, Natural Resources Canada: Ottawa, ON, Canada, 2011.

51. Natural Resources Canada Canada-Permafrost. *National Atlas of Canada*; Geological Survey of Canada, Natural Resources Canada: Ottawa, ON, Canada, 1993.

52. Key, C.H.; Benson, N.C. *Landscape Assessment (LA): Sampling and Analysis Methods*; Tech. Rep. RMRS-GTR-164-CD; SDA Forest Service: Fort Collins, CO, USA, 2006; pp. LA-1–LA-51.

53. Hall, R.J.; Freeburn, J.T.; De Groot, W.J.; Pritchard, J.M.; Lynham, T.J.; Landry, R. Remote sensing of burn severity: Experience from western Canada boreal fires. *Int. J. Wildland Fire* **2008**, *17*, 476–489. [CrossRef]

54. Beckingham, J.D.; Archibald, J.H. *Field Guide to Ecosites of Northern Alberta*; Natural Resources Canada, Canadian Forest Service, Northern Forestry Centre: Edmonton, AB, Canada, 1996.

55. Mitchell, K. *Quantitative Analysis by the Point-Centered Quarter Method*; Department of Mathematics and Computer Science, Hobart and William Smith Colleges: Geneva, NY, USA, 2015; pp. 1–56.

56. Cottam, G.; Curtis, J.T.; Hale, B.W. Some sampling characteristics of a population of randomly dispersed individuals. *Ecology* **1953**, *34*, 741–757. [CrossRef]

57. Cybis Elektronik & Data AB. CooRecorder v.9.1. 2013. Available online: http://www.cybis.se/forfun/dendro/index.htm (accessed on 10 February 2018).

58. Loboda, T.V.; French, N.H.F.; Hight-Harf, C.; Jenkins, L.; Miller, M.E. Mapping fire extent and burn severity in Alaskan tussock tundra: An analysis of the spectral response of tundra vegetation to wildland fire. *Remote Sens. Environ.* **2013**, *134*, 194–209. [CrossRef]

59. Dyrness, C.T.; Norum, R.A. The effects of experimental fires on black spruce forest floors in interior Alaska. *Can. J. For. Res.* **1983**, *17*, 1207–1212. [CrossRef]

60. Kasischke, E.S.; O'Neill, K.P.; French, N.H.F.; Bourgeau-Chavez, L.L. Controls on Patterns of Biomass Burning in Alaskan Boreal Forests. In *Fire, Climate Change, and Carbon Cycling in the Boreal Forest*; Kasischke, E.S., Stocks, B.J., Eds.; Springer: New York, NY, USA, 2000; pp. 173–196.

61. Moss, E.H. *Flora of Alberta*, 2nd ed.; Packer, J.G., Ed.; University of Toronto Press Inc.: Buffalo, NY, USA, 1994; ISBN 0-8020-2508-0.

62. Cody, W.J. *Flora of the Yukon Territory*, 2nd ed.; NRC Research Press: Ottawa, ON, Canada, 2000; ISBN 978-0-660-18110-3.

63. Daly, C.; Gibson, W.P.; Taylor, G.H.; Johnson, G.L.; Pasteris, P. A knowledge-based approach to the statistical mapping of climate. *Clim. Res.* **2002**, *22*, 99–113. [CrossRef]

64. Natural Resources Canada. *Canadian Digital Elevation Model*; Natural Resources Canada: Ottawa, ON, Canada, 2016. Available online: https://open.canada.ca/data/en/dataset/7f245e4d-76c2-4caa-951a-45d1d2051333/ (accessed on 03 September 2017).

65. R Core Team R: A Language and Environment for Statistical Computing. v 3.4.3. 2017. Available online: https://www.R-project.org/ (accessed on 20 December 2017).

66. Bates, D.; Mächler, M.; Bolker, B.; Walker, S. Fitting Linear Mixed-Effects Models Using lme4. *J. Stat. Softw.* **2015**, *67*, 1–48. [CrossRef]

67. Kuznetsova, A.; Brockhoff, P.B.; Christensen, R.H.B. lmerTest Package: Tests in Linear Mixed Effects Models. *J. Stat. Softw.* **2017**, *82*, 1–26. [CrossRef]

68. Luke, S.G. Evaluating significance in linear mixed-effects models in R. *Behav. Res. Methods* **2017**, *49*, 1494–1502. [CrossRef] [PubMed]

69. Lenth, R.V. Least-Squares Means: The R Package lsmeans. *J. Stat. Softw.* **2016**, *69*, 1–33. [CrossRef]

70. Oksanen, J.; Blanchet, G.; Friendly, M.; Kindt, R.; Legendre, P.; McGlinn, D.; Minchin, P.R.; O'Hara, R.B.; Simpson, G.L.; Solymos, P.; et al. Vegan: Community Ecology Package R Package v.2.4-4. 2017. Available online: https://CRAN-R-project.org/package=vegan (accessed on 20 December 2017).

71. De Caceres, M.; Legendre, P. Associations between species and groups of sites: Indices and statistical inference. *Ecology* **2009**, *90*, 3566–3574. [CrossRef] [PubMed]
72. Van den Boogaart, K.G.; Tolosana, R.; Bren, M. compositions R package v.1.40-1. 2014. Available online: https://CRAN.R-project.org/package=compositions (accessed on 2 February 2018).
73. Ripley, B. Tree: Classification and Regression Trees R Package v.1.0-37. 2016. Available online: https://CRAN.R-project.org/package=tree (accessed on 30 January 2018).
74. Andrews, J.T. Present and postglacial rates of uplift for glaciated northern and eastern North America derived from postglacial uplift curves. *Can. J. Earth Sci.* **1970**, *7*, 703–715. [CrossRef]
75. Portenga, E.W.; Bierman, P.R. Understanding earth's eroding surface with 10Be. *GSA Today* **2011**, *21*, 4–10. [CrossRef]
76. Loarie, S.R.; Duffy, P.B.; Hamilton, H.; Asner, G.P.; Field, C.B.; Ackerly, D.D. The velocity of climate change. *Nature* **2009**, *462*, 1052–1055. [CrossRef] [PubMed]
77. Waddington, J.M.; Morris, P.J.; Kettridge, N.; Granath, G.; Thompson, D.K.; Moore, P.A. Hydrological feedbacks in northern peatlands. *Ecohydrology* **2015**, *8*, 113–127. [CrossRef]
78. Schneider, R.R.; Devito, K.; Kettridge, N.; Bayne, E. Moving beyond bioclimatic envelope models: Integrating upland forest and peatland processes to predict ecosystem transitions under climate change in the western Canadian boreal plain. *Ecohydrology* **2016**, *9*, 899–908. [CrossRef]
79. Messier, C.; Parent, S.; Bergeron, Y. Effects of Overstory and Understory Vegetation on the Understory Light Environment in Mixed Boreal Forests. *J. Veg. Sci.* **1998**, *9*, 511–520. [CrossRef]
80. Nilsson, M.; Wardle, D. Understory vegetation as a forest ecosystem driver: Evidence from the northern Swedish boreal forest. *Front. Ecol. Environ.* **2005**, *3*, 421–428. [CrossRef]
81. Tsuyuzaki, S.; Narita, K.; Sawada, Y.; Kushida, K. The establishment patterns of tree seedlings are determined immediately after wildfire in a black spruce (*Picea mariana*) forest. *Plant Ecol.* **2014**, *215*, 327–337. [CrossRef]
82. Grosse, G.; Harden, J.; Turetsky, M.; McGuire, A.D.; Camill, P.; Tarnocai, C.; Frolking, S.; Schuur, E.A.G.; Jorgenson, T.; Marchenko, S.; et al. Vulnerability of high-latitude soil organic carbon in North America to disturbance. *J. Geophys. Res. Biogeosci.* **2011**, *116*, 1–23. [CrossRef]
83. Neff, J.C.; Harden, J.W.; Gleixner, G. Fire effects on soil organic matter content, composition, and nutrients in boreal interior Alaska. *Can. J. For. Res.* **2005**, *35*, 2178–2187. [CrossRef]
84. Devito, K.; Creed, I.; Gan, T.; Mendoza, C.; Petrone, R.; Silins, U.; Smerdon, B. A framework for broad-scale classification of hydrologic response units on the Boreal Plain: Is topography the last thing to consider? *Hydrol. Process.* **2005**, *19*, 1705–1714. [CrossRef]
85. Walker, X.; Baltzer, J.; Cumming, S.; Day, N.; Johnstone, J.; Rogers, B.; Solvik, K.; Turetsky, M.; Mack, M. Soil organic layer combustion in black spruce and jack pine stands of the Northwest Territories, Canada. *Int. J. Wildland Fire* **2018**, *27*, 125–134. [CrossRef]
86. Peters, V.S.; Macdonald, S.E.; Dale, M.R.T. The Interaction Between Masting and Fire is Key to White Spruce Regeneration. *Ecology* **2005**, *86*, 1744–1750. [CrossRef]
87. Peters, V.S.; Macdonald, S.E.; Dale, M.R.T. Patterns of initial versus delayed regeneration of white spruce in boreal mixedwood succession. *Can. J. For. Res.* **2006**, *36*, 1597–1609. [CrossRef]
88. Searle, E.B.; Chen, H.Y.H. Persistent and pervasive compositional shifts of western boreal forest plots in Canada. *Glob. Chang. Biol.* **2017**, *23*, 857–866. [CrossRef] [PubMed]
89. Greene, D.F.; Zasada, J.C.; Sirois, L.; Kneeshaw, D.; Morin, H.; Charron, I.; Simard, M.J. A review of the regeneration dynamics of North American boreal forest tree species. *Can. J. Bot.* **1999**, *29*, 824–839. [CrossRef]
90. Bergeron, Y.; Chen, H.Y.H.; Kenkel, N.C.; Leduc, A.L.; Macdonald, S.E. Boreal mixedwood stand dynamics: Ecological processes underlying multiple pathways. *For. Chron.* **2014**, *90*, 202–213. [CrossRef]
91. Ilisson, T.; Chen, H.Y.H. The direct regeneration hypothesis in northern forests. *J. Veg. Sci.* **2009**, *20*, 735–744. [CrossRef]
92. Walker, X.J.; Mack, M.C.; Johnstone, J.F. Predicting Ecosystem Resilience to Fire from Tree Ring Analysis in Black Spruce Forests. *Ecosystems* **2017**, *20*, 1137–1150. [CrossRef]
93. Johnstone, J.F.; Hollingsworth, T.N.; Chapin, F.S., III; Mack, M.C. Changes in fire regime break the legacy lock on successional trajectories in Alaskan boreal forest. *Glob. Chang. Biol.* **2010**, *16*, 1281–1295. [CrossRef]
94. White, J.C.; Wulder, M.A.; Hermosilla, T.; Coops, N.C.; Hobart, G.W. A nationwide annual characterization of 25 years of forest disturbance and recovery for Canada using Landsat time series. *Remote Sens. Environ.* **2017**, *194*, 303–321. [CrossRef]

95. Hogg, E.H.; Wein, R.W. Impacts of drought on forest growth and regeneration following fire in southwestern Yukon, Canada. *Can. J. For. Res.* **2005**, *35*, 2141–2150. [CrossRef]

96. Barber, V.A.; Juday, G.P.; Finney, B.P. Reduced growth of Alaskan white spruce in the 20th century from temperature-induced drought stress. *Nature* **2000**, *405*, 668–673. [CrossRef] [PubMed]

97. Michaelian, M.; Hogg, E.H.; Hall, R.J.; Arsenault, E. Massive mortality of aspen following severe drought along the southern edge of the Canadian boreal forest. *Glob. Chang. Biol.* **2011**, *17*, 2084–2094. [CrossRef]

98. Stralberg, D.; Wang, X.; Parisien, M.-A.; Robinne, F.-N.; Sólymos, P.; Mahon, L.C.; Nielsen, S.E.; Bayne, E.M. Wildfire-mediated vegetation change in boreal forests of Alberta, Canada. *Ecosphere* **2018**, in press.

99. Forestry Canada Fire Danger Group. *Development of the Canadian Forest Fire Behavior Prediction System*; Information Report ST-X-3; Forestry Canada: Ottawa, ON, Canada, 1992; pp. 1–63. Available online: https://cfs.nrcan.gc.ca/publications?id=10068/ (accessed on 01 September 2018).

100. Cumming, S.G. Forest type and wildfire in the Alberta Boreal Mixedwood: What do fires burn? *Ecol. Appl.* **2001**, *11*, 97–110. [CrossRef]

101. Euskirchen, E.S.; Bennett, A.P.; Breen, A.L.; Genet, H.; Lindgren, M.A.; Kurkowski, T.A.; McGuire, A.D.; Rupp, T.S. Consequences of changes in vegetation and snow cover for climate feedbacks in Alaska and northwest Canada. *Environ. Res. Lett.* **2016**, *11*. [CrossRef]

102. Terrier, A.; Girardin, M.P.; Périé, C.; Legendre, P.; Bergeron, Y. Potential changes in forest composition could reduce impacts of climate change on boreal wildfires. *Ecol. Appl.* **2013**, *23*, 21–35. [CrossRef] [PubMed]

103. Dai, A. Increasing drought under global warming in observations and models. *Nat. Clim. Chang.* **2013**, *3*, 52–58. [CrossRef]

104. Hogg, E.H.; Bernier, P.Y. Climate change impacts on drought-prone forests in western Canada. *For. Chron.* **2005**, *81*, 675–682. [CrossRef]

105. Way, D.A.; Crawley, C.; Sage, R.F. A hot and dry future: Warming effects on boreal tree drought tolerance. *Tree Physiol.* **2013**, *33*, 1003–1005. [CrossRef] [PubMed]

106. Darlington, A.B.; Halinska, A.; Dat, J.F.; Blake, T.J. Effects of increasing saturation vapour pressure deficit on growth and ABA levels in black spruce and jack pine. *Trees* **1997**, *11*, 223–228. [CrossRef]

107. Parks, S.A.; Parisien, M.-A.; Miller, C.; Holsinger, L.M.; Baggett, L.S. Fine-scale spatial climate variation and drought mediate the likelihood of reburning. *Ecol. Appl.* **2017**, *28*, 573–586. [CrossRef] [PubMed]

108. Greene, D.F.; Noël, J.; Bergeron, Y.; Rousseau, M.; Gauthier, S. Recruitment of *Picea mariana*, *Pinus banksiana*, and *Populus tremuloides* across a burn severity gradient following wildfire in the southern boreal forest of Quebec. *Can. J. For. Res.* **2004**, *34*, 1845–1857. [CrossRef]

109. Greene, D.F.; Macdonald, S.E.; Haeussler, S.; Domenicano, S.; Noël, J.; Jayen, K.; Charron, I.; Gauthier, S.; Hunt, S.; Gielau, E.T.; et al. The reduction of organic-layer depth by wildfire in the North American boreal forest and its effect on tree recruitment by seed. *Can. J. For. Res.* **2007**, *37*, 1012–1023. [CrossRef]

110. Barrett, K.; McGuire, A.D.; Hoy, E.E.; Kasischke, E.S. Potential shifts in dominant forest cover in interior Alaska driven by variations in fire severity. *Ecol. Appl.* **2011**, *21*, 2380–2396. [CrossRef] [PubMed]

111. Kemball, K.J.; Wang, G.G.; Westwood, A.R. Are mineral soils exposed by severe wildfire better seedbeds for conifer regeneration? *Can. J. For. Res.* **2006**, *36*, 1943–1950. [CrossRef]

112. Barrett, K.; Kasischke, E.S.; McGuire, A.D.; Turetsky, M.R.; Kane, E.S. Modeling fire severity in black spruce stands in the Alaskan boreal forest using spectral and non-spectral geospatial data. *Remote Sens. Environ.* **2010**, *114*, 1494–1503. [CrossRef]

forests

Article

Overstory Structure and Surface Cover Dynamics in the Decade Following the Hayman Fire, Colorado

Paula J. Fornwalt [1,*], Camille S. Stevens-Rumann [2] and Byron J. Collins [1]

[1] USDA Forest Service, Rocky Mountain Research Station, 240 West Prospect Road, Fort Collins, CO 80526, USA; byroncollins@outlook.com

[2] Department of Forest and Rangeland Stewardship, Colorado State University, Fort Collins, CO 80523, USA; c.stevens-rumann@colostate.edu

*   Correspondence: pfornwalt@fs.fed.us; Tel.: +1-970-498-2581

Received: 16 February 2018; Accepted: 15 March 2018; Published: 17 March 2018

Abstract: The 2002 Hayman Fire burned with mixed-severity across a 400-ha dry conifer study site in Colorado, USA, where overstory tree and surface cover attributes had been recently measured on 20 0.1-ha permanent plots. We remeasured these plots repeatedly during the first post-fire decade to examine how the attributes changed through time and whether changes were influenced by fire severity. We found that most attributes were temporally dynamic and that fire severity shaped their dynamics. For example, low-severity plots experienced a modest reduction in live overstory density due to both immediate and delayed tree mortality, and no change in live overstory basal area through time; in contrast, high-severity plots experienced an immediate and total loss of live overstory density and basal area. Large snag density in low-severity plots did not vary temporally because snag recruitment balanced snag loss; however, in high-severity plots large snag density increased markedly immediately post-fire and then declined by about half by post-fire year ten as snags fell. Mineral soil cover increased modestly immediately post-fire in low-severity plots and substantially immediately post-fire in high-severity plots, but changed little in ensuing years for either severity class. By incorporating pre-fire and repeatedly-measured post-fire data for a range of severities, our study uniquely contributes to the current understanding of wildfire effects in dry conifer forests and should be of interest to managers, researchers, and others.

Keywords: Colorado; USA; delayed tree mortality; Douglas-fir (*Pseudotsuga menziesii* (Mirb.) Franco); Hayman Fire; ponderosa pine (*Pinus ponderosa* Lawson & C. Lawson); snag; surface cover

1. Introduction

Wildfires have long regulated dry conifer forests of the western USA. While wildfire activity for these forests was diminished relative to historical levels for most of the 20th century due to fire suppression, logging, grazing, and other land-use practices [1–3], in recent decades it has increased markedly [4–6]. The recent increase has likewise been borne out of past land-use practices, which allowed forests to become denser and more homogeneous [1,3,7], as well as out of a changing climate [4–6]. The resurgence of wildfires in western dry conifer forests thus makes it important that managers, researchers, and others thoroughly understand how forests are directly affected by fire and how they subsequently develop through time.

For forests, the direct effect of fire on overstory trees and organic surface material is commonly captured by the term "fire severity" [8]. Many of the recent wildfires in dry conifer forests of the west burned with uncharacteristic high-severity crown fires across fair portions of their area, creating patches where all or nearly all trees were killed and where most of the tree crown and surface organic material were consumed [9–13]. Yet even the most severe recent wildfires typically burned with a mix of severities, with low- to moderate-severity patches also comprising much of their area [9–13]. In these patches, where fire severity is generally more in line with historical fire severity [1–3], some or even many of the trees survived burning and tree crown and surface organic material were at most only partially consumed.

Overstory structure can change dramatically in the years after wildfire, with fire severity influencing the magnitude and timing of temporal change. Few studies have quantified time- and severity-related overstory dynamics simultaneously, however, with most focusing instead on either the effects of time in high-severity areas or on the effects of severity for one point in time. Low- and moderate-severity areas can experience additional declines in live tree abundance due to delayed mortality, particularly if trees were badly fire-injured [14–16]. The abundance of dead trees, or snags, can probably also be temporally variable, depending on the degree and timing of snag recruitment relative to snag fall. High-severity areas, on the other hand, experience complete or nearly complete tree mortality due to fire, so mortality in subsequent years is nominal. The transformation of all or nearly all live trees to snags can dramatically increase snag abundance in the short-term, but snag abundance can decline through time as they fall [17–19]. These variations in post-fire overstory structure across space and time can in turn have important implications for tree regeneration [15,20,21] and wildlife use [22–24], among other things.

The amount and type of organic material covering the mineral soil surface following wildfire influences hillslope runoff and erosion [25–27], fire behavior and severity during a reburn [28–30], and a host of other ecological properties and processes. As with post-fire overstory conditions, post-fire surface cover conditions can vary considerably with both time since fire and fire severity, but few holistic examinations of these factors have been conducted. Modest amounts of litter, duff, and wood can accumulate on the ground in low- and moderate-severity areas as scorched needles and dead tree branches and boles fall, augmenting unconsumed pre-fire material [15,31]. If light, water, and nutrient availability were boosted enough in low- and moderate-severity areas to promote understory plant growth, then herbs and shrubs can also modestly increase [31,32]. In contrast, in high-severity areas, the amount of litter and duff can remain minimal in ensuing years due to a lack of foliage in the crowns of overstory snags [15,31,33]. The amount of wood in high-severity areas, meanwhile, can increase substantially as the numerous overstory snags break and fall [15,18,33], and herbs and shrubs can be substantially promoted due to the greatly elevated levels of light, water, and nutrients [31–33].

Wildfires were largely excluded from dry conifer forests of the Colorado Front Range for much of the 20th century, but they have become increasingly frequent as of late [34]. The largest recent fire, the 2002 Hayman Fire, burned more than 52,000 ha of forest comprised predominately of ponderosa pine (*Pinus ponderosa* Lawson & C. Lawson) and Douglas-fir (*Pseudotsuga menziesii* (Mirb.) Franco) [9]. The Hayman Fire also burned a 400-ha study site where overstory tree, surface cover, and other measurements had been made a few years prior in 20 0.1-ha permanent plots [35–37]. While much of the area within the Hayman Fire footprint burned as an uncharacteristically severe crown fire [9,38], our study site and study plots burned more heterogeneously [32,39,40]. We took advantage of this serendipitous opportunity by remeasuring the plots repeatedly in the first post-fire decade [32,39,40]. Here, we report on how a variety of live overstory structure, dead overstory structure, and surface cover attributes changed through time as a result of fire, and whether fire severity influenced these changes. By incorporating pre-fire and repeatedly-measured post-fire data for a gradient of severities, our study contributes a unique perspective to the current understanding of wildfire effects in western dry conifer forests, and should be of interest to the management, scientific, and other communities.

2. Materials and Methods

2.1. Study Site and Study Plots

Our 400-ha study site is approximately 60 km southwest of Denver, Colorado, on Pike National Forest lands within the Hayman Fire perimeter (39.14° N, 105.24° W; Figure 1) [35–37]. Elevations range from about 2300 to 2500 m. The climate is warm and dry, with average January and July temperatures of about 4 °C and 17 °C, respectively, and average annual precipitation of about 38 cm [41]. The poorly-developed, gravelly soils are derived from Pikes Peak granite [42]. Vegetation at the site is characteristic of Front Range dry conifer forests. Overstories are dominated primarily by ponderosa pine and secondarily by Douglas-fir [35]. Understories are diverse communities of graminoids, forbs, and relatively low-statured shrubs [36,37]. The site's disturbance history is also characteristic of Front Range dry conifer forests. Prior to Euro-American settlement in the mid to late 1800s, the wildfire regime was one of mixed-severity, with fire intervals for individual stands varying from very short (<10 years) to very long (>100 years), and fire severity varying from low to high [43]. The site experienced very few wildfires beginning in the late 1800s (that is, until the 2002 Hayman Fire), likely due in large part to the fire suppression policy that began in the early 1900s [43]. Logging and grazing are thought to have been rampant at the site from the late 1800s and early 1900s, and also may have contributed to the general lack of wildfires during this period [35–37,43]. To our knowledge, logging and grazing have not occurred since this time.

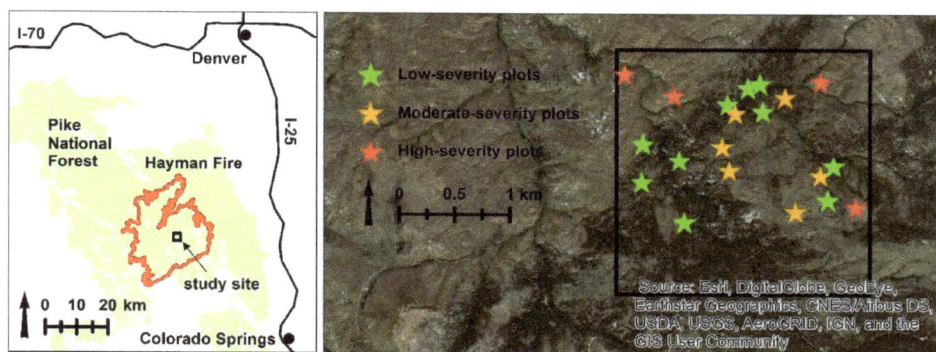

Figure 1. Location of the 400-ha study site within the Hayman Fire, Colorado, USA (**left**). Location and severity of the 20 0.1-ha plots within the study site, with post-fire aerial imagery in the background (**right**).

In 1996–1997, 20 randomly-located, upland, 0.1 ha (20 × 50 m) permanent plots were established at the study site (Figure 1) [35–37]. The plots subsequently burned in the Hayman Fire. This fire was human-ignited on June 8, 2002 [9]. Low fuel moistures, heavy and continuous fuel loads, and strong winds enabled the fire to burn approximately 24,000 ha on June 9, largely as a high-severity crown fire. The two plots in the northwest corner of our study site are thought to have been impacted by the fire on this day. While historical fires sometimes contained a high-severity component, the size of the high-severity patches created by the Hayman Fire on June 9 appear to be unprecedented over at least the last four centuries [38,43,44]. Less extreme weather conditions arrived on June 10 and persisted for much of the next three weeks, allowing the fire to burn somewhat more heterogeneously. The remaining 18 plots are thought to have burned early during this period. The Hayman Fire was contained on July 2, after having impacted more than 52,000 ha. In 2003, one year post-fire, we successfully reestablished all original plots [32,39,40]. Plot locations were precisely identified using pre-fire data such as plot coordinates and overstory tree stem maps, as well as from finding the remains of plot corner stakes and tree tags.

2.2. Data Collection

Overstory tree data in the 20 plots were first collected in 1996–1997 [35]. At this time, each tree taller than breast height (1.4 m) was tagged with a numbered aluminum tag and its location was mapped with a survey laser. Diameter at breast height (DBH), species, and live or dead status were also recorded for each tree. In 2003, one year post-fire, we relocated all pre-fire overstory trees, retagged them as necessary, and noted their status as either live, dead and standing, dead and downed, or dead and missing [32,39,40]. Dead and downed trees were those that were either uprooted or broken below breast height. Dead and missing trees were those that could not be relocated; these were typically trees that were dead before the fire and that were likely incinerated. Also in 2003, we assessed direct fire effects on trees by visually estimating the percent of the pre-fire live crown volume that was undamaged, scorched, and consumed. Status of all trees was recorded again in 2004, 2005, 2006, 2007, and 2012. DBH was measured again in 2004 and 2012. A small number of trees, mostly quaking aspen (*Populus tremuloides* Michx.), grew into the overstory of some plots during the course of this study. These trees had high turnover and were not tracked.

Surface cover data in the 20 plots were likewise first collected in 1996–1997, as part of an understory plant community survey [36,37]. Cover was visually estimated to the nearest percent within ten systematically-placed 1-m^2 subplots per plot, with separate estimates made for mineral soil, rock, combined litter and duff, wood (including stumps and boles of standing dead trees), and individual live herb and shrub species. Total cover could therefore exceed 100% due to overlap among these elements. Measurements were repeated in 2003, 2004, 2005, 2006, 2007, and 2012 [32,39,40]. In 2003, we also assessed direct fire effects on organic surface material by noting the degree of scorch and consumption for pre-fire litter, duff, and wood in the subplots.

2.3. Data Calculations and Analyses

We used our overstory tree and surface cover data to produce plot-level estimates for ten attributes. Attributes describing live overstory structure were (1) density (stems ha^{-1}) and (2) basal area (m^2 ha^{-1}). Because DBH measurements were only taken in 1996–1997, 2004, and 2012, we estimated the DBH of live trees for other study years prior to calculating basal area. For trees that were alive during the entire post-fire period, we accomplished this by determining the tree's average annual DBH growth from 2004 to 2012 and linearly adjusting 2004 DBH values as necessary. For trees that died between 2004 and 2012, we calculated DBH assuming the average annual DBH growth rate for all trees alive during the entire post-fire period (0.1 cm year^{-1}). Attributes describing dead overstory structure were (3) density (stems ha^{-1}) and (4) persistence (percent standing) of large (>20 cm DBH) snags. We focused on large snags (sensu [45]) because they provide the most valuable wildlife habitat [22,46,47]. For persistence calculations, we only included trees that were alive pre-fire and dead in the first post-fire year, such that time since fire was synonymous with time since death. Persistence was the percent of these trees standing in each year. Surface cover attributes were percent (5) mineral soil; (6) rock; (7) litter and duff; (8) wood; (9) herb; and (10) shrub cover. The first four attributes were calculated by averaging cover across the ten subplots per plot. The latter two attributes were calculated by summing the cover of all relevant species in each subplot and then averaging across subplots.

We classified each plot as burning with low-, moderate-, or high-severity by utilizing our direct fire effects data [32,39,40]. We categorized plots where <50% of the overstory trees were killed and where tree crown and organic surface material consumption were generally slight as burning with low-severity (i.e., burned by light surface fire; ten plots). Moderate-severity plots experienced >50% overstory mortality and had modest tree crown and organic surface material consumption (i.e., burned by moderate to severe surface fire; six plots). High-severity plots experienced 100% overstory mortality and complete or nearly complete tree crown and organic surface material consumption (i.e., burned by severe crown fire; four plots).

We used generalized linear mixed models in SAS 9.4 (SAS Institute Inc., Cary, NC, USA) to examine how the ten attributes changed through time, and how temporal changes were influenced by fire severity. We modeled each against fire severity, time since fire, and fire severity × time since fire. Fire severity had three levels (low, moderate, and high) and time since fire had seven levels (pre-fire and one, two, three, four, five, and ten years post-fire). The appropriate distribution and link function for the attributes were defined in the models (e.g., a negative binomial distribution with a logarithmic link function for live overstory density and large snag density, and a beta distribution with a logit link function for all surface cover attributes). Models accounted for the repeated measurement of plots through time using the spatial power covariance structure. This structure was the one best able to account for the higher degree of correlation between measurements closer in time than between measurements further apart in time and for the unevenly-spaced time intervals between measurements. For models where severity or time was significant, we examined pairwise differences between levels using least squares means and Tukey-Kramer *p*-value adjustments for multiple comparisons. For models where severity × time was significant, we also examined pairwise differences using least squares means and Tukey-Kramer adjustments, but we "sliced" the interaction term to limit comparisons to those of levels of time within levels of severity and to those of levels of severity within levels of time. Significance was assessed at $\alpha = 0.050$. We note that a similar analysis of shrub cover was conducted by Abella and Fornwalt [32], but it differs from ours in that they also incorporated data from unburned plots and they employed a slightly different model structure.

3. Results

3.1. Live Overstory Structure

Live overstory density varied with fire severity, time since fire, and fire severity × time since fire, indicating that the severity classes experienced different temporal density patterns (Figure 2). Before the Hayman Fire, live overstory density averaged 572 stems ha^{-1} across all 20 plots, with no differences among fire severity classes. Trees were generally more abundant in the smaller than the larger size classes pre-fire, with 34% of trees in the 0–10 cm DBH class, 25% in the 10–20 cm DBH class, 24% in the 20–30 cm DBH class, and 17% in the 30+ cm DBH class. Ponderosa pine constituted 61% of pre-fire trees and Douglas-fir constituted 36%. Live overstory density in low-severity plots decreased 21% in post-fire year one, to an average of 371 stems ha^{-1}. In post-fire year two, density in low-severity plots decreased an additional 5% due to delayed tree mortality. Density continued to decrease non-significantly thereafter, such that by post-fire year ten, it had been reduced by a total of 34%, to an average of 308 stems ha^{-1}. Trees with the smallest DBHs experienced the greatest overall mortality, with 60% of 0–10 cm DBH trees killed, 35% of 10–20 cm DBH trees killed, 21% of 20–30 DBH trees killed, and 24% of 30+ cm DBH trees killed (Figure 3). Moreover, Douglas-fir experienced greater overall mortality than ponderosa pine, with 39% of the former and 25% of the latter killed. Meanwhile, the post-fire year one decrease in live overstory density was more substantial in moderate-severity plots; in these plots, density dropped 68%, to an average of 260 stems ha^{-1}. Density decreased an additional 5% in post-fire year two and 2% in post-fire year three. By the time a decade had passed since the Hayman Fire, mean live overstory density in moderate-severity plots averaged 177 stems ha^{-1}, 79% lower than pre-fire. Overall, 93% of trees in the 0–10 cm DBH class were dead, while 82% of trees in the 10–20 cm DBH class, 63% of trees in the 20–30 cm DBH class, and 50% of the trees in the 30+ cm DBH class were dead. Overall mortality was 86% for Douglas-fir and 66% for ponderosa pine. In high-severity plots, all pre-fire live trees were dead in the first post-fire year.

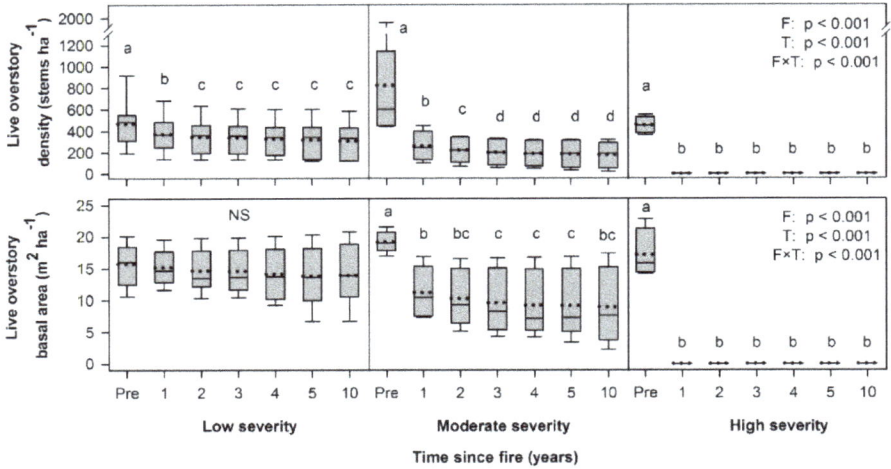

Figure 2. Live overstory (trees > 1.4 m tall) density (**top**) and basal area (**bottom**) with respect to fire severity and time since fire. Boxes represent 25th, 50th, and 75th percentiles, whiskers represent 10th and 90th percentiles, and dotted lines represent means. The *p*-values are model results for the effects of fire severity (F), time since fire (T), and their interaction (F × T). Within fire severity classes, letters separate means though time; NS indicates that means did not differ.

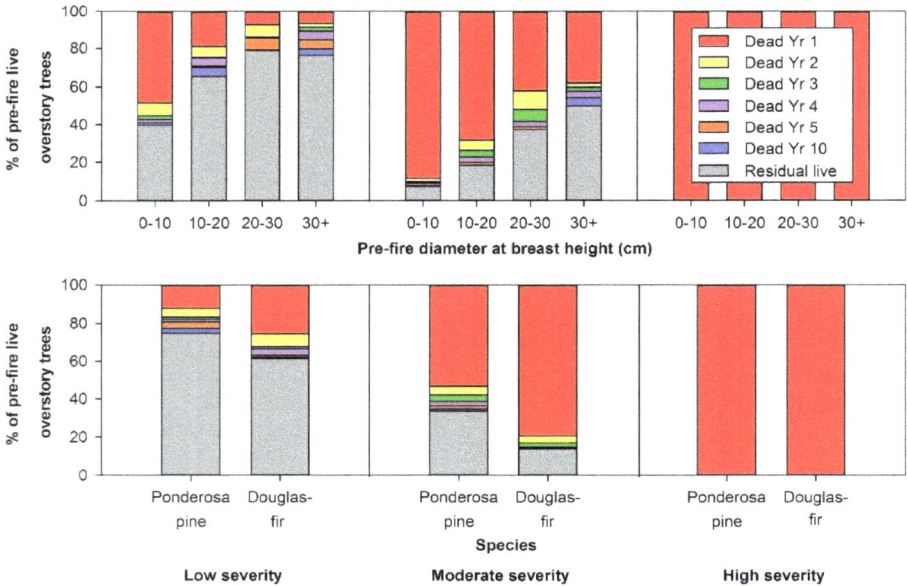

Figure 3. Mean percent incremental post-fire mortality of pre-fire live overstory (>1.4 m tall) trees, by pre-fire diameter and fire severity (**top**) and by species and fire severity (**bottom**).

Live overstory basal area also varied with fire severity, time since fire, and their interaction (Figure 2). Live overstory basal area prior to the fire did not differ among severity classes, averaging 17 m² ha⁻¹ across all plots. Basal area was reduced by fire in moderate- and high-severity plots but not in low-severity plots. In moderate-severity plots, basal area declined 41% immediately following fire, to an average of 11 m^2 ha^{-1} in post-fire year one. An additional small decline was also observed in the third post-fire year due to delayed tree mortality. By the end of the study, basal area in moderate-severity plots averaged 9 m^2 ha^{-1}, 53% lower than pre-fire basal area. In high-severity plots, basal area declined to 0 m^2 ha^{-1} in post-fire year one, where it remained.

3.2. Dead Overstory Structure

Time since fire was a strong predictor of whether large snags created by the Hayman Fire (i.e., trees that were alive before the fire and dead in post-fire year one) were standing (Figure 4). On average, >90% of the large fire-created snags remained standing through post-fire year three. By post-fire year five, an average of 76% of the large fire-created snags were standing; by post-fire year ten, 43% were standing. Moreover, ten years post-fire, an average of 67% of the large fire-created Douglas-fir snags remained standing, in contrast to 15% of the ponderosa pine snags.

Time since fire was also a strong predictor of large snag density, as were fire severity and severity × time (Figure 5). Prior to the fire, large snag density averaged 15 stems ha^{-1}, with no differences among the severity classes. Following the fire, snag density did not deviate from pre-fire levels in low-severity plots. In contrast, moderate- and high-severity plots saw a sustained post-fire increase in snag density over pre-fire levels for the first five years; the magnitude of this increase was 9-fold in moderate-severity plots and 11-fold in high-severity plots. By post-fire year ten, snag density in moderate-severity plots had returned to pre-fire levels. Snag density also declined by post-fire year ten in high-severity plots but was still 6-fold greater than pre-fire.

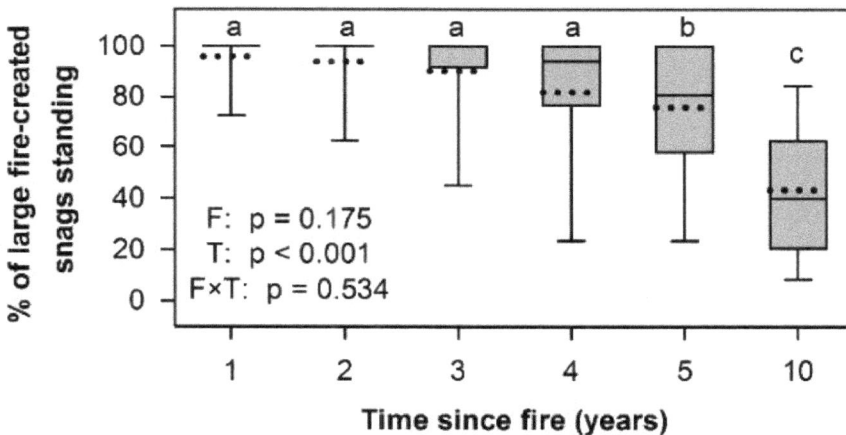

Figure 4. Persistence of large (>20 cm diameter at breast height (DBH)) fire-created snags with respect to time since fire. Boxes represent 25th, 50th, and 75th percentiles, whiskers represent 10th and 90th percentiles, and dotted lines represent means. The *p*-values are model results for the effects of fire severity (F), time since fire (T), and their interaction (F × T). Letters separate means though time.

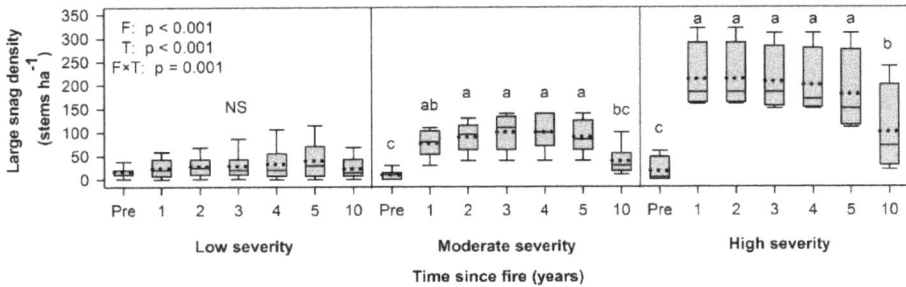

Figure 5. Large (>20 cm DBH) snag density with respect to fire severity and time since fire. Boxes represent 25th, 50th, and 75th percentiles, whiskers represent 10th and 90th percentiles, and dotted lines represent means. The *p*-values are model results for the effects of fire severity (F), time since fire (T), and their interaction (F × T). Letters separate means though time within severity classes; NS indicates that means did not differ.

3.3. Surface Cover

The Hayman Fire caused an increase in mineral soil cover, but the magnitude of the increase was severity-dependent, and to a lesser extent, time-dependent (Figure 6). Relative to pre-fire conditions, mineral soil cover increased 43% in low-severity plots in post-fire year one, while it increased 125% in moderate-severity plots. Mineral soil cover in high-severity plots increased >300% in post-fire year one, with mineral soil exposed across 86% of the ground surface on average. Mineral soil cover in low- and moderate-severity plots did not change in ensuing years relative to post-fire year one conditions but declined somewhat in high-severity plots.

The combined cover of litter and duff was likewise shaped by the interaction of fire severity and time since fire (Figure 6). For all severity classes, litter and duff covered 50% or more of the ground surface prior to the fire. Litter and duff cover were relatively unaffected by low-severity burning for the entire post-fire decade. Moderate-severity burning caused a sustained decrease in litter and duff cover that ranged from 34–45% through time, although interestingly this decrease did not manifest until the second post-fire year. High-severity burning caused an immediate and sustained decrease in litter and duff cover that ranged from 67–78% through time.

Wood cover was temporally dynamic, but these dynamics were not dependent on fire severity (Figure 6). Burning did not have an immediate effect on wood cover; average pre-fire wood cover was 5%, on par with the average post-fire year one cover of 3%. Wood cover gradually increased during the post-fire observation period but did not exceed pre-fire levels until the tenth post-fire year, when it averaged nearly 10%.

Finally, herb cover varied with fire severity, time since fire, and their interaction, while shrub cover varied solely with time (Figure 6). In low-severity plots, pre-fire herb cover was comparable to post-fire herb cover for all post-fire years, although some differences among post-fire years were evident. Across all years, herb cover averaged 14% in low-severity plots. In moderate- and high-severity plots, pre-fire herb cover was comparable to post-fire herb cover for the first four post-fire years, averaging 14% in the former and 12% in the latter across these years. Herb cover in moderate-severity plots increased over pre-fire levels in post-fire year five (to an average of 31%), while in high-severity plots it increased over pre-fire levels in post-fire years five (to an average of 37%) and ten (to an average of 28%). As Abella and Fornwalt [32] similarly show, shrub cover was reduced 93% immediately following burning, from an average of 9% pre-fire to 1% in post-fire year one. Shrub cover gradually increased in later years but always remained below pre-fire levels, peaking at 4% ten years post-fire.

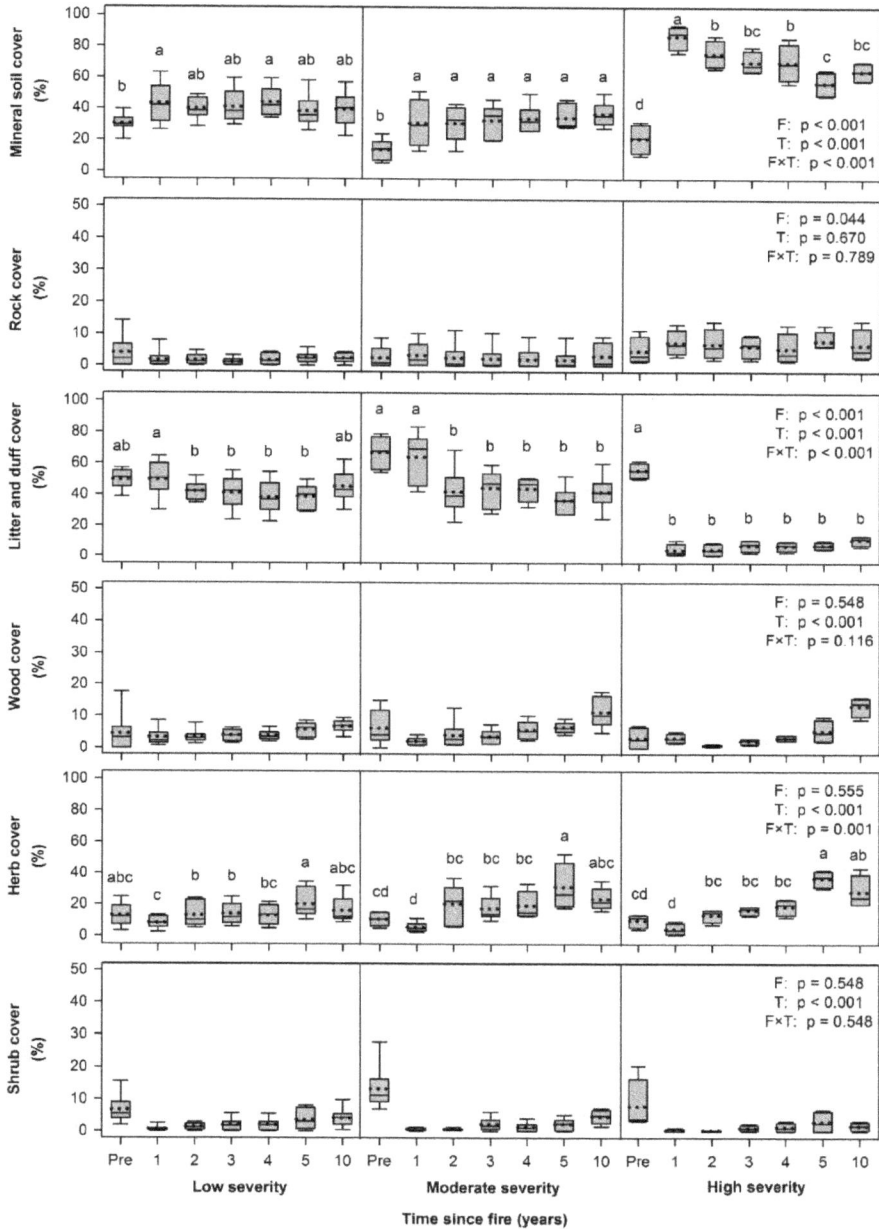

Figure 6. Mineral soil, rock, litter and duff, wood, herb, and shrub cover (top to bottom) with respect to fire severity and time since fire. Boxes represent 25th, 50th, and 75th percentiles, whiskers represent 10th and 90th percentiles, and dotted lines represent means. The *p*-values are model results for the effects of fire severity (F), time since fire (T), and their interaction (F × T). For attributes where F × T was significant, letters separate means though time within severity classes.

4. Discussion

Dry conifer forests of the western USA have experienced a recent increase in wildfire activity [4–6], making it important that the direct and longer-term consequences of wildfires be thoroughly understood. Our unique study utilized pre-fire and repeatedly-collected post-fire data to examine how one recent wildfire, Colorado's 2002 Hayman Fire, affected several live overstory structure, dead overstory structure, and surface cover attributes over the first post-fire decade, and whether temporal patterns were contingent on fire severity. We found that nearly all attributes changed through time as a result of the fire. Moreover, we found that for the majority of attributes, the magnitude and timing of changes depended on the severity with which the fire burned.

4.1. Live Overstory Structure

The Hayman Fire's net effect on live overstory structure was wide-ranging at our study site. Low-severity areas underwent little structural change. Density was reduced by about a third by the tenth post-fire year, but because mortality was concentrated in the 0–10 cm DBH class, basal area was not reduced at all. These patterns resembled those documented for other western dry conifer forests experiencing low-severity wildfire [15,31], as well as for forests experiencing prescribed fire treatments and light hand and mechanical thinning treatments [48–50]. More substantial change occurred in moderate-severity areas, with density reduced by over 75% and basal area reduced by over 50% by the tenth post-fire year. The vast majority of trees in both the 0–10 and the 10–20 cm DBH classes were killed, as were the vast majority of Douglas-fir trees. The changes we observed in moderate-severity areas were in line with those brought about elsewhere by moderate-severity wildfire [15,31], and by aggressive hand and mechanical thinning treatments [48,50,51]. Meanwhile, high-severity areas experienced the greatest change in live overstory structure; in these areas, burning transformed dry conifer forests into herb-dominated openings devoid of overstory trees, just as it has done across the west [15,18,31].

Many managers, researchers, and others are interested in examining to what extent recent wildfires are advancing restoration goals in overly-dense, fire-excluded western dry conifer forests [31,52–55]. Restoration treatments in these forests primarily aim to increase their resilience to future burning (sensu [56]) by creating open and heterogeneous overstory conditions that are unlikely to carry large-scale high-severity crown fire [34,57,58]. Restoration objectives for a given area are usually informed by the range of conditions that occurred there historically (i.e., the historical range of variability (HRV)), as they represent the fire-resilient conditions that existed prior to post-settlement land-use practices like fire suppression, logging, and grazing [34,57,58]. Regarding live overstory structure, Battaglia et al. [7] estimated that historical density in dry conifer forests of the Front Range ranged from about 62 to 214 stems ha^{-1} and that historical basal area ranged from about 5 to 11 $m^2\ ha^{-1}$. Comparing these values with our post-fire year ten values suggests that live overstory structure was generally restored to within HRV in moderate-severity portions of our study site. In contrast, post-fire live overstory structure in low-severity areas generally remained above HRV, while in high-severity areas it was well below HRV. That moderate-severity burning was most effective at moving structural attributes to within HRV has also been documented for other dry conifer forests in the west [15,31,54].

Repeatedly assessing the status of all pre-fire overstory trees in our low- and moderate-severity plots allowed us to quantify delayed tree mortality during the first post-fire decade. Similar to others [14–16], we detected small but significant reductions in live overstory structural attributes due to delayed mortality in post-fire year two in low-severity areas and in post-fire years two and three in moderate-severity areas. We suspect that this delayed mortality was primarily driven by the degree of fire-caused crown and bole injury [16,59,60] and exacerbated overall by the below-average annual precipitation conditions that began in 2000 and persisted for the entire post-fire sampling period [32,61]. We do not think that insects were also a primary driver of delayed mortality, as others have found [16,59,60], because we did not observe many signs of insect activity. While we also observed mortality in the other post-fire years, it was not substantial enough to elicit changes in live overstory structure. We anticipate that mortality in future years and decades will likewise have a negligible impact on structure in the absence of additional disturbance.

We anticipate that future live overstory structure at our study site will, however, be strongly shaped by patterns of conifer regeneration. We did not quantify conifer regeneration, but several authors have quantified it elsewhere in areas affected by the Hayman Fire and other Front Range wildfires [20,62–65]. Collectively, they found that low- and moderate-severity areas generally had some or even ample conifer regeneration, while high-severity areas generally had little or even no regeneration, particularly where surviving conifers were distant. This is because ponderosa pine, Douglas-fir, and most other co-occurring conifers rely on seeds from nearby surviving overstory trees to regenerate after fire. The regeneration patterns documented by these authors suggest that live overstory structure may ultimately come to resemble that found prior to the fire in low- and moderate-severity portions of our site, assuming that they are not further disturbed. However, in high-severity portions of our site, a return to pre-fire conditions is unlikely to occur naturally for centuries, if at all.

4.2. Dead Overstory Structure

Large snags provide nesting, foraging, and roosting habitat for numerous species of birds, bats, and other wildlife, but they can be rare in undisturbed western dry conifer forests [46,47,66]. Our results illustrate the striking influence that fire severity and time since fire can have on large snag structure, via their influence on snag recruitment and snag persistence. Low-severity areas did not experience a post-fire increase in large snag density. This stasis was not because such areas failed to recruit new large snags post-fire; rather, it was because recruitment was balanced by loss through incineration and snag fall. In contrast, large snag density increased following fire in moderate- and high-severity areas, at least temporarily. Large snag density in these areas peaked in the first post-fire year, driven by high rates of recruitment and low rates of loss in that year relative to others. By ten years post-fire, large snag densities had returned to pre-fire levels in moderate-severity areas. In high-severity areas, large snag densities were still elevated but nonetheless were in decline. We expect that densities in high-severity areas will return to pre-fire levels in upcoming years as the bulk of the remaining snags, most of which are Douglas-fir, fall [18,19,33]. We therefore also expect that any increased use by wildlife species dependent on large snags will be transient in our study site [67].

4.3. Surface Cover

Litter and duff typically blanket most of the ground surface in dry conifer forests of the west [21,51,68], and prior to the Hayman Fire, our study site was no exception. Although the Hayman Fire consumed some of the pre-fire litter and duff in low-severity areas, it did not alter their cover in the first post-fire year relative to pre-fire levels. This was primarily due to the rapid casting of scorched needles, the cover of which augmented residual litter and duff cover. Cover from scorched needle cast also augmented residual litter and duff cover in moderate-severity areas one year following fire. A different scenario unfolded in high-severity areas. Here the Hayman Fire consumed nearly all pre-fire litter and duff, and because it also consumed nearly all needles in the tree crowns, scorched needle cast contributed negligible new cover in the first post-fire year. Moreover, litter and duff cover in high-severity areas did not change in subsequent years. We think that litter and duff cover in high-severity areas will begin to show signs of recovery soon due to the accumulation of sloughing bark and detached herbaceous material, the most likely sources of new material [18,19,33]. However, pre-fire levels are unlikely to be attained for decades.

Several studies have examined how fire severity and time since fire influence the amount of wood in burned western dry conifer forests [15,18,19,31,33], although to our knowledge, only Keyser et al. [15] also examined these factors simultaneously. These latter authors found that substantial fine wood (<7.6 cm diameter) and coarse wood (>7.6 cm diameter) biomass accumulated in areas that burned with moderate- and high-severity, but not in areas that burned with low-severity, in the five years following a South Dakota wildfire. We were therefore surprised to find that temporal wood cover trajectories also did not vary with fire severity at our study site. Like Keyser et al., however, we did find that wood cover increased through the post-fire period; at our site, wood cover values ten years post-fire were approximately double what they were pre-fire and triple what they were one year post-fire. These increases undoubtedly reflect temporal snag fall dynamics, as discussed earlier, as well as temporal snag break-up dynamics due to the shedding of twigs, branches, and upper boles [18,19,33]. Moreover, we expect that fire severity will begin to influence temporal wood cover trajectories in the near future as the large number of snags still standing in high-severity areas continue to break apart and fall.

Many herb and shrub species in western dry conifer forests are considered to be fire-adapted due to their ability to rapidly establish following fire via sprouting or germinating from seeds [69]. Our herb cover results are consistent with this generalization; herb cover in the first post-fire year was comparable with pre-fire cover for all severity classes. Moreover, in later post-fire years herb cover more than doubled relative to pre-fire levels in moderate- and high-severity classes, highlighting the ability of many herb species to also rapidly expand following fire and other disturbances so long as overstory tree mortality is sufficient [31,40,70]. Herb cover appeared to more-or-less stabilize by the last five years of our study, suggesting it had equilibrated with the new overstory conditions. Thus, we predict that herb cover will probably not change appreciably in upcoming years. On the other hand, our shrub cover results indicate that shrub cover failed to reach pre-fire levels by the tenth post-fire year, regardless of fire severity (also see [32]). This is probably because kinnikinnick (*Arctostaphylos uva-ursi* (L.) Spreng.), by far the most common shrub species at our site prior to the fire, sprouts poorly following fire and does not establish readily from seeds [71]. It will likely take several more years for shrubs, especially kinnikinnick, to recover from burning.

The above surface cover results have a host of ecological implications. For example, the amount of litter, duff, wood, and other organic surface material affects fire behavior and fire effects during a reburn [72]. The amount of coarse wood is often of particular concern, as it strongly influences fire hazard and soil heating [28]. The relatively low wood cover values that we observed ten years post-fire, coupled with relatively low coarse wood biomass values (average of 15 Mg ha^{-1} and range of 0–48 Mg ha^{-1}; P. Fornwalt unpublished data [73]), suggests that the amount of coarse wood currently does not surpass recommended upper thresholds from either a fire hazard or soil heating perspective [28]. However, it may surpass them in high-severity areas in future years as snags fall.

Despite post-fire increases in wood and herbs through time in high-severity portions of our study site, exposed mineral soil was still abundant ten years after the Hayman Fire, averaging around 60% cover. High levels of mineral soil cover can persist for several years following high-severity burning, particularly in relatively unproductive dry conifer forests like those studied here [26,74,75]. It should be noted that Robichaud et al. [74,75] also measured mineral soil cover in high-severity portions of the Hayman Fire, and found that it averaged approximately 30% to 45% seven years post-fire. The discrepancy between their values and ours may be partly due to measurement technique. Whereas Robichaud et al. estimated mineral soil cover as the amount of soil that did not have wood, herb, shrub, or other ground cover elements overlapping it, we estimated it as the amount of soil that was visible, irrespective of other overlapping elements. Regardless, it is clear that the amount of mineral soil cover remains greatly elevated in areas where the Hayman Fire burned most severely. This in turn suggests that post-fire sediment production, which increases as mineral soil cover increases, may also be greatly elevated in such areas [25–27]. Indeed, Rhoades et al. [76] found that drainage basins experiencing extensive high-severity burning in the Hayman Fire had considerably more streamwater-suspended sediment than unburned basins even into the fifth post-fire year, the last year of their study; likewise, Robichaud et al. [74] reported that high-severity hillslopes in the Hayman Fire were producing considerably more sediment than unburned hillslopes even into the seventh post-fire year.

4.4. Study Design Considerations

Our study design had some inescapable limitations owing to its opportunistic nature. Primary among them is that sampling was restricted to one study site within the very large Hayman Fire, constraining our ability to make inferences about the effects of fire severity and time since fire to elsewhere in this wildfire or to other wildfires. Another limitation is that our sample sizes were low, most notably for the high-severity class. This may have affected our ability to accurately characterize overstory and surface cover attributes with respect to fire severity and time since fire, and to detect changes in them due to these factors.

Yet our study design also allowed us to overcome some of the limitations inherent to many wildfire studies. First, we were able to utilize pre-fire data to assess change due to wildfire. Wildfires are unplanned events and pre-fire data are usually not available; thus, studies evaluating their effects tend to collect data in both unburned and burned sites and assume that these sites were comparable prior to burning [15,21,31]. Second, we were able to utilize repeatedly-measured data to investigate a decade of post-fire temporal dynamics. Such dynamics are more commonly evaluated using a chronosequence approach [18,19,33], which substitutes space for time and requires an assumption that differences due to site are small relative to differences due to the passage of time.

5. Conclusions and Recommendations

Our results highlight the considerable variability that can result from recent wildfires in western dry conifer forests due to gradients of both fire severity and time. We therefore suggest that management actions, if undertaken for our Hayman Fire study site or for similar post-fire sites, factor in how and when burning occurred. We also suggest that management actions aim to promote ecologically-appropriate conditions that will be resilient to future wildfires and other disturbances, such as the conditions found historically. Low-severity burning caused the least amount of change. Areas burned in this manner experienced a small reduction in live overstory density, but not a concurrent reduction in live overstory basal area. They also experienced no change in large snag density and little to no change in surface cover attributes like mineral soil cover, litter and duff cover, and herb cover. As in unburned areas, thinning or prescribed fire treatments could be implemented in low-severity areas in the first few post-fire years or decades to move live overstory structure closer to HRV [48,50,51]. Such treatments could also promote large snag and herb abundance. Moderate-severity burning caused significant reductions in live stand density and basal area, generally moving areas to within HRV; it also reduced the abundance of litter and duff and increased the abundance of herbs. Prescribed fire could be utilized in moderate-severity areas in future decades to maintain these ecological benefits [77]. Prescribed fire could also be used to create new large snags, as we found that elevated post-fire snag levels in moderate-severity areas were relatively short-lived. High-severity burning clearly caused the greatest ecological transformation. Notably, areas experiencing high-severity burning underwent an immediate, total, and likely long-lived [15,20,21] reduction in live overstory density and basal area. Tree planting could be conducted to hasten the return of a forested condition. In high-severity areas, we observed a marked increase in large snag density immediately post-fire, but it declined toward pre-fire levels by post-fire year ten due to snag fall. The bulk of the remaining snags will probably fall in upcoming years [18,19,33], potentially creating undesirable coarse wood loads [28]. Prescribed fire conducted during cool weather conditions could potentially be used before this point is reached to remove some of the coarse wood while minimizing adverse effects [28]. Finally, we also suggest that additional research be conducted in recently-burned western dry conifer forests so that the direct and longer-term implications of wildfires, and their relationship to fire severity, can be further clarified. Such research should help improve management practices for burned forests as they become ever more prevalent across the west.

Acknowledgments: This study was funded by the Joint Fire Science Program (projects 03-2-3-08, 04-2-1-118, and 11-1-1-5) and the Rocky Mountain Research Station (project NFP-13-16-FWE-38). We thank Stephanie Asherin, Marin Chambers, Allison Grow, Beckie Hemmerling, Lucas Herman, Chris Hines, Chris Martin, Jill Oropeza, Bill Palm, Chris Peltz, Sean Rielly, Jenny Ventker, and Chris Welker for assisting with field and lab work; Scott Baggett for providing statistical advice; and Mike Battaglia for commenting on earlier versions of the manuscript. Lodging, lab, and office facilities during field campaigns were provided by the Rocky Mountain Research Station's Manitou Experimental Forest.

Author Contributions: P.J.F. conceived the study; P.J.F. and B.J.C. collected the data; P.J.F. and C.S.S. analyzed the data; P.J.F., C.S.S., and B.J.C. wrote the paper.

Conflicts of Interest: The authors declare no conflict of interest. The funding sponsors had no role in the design of the study; in the collection, analyses, or interpretation of data; in the writing of the manuscript; or in the decision to publish the results.

References

1.   Keane, R.E.; Ryan, K.C.; Veblen, T.T.; Allen, C.D.; Logan, J.; Hawkes, B. *Cascading Effects of Fire Exclusion in Rocky Mountain Ecosystems: A Literature Review*; General Technical Report RMRS-GTR-91; US Department of Agriculture, Forest Service, Rocky Mountain Research Station: Fort Collins, CO, USA, 2002; p. 24.

2.   Swetnam, T.W.; Baisan, C.H. Historical fire regime patterns in the southwestern United States since AD 1700. In *Fire Effects in Southwestern Forests: Proceedings of the Second La Mesa Fire Symposium*; Allen, C.D., Ed.; General Technical Report RM-286; US Department of Agriculture, Forest Service, Rocky Mountain Research Station: Fort Collins, CO, USA, 1996; pp. 11–32.

3.   Hessburg, P.F.; Agee, J.K.; Franklin, J.F. Dry forests and wildland fires of the inland Northwest USA: Contrasting the landscape ecology of the pre-settlement and modem eras. *For. Ecol. Manag.* **2005**, *211*, 117–139. [CrossRef]

4.   Westerling, A.L.; Hidalgo, H.G.; Cayan, D.R.; Swetnam, T.W. Warming and earlier spring increase western US forest wildfire activity. *Science* **2006**, *313*, 940–943. [CrossRef] [PubMed]

5.   Litschert, S.E.; Brown, T.C.; Theobald, D.M. Historic and future extent of wildfires in the Southern Rockies Ecoregion, USA. *For. Ecol. Manag.* **2012**, *269*, 124–133. [CrossRef]

6.   Westerling, A.L. Increasing western US forest wildfire activity: Sensitivity to changes in the timing of spring. *Philos. Trans. Royal Soc. B-Biol. Sci.* **2016**, *371*, 20150178. [CrossRef] [PubMed]

7.   Battaglia, M.A.; Gannon, B.; Brown, P.M.; Fornwalt, P.J.; Cheng, A.S.; Huckaby, L.S. Changes in forest structure since 1860 in ponderosa pine dominated forests of the Colorado and Wyoming Front Range, USA. *For. Ecol. Manag.* **2018**, submitted.

8.   Keeley, J.E. Fire intensity, fire severity and burn severity: A brief review and suggested usage. *Int. J. Wildland Fire* **2009**, *18*, 116–126. [CrossRef]

9.   Graham, R.T. *Hayman Fire Case Study*; General Technical Report RMRS-GTR-114; US Department of Agriculture, Forest Service, Rocky Mountain Research Station: Ogden, UT, USA, 2003; p. 396.

10. Lentile, L.B.; Smith, F.W.; Shepperd, W.D. Patch structure, fire-scar formation, and tree regeneration in a large mixed-severity fire in the South Dakota Black Hills, USA. *Can. J. For. Res.* **2005**, *35*, 2875–2885. [CrossRef]

11. Kokaly, R.F.; Rockwell, B.W.; Haire, S.L.; King, T.V.V. Characterization of post-fire surface cover, soils, and burn severity at the Cerro Grande Fire, New Mexico, using hyperspectral and multispectral remote sensing. *Remote Sens. Environ.* **2007**, *106*, 305–325. [CrossRef]

12. Hayes, J.J.; Robeson, S.M. Relationships between fire severity and post-fire landscape pattern following a large mixed-severity fire in the Valle Vidal, New Mexico, USA. *For. Ecol. Manag.* **2011**, *261*, 1392–1400. [CrossRef]

13. Casas, A.; Garcia, M.; Siegel, R.B.; Koltunov, A.; Ramirez, C.; Ustin, S. Burned forest characterization at single-tree level with airborne laser scanning for assessing wildlife habitat. *Remote Sens. Environ.* **2016**, *175*, 231–241. [CrossRef]

14. Harrington, M.G. Predicting *Pinus ponderosa* mortality from dormant season and growing season fire injury. *Int. J. Wildland Fire* **1993**, *3*, 65–72. [CrossRef]

15. Keyser, T.L.; Lentile, L.B.; Smith, F.W.; Shepperd, W.D. Changes in forest structure after a large, mixed-severity wildfire in ponderosa pine forests of the Black Hills, South Dakota, USA. *For. Sci.* **2008**, *54*, 328–338.

16. Hood, S.M.; Smith, S.L.; Cluck, D.R. Predicting mortality for five California conifers following wildfire. *For. Ecol. Manag.* **2010**, *260*, 750–762. [CrossRef]

17. Dunn, C.J.; Bailey, J.D. Temporal dynamics and decay of coarse wood in early seral habitats of dry mixed conifer forests in Oregon's eastern Cascades. *For. Ecol. Manag.* **2012**, *276*, 71–81. [CrossRef]

18. Roccaforte, J.P.; Fulé, P.Z.; Chancellor, W.W.; Laughlin, D.C. Woody debris and tree regeneration dynamics following severe wildfires in Arizona ponderosa pine forests. *Can. J. For. Res.* **2012**, *42*, 593–604. [CrossRef]

19. Passovoy, M.D.; Fulé, P.Z. Snag and woody debris dynamics following severe wildfires in northern Arizona ponderosa pine forests. *For. Ecol. Manag.* **2006**, *223*, 237–246. [CrossRef]

20. Chambers, M.E.; Fornwalt, P.J.; Malone, S.L.; Battaglia, M.A. Patterns of conifer regeneration following high severity wildfire in ponderosa pine - dominated forests of the Colorado Front Range. *For. Ecol. Manag.* **2016**, *378*, 57–67. [CrossRef]

21. Welch, K.R.; Safford, H.D.; Young, T.P. Predicting conifer establishment post wildfire in mixed conifer forests of the North American Mediterranean-climate zone. *Ecosphere* **2016**, *7*, e01609. [CrossRef]

22. Chambers, C.; Mast, J. Ponderosa pine snag dynamics and cavity excavation following wildfire in northern Arizona. *For. Ecol. Manag.* **2005**, *216*, 227–240. [CrossRef]

23. Koprowski, J.L.; Leonard, K.M.; Zugmeyer, C.A.; Jolley, J.L. Direct effects of fire on endangered Mount Graham red squirrels. *Southwest. Nat.* **2006**, *51*, 59–63. [CrossRef]

24. Kotliar, N.B.; Reynolds, E.W.; Deutschman, D.H. American three-toed woodpecker response to burn severity and prey availability at multiple spatial scales. *Fire Ecol.* **2008**, *4*, 26–45. [CrossRef]

25. Johansen, M.P.; Hakonson, T.E.; Breshears, D.D. Post-fire runoff and erosion from rainfall simulation: Contrasting forests with shrublands and grasslands. *Hydrol. Process.* **2001**, *15*, 2953–2965. [CrossRef]

26. Benavides-Solorio, J.d.D.; MacDonald, L.H. Measurement and prediction of post-fire erosion at the hillslope scale, Colorado Front Range. *Int. J. Wildland Fire* **2005**, *14*, 457–474. [CrossRef]

27. Wagenbrenner, J.W.; Robichaud, P.R. Post-fire bedload sediment delivery across spatial scales in the interior western United States. *Earth Surf. Process. Landf.* **2014**, *39*, 865–876. [CrossRef]

28. Brown, J.K.; Reinhardt, E.D.; Kramer, K.A. *Coarse Woody Debris: Managing Benefits and Fire Hazard in the Recovering Forest*; General Technical Report RMRS-GTR-105; US Department of Agriculture, Forest Service, Rocky Mountain Research Station: Ogden, UT, USA, 2003; p. 16.

29. Coppoletta, M.; Merriam, K.E.; Collins, B.M. Post-fire vegetation and fuel development influences fire severity patterns in reburns. *Ecol. Appl.* **2016**, *26*, 686–699. [CrossRef] [PubMed]

30. Stevens-Rumann, C.; Morgan, P. Repeated wildfires alter forest recovery of mixed-conifer ecosystems. *Ecol. Appl.* **2016**, *26*, 1842–1853. [CrossRef] [PubMed]

31. Stevens-Rumann, C.S.; Sieg, C.H.; Hunter, M.E. Ten years after wildfires: How does varying tree mortality impact fire hazard and forest resiliency? *For. Ecol. Manag.* **2012**, *267*, 199–208. [CrossRef]

32. Abella, S.R.; Fornwalt, P.J. Ten years of vegetation assembly after a North American mega fire. *Glob. Chang. Biol.* **2015**, *21*, 789–802. [CrossRef] [PubMed]

33. Dunn, C.J.; Bailey, J.D. Temporal fuel dynamics following high-severity fire in dry mixed conifer forests of the eastern Cascades, Oregon, USA. *Int. J. Wildland Fire* **2015**, *24*, 470–483. [CrossRef]

34. Addington, R.N.; Aplet, G.H.; Battaglia, M.A.; Briggs, J.S.; Brown, P.M.; Cheng, T.S.; Dickinson, Y.; Feinstein, J.A.; Fornwalt, P.J.; Gannon, B.; et al. *Principles and Practices for the Restoration of Ponderosa Pine and Dry Mixed-Conifer Forests of the Colorado Front Range*; General Technical Report RMRS-GTR-373, US Department of Agriculture, Forest Service, Rocky Mountain Research Station: Fort Collins, CO, USA, 2018; p. 121.

35. Kaufmann, M.R.; Regan, C.M.; Brown, P.M. Heterogeneity in ponderosa pine/Douglas-fir forests: Age and size structure in unlogged and logged landscapes of central Colorado. *Can. J. For. Res.* **2000**, *30*, 698–711. [CrossRef]

36. Fornwalt, P.J.; Kaufmann, M.R.; Huckaby, L.S.; Stoker, J.A.; Stohlgren, T.J. Non-native plant invasions in managed and protected ponderosa pine/Douglas-fir forests of the Colorado Front Range. *For. Ecol. Manag.* **2003**, *177*, 515–527. [CrossRef]

37. Fornwalt, P.J.; Kaufmann, M.R.; Huckaby, L.S.; Stohlgren, T.J. Effects of past logging and grazing on understory plant communities in a montane Colorado forest. *Plant Ecol.* **2009**, *203*, 99–109. [CrossRef]

38. Fornwalt, P.J.; Huckaby, L.S.; Alton, S.K.; Kaufmann, M.R.; Brown, P.M.; Cheng, A.S. Did the 2002 Hayman Fire, Colorado, USA, burn with uncharacteristic severity? *Fire Ecol.* **2016**, *12*, 117–132. [CrossRef]

39. Fornwalt, P.J.; Kaufmann, M.R.; Stohlgren, T.J. Impacts of mixed severity wildfire on exotic plants in a Colorado ponderosa pine—Douglas-fir forest. *Biol. Invasions* **2010**, *12*, 2683–2695. [CrossRef]

40. Fornwalt, P.J.; Kaufmann, M.R. Understorey plant community dynamics following a large, mixed severity wildfire in a *Pinus ponderosa-Pseudotsuga menziesii* forest, Colorado, USA. *J. Veg. Sci.* **2014**, *25*, 805–818. [CrossRef]

41. US Department of Agriculture, Forest Service, Rocky Mountain Research Station. *Long-Term Climate Data for Manitou Experimental Forest*; US Department of Agriculture, Forest Service, Rocky Mountain Research Station: Woodland Park, CO, USA, 2018, unpublished.

42. Moore, R. *Soil Survey of Pike National Forest, Eastern Part, Colorado, Parts of Douglas, El Paso, Jefferson, and Teller Counties*; US Department of Agriculture, Forest Service; US Department of Agriculture, Soil Conservation Service; Colorado Agricultural Experiment Station: Washington, DC, USA, 1992; p. 106.

43. Brown, P.M.; Kaufmann, M.R.; Shepperd, W.D. Long-term, landscape patterns of past fire events in a montane ponderosa pine forest of central Colorado. *Landsc. Ecol.* **1999**, *14*, 513–532. [CrossRef]

44. Huckaby, L.S.; Kaufmann, M.R.; Stoker, J.M.; Fornwalt, P.J. Landscape patterns of montane forest age structure relative to fire history at Cheesman Lake in the Colorado Front Range. In *Proceedings of the Ponderosa Pine Ecosystems Restoration and Conservation: Steps toward Stewardship Conference, Proceedings RMRS-P-22*; Vance, R.K., Covington, W.W., Edminster, C.B., Eds.; US Department of Agriculture, Forest Service, Rocky Mountain Research Station: Fort Collins, CO, USA, 2001; pp. 19–27.

45. Hessburg, P.F.; Povak, N.A.; Salter, R.B. Thinning and prescribed fire effects on snag abundance and spatial pattern in an eastern Cascade Range dry forest, Washington, USA. *For. Sci.* **2010**, *56*, 74–87.

46. Rabe, M.J.; Morrell, T.E.; Green, H.; deVos, J.C.; Miller, C.R. Characteristics of ponderosa pine snag roosts used by reproductive bats in northern Arizona. *J. Wildl. Manag.* **1998**, *62*, 612–621. [CrossRef]

47. Ganey, J.; Vojta, S. Characteristics of snags containing excavated cavities in northern Arizona mixed-conifer and ponderosa pine forests. *For. Ecol. Manag.* **2004**, *199*, 323–332. [CrossRef]

48. Korb, J.E.; Fulé, P.Z.; Stoddard, M.T. Forest restoration in a surface fire-dependent ecosystem: An example from a mixed conifer forest, southwestern Colorado, USA. *For. Ecol. Manag.* **2012**, *269*, 10–18. [CrossRef]

49. Youngblood, A.; Metlen, K.L.; Coe, K. Changes in stand structure and composition after restoration treatments in low elevation dry forests of northeastern Oregon. *For. Ecol. Manag.* **2006**, *234*, 143–163. [CrossRef]

50. Roccaforte, J.P.; Huffman, D.W.; Fule, P.Z.; Covington, W.W.; Chancellor, W.W.; Stoddard, M.T.; Crouse, J.E. Forest structure and fuels dynamics following ponderosa pine restoration treatments, White Mountains, Arizona, USA. *For. Ecol. Manag.* **2015**, *337*, 174–185. [CrossRef]

51. Briggs, J.S.; Fornwalt, P.J.; Feinstein, J.A. Short-term ecological consequences of collaborative restoration treatments in ponderosa pine forests of Colorado. *For. Ecol. Manag.* **2017**, *395*, 69–80. [CrossRef]

52. Fulé, P.Z.; Laughlin, D.C. Wildland fire effects on forest structure over an altitudinal gradient, Grand Canyon National Park, USA. *J. Appl. Ecol.* **2007**, *44*, 136–146. [CrossRef]

53. North, M.P.; Stephens, S.L.; Collins, B.M.; Agee, J.K.; Aplet, G.; Franklin, J.F.; Fulé, P.Z. Reform forest fire management. *Science* **2015**, *349*, 1280–1281. [CrossRef] [PubMed]

54. Huffman, D.W.; Sanchez Meador, A.J.; Stoddard, M.T.; Crouse, J.E.; Roccaforte, J.P. Efficacy of resource objective wildfires for restoration of ponderosa pine (*Pinus ponderosa*) forests in northern Arizona. *For. Ecol. Manag.* **2017**, *389*, 395–403. [CrossRef]

55. Hunter, M.E.; Iniguez, J.M.; Lentile, L.B. Short- and long-term effects on fuels, forest structure and wildfire potential from prescribed fire and resource benefit fire in southwestern forests, USA. *Fire Ecol.* **2011**, *7*, 108–121. [CrossRef]

56. Lake, P.S. Resistance, resilience and restoration. *Ecol. Manag. Restor.* **2013**, *14*, 20–24. [CrossRef]

57. Allen, C.D.; Savage, M.; Falk, D.A.; Suckling, K.F.; Swetnam, T.W.; Schulke, T.; Stacey, P.B.; Morgan, P.; Hoffman, M.; Klingel, J.T. Ecological restoration of southwest ponderosa pine ecosystems: A broad perspective. *Ecol. Appl.* **2002**, *12*, 1418–1433. [CrossRef]

58. Hessburg, P.F.; Churchill, D.J.; Larson, A.J.; Haugo, R.D.; Miller, C.; Spies, T.A.; North, M.P.; Povak, N.A.; Belote, R.T.; Singleton, P.H.; et al. Restoring fire-prone inland Pacific landscapes: Seven core principles. *Landsc. Ecol.* **2015**, *30*, 1805–1835. [CrossRef]

59. Sieg, C.H.; McMillin, J.D.; Fowler, J.F.; Allen, K.K.; Negron, J.F.; Wadleigh, L.L.; Anhold, J.A.; Gibson, K.E. Best predictors for postfire mortality of ponderosa pine trees in the intermountain west. *For. Sci.* **2006**, *52*, 718–728.

60. Ganio, L.M.; Progar, R.A. Mortality predictions of fire-injured large Douglas-fir and ponderosa pine in Oregon and Washington, USA. *For. Ecol. Manag.* **2017**, *390*, 47–67. [CrossRef]

61. Koepke, D.F.; Kolb, T.E.; Adams, H.D. Variation in woody plant mortality and dieback from severe drought among soils, plant groups, and species within a northern Arizona ecotone. *Oecologia* **2010**, *163*, 1079–1090. [CrossRef] [PubMed]

62. Ziegler, J.P.; Hoffman, C.M.; Fornwalt, P.J.; Sieg, C.H.; Battaglia, M.A.; Chambers, M.E.; Iniguez, J.M. Tree regeneration spatial patterns in ponderosa pine forests following stand-replacing fire: Influence of topography and neighbors. *Forests* **2017**, *8*, 391. [CrossRef]

63. Malone, S.L.; Fornwalt, P.J.; Battaglia, M.A.; Chambers, M.E.; Iniguez, J.M.; Sieg, C.H. Mixed-severity fire fosters heterogeneous spatial patterns of conifer regeneration in a dry conifer forest. *Forests* **2018**, *9*, 45. [CrossRef]

64. Rother, M.T.; Veblen, T.T. Limited conifer regeneration following wildfires in dry ponderosa pine forests of the Colorado Front Range. *Ecosphere* **2016**, *7*, e01594. [CrossRef]

65. Rother, M.T.; Veblen, T.T. Climate drives episodic conifer establishment after fire in dry ponderosa pine forests of the Colorado Front Range, USA. *Forests* **2017**, *8*, 159. [CrossRef]

66. Spiering, D.J.; Knight, R.L. Snag density and use by cavity-nesting birds in managed stands of the Black Hills National Forest. *For. Ecol. Manag.* **2005**, *214*, 40–52. [CrossRef]

67. Saab, V.A.; Russell, R.E.; Dudley, J.G. Nest-site selection by cavity-nesting birds in relation to postfire salvage logging. *For. Ecol. Manag.* **2009**, *257*, 151–159. [CrossRef]

68. Kerns, B.K.; Thies, W.G.; Niwa, C.G. Season and severity of prescribed burn in ponderosa pine forests: Implications for understory native and exotic plants. *Ecoscience* **2006**, *13*, 44–55. [CrossRef]

69. Brown, J.K.; Smith, J.K. *Wildland Fire in Ecosystems: Effects of Fire on Flora*; General Technical Report RMRS-GTR-42-2; US Department of Agriculture, Forest Service, Rocky Mountain Research Station: Ogden, UT, USA, 2000; p. 257.

70. Laughlin, D.C.; Moore, M.M.; Bakker, J.D.; Casey, C.A.; Springer, J.D.; Fulé, P.Z.; Covington, W.W. Assessing targets for the restoration of herbaceous vegetation in ponderosa pine forests. *Restor. Ecol.* **2006**, *14*, 548–560. [CrossRef]

71. Ferguson, D.E.; Byrne, J.C. *Shrub Succession on Eight Mixed-Severity Wildfires in Western Montana, Northeastern Oregon, and Northern Idaho*; Research Paper RMRS-RP-106; US Department of Agriculture, Forest Service, Rocky Mountain Research Station: Fort Collins, CO, USA, 2016.

72. Anderson, H.E. *Aids to Determining Fuel Models for Estimating Fire Behavior*; General Technical Report INT-122; US Department of Agriculture, Forest Service, Intermountain Forest and Range Experiment Station: Fort Collins, CO, USA, 1982; p. 22.

73. Fornwalt, P.J. *Fuel Biomass Data for 20 0.1-ha Hayman Fire Plots*; US Department of Agriculture, Forest Service, Rocky Mountain Research Station: Fort Collins, CO, USA, 2016, unpublished data.

74. Robichaud, P.R.; Lewis, S.A.; Wagenbrenner, J.W.; Ashmun, L.E.; Brown, R.E. Post-fire mulching for runoff and erosion mitigation part I: Effectiveness at reducing hillslope erosion rates. *Catena* **2013**, *105*, 75–92. [CrossRef]

75. Robichaud, P.R.; Wagenbrenner, J.W.; Lewis, S.A.; Ashmun, L.E.; Brown, R.E.; Wohlgemuth, P.M. Post-fire mulching for runoff and erosion mitigation part II: Effectiveness in reducing runoff and sediment yields from small catchments. *Catena* **2013**, *105*, 93–111. [CrossRef]

76. Rhoades, C.C.; Entwistle, D.; Butler, D. The influence of wildfire extent and severity on streamwater chemistry, sediment and temperature following the Hayman Fire, Colorado. *Int. J. Wildland Fire* **2011**, *20*, 430–442. [CrossRef]

77. North, M.; Collins, B.M.; Stephens, S. Using fire to increase the scale, benefits, and future maintenance of fuels treatments. *J. For.* **2012**, *110*, 392–401. [CrossRef]

![forests logo] *forests*

MDPI

Article

Temporal Patterns of Wildfire Activity in Areas of Contrasting Human Influence in the Canadian Boreal Forest

Rodrigo Campos-Ruiz [1,2,*], **Marc-André Parisien** [2,3] **and Mike D. Flannigan** [1,2]

[1] Department of Renewable Resources, University of Alberta, Edmonton, AB T6G 2R3, Canada; mike.flannigan@ualberta.ca

[2] Canadian Partnership for Wildland Fire Science, Department of Renewable Resources, University of Alberta, Edmonton, AB T6G 2R3, Canada

[3] Northern Forestry Centre, Canadian Forest Service, Natural Resources Canada, Edmonton, AB T6H 3S5, Canada; marc-andre.parisien@nrcan-rncan.gc.ca

* Correspondence: rcampos@ualberta.ca

Received: 15 February 2018; Accepted: 19 March 2018; Published: 21 March 2018

Abstract: The influence of humans on the boreal forest has altered the temporal and spatial patterns of wildfire activity through modification of the physical environment and through fire management for the protection of human and economic values. Wildfires are actively suppressed in areas with higher human influence, but, paradoxically, these areas have more numerous ignitions than low-impact ones because of the high rates of human-ignited fires, especially during the springtime. The aim of this study is to evaluate how humans have altered the temporal patterns of wildfire activity in the Canadian boreal forest by comparing two adjacent areas of low and high human influence, respectively: Wood Buffalo National Park (WBNP) and the Lower Athabasca Plains (LAP). We carried out Singular Spectrum Analysis to identify trends and cycles in wildfires from 1970 to 2015 for the two areas and examined their association with climate conditions. We found human influence to be reflected in wildfire activity in multiple ways: (1) by dampening (i.e., for area burned)—and even reversing (i.e., for the number of fires)—the increasing trends of fire activity usually associated with drier and warmer conditions; (2) by shifting the peak of fire activity from the summer to the spring; (3) by altering the fire-climate association; and (4) by exhibiting more recurrent (<8 year periodicities) cyclical patterns of fire activity than WBNP (>9 years).

Keywords: Wildfire; Wildland fire; forest fire; boreal forest; fire management; human influence; climate

1. Introduction

Wildfire is a critical phenomenon maintaining the ecological processes and integrity of the Canadian boreal forest [1]. Although most fire regimes in Canada are characterized by infrequent, high-intensity, extensive fires occurring mostly between May and August, there is great variability in the components of the fire regime across spatial and temporal scales [2]. Understanding the spatio-temporal patterns of fire occurrence and area burned is thus of foremost interest in Canada in ecological, social, and economic terms. Wildland fires burn on average between 1 to 3 million ha annually in Canada, affecting biological diversity [3,4], ecological services [5,6], and forest resources [7], and require costly fire management strategies for infrastructure and community protection [8,9].

Fire activity, which is usually measured as the number of fires and the total area burned, is driven by weather, climate [10,11], vegetation type [12,13], topography, human activities [8,14,15], and complex interactions among these factors. Globally, wildfire dynamics have been altered by human activities for millennia through people setting (accidental or deliberate) ignitions that add to the ones caused naturally by lightning, and by modifying the landscape through their activities, increasing access to wildlands, and by altering the arrangement, continuity, amount, structure, and distribution of fuels (i.e., flammable biomass). These changes, combined with climate change, have modified the components of the fire regime: severity, likelihood, seasonality, size, frequency, and intensity of fires [13,16–20].

Evidence suggests that fire suppression practices also have an important effect on the fire–weather relationship, potentially undermining the reliability of weather-based predictions. Current ignition rates in France, for example, are not as high as they were during the pre-suppression period, even under similar weather conditions [21]. Fire regimes in Catalonia, Spain, cannot be efficiently predicted unless fire suppression activities, in addition to climatic variables, are taken into account [22]. In the U.S. Rocky Mountains, alternating periods of fire–climate relationship strength were reported for the past century, suggesting that the interaction of climatic and non-climatic factors, such as fire suppression, is highly complex and needs to be assessed to improve our understanding of fire activity [23].

Over the last century, fire management activities in the Canadian boreal forest have shaped fire activity through aggressive suppression efforts and preventive measures. However, fire management has not been uniform over space and time because deploying resources to prevent and attack fires involves balancing potential economic, social, and ecological impacts, which can sometimes be conflicting [24–27]. To better reflect regional priorities, most of Alberta's forested land surface has been divided into ten Wildfire Management Areas [28] that are managed by the provincial government. Fire management in Alberta has increased the containment (i.e., extinguishment before they reach 2 ha in size) of fires from 75% in 1998 to 93% in 2015 [29]. Regardless of technological advances and preventive measures, large fire events still occur in Alberta. Such is the case of the Horse River fire in Fort McMurray, Alberta (spring 2016; 580,633 ha burned), which caused the evacuation of 80,000 people and was the costliest natural disaster in Canada, with an estimated damage worth of CAN$10.9 billion [30,31]. In contrast, National Parks within Alberta are not included in the provincial Wildfire Management Areas because the administration falls under federal jurisdiction. In these parks, unlike the rest of forested Alberta, the ecological role of fires is favored and wildfires are only supressed when they pose a risk to surrounding inhabited areas and areas containing rare natural resources [32].

Although it is well recognized that human influence can alter fire activity in the boreal forest, the magnitude and direction of recent changes remain largely undocumented. The main goal of this study is to assess how human influence has affected the fire regime of two contrasting areas in the Canadian boreal forest over the past few decades. To achieve this goal, we analyzed the changes in the number of fires and area burned (fire activity) over time and their relationships with climate in two contiguous regions of Alberta with contrasting human influence (fire management and human land use). First, to establish whether fire activity has changed in each region, we analyzed annual trends of the number of fires and area burned, from 1970 to 2015, and compared them while distinguishing natural from anthropogenic fires. Secondly, to determine if the fire–climate relationships remain coherent despite the dissimilar human influence, we compared these relationships between regions. Finally, to better understand the correlations between fire activity and climate, we characterized the cyclical patterns of the number of fires, area burned, and climate, and compared them between regions in terms of the duration of their periodicities.

2. Materials and Methods

2.1. Study Areas

We chose two adjacent study regions located in the northeastern corner of the province of Alberta and a southern portion of Northwest Territories, Canada, that differ in their level of human influence, including wildfire management (Figure 1): Wood Buffalo National Park (WBNP, 4.3 Mha) and the Lower Athabasca Plains (LAP, 7.9 Mha). Both regions are located within the Boreal Forest Natural

Region (BFNR), which covers approximately 58% of Alberta. The topography is mostly flat to gently hilly, and the vegetation is represented by four dominant types: upland deciduous, coniferous, mixed forests, and wetlands. The most common tree species found are black spruce (*Picea mariana* (Mill.) B.S.P.), white spruce (*P. glauca* (Moench) Voss), jack pine (*Pinus banksiana* Lamb.), trembling aspen (*Populus tremuloides* (Michx.)), balsam poplar (*P. balsamifera* L.), and eastern larch (*Larix laricina* (Du Roi) K. Koch) [33].

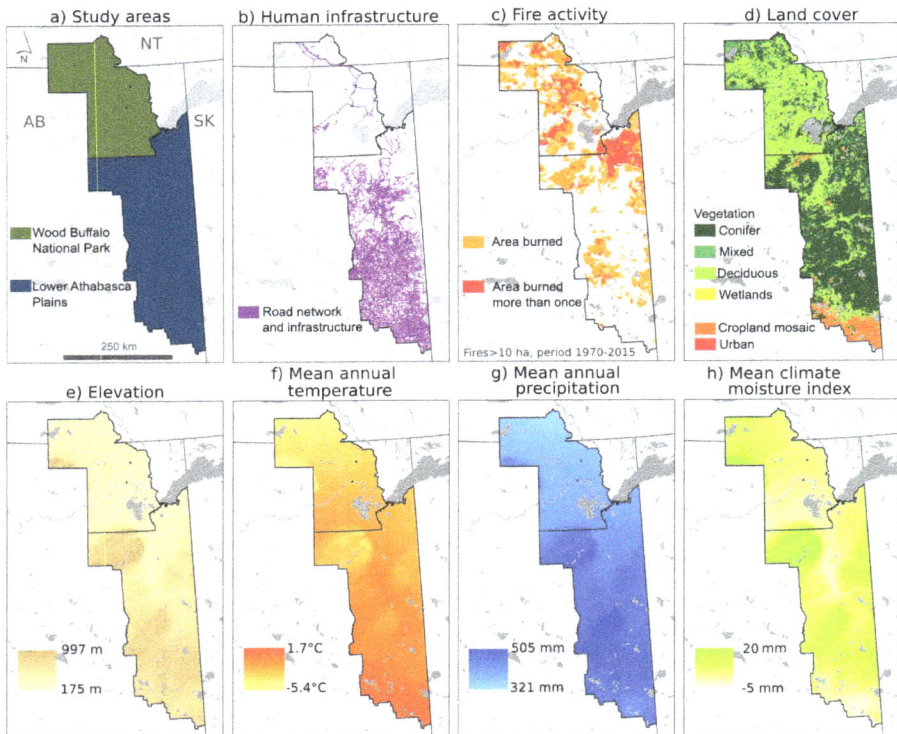

Figure 1. Maps showing (**a**) the location of the areas of interest (53.63° N, 110° W, 60.70° N, 115.60° W): Wood Buffalo National Park, which occupies a portion of Alberta (AB) and Northwest Territories (NT), and Lower Athabasca Plains, which borders with the province of Saskatchewan (SK), (**b**) Human infrastructure cover (30-metre-resolution, Landsat; [34]), (**c**) Area burned from 1970–2015 (one-kilometer resolution), (**d**) Land cover (Global Land Cover Characterization Project, one-kilometer resolution, AVHRR; [35]), (**e**) Elevation, (**f**) Mean annual temperature, (**g**) Mean annual precipitation, and (**h**) Mean climate moisture index (precipitation minus potential evapotranspiration, mm, 1° latitude by 1° longitude resolution; [36]).

Although WBNP and LAP are very similar in terms of topography and vegetation composition, they are subjected to contrasting levels of human influence. Established in 1922, WBNP is a UNESCO (United Nations Educational, Scientific and Cultural Organization) world heritage site where the human influence is remarkably low; less than one percent of its area has been modified by human activities, which mainly consist of two roads (Table 1, Figure 1b). In contrast, LAP has almost 10 percent of its area destined to agriculture, tree harvesting, oil-and-gas exploration, and other rural and industrial activities. Comparatively, LAP also possesses a more extensive road network and more human settlements (Table 1, Figure 1b) [33,34]. Despite belonging to the same land management unit, we excluded the portion of LAP located on the north side of Lake Athabasca, because it is located in the Canadian Shield, which represents a topo-edaphic setting that differs substantially from the rest of the study area and WBNP, both located in the Boreal Plain.

Table 1. Area occupied or modified by the leading human activities in Wood Buffalo National Park (WBNP) and Lower Athabasca Plains (LAP).

Region	WBNP Area kha (%)		LAP Area kha (%)	
Total	4266	(100)	7932	(100)
Cultivation	0		134.80	(3.16)
Harvested (cut blocks)	0		111.33	(2.61)
Mining	0		32.42	(0.76)
Seismic lines	0		29.86	(0.70)
Industrial-rural	0		23.90	(0.56)
Roads and vegetated margins	0.85	(0.02)	15.36	(0.36)
Urban	0		2.58	(0.06)
Total human-modified area	0.85	(0.02)	350.25	(8.21)

Fire management also differs between these regions. Fires in WBNP are not actively suppressed unless they pose an imminent threat to infrastructure within the park or neighbouring communities, leaving wildfires to fulfill their "natural" role. In contrast, LAP includes the community of Fort McMurray and most of the Lac La Biche Alberta Wildfire Management areas, where there is a strict policy to suppress and prevent fires. On average, during the period 1970–2015, lightning-caused fires in WBNP are larger than the ones in LAP (Table A1), whereas anthropogenic ones are larger in LAP. Human contribution to area burned is so prominent in LAP that the area burned by human-caused fires exceeds the area burned by lightning-caused fires in this region (Table A1). Summer is the season with the highest fire activity (number of fires and area burned; Figure 2) in both regions, but the human contribution to fires is considerably higher during the spring in LAP than in WBNP.

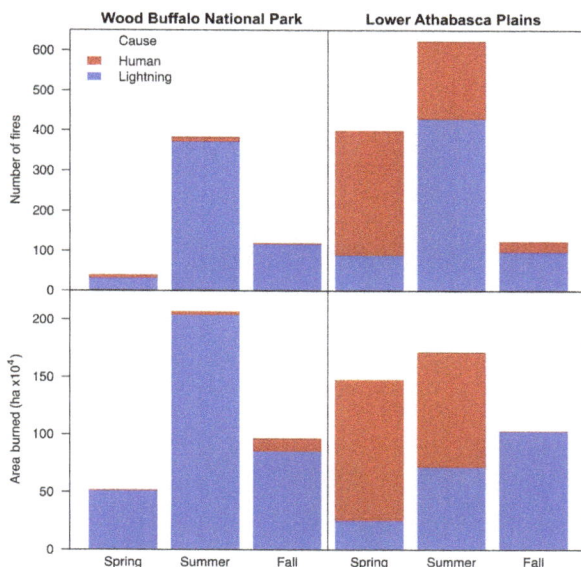

Figure 2. Number of fires and area burned in Wood Buffalo National Park and Lower Athabasca Plains by season (spring, summer, and fall) and cause of ignition (lightning or human). Information based on fires ≥10 ha from 1970–2015.

2.2. Fire Variables

We obtained the historical fire records from the Canadian National Fire Database [37] from which we derived the number of fires and area burned (ha) during the fire season (1 March to 31 October) [38] for the period 1970–2015. Although the database extends earlier than 1970, the analysis was limited to this time period because the data recording and methods of detection during the contemporary era were more reliable and consistent than those of the earlier period [24,27,39]. In addition, we only included fires ≥10 ha because they are less likely to go undetected or unreported than fires <10 ha. The fires were stratified by area (WBNP or LAP) and by cause (human or lightning) and summarized into annual time series using the total number of fires and total area burned (log transformed) during the fire season.

2.3. Fire-Climate Variables

In order to uncover possible relationships with the fire activity variables (area burned and number of fires), we built mean annual time series of climate and fire danger indexes from the Canadian Fire Weather Index System (FWIS) [40] (Table 2). We used the records from 67 weather stations located in WBNP and LAP for the period 1970–2015 [41]. Climate variables include annual means of temperature (°C), relative humidity (%), wind speed (km/h), and 24-h precipitation (mm), as well as codes and indices from the FWIS [40]: Fine Fuel Moisture Code (FFMC), Duff Moisture Content (DMC), Drought Code (DC), Initial Spread Index (ISI), Buildup Index (BUI), Fire Weather Index (FWI), and Daily Severity Rating (DSR). Indices and codes are calculated based on weather measurements taken at noon local standard time (LST) and the FWIS moisture codes from the previous day (Table 2).

Table 2. Climate variables and Fire Weather Index System codes and indices used (Modified from Van Wagner [40]).

Acronym	Name	Units	Description
TEMP	Temperature	°C	A measure of heat present in the air
PRECIP	Precipitation	mm	A form of water, such as rain, snow etc. that condenses from the atmosphere and fall to the Earth
RH	Relative Humidity	%	Amount of water vapor present in air
WINS	Wind Speed	km/h	Velocity of air flow
FFMC	Fine Fuel Moisture Code	unitless	Moisture contained in the upper soil layer (litter and fine fuels)
DMC	Duff Moisture Code	unitless	Moisture for the loose organic layers of the soil, including medium-sized woody debris
DC	Drought Code	unitless	Moisture in deep compacted organic layer and large woody debris
ISI	Initial Spread Index	unitless	The expected rate of spread based on FFMC and wind speed
BUI	Buildup Index	unitless	Proxy for the fuel load available for combustion. Based on DMC and DC
FWI	Fire Weather Index	unitless	Reflects fire intensity and fire danger in forested areas. Based on ISI and BUI
DSR	Daily Severity Rating	unitless	Exponential transformation of FWI indicating severe conditions when DSR >2

2.4. Statistical Analysis

We searched for linear trends in the time series to identify overall changes in the number of fires and area burned during the 1970–2015 period. By decomposing the time series, we were able to separate them into their additive components: non-linear trend, oscillations (cyclical), and white noise. The non-linear trend was subtracted from the original time series as a "pre-whitening" step, which reduces the occlusion of the remaining components; in this way, we were able to identify oscillation frequencies and calculate their periodicities (years between peaks). Finally, we ran correlation tests between the fire-climate and fire activity time series in order to identify potential relationships between them. If a relationship was identified, we described its nature (positive, negative), strength, and periodicity (through their regular oscillations). All analyses were performed using R [42].

2.4.1. Linear Trend Detection

We tested the time series for autocorrelation processes, after which we ran a phase-randomized version of the Mann-Kendall trend test (slope different from 0: $p \leq 0.05$, H_0: $\beta_1 = \beta_2$, H_a: $\beta_1 \neq \beta_2$) to find a significant linear trend. The Theil-Sen slope method was employed to calculate the slope value (β).

Some time series showed autocorrelation processes (aka. "red noise"), for which the magnitude and order were calculated. This was achieved by means of the autocorrelation and partial autocorrelation functions using the *Acf* function of the "forecast" R package [43] (Figures A1 and A2). We found autocorrelation processes in the number of fires and area burned for both lightning-caused fires (first order autoregressive model, AR1 = 0.34) and human-caused fires (autoregressive-moving-average model, ARMA (3, 0), $p = 0.42$) in WBNP. In LAP, only area burned by human-caused fires and the total (L + H) showed autocorrelation ARMA (6, 0). To account for the serial correlation, we employed a phase-randomization method for the hypothesis testing of the trend, which is suitable for non-normal, autocorrelated data, and is robust against outliers (i.e., influential or extreme data). This method consists of the creation of surrogate time series (randomized versions of the original) to create a distribution which is then compared to the original time series to determine its significance. By doing so, we avoided spurious regressions due to the lack of error independence and unequal variances [44–46] and obtained a robust estimate of parameters of the regression by bootstrapping [47]. We employed the *MKcorr.test* function contained in the "MKCorr R" package [48].

2.4.2. Time Series Decomposition

Next, we separated the time series (number of fires, area burned, climate and FWIS indices) into their additive components: non-linear trend, oscillations (regular cycles), and white noise (random signal; see example in the appendix) by using an iterative Singular Spectrum Analysis (SSA, [49]) included in the "Rssa" package [50] in R. SSA is an adaptive non-parametric method ideal for short, noisy time series. This method does not require a priori knowledge of the model to be fitted (e.g., linearity, normality, and stationarity of the residuals, or the number and value of the contained periodicities), making it an advantageous technique to explore and analyze data when the parameters are unknown.

The separation of the components was achieved by running the decomposition process twice (hence the term "iterative"): first, we ran the analysis to extract the non-linear trend and, second, we subtracted this trend (detrending) from the original time series and ran the decomposition process again to extract the oscillatory (cyclical) components. This step also reduces the red noise significantly, so that the oscillations detected are unlikely to be autocorrelation processes [51]. Each iteration requires "windows" (adjacent values) of different lengths (number of values) to establish the resolution and minimum periodicity to detect. We used a small window ($L = 12$) to extract the non-linear trend, and a larger window ($L = 24$) on the detrended time series to extract the cyclical components [49]. We calculated the periodicities of the cyclical components originating from the eigenvectors that explained most variance (>20%) according to the SSA. This was achieved through the Estimation of Signal Parameters via Rotational Invariance Techniques (ESPRIT; [52]) using the function *parestimate* in the "Rssa" package and confirmed with the spectrum (*mvspec* function, "astsa" package [53]). As a result, we obtained a new set of time series produced for the two study areas (WBNP and LAP), the three ignition causes (lightning, human, lightning + human), the two fire variables (number of fires and area burned), the eleven fire-climate variables, and the two SSA derived outputs (one non-linear-trend and one detrended time series). Non-linear trends, unlike linear ones, may show changes over time according to different rates that might not be detected with the Mann-Kendall test, but they may still represent a relevant pattern in the time series.

2.4.3. Cross-Correlations of Time Series

In order to evaluate the correlations of fire activity time series between areas and fire activity series with fire-climate variables, we calculated the Kendall rank correlation coefficients and their significance from the detrended time series. We carried out correlations for two different time lags (zero and one year) to test for current and delayed effects of climate on fire, respectively.

We used a non-parametric "randomized-phase surrogate" technique for significance testing to reduce type-1 errors. This test employs the Fourier transform to generate a large number of random time series (called surrogates) with the same spectral properties as one of original (and thus, the same autocorrelation, if present) but with random phases. Then, the correlation between the two original time series is compared to a distribution of correlations produced by the surrogate series to obtain the statistical significance [54,55]. This analysis was performed using the *surrogateCor* function contained in the package "astrochron" [56] in R with 2000 random surrogate series.

3. Results

Fire activity from 1970 to 2015 has increased in WBNP in both the number of fires ($\beta = 0.30$, $p < 0.01$, Figure 3a) and area burned ($\beta = 0.14$, $p < 0.01$, Figure 3c), while in LAP, there is a statistically significant decrease in the number of fires ($\beta = -0.21$, $p = 0.01$, Figure 3b), and no overall change in area burned ($\beta = 0.02$, $p > 0.05$; Figure 3d). Lightning-caused fires were responsible for raising the number of fires in WBNP ($\beta = 0.28$, $p < 0.05$), whereas anthropogenic fires remained low ($\beta = 0$, $p > 0.05$) and their area burned unchanged ($\beta = 0$, $p > 0.05$; Figure 3a,c). Fire activity in WBNP exhibited nonlinear trends, indicating that the increase was faster during the second half of the time period than the first half. Although the non-linear trends suggest a decline in the number of fires caused by lightning or humans in LAP, neither of them were statistically significant ($\beta = 0$, $p > 0.05$ and $\beta = -0.9$, $p > 0.05$ respectively), but their sum (L + H) was (Figure 3b).

In general, a warming and drying climate trend was found over the 46 years studied in both WBNP and LAP, with faster changes in WBNP (Table 3). Most of the fire-climate time series showed some change (linear trend) over the same period that would lead to an increased wildfire activity, but this is only observed in WBNP. In general, for both WBNP and LAP, we found a declining trend in PRECIP, RH, and WINS, and an increasing trend for FFMC, DMC, DC, and BUI. Only ISI did not change overall in any of the areas. Although we did not find a linear trend for temperature in LAP, the non-linear trend extracted by SSA does show an increase (Figure A3), as well as for FWI and DSR in WBNP. The most drastic increase in both areas was observed in the Drought Code, which indicates moisture deficits in the deeper soil levels (Table 3).

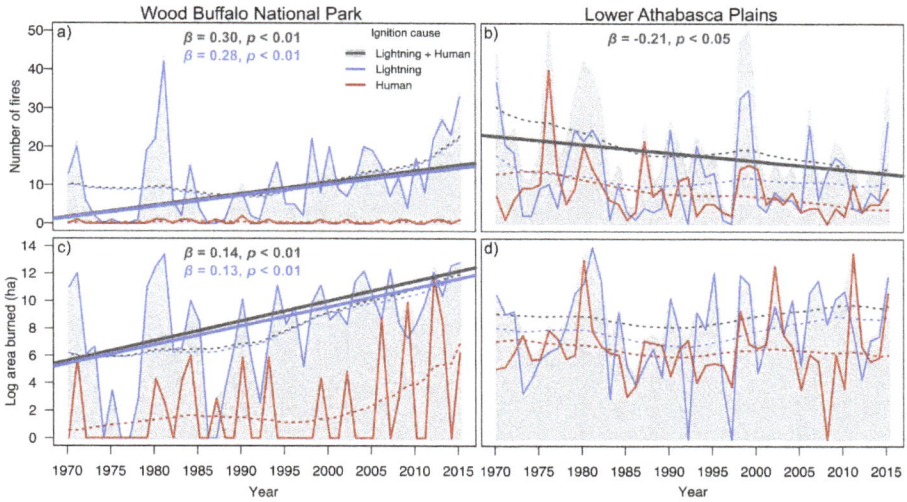

Figure 3. Time series representing the number of fires (**a,b**) and area burned (**c,d**) for the period 1970–2015 in Wood Buffalo National Park (**a,c**) and Lower Athabasca Plains (**b,d**). Thin solid lines and shaded background indicate raw time series, dotted lines show the non-linear trend extracted by Singular Spectrum Analysis, and solid straight lines display the significant linear trends obtained by the Mann-Kendall trend test. Different colors indicate the cause of ignition. Theil-Sen's slopes (β) and trend significance by Mann-Kendall are shown.

Table 3. Theil-Sen slope values (β) and significance by a random-phase Mann-Kendall trend test (* $\leq$0.05, ** $\leq$0.005) for the fire-climate variables in Wood Buffalo National Park (WBNP) and Lower Athabasca Plains (LAP). Values were calculated from annual time series.

Acronym	Name	WBNP	LAP
TEMP	Temperature	0.03 **	0.01
PRECIP	Precipitation	<−0.01 *	−0.01 **
RH	Relative Humidity	−0.15 **	−0.06 **
WINS	Wind Speed	−0.10 **	−0.03 *
FFMC	Fine Fuel Moisture Code	0.07 *	0.07 **
DMC	Duff Moisture Code	0.30 **	0.23 **
DC	Drought Code	2.83 **	3.23 **
ISI	Initial Spread Index	0	0
BUI	Buildup Index	0.44 **	0.42 **
FWI	Fire Weather Index	0.04	0.06 **
DSR	Daily Severity Rating	0.02	0.02 *

Cross-correlations of detrended time series between regions indicated a similar pattern of fire activity regardless of the differences in human influence. Lightning-caused fire time series correlations (for both number of fires and area burned) showed the highest coefficients, whereas human-caused fires showed the lowest (Table 4).

Table 4. Correlation coefficient table for the number of fires (upper-right) and area burned (lower left, shaded) of the fire activity time series in Wood Buffalo National Park (WBNP) and Lower Athabasca Plains (LAP). Significance was calculated by a random-phase test (* ≤0.05, ** ≤0.005). Letters following area name indicate ignition cause: Lightning (L) and human (H).

Region			WBNP			LAP		
	Cause	L + H	L	H	L + H	L	H
WBNP	L + H	1	0.96 **	0.17 *	0.31 **	0.36 **	0.08
	L	0.94 **	1	0.13	0.30 **	0.35 **	0.09
	H	0.09	0.04	1	0.18	0.24 *	−0.07
LAP	L + H	0.26 *	0.25 *	0.10	1	0.68 **	0.37 **
	L	0.33 **	0.32 **	0.09	0.76 **	1	0.06
	H	−0.01	−0.01	0	0.29 *	0.09	1

Although the association of fire activity with fire climate was similar in both regions (we observed significant correlations with similar coefficient values for both the number of fires and area burned time series; Table 5), we noted that: (1) mean annual temperature was only correlated with lightning-caused fires in WBNP and human-caused fires and area burned in LAP, (2) mean annual precipitation was associated with lightning-caused fire activity but not human-caused fires in both areas, (3) human-caused fires in WBNP do not have any association with fire-climate, and (4) lightning-caused fires only correlated with temperature and relative humidity in WBNP. We only found three one-year lagged correlations: between the number of human-caused fires in LAP with wind speed ($\tau = -0.30$, $p < 0.005$), and area burned by lightning ignitions with precipitation in WBNP ($\tau = -0.26$, $p < 0.05$) and temperature in LAP ($\tau = 0.29$, $p < 0.005$).

Table 5. Kendall-correlation coefficient table for the relationship between the number of fires and area burned with fire-climate time series at lag 0 (i.e., current year) in Wood Buffalo National Park (WBNP) and Lower Athabasca Plains (LAP). Significance was calculated using a random-phase test (* ≤0.05, ** ≤0.005).

	Number of Fires						Area Burned					
Region	WBNP			LAP			WBNP			LAP		
Ignition cause	L + H	L	H	L + H	L	H	L + H	L	H	L + H	L	H
Temperature	0.20 *	0.19 *	0.17	0.20	0.12	0.23 *	0.07	0.07	0.1	0.10	0.09	0.17 *
Precipitation	−0.29 *	−0.27 *	−0.02	−0.22 *	−0.21 *	−0.14	−0.23 *	−0.25 *	0.07	−0.20 *	−0.21 *	−0.14
Relative humidity	−0.24 *	−0.23 *	−0.05	−0.20	−0.13	−0.18 *	−0.21	−0.22 *	−0.03	−0.27 *	−0.24 *	-0.22 *
Wind speed	−0.14	−0.15	−0.05	−0.18	−0.07	−0.2 *	−0.27	−0.23	−0.09	−0.02	−0.01	−0.12
FFMC	0.26 *	0.24 *	0.1	0.23 *	0.22 *	0.15	0.20	0.20 *	0.05	0.29 *	0.27 *	0.19 *
DMC	0.33 *	0.32 **	−0.01	0.27 *	0.03 *	0.21 *	0.37 **	0.40 **	−0.06	0.40 **	0.34 **	0.30 **
DC	0.34	0.34 **	0.04	0.23 *	0.25 *	0.18 *	0.36 *	0.40 **	−0.02	0.24 *	0.26 **	0.27 *
BUI	0.20 *	0.35 **	0	0.30 *	0.26 *	0.2 *	0.37 **	0.40 **	−0.03	0.40 **	0.35 **	0.30 *
ISI	0.21 *	0.20 *	0.10	0.28 *	0.30 *	0.1	0.23 *	0.26 *	0.07	0.40 **	0.32 **	0.21 *
FWI	0.30 *	0.30 *	0.06	0.31 *	0.30 *	0.16	0.36 **	0.40 **	0.01	0.44 **	0.36 **	0.30 **
DSR	0.28 *	0.26 *	0.01	0.34 **	0.30 **	0.20	0.37 **	0.40 **	0.02	0.44 **	0.35 **	0.31 **

Fire activity in WBNP was characterized by longer and more acute periodicities than the ones found in LAP. We observed that the number of fires and area burned by lightning in WBNP showed strong activity peaks every ~12 years, whereas in LAP, weaker oscillations (i.e., under eight years) were most frequent. Anthropogenic ignitions in both areas mostly displayed high frequency (short periods) oscillations (~3–6 years); however, area burned by humans also showed strong oscillations every ~11 years (Figure 4).

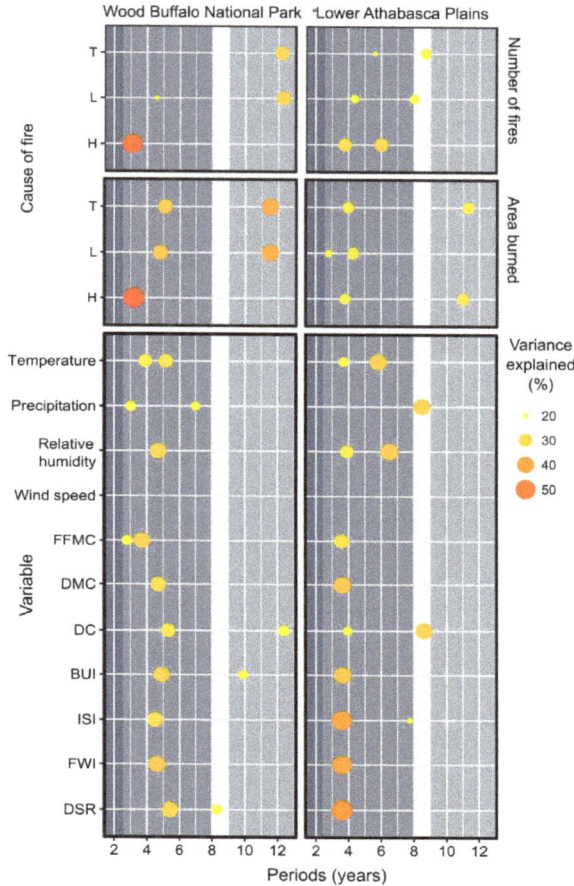

Figure 4. Periodicities (i.e., duration of oscillations) of the fire activity and fire-climate variables, calculated for the cyclical components found in the time series. Size and color of the points indicate the variance explained by the eigenvectors from which the oscillations were extracted during SSA. Shading corresponds to the "typical" climatic teleconnection domains: Quasi-Biennial Oscillation (2–2.5 years, dark grey), El Niño Southern Oscillation (2.5–8 years, moderate grey), and a quasi-decadal oscillation (9–13 years, light grey).

Fire-climate oscillations were very similar between regions, with the strongest periodicities (i.e., higher variance explained) falling under the ENSO domain (2.5–8 years) and aligning with many of the periodicities of fire activity (between 3–6 years). Most of the strongest fire-climate periodicities were found at around four years. With the exception of DC (an index of drought involving precipitation and temperature) and BUI (calculated through DC and DMC), we did not detect strong long periodicities (over nine years) from the fire-climate time series (Figure 4).

4. Discussion

Human activities have been continuously altering the dynamics of the boreal forest in Alberta over the past few decades, and have generated recognizable spatial and temporal patterns of wildfire activity. In the 1970–2015 period, we found increasing fire-conducive climatic trends in both study areas that were reflected in occurrence and area burned increases only in the region with the lowest human interference (WBNP), which agrees with findings reported across Canada [57–59]. In contrast, we observed a dampening effect of fire activity in the region with the most human impact (LAP), resulting in a decline in the number of fires and no overall change (no trend) in the area burned. Previous studies also suggest that although area burned has increased due to the changing climate in the boreal forest, this is not necessarily the case in areas of higher human influence [60,61].

The main causes limiting the potential area burned under high human influence are: (1) the strong prevention and fire suppression policies, (2) improved accessibility, (3) the location of the human ignitions (i.e., closer to human infrastructures), and (4) land-use change. The first cause is a consequence of two of the objectives of fire suppression in Alberta: reducing the spread of fires before 10 AM of the following day and preventing them from attaining 2 ha in size [28], which results in fewer escaped fires (i.e., defined here as fires $\geq$10 ha). In addition, the widespread road network facilitates access for fire-management activities to take place. Although human ignitions might increase over time, they also have a tendency to cluster in the wildland-urban interface (WUI; [61,62]), where fire detection and initial attack are more efficient, impeding their further growth in spite of fire-prone weather conditions. Finally, conversion of forested areas to agricultural, urban, and petrochemical mining land uses has altered the vegetation's structure and composition. These changes, in turn, have impacted the burn rates and ignition likelihood, and the increased fragmentation has reduced vegetation continuity, countering potential area burned [15,62–67].

Wildfire suppression practices, in conjunction with other human-induced changes, have the potential to not just slow down, but also to reverse trends in fire activity; hence, non-climatic factors have the potential to alter fire-climate relationships. Such cases have been reported in France, where recent fire activity stopped tracking climatic trends and decreased along with major changes in fire suppression policy [21,68], and in South Africa, where land-use changes mediated the relationship between climate and area burned [69].

Unlike trends, overall annual fluctuation patterns of wildfire activity exhibited some similarities between regions, regardless of the level of human influence. This is because climate and lightning still persist as dominant factors regulating the totality of fire activity in the boreal forest [70–72]. We further support this observation, given that wildfire activity was similarly related to climate in both regions. In general, peaks of fire activity tracked a drier and warmer climate in both areas. We also found that temperature and relative humidity did not correlate to the number of lightning-caused fires in LAP, whereas they did in WBNP, suggesting that non-climatic factors (i.e., fire management, land-use change, road density) might have interfered with those relationships [21,73].

Anthropogenic fire activity is associated with the same climatic variables as lightning-caused fire activity, with only very few exceptions. Most notably, we observed a lack of association between anthropogenic fires and precipitation that might indicate that more of these fires may occur in years with higher soil moisture conditions than lightning-caused fires. This observation has also been reported before in the U.S. [74], where the authors concluded that anthropogenic ignitions can occur in a broader range of moisture environments than lightning-caused fires, thereby resulting in a wider wildfire "niche". In addition, the shift of the peak of anthropogenic fire activity from the summer to the spring, accompanied by a longer fire season length [20,74–76], might have caused anthropogenic and lightning fires associated with different climatic conditions within the year. This means that even if precipitation was higher overall during the year, the actual precipitation events might have been clustered to only the season where they could limit lightning-caused fires (summer), but not anthropogenic fires (spring).

We found periodicities that suggest a match with different teleconnections' oscillatory patterns. Large-scale climatic patterns (teleconnections) and their interactions influence weather and local climate, and consequently, fire activity in Canada [77,78]. Wildfire cycles under higher human influence (LAP), as well as most fire-climate periodicities, were predominantly characterized by shorter periods (<8 years) compared to longer periods under low human influence (WBNP; >9 years).

The short periodicities of fire activity under high human influence suggest a higher susceptibility to Quasi-biennial and el Niño Southern Oscillations (QBO and ENSO, 2–2.5 and 2.5–8 years), whereas under lower human influence (WBNP), they seem to respond to ENSO and a quasi-decadal oscillation (Pacific Decadal Oscillation + sunspot cycle, 9–13 years) [79]. We also found traces of larger oscillations in WBNP (<15 years, not shown) that are usually associated with the PDO and IPO (Pacific Decadal and Pacific Interdecadal Oscillations), but due to the short time series we used, the signals were weak and possibly spurious. The absence of long fire-climate periodicities explaining area burned by humans in LAP at 10–12 years, might be a result of coinciding peaks of short oscillations of different periodicity that may create longer, stronger oscillations. These kinds of interactions have been documented for longer-term climatic patterns, when negative phases of ENSO and PDO concur with positive AMO phase, increasing the occurrence of fires in Colorado [80], or when positive ENSO and PDO phases coincide in the Rocky Mountains [81] in the U.S. In order to support these partial observations, future research with longer, seasonal time series are required.

The creation of Wood Buffalo National Park almost a century ago gave us the opportunity to compare this area of very low human impact with the adjacent area under a strong anthropogenic transformation in the same ecological region, avoiding the conflation of human influence with other factors. Furthermore, we used a temporal and spatial extent that allowed us to distinguish more directly the effect of human influence on fire activity in the short term (years to decades) [57,62,70], which generates useful information for land managers. Understanding the temporal patterns of fire activity helps fire management agencies assign and efficiently distribute material and human resources to fight and prevent fires. For example, in the province of Alberta, increasing attention is being given to the earliest part of the fire season (i.e., spring), when numerous human ignitions often coincide with the early onset of warm weather due to a lengthening of the fire season that has resulted in large and destructive wildfires (e.g., the Fort McMurray fire of 2016) [82].

5. Conclusions

Over the 46-year period studied (1970–2015), we observed how wildfire activity patterns in the boreal forest have been shaped by the continuously increasing influence of humans, potentially creating a novel fire regime through the modification of the seasonality, size, and frequency of fires. In our area of study, under high human influence, fire activity (area burned and number of fires) peaks in the spring instead of the summer, burning rates are lower, on average, and fewer fires over 10 ha occur than in the more natural area. Analyses used mostly non-parametric statistical techniques that are suitable for the highly variable and stochastic nature of the data. We showed how human influence affects fire activity by changing its trends and cyclical patterns, and how anthropogenic wildfire activity generates temporal patterns and associations with climate distinctive (albeit similar) from those associated with lightning wildfire activity. In general, although northern Alberta is subjected to drier and warmer climatic conditions, in areas of a high anthropogenic footprint, human influence appears to dampen and reverse the expected fire activity trends and affect the cyclical nature of fire occurrence. These rapid changes pose a new set of challenges for managers and researchers who try to understand and predict the impact of altered fire regimes on the diversity, structure, and future fire activity of a boreal forest. Our results further emphasize the importance of explicitly incorporating the multi-faceted human impact to improve our understanding of fire activity, how it is affecting the fire regime at different spatial and temporal scales, and to produce more accurate predictive models of fire activity.

Acknowledgments: Support for this research came from the Canadian Partnership for Wildland Fire, Mike Flannigan NSERC Discovery Grant and CONACYT Mexico (Scholarship 312318) and the Department of Renewable Resources of the University of Alberta. We would also like to thank Richard Carr for providing the fire-climate data, Piyush Jain for introducing us to the "randomized-phase-surrogate" techniques for trend detections and correlations and sharing his R package for the Mann-Kendall trend test, John Little for his advice on using and accessing the National Fire Database, Ellen Whitman for her assistance in building the land cover map, Karen Blouin for proofreading, and Filmer Chu for our long discussions over the potential applications of time series analysis.

Author Contributions: M.D.F. conceived the original idea and supervised this project. R.C.-R. designed and performed the data analysis; R.C.-R. and M.-A.P. interpreted the results and wrote the paper.

Conflicts of Interest: The authors declare no conflict of interest. The founding sponsors had no role in the design of the study; in the collection, analyses, or interpretation of data; in the writing of the manuscript, and in the decision to publish the results.

Appendix

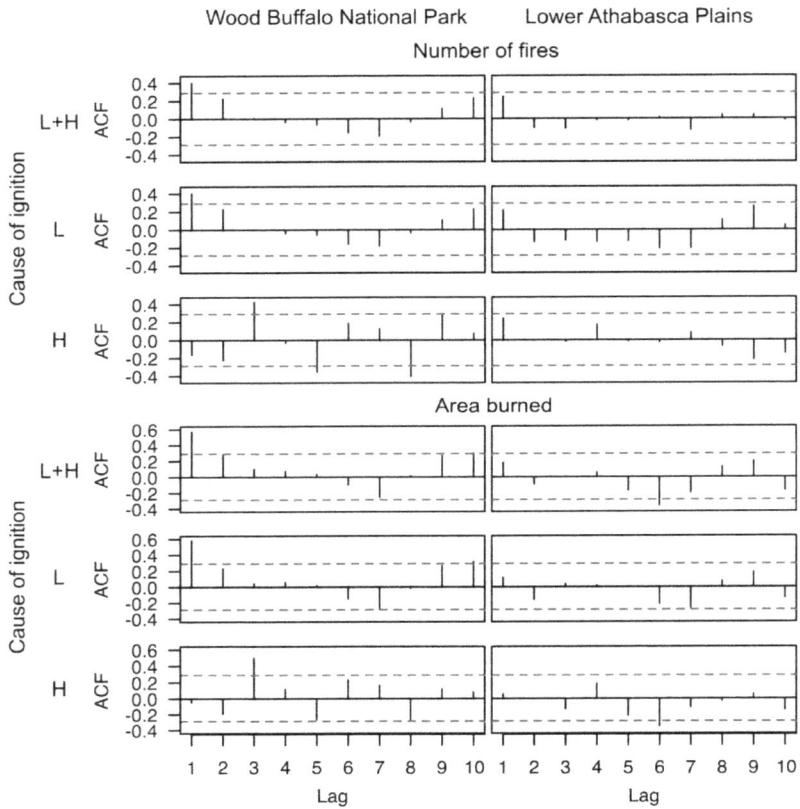

Figure A1. Autocorrelation plots for the number of fires and area burned (log transformed) time series in Wood Buffalo National Park and Lower Athabasca Plains.

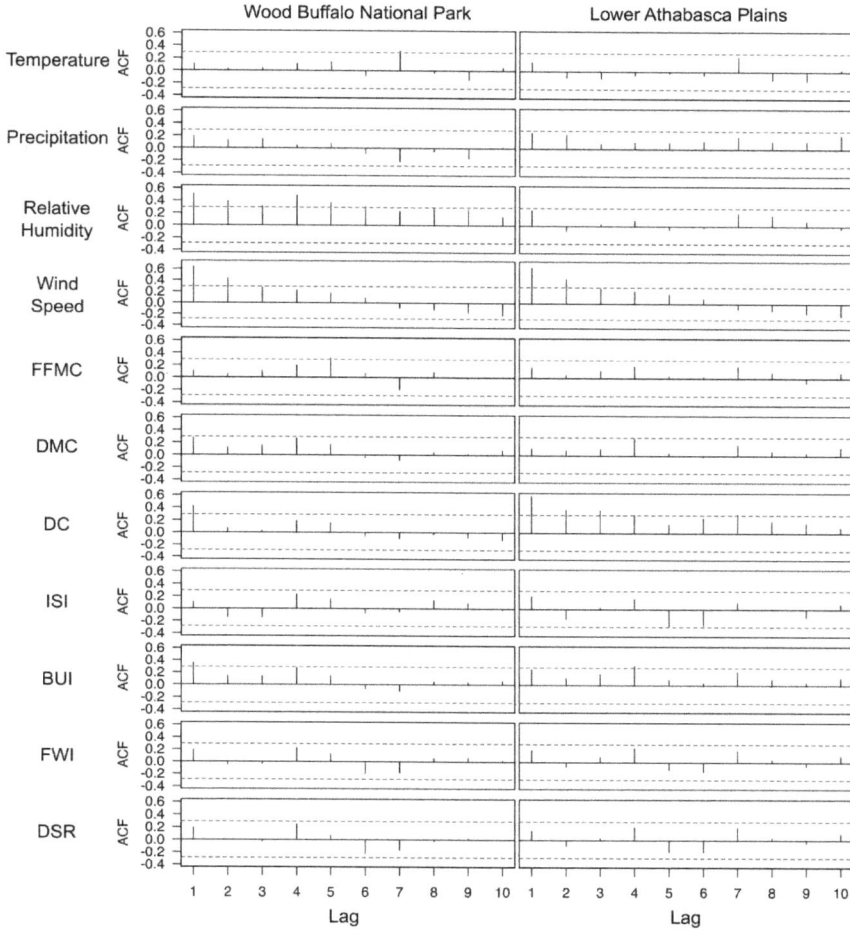

Figure A2. Autocorrelation plots for the fire-climate (climate and Fire Weather Index System) time series in Wood Buffalo National Park and Lower Athabasca Plains.

Table A1. Fire statistics in Wood Buffalo National Park (WBNP) and Lower Athabasca Plains (LAP) from 1970 to 2015. Letters L and H indicate lightning and human-caused fires, respectively.

Region	WBNP			LAP		
Cause	L + H	L	H	L + H	L	H
Number of fires	541 (100%)	518 (95.75%)	23 (4.25%)	1146 (100%)	614 (53.60%)	532 (46.40%)
Proportional number of fires (fires per 100 kha)	12.70	12.14	0.54	14.44	7.74	6.70
Area burned (ha × 10^4)	355.08 (100%)	339.87 (95.70%)	15.21 (4.30%)	422.39 * (100%)	199.56 (47.25%)	222.83 (52.75%)
Mean fire size (ha)	7332.51	6978.90	353.60	7384.40	4411.42	2972.98

Note: Information shown here is for fires ≥10 ha. Based on the National Fire Database [37]. * The Horse River fire (Fort McMurray) in 2016, increased this value by 13% in just one year.

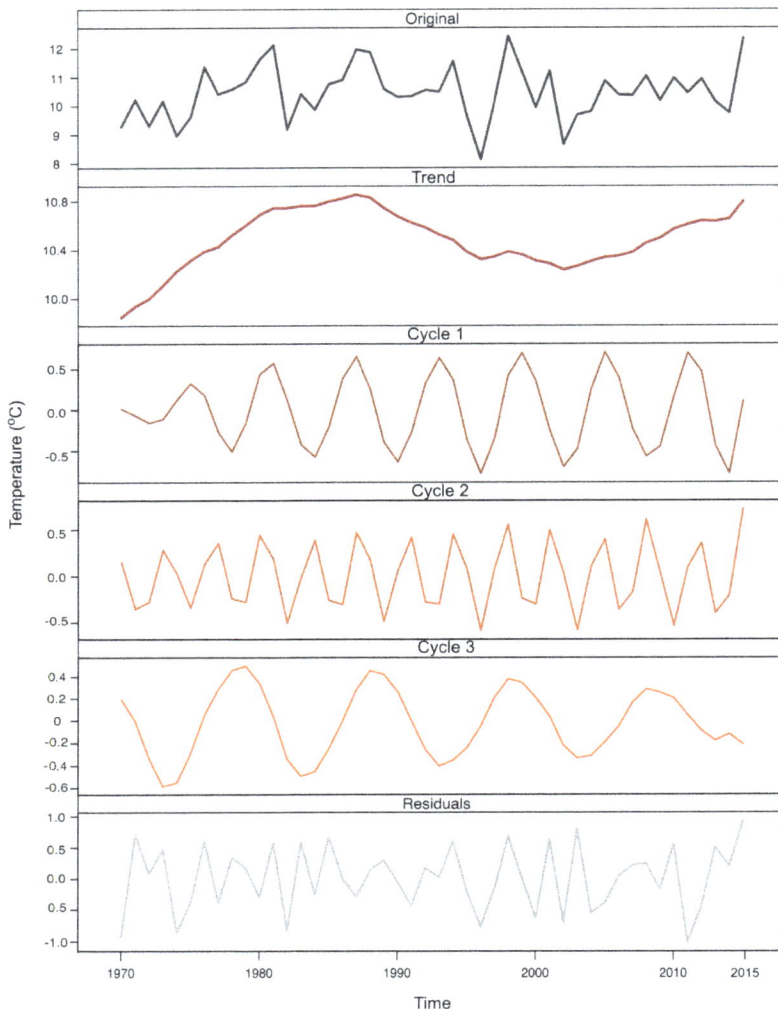

Figure A3. Example of decomposition by Singular Spectrum Analysis for temperature time series in Lower Athabasca Plains, showing its additive components: trend, cycles (oscillations), and residuals (white noise, random). In this case, the trend is upward non-monotonic and each cycle has a distinctive periodicity of 5.8, 3.7, and 9.5 years, respectively.

References

1. Weber, M.G.; Stocks, B.J. Forest Fires and Sustainability in the Boreal Forest of Canada. *Ambio* **1998**, *27*, 545–550.
2. Stocks, B.J.; Mason, J.A.; Todd, J.B.; Bosch, E.M.; Wotton, B.M.; Amiro, B.D.; Flannigan, M.D.; Hirsch, K.G.; Logan, K.A.; Martell, D.L.; et al. Large forest fires in Canada, 1959–1997. *J. Geophys. Res.* **2002**, *108*. [CrossRef]
3. Bergeron, Y.; Leduc, A.; Harvey, B.; Gauthier, S. Natural fire regime: A guide for sustainable management of the Canadian boreal forest. *Silva Fenn.* **2002**, *36*, 81–95. [CrossRef]
4. Johnstone, J.F. Response of boreal plant communities to variations in previous fire-free interval. *Int. J. Wildl. Fire* **2006**, *15*, 497–508. [CrossRef]

5. Kurz, W.; Apps, M. A 70-year retrospective analysis of carbon fluxes in the Canadian forest sector. *Ecol. Appl.* **1999**, *9*, 526–547. [CrossRef]
6. Amiro, B.D.; Stocks, B.J.; Alexander, M.E.; Flannigan, M.D.; Wotton, B.M. Fire, climate change, carbon and fuel management in the Canadian boreal forest. *J. Wildl. Fire* **2001**, *10*, 405–413. [CrossRef]
7. Martell, D.L. The impact of fire on timber supply in Ontario. *For. Chron.* **1994**, *70*, 164–173. [CrossRef]
8. Bowman, D.M.J.S.; Balch, J.; Artaxo, P.; Bond, W.J.; Cochrane, M.A.; D'Antonio, C.M.; Defries, R.; Johnston, F.H.; Keeley, J.E.; Krawchuk, M.A.; et al. The human dimension of fire regimes on Earth. *J. Biogeogr.* **2011**, *38*, 2223–2236. [CrossRef] [PubMed]
9. Hope, E.S.; McKenney, D.W.; Pedlar, J.H.; Stocks, B.J.; Gauthier, S. Wildfire Suppression Costs for Canada under a Changing Climate. *PLoS ONE* **2016**, *11*, 2223–2236. [CrossRef] [PubMed]
10. Weber, M.G.; Flannigan, M.D. Canadian boreal forest ecosystem structure and function in a changing climate: Impact on fire regimes. *Environ. Rev.* **1997**, *5*, 145–166. [CrossRef]
11. Flannigan, M.; Wotton, B. Climate, weather, and area burned. In *Forest Fires: Behavior and Ecological Effects*; Johnson, E.A., Miyanishi, K., Eds.; Academic Press: New York, NY, USA, 2001; pp. 351–373.
12. Cumming, S.G. A parametric model of the fire-size distribution. *Can. J. For. Res.* **2001**, *31*, 1297–1303. [CrossRef]
13. Krawchuk, M.A.; Cumming, S.G. Effects of biotic feedback and harvest management on boreal forest fire activity under climate change. *Ecol. Appl.* **2011**, *21*, 122–136. [CrossRef] [PubMed]
14. Podur, J. Spatial and temporal patterns of forest fire activity in Canada. Ph.D. Thesis, University of Toronto, Toronto, ON, Canada, 2001.
15. Gralewicz, N.J.; Nelson, T.A.; Wulder, M.A. Factors influencing national scale wildfire susceptibility in Canada. *For. Ecol. Manag.* **2012**, *265*, 20–29. [CrossRef]
16. Marlon, J.R.; Bartlein, P.J.; Carcaillet, C.; Gavin, D.G.; Harrison, S.P.; Higuera, P.E.; Joos, F.; Power, M.J.; Prentice, I.C. Climate and human influences on global biomass burning over the past two millennia. *Nat. Geosci.* **2008**, *1*, 697–702. [CrossRef]
17. Gustafson, E.J.; Zollner, P.A.; Sturtevant, B.R.; He, H.S.; Mladenoff, D.J. Influence of forest management alternatives and land type on susceptibility to fire in northern Wisconsin, USA. *Landsc. Ecol.* **2004**, *19*, 327–341. [CrossRef]
18. Parisien, M.; Peters, V.; Wang, Y. Spatial patterns of forest fires in Canada, 1980–1999. *J. Wildl. Fire* **2006**, *15*, 361–374. [CrossRef]
19. Pechony, O.; Shindell, D.T. Driving forces of global wildfires over the past millennium and the forthcoming century. *Proc. Natl. Acad. Sci. USA* **2010**, *107*, 19167–19170. [CrossRef] [PubMed]
20. Flannigan, M.; Cantin, A.S.; de Groot, W.J.; Wotton, M.; Newbery, A.; Gowman, L.M. Global wildland fire season severity in the 21st century. *For. Ecol. Manag.* **2013**, *294*, 54–61. [CrossRef]
21. Ruffault, J.; Mouillot, F. How a new fire-suppression policy can abruptly reshape the fire-weather relationship. *Ecosphere* **2015**, *6*, art199. [CrossRef]
22. Brotons, L.; Aquilué, N.; de Cáceres, M.; Fortin, M.-J.; Fall, A. How fire history, fire suppression practices and climate change affect wildfire regimes in Mediterranean landscapes. *PLoS ONE* **2013**, *8*, e62392. [CrossRef] [PubMed]
23. Higuera, P.E.; Abatzoglou, J.T.; Littell, J.S.; Morgan, P. The changing strength and nature of fire-climate relationships in the northern Rocky Mountains, USA, 1902–2008. *PLoS ONE* **2015**, *10*, e0127563. [CrossRef] [PubMed]
24. Cumming, S.G. Effective fire suppression in boreal forests. *Can. J. For. Res.* **2005**, *786*, 772–786. [CrossRef]
25. Martell, D.L.; Sun, H. The impact of fire suppression, vegetation, and weather on the area burned by lightning-caused forest fires in Ontario. *Can. J. For. Res.* **2008**, *38*, 1547–1563. [CrossRef]
26. Magnussen, S.; Taylor, S.W. Inter- and intra-annual profiles of fire regimes in the managed forests of Canada and implications for resource sharing. *Int. J. Wildl. Fire* **2012**, *21*, 328–341. [CrossRef]
27. Murphy, P. Methods for evaluating the effects of forest fire management in Alberta. Ph.D. Thesis, University of British Columbia, Vancouver, BC, Canada, 1985.
28. Alberta Agriculture and Forestry Forest Areas of Alberta. Available online: http://wildfire.alberta.ca/resources/maps-data/documents/ForestAreasAlberta-May03-2017.pdf (accessed on 4 November 2016).
29. Alberta Environment and Parks Alberta Wildfire 2015 Statistics. Available online: http://aep.alberta.ca/files/PREV_AlbertaWildfire_2015_Infographic_.pdf (accessed on 1 January 2016).

30. Alam, R.; Islam, S.; Mosely, E.; Thomas, S.; Dodwell, V.; Doel, D. Rapid Impact Assessment of Fort McMurray Wildfire. Available online: https://www.iclr.org/images/AlamIslam_QuickResponseSummary-ICLR.pdf (accessed on 7 January 2017).
31. KPMG May 2016 Wood Buffalo Wildfire, Post-Incident Assessment Report. Available online: https://www.alberta.ca/assets/documents/Wildfire-KPMG-Report.pdf (accessed 7 January 2017).
32. Parks Canada Agency Fire Management Zones. Available online: https://www.pc.gc.ca/en/nature/science/conservation/feu-fire/feuveg-fireveg/incendies-wildfire (accessed on 1 January 2017).
33. Natural Regions Committee. *Natural Regions and Subregions of Alberta*; Downing, D.J., Pettapiece, W.W., Eds.; Government of Alberta: Edmonton, AB, Canada, 2006; Pub. No. T/852.
34. Alberta Biodiversity Monitoring Institute ABMI Wall-to-wall Land Cover Map. 2012. Available online: http://abmi.ca/home/data-analytics/da-top/da-product-overview/GIS-Land-Surface/Land-Cover.html (accessed on 10 April 2017).
35. US Geological Survey USGS. Global Visualization Viewer. Available online: http://glovis.usgs.gov (accessed on 1 April 2017).
36. Wang, T.; Hamann, A.; Spittlehouse, D.L.; Murdock, T.Q. ClimateWNA—High-resolution spatial climate data for western North America. *J. Appl. Meteorol. Climatol.* **2012**, *51*, 16–29. [CrossRef]
37. Canadian Forest Service. Canadian National Fire Database—Agency Fire Data. 2016. Available online: http://cwfis.cfs.nrcan.gc.ca/datamart (accessed on 15 November 2016).
38. Government of Alberta. Forest and Prairie Protection Act. Available online: http://www.qp.alberta.ca/574.cfm?page=F19.cfm&leg_type=Acts&isbncln=9780779726554%5CnR:%5CAdmin%5COffice%5CRefMan%5CPrairieOffices%5CCalgaryOffice%5C1322-EnvironmentalAssessment%5C1332-Biophysical%5CReferences (accessed on 20 July 2016).
39. Armstrong, G.W. A stochastic characterisation of the natural disturbance regime of the boreal mixedwood forest with implications for sustainable forest management. *Can. J. For. Res.* **1999**, *29*, 424–433. [CrossRef]
40. Van Wagner, C.E. *Development and Structure of the Canadian Forest Fire Weather Index System*; Forestry Technical Report 35; Canadian Forestry Service: Otawa, ON, Canada, 1987.
41. Natural Resources Canada. *Canadian Forest Fire Weather Index System from the Canadian Wildland Fire Information System*; Version 3.0; Canadian Forest Service, Northern Forestry Center: Edmonton, AB, Canada, 2016.
42. R Development Core Team. *R: A Language and Environment for Statistical Computing*; R Foundation for Statistical Computing: Vienna, Austria. Available online: https://www.R-project.org/ (accessed on 4 June 2016).
43. Hyndman, R.J.; Khandakar, Y. Automatic time series forecasting: The forecast package for R. *J. Stat. Softw.* **2008**, *27*, C3. [CrossRef]
44. Metcalfe, A.V.; Cowpertwait, P.S.P. *Introductory Time Series with R*; Springer New York: New York, NY, USA, 2009; ISBN 978-0-387-88697-8.
45. Crawley, M.J. *The R Book*, 2nd ed.; John Wiley and Sons: Chichester, UK, 2013; ISBN 9780470973929.
46. Fox, J.; Weisberg, S. *An {R} Companion to Applied Regression*; Sage Publications: Thousand Oaks, CA, USA, 2002; ISBN 9781412975148.
47. Harrell, F.E. *Regression Modeling Strategies*, 2nd ed.; Springer International Publishing: Cham, Switzerland, 2015; ISBN 978-3-319-19424-0.
48. Jain, P. Package "MKcorrR": Mann-Kendall Test with Autocorrelated Data. Unpublished work. 2017.
49. Golyandina, N.; Korobeynikov, A. Basic singular spectrum analysis and forecasting with R. *Comput. Stat. Data Anal.* **2014**, *71*, 934–954. [CrossRef]
50. Golyandina, N.; Korobeynikov, A.; Shlemov, A.; Usevich, K. Multivariate and 2D extensions of singular spectrum analysis with the Rssa package. *J. Stat. Softw.* **2014**, *67*, 1–78. [CrossRef]
51. Yue, S.; Wang, C.Y. Assessment of the significance of sample serial correlation by the bootstrap test. *Water Resour. Manag.* **2002**, *16*, 23–35. [CrossRef]
52. Roy, R.; Kailath, T. ESPRIT-estimation of signal parameters via rotational invariance techniques. *IEEE Trans. Acoust.* **1989**, *37*, 984–995. [CrossRef]
53. Stoffer, D. Astsa: Applied Statistical Time Series Analysis. 2016. Available online: https://CRAN.R-project.org/package=astsa (accessed on 20 January 2017).

54. Ebisuzaki, W.; Ebisuzaki, W. A method to estimate the statistical significance of a correlation when the data are serially correlated. *J. Clim.* **1997**, *10*, 2147–2153. [CrossRef]
55. Baddouh, M.; Meyers, S.R.; Carroll, A.R.; Beard, B.L.; Johnson, C.M. Lacustrine ^{87}Sr/^{86}Sr as a tracer to reconstruct Milankovitch forcing of the Eocene hydrologic cycle. *Earth Planet. Sci. Lett.* **2016**, *448*, 62–68. [CrossRef]
56. Meyers, S.R. astrochron: An R Package for Astrochronology. 2014. Available online: https://cran.r-project.org/package=astrochron (accessed on 15 December 2016).
57. Tymstra, C.; Flannigan, M.D.; Armitage, O.B.; Logan, K. Impact of climate change on area burned in Alberta's boreal forest. *Int. J. Wildl. Fire* **2007**, *16*, 153–160. [CrossRef]
58. Girardin, M.P. Interannual to decadal changes in area burned in Canada from 1781 to 1982 and the relationship to Northern Hemisphere land temperatures. *Glob. Ecol. Biogeogr.* **2007**, *16*, 557–566. [CrossRef]
59. Jolly, W.M.; Cochrane, M.A.; Freeborn, P.H.; Holden, Z.A.; Brown, T.J.; Williamson, G.J.; Bowman, D.M.J.S. Climate-induced variations in global wildfire danger from 1979 to 2013. *Nat. Commun.* **2015**, *6*, 7537. [CrossRef] [PubMed]
60. Parisien, M.-A.; Miller, C.; Parks, S.A.; DeLancey, E.R.; Robinne, F.-N.; Flannigan, M.D. The spatially varying influence of humans on area burned in North America. *Environ. Res. Lett.* **2016**, *11*, 075005. [CrossRef]
61. Robinne, F.-N.; Parisien, M.-A.; Flannigan, M.D. Anthropogenic influence on wildfire activity in Alberta, Canada. *Int. J. Wildl. Fire* **2016**, *25*, 1131–1143. [CrossRef]
62. Gralewicz, N.J.; Nelson, T.A.; Wulder, M.A. Spatial and temporal patterns of wildfire ignitions in Canada from 1980 to 2006. *Int. J. Wildl. Fire* **2012**, *21*, 230. [CrossRef]
63. Podur, J.; Martell, D.; Knight, K. Statistical quality control analysis of forest fire activity in Canada. *Can. J. For.* **2002**, *205*, 195–205. [CrossRef]
64. Ryu, S.-R.; Chen, J.; Zheng, D.; Bresee, M.K.; Crow, T.R. Simulating the effects of prescribed burning on fuel loading and timber production (EcoFL) in managed northern Wisconsin forests. *Ecol. Modell.* **2006**, *196*, 395–406. [CrossRef]
65. Krawchuk, M.A.; Cumming, S.G.; Flannigan, M.D.; Wein, R.W. Biotic and abiotic regulation of lightning fire initiation in the mixedwood boreal forest. *Ecology* **2006**, *87*, 458–468. [CrossRef] [PubMed]
66. Cumming, S. Forest type and wildfire in the Alberta boreal mixedwood: What do fires burn? *Ecol. Appl.* **2001**, *11*, 97–110. [CrossRef]
67. Parisien, M.-A.; Parks, S.A.; Krawchuk, M.A.; Flannigan, M.D.; Bowman, L.M.; Moritz, M.A. Scale-dependent controls on the area burned in the boreal forest of Canada, 1980-2005. *Ecol. Appl.* **2011**, *21*, 789–805. [CrossRef] [PubMed]
68. Fréjaville, T.; Curt, T. Seasonal changes in the human alteration of fire regimes beyond the climate forcing. *Environ. Res. Lett.* **2017**, *12*, 35006. [CrossRef]
69. Archibald, S.; Roy, D.P.; van Wilgen, B.W.; Scholes, R.J. What limits fire? An examination of drivers of burnt area in Southern Africa. *Glob. Chang. Biol.* **2009**, *15*, 613–630. [CrossRef]
70. Wang, Y.; Anderson, K. An evaluation of spatial and temporal patterns of lightning-and human-caused forest fires in Alberta, Canada, 1980-2007. *Int. J. Wildl. Fire* **2011**, *19*, 1059–1072. [CrossRef]
71. Veraverbeke, S.; Rogers, B.M.; Goulden, M.L.; Jandt, R.R.; Miller, C.E.; Wiggins, E.B.; Randerson, J.T. Lightning as a major driver of recent large fire years in North American boreal forests. *Nat. Clim. Chang.* **2017**, *7*, 529–534. [CrossRef]
72. Erni, S.; Arseneault, D.; Parisien, M.A.; Bégin, Y. Spatial and temporal dimensions of fire activity in the fire-prone eastern Canadian taiga. *Glob. Chang. Biol.* **2017**, *23*, 1152–1166. [CrossRef] [PubMed]
73. Arienti, M.C.; Cumming, S.G.; Krawchuk, M.A.; Boutin, S. Road network density correlated with increased lightning fire incidence in the Canadian western boreal forest. *Int. J. Wildl. Fire* **2009**, *18*, 970–982. [CrossRef]
74. Balch, J.K.; Bradley, B.A.; Abatzoglou, J.T.; Nagy, R.C.; Fusco, E.J. Human-started wildfires expand the fire niche across the United States. *Proc. Natl. Acad. Sci. USA* **2017**, *114*. [CrossRef] [PubMed]
75. Albert-Green, A.; Dean, C.B.; Martell, D.L.; Woolford, D.G. A methodology for investigating trends in changes in the timing of the fire season with applications to lightning-caused forest fires in Alberta and Ontario, Canada. *Can. J. For. Res.* **2013**, *43*, 39–45. [CrossRef]
76. Wang, X.; Thompson, D.K.; Marshall, G.A.; Tymstra, C.; Carr, R.; Flannigan, M.D. Increasing frequency of extreme fire weather in Canada with climate change. *Clim. Chang.* **2015**, *130*, 573–586. [CrossRef]

77. Fauria, M.M.; Johnson, E.A. Large-scale climatic patterns control large lightning fire occurrence in Canada and Alaska forest regions. *J. Geophys. Res. Biogeosci.* **2006**, *111*, G04008. [CrossRef]

78. Mori, A.S. Climatic variability regulates the occurrence and extent of large fires in the subalpine forests of the Canadian Rockies. *Ecosphere* **2011**, *2*, 1–20. [CrossRef]

79. Bridgman, H.A.; Oliver, J.; Glantz, M. *The Global Climate System: Patterns, Processes, and Teleconnections*; Cambridge University Press: Cambridge, UK, 2006; ISBN 052182642X.

80. Schoennagel, T.; Veblen, T.T.; Kulakowski, D.; Holz, A. Multidecadal climate variability and climate interactions affect subalpine fire occurrence, Western Colorado (USA). *Ecology* **2007**, *88*, 2891–2902. [CrossRef] [PubMed]

81. Schoennagel, T.; Veblen, T.T.; Romme, W.H.; Sibold, J.S.; Cook, E.R. ENSO and PDO variability affect drought-induced fire occurrence in rocky mountain subalpine forests. *Ecol. Appl.* **2005**, *15*, 2000–2014. [CrossRef]

82. Pickell, P.D.; Coops, N.C.; Ferster, C.J.; Bater, C.W.; Blouin, K.D.; Flannigan, M.D.; Zhang, J. An early warning system to forecast the close of the spring burning window from satellite-observed greenness. *Sci. Rep.* **2017**, *1*, 14190. [CrossRef] [PubMed]

Article

What Drives Low-Severity Fire in the Southwestern USA?

Sean A. Parks [1,*], Solomon Z. Dobrowski [2] and Matthew H. Panunto [3]

[1] Aldo Leopold Wilderness Research Institute, Rocky Mountain US Forest Service, 790 E. Beckwith Ave., Missoula, MT 59801, USA
[2] W.A. Franke College of Forestry and Conservation, Department of Forest Management, University of Montana, 32 Campus Dr., Missoula, MT 59812, USA; solomon.dobrowski@umontana.edu
[3] Missoula Fire Sciences Laboratory, Rocky Mountain Research Station, US Forest Service, 5775 Hwy 10 W, Missoula, MT 59808, USA; mpanunto@fs.fed.us
*   Correspondence: sean_parks@fs.fed.us; Tel.: +1-406-542-4182

Received: 27 February 2018; Accepted: 21 March 2018; Published: 24 March 2018

Abstract: Many dry conifer forests in the southwestern USA and elsewhere historically (prior to the late 1800's) experienced fairly frequent surface fire at intervals ranging from roughly five to 30 years. Due to more than 100 years of successful fire exclusion, however, many of these forests are now denser and more homogenous, and therefore they have a greater probability of experiencing stand-replacing fire compared to prior centuries. Consequently, there is keen interest in restoring such forests to conditions that are conducive to low-severity fire. Yet, there have been no regional assessments in the southwestern USA that have specifically evaluated those factors that promote low-severity fire. Here, we defined low-severity fire using satellite imagery and evaluated the influence of several variables that potentially drive such fire; these variables characterize live fuel, topography, climate (30-year normals), and inter-annual climate variation. We found that live fuel and climate variation (i.e., year-of-fire climate) were the main factors driving low-severity fire; fuel was ~2.4 times more influential than climate variation. Low-severity fire was more likely in settings with lower levels of fuel and in years that were wetter and cooler than average. Surprisingly, the influence of topography and climatic normals was negligible. Our findings elucidate those conditions conducive to low-severity fire and provide valuable information to land managers tasked with restoring forest structures and processes in the southwestern USA and other regions dominated by dry forest types.

Keywords: fire severity; burn severity; wildland fire; forests; fire regime; fire refugia

1. Introduction

Wildland fire is an integral component of most dry conifer forest ecosystems in the southwestern USA and elsewhere [1]. Analyses of fire scarred trees indicate that most dry conifer forests in the southwest USA historically (i.e., prior to the late 19th century) experienced frequent surface fire and less frequent mixed-severity fire at intervals ranging from roughly five to thirty years [2–4]. However, as a result of fire exclusion policies that reduced fire frequency and area burned after the late 19th century [5,6], many dry conifer forests in the southwestern USA are denser and more homogenous compared to the pre-settlement era [7,8]. Consequently, there is growing concern that some dry forests are at risk of burning at higher severities (i.e., stand-replacing) than occurred in past centuries [9,10]. Recent research suggests this is indeed the case [11–13].

Stand-replacing fire in dry conifer forests has caused substantial concern about enduring conversions to non-forest. It is evident, for example, that the regeneration of dry conifer species (e.g., ponderosa pine) becomes more limited with increasing fire severity, increasing distance to seed source, and at sites with drier biophysical characteristics [14–16]. Short-interval high-severity fire (i.e., reburning at high-severity) in some dry forests also leads to post-fire successional trajectories that substantially differ from the pre-fire conditions, raising additional concern about altered successional trajectories and conversion to non-forest [17–19]. Although the drivers and consequences of high-severity fire are being increasingly studied, little to no research has been conducted that specifically focuses on the factors that promote low-severity fire, particularly in regions dominated by dry conifer forest that historically experienced frequent surface fire. A better understanding of those factors promoting low-severity fire could assist managers interested in reintroducing such fire to dry conifer forests in the southwest USA and elsewhere. Furthermore, identifying factors that promote low-severity fire could help identify biophysical settings in need of restoration treatments (e.g., prescribed fire and mechanical thinning) that will increase the likelihood of surface fire, thereby lowering the likelihood of stand-replacing fire and potential fire-facilitated conversions to non-forest.

Indeed, many dry conifer forests in the southwestern USA are in need of restoration in order to increase their resilience (i.e., reduce the probability of stand-replacing fire and associated transition to non-forest) [20,21]. Restoration treatments usually refer to mechanical thinning and prescribed fire [22], but it has been pointed out that the pace and scale of such treatments are inadequate in addressing the large area in need of restoration due to logistical, legal, and physical (i.e., topography) constraints [23]. However paradoxical it may seem, wildland fire itself has also been espoused as an effective method for increasing the resilience of dry conifer forests [24,25]. Reintroducing stand-replacing fire is obviously counterproductive for dry conifer forests, and consequently, Allen et al. [26] recommend, among other restoration treatments, the reintroduction of low-severity fire in such forests. This said, uncertainty about the biophysical settings in which low-severity fire is probable, and under what weather conditions, likely precludes the reintroduction of such fire in most cases (cf. [27]). This is a substantial knowledge gap given that low-severity fire was common in such forests prior to European settlement and the growing interest in restoring surface fire to dry conifer forests. Excluding studies involving fire refugia, which focus on unburned or low-severity patches within a matrix of moderate- to high-severity fire [28,29], little-to-no research has been conducted that specifically focuses on the drivers of low-severity fire in dry conifer forests such as those found in the southwestern USA.

The overarching goal of our study was to identify the most important factors driving low-severity fire in the southwestern USA. We measured fire severity using a satellite-inferred metric of fire-induced change, the relativized burn ratio [30]. We evaluated the relative influence of several factors driving low-severity fire including live fuel, topography, climate (30-year normals), and inter-annual climate variation (i.e., year-of-fire climate). We were also interested in functional relationships between important variables and low-severity fire, thereby providing managers with information pertaining to the biophysical and year-of-fire climatic conditions that promote low-severity fire. Consequently, our results will be highly relevant and timely to land managers interested in restoring fire regimes in the southwestern USA and other regions dominated by dry conifer forest.

2. Materials and Methods

2.1. Study Area

We conducted our study in the southwestern USA because of the high prevalence of dry conifer forest and the historical dominance of frequent, low-severity fire [31]. Specifically, we focused on the Arizona—New Mexico ecoregion (plus a 10-km buffer; 150,747 km^2) as defined by The Nature Conservancy [32] (Figure 1). Elevation ranges from 1053 to 3756 m (mean across ecoregion = 1986 m). The ecoregion is climatically diverse; mean annual temperature ranges from 0.5 to 17.2 °C (mean = 11.1 °C) and mean annual precipitation from 16.7 to 121.1 cm/year (mean = 40.6 cm/year) [33]. Almost half (48%) of the precipitation occurs in the summer (July–September) due to monsoonal storms [34]. The vegetation is also diverse; dominant forest types include pinyon-juniper woodland (22.4% of study area) and ponderosa pine woodland and savannah (12.7%) [31]. Other forest types such as mixed conifer, spruce-fir, and conifer-oak represent a fairly small proportion of the study area. Our study does not include non-forested vegetation (see below) and is therefore not described here. The proportional coverage of vegetation communities within the burned areas can be characterized as follows: ponderosa pine = 52%, pine-oak types = 20% (includes Arizona pine, alligator juniper, and Emory oak), mixed-conifer types = 15% (includes Douglas fir and white fir), subalpine types = 5% (includes Engelmann spruce and subalpine fir), riparian = 5% (includes black cottonwood), and pinyon-juniper = 4% [31]. The fire season runs from early May through late-August (USDA Forest Service 2013), although fires are less likely after early July due to rains associated with monsoonal storms from the Gulf of Mexico [35,36]. Fires in this region were generally characterized as occurring frequently and at a low-severity prior to European settlement, although it is recognized that fire severity varies with elevation and topography [5,37]. Extensive cattle and sheep grazing began in the 1880s, which substantially reduced fine fuel amount and continuity and caused a decrease in fire frequency [38]. Continued fire exclusion via direct fire suppression has contributed to increases in tree density and shade-tolerant species, thereby heightening concern about uncharacteristically severe fire and altered post-fire successional trajectories [20,39,40].

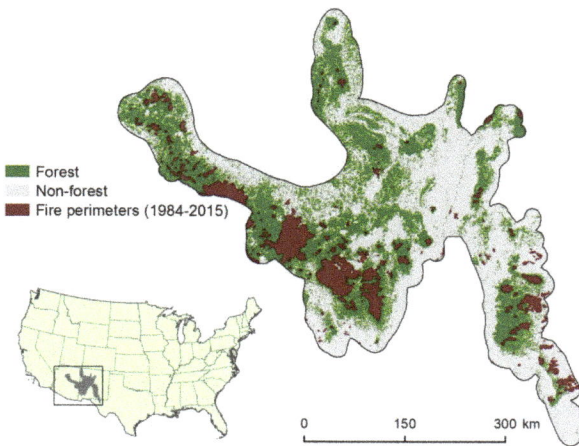

Figure 1. Study area map shows the distribution of forest, non-forest, and fire in our study area (the Arizona-New Mexico Mountains ecoregion). Inset shows this ecoregion's location in the context of the contiguous USA.

2.2. Data

Fire severity was measured using the relativized burn ratio (RBR), an index (resolution: 30-m) that quantifies the difference between pre- and post-fire Landsat thematic mapper (TM), enhanced thematic mapper plus (ETM+), and operational land imager (OLI) satellite data. The RBR has a high correspondence to field-based measures of severity such as the composite burn index (CBI; $r^2 = 0.71$) [30]. We classified the RBR data into binary categories representing low-severity (RBR $\leq$ 116) and other severity (RBR > 116) (Figure 2b). The RBR = 116 value corresponds to the average threshold between low and moderate severity for the nine fires analyzed in the southwestern USA by Parks et al. [30]; a similar thresholding approach was used by Dillon et al. [41] in their analysis involving high-severity fire. Satellite imagery used to generate RBR was obtained from the Monitoring Trends in Burn Severity program (MTBS) [42], which distributes fire and satellite data for fires $\geq$400 ha for the years 1984–2015. RBR was calculated using the 'dNBR offset', which accounts for differences due to phenology or precipitation between the pre- and post-fire imagery by subtracting the average delta normalized burn ratio (dNBR) of pixels outside the burn perimeter [43]; this can be important when comparing severity among fires [30].

We evaluated 13 explanatory variables in describing low-severity fire that can be categorized into four groups characterizing live fuel, topography, climate (30-year normals), and inter-annual climate variation (i.e., year-of-fire climate) (Table 1). The fuel group is comprised of three vegetation indices derived from satellite data: NDVI, NDMI, and EVI (Table 1) (resolution = 30-m). These indices were generated using pre-fire imagery distributed by MTBS. NDVI is an index of vegetation productivity and biomass [44]. NDMI is a measure of vegetation moisture and is frequently used in drought monitoring, and because of its sensitivity, it is also key in assessing wildfire potential and severity [45,46]. EVI is an alternative index of vegetation productivity, but, whereas NDVI is chlorophyll sensitive, EVI is more responsive to canopy structural variations (i.e., leaf area index, canopy type, plant physiognomy, and canopy architecture) [47] (Figure 2).

Table 1. Variables evaluated as predictors in modeling the probability of low-severity fire in forests of the southwestern USA.

Group	Variable Name	Description	Source
Live fuel	NDVI	Normalized differenced vegetation index. Calculated using pre-fire imagery distributed by the Monitoring Trends in Burn Severity (MTBS) program [41].	Pettorelli et al. [44]
	NDMI	Normalized differenced moisture index. Calculated using pre-fire imagery distributed by MTBS [41].	McDonald et al. [46]
	EVI	Enhanced vegetation index. Calculated using pre-fire imagery distributed by MTBS [41].	Huete [47]
Topography	DISS	Dissection index with a 450 m radius. DISS is a measure of topographic complexity.	Evans [48]
	TPI	Topographic position index. TPI is a measure of valley bottom vs. ridge top and measures the elevational difference (meters) between each pixel and an annulus with a 2000-m radius.	NA
	SRAD	Potential solar radiation, as calculated using the SOLPET6 model.	Flint et al. [49]
	Slope	Slope angle	NA
Climate	CMD	Climatic moisture deficit [49]. Mean over the 1981–2010 time period.	Wang et al. [50]; https://adaptwest.databasin.org/
	ET	Evapotranspiration (i.e., Eref-CMD). Mean over the 1981–2010 time period.	
	MAT	Mean annual temperature. Mean over the 1981–2010 time period.	
Inter-annual climate variation	Temp.z	Mean June temperature for the year in which the fire occurred. Converted to a z-score.	ClimateNA software package; Wang et al. [50]
	ET.z	Mean June evapotranspiration for the year in which the fire occurred. Converted to a z-score.	
	CMD.z	Mean June climatic moisture deficit for the year in which the fire occurred. Converted to a z-score.	

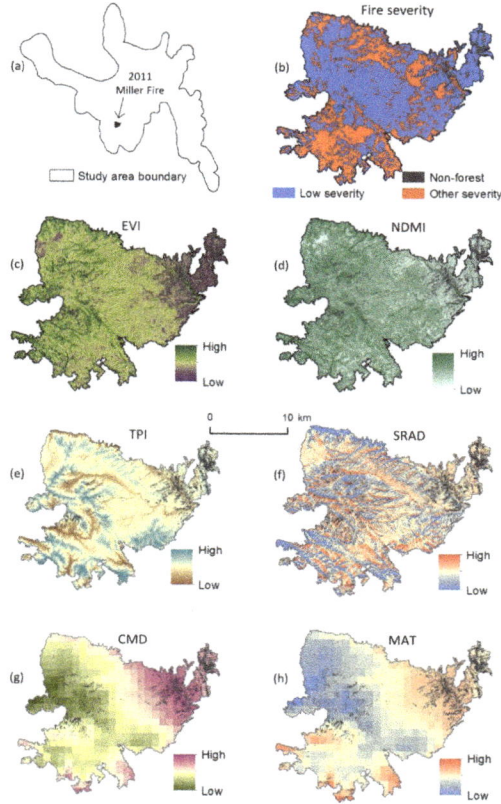

Figure 2. Example shows one of the >400 fires evaluated. Location of the 2011 Miller fire within the study area (**a**). Fire severity for the 2011 Miller Fire (**b**). Examples of the variables we used to represent pre-fire fuel (**c,d**), topography (**e,f**), and climate (**g,h**) for the 2011 Miller Fire. Inter-annual climate variation is not shown here because such variables are more indicative of temporal variability as opposed to spatial variability for individual fires. EVI: enhanced vegetation index; NDMI: normalized differenced moisture index; TPI: topographic position index; SRAD: solar radiation; CMD: climatic moisture deficit; MAT: mean annual temperature.

Climate is represented by three variables (resolution = 1-km): climatic moisture deficit (CMD), reference evapotranspiration minus CMD, hereafter referred to as evapotranspiration (ET), and mean annual temperature (MAT) [50] (Table 1; Figure 2). These variables characterize spatial variability and represent climate normals over the 1981–2010 time period (they do not vary annually) and have been identified as predictors of wildland fire in several studies [51–53].

Inter-annual climate variation is represented by three 'year-of-fire' variables: Temp.z, CMD.z, and ET.z (Table 1). These variables represent the z-scores for the month of June in the year in which each fire burned; June experiences the highest fire activity on average in the southwestern USA [54]. As such, Temp.z represents mean temperature for the month of June in the year in which the fire burned. CMD.z represents climatic moisture deficit and ET.z represents evapotranspiration for the month of June in the year in which the fire burned. These variables (resolution = 1-km) were generated using the ClimateNA software package (version 5.10) [50]. Recent studies have used similar variables representing climate variation in evaluations of fire severity [55,56]. All variables representing climate variation were converted to z-scores using the per-pixel mean and standard deviation for the month of June over a 30-year time period (1986–2015). Z-scores therefore represent the value in the month of June in terms of standard deviations away from the June mean.

2.3. Sampling Design and Statistical Model

We sampled fires that occurred from 1984–2015. We only sampled pixels identified as forest (i.e., forest, woodland, and savanna), as defined by a combination of landscape level vegetation products that include Landfire's [31] Existing Vegetation Cover (EVC), Environmental Site Potential (ESP), and the Landsat Time Series Stacks–Vegetation Change Tracker (LTSS-VCT) [57]. From the full set of burned forested pixels, we generated an initial 5% random sample, but then removed all pixels <100 m from the fire perimeter to reduce edge effects common at fire boundaries [58]. Although predictor variables ranged in resolution from 30-m to 1-km, all extractions were conducted using the native resolution of the response variable (30-m).

We produced a logistic regression model (family = binomial) describing low-severity fire (binary response) as a function of the 13 variables representing live fuel, topography, climate, and inter-annual climate variation (Table 1). We used a five-fold cross-validated procedure in which 80% of the fires (not the samples/pixels) were used to build a model and the remaining 20% of the fires were used to test the model; this ensures our cross-validation was spatially and temporally structured and that our model validation and inferences are not a result of autocorrelation common in satellite-inferred severity data [58–60]. For each of the five folds, we calculated the area under curve (AUC) statistic derived from the receiver operating characteristic curve of the full model (includes all 13 explanatory variables). We then compared this AUC to the AUC of additional models in which each variable was excluded. The AUC using the test data was averaged over the five folds. If the cross-validated AUC *increased* when a variable was removed, it was an indication that the variable did not provide unique information that improved model fit. As such, we removed the variable that resulted in the largest AUC increase when it was removed from the model. We then repeated this procedure until all variables resulted in a decrease in the cross-validated AUC when they were individually removed from the model. All statistical analyses were conducted using the *R* statistical program [61]. The cross-validation and stepwise variable selection procedures follow that of Parks et al. [62].

The cross-validated stepwise procedure we employed has some advantages compared to approaches that do not hold out independent data. For example, this procedure reduces the possibility of model overfitting and avoids falsely inflating our model skill (i.e., AUC statistic). Because our test data are independent—data from fires used to build the model (i.e., training data) were not used for model validation and variable selection (i.e., testing data)—our models are spatially and temporally transferable. Variables are retained based solely on whether or not they improve model fit; even if retained variables are correlated, they still possess unique information that improves the model.

Once the final set of variables was identified using the procedure described above, we calculated the relative influence of each variable group (fuel, topography, climate, and climate variation). This was achieved using a five-fold cross validation while excluding each group of variables. Specifically, we compared the five-fold cross validated AUC of the final model to models that excluded variables characterizing fuel, topography, climate, and inter-annual climate variation. Small decreases in AUC (compared to the final model) for any particular variable group are interpreted as having little influence, whereas sizeable decreases in AUC are interpreted as having large influence. The specific equation is as follows:

$$Relative\ influence_i = \frac{AUC.full - AUC.no.var_i}{\sum_{i=1}^{i=4}(AUC.full - AUC.no.var_i)} \times 100$$

where *AUC.full* is the AUC of the full model, *AUC.no.var_i* is the AUC of the model excluding any particular variable group, and *i* represents one of the variable groups.

We produced response curves describing the probability of low-severity fire as a function of all variables retained in the final model. To do so, we built individual logistic regression models (family = binomial) for each variable and plotted the response curves.

3. Results

We included data from over 400 fires that burned over 12,000 km^2 of forest to inform our model describing the probability of low-severity fire. The spatially and temporally cross-validated AUC was 0.701. Live fuel was the most influential factor driving low-severity fire (relative influence = 70.0%). This was followed by inter-annual climate variation (relative influence = 28.6%). The influence of topography and climate was negligible (0.9% and 0.5%, respectively). Our final model included eight variables that remained after the cross-validated stepwise procedure: EVI, NDMI, TPI, SRAD, ET, TEMP.z, ET.z, and CMD.z.

The response curves show a negative relationship between low-severity fire and both measures of fuel; that is, the probability of low-severity fire decreases with increasing fuel (Figure 3). Low-severity fire has a negative relationship with both Temp.z and CMD.z, so low-severity fire is more likely in years in which the June temperature and climatic moisture deficit are lower than average (i.e., z-scores < 0) compared to higher than average (z-score > 0). Finally, the relationship between low-severity fire and ET.z is positive, meaning the probability of low-severity fire increased with June evapotranspiration. We do not show the functional relationships with SRAD, TPI, and ET because the relative influence of these variables is less than 1% each.

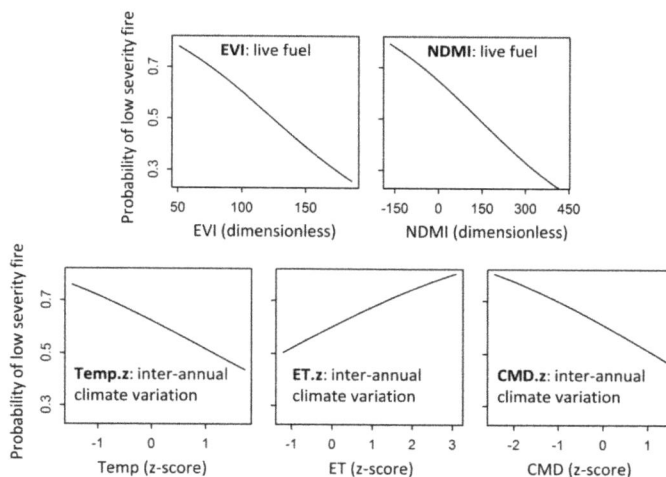

Figure 3. Functional relationships depict the probability of low-severity fire as a function of live fuels and inter-annual climate variation. Each of these was produced with a logistic regression with only the variable of interest. EVI: enhanced vegetation index; NDMI: normalized differenced moisture index; Temp.z: temperature z-score; ET.z: evapotranspiration z-score; CMD.z: climatic moisture deficit z-score. Functional relationships for TPI, SRAD, and ET are not shown since the relative influence of these variables is less than 1%.

4. Discussion

Our study pertains to those factors responsible for low-severity fire, thereby providing a different lens with which to view fire compared to the numerous studies that focus on the drivers and distribution of high-severity fire [37,41,56,62–64]. Specifically, because our study identifies the drivers of, and their relationship to, low-severity fire, we fill a critical information gap for dry forested regions in which prescribed fire and wildland fire managed for resource benefit are often espoused as forest restoration strategies [26,65,66]. This contrasts from those evaluations of high-severity fire, which often underscore the legitimate negative ecological and social impacts of such fire including the potential for altered successional trajectories and conversion to non-forest, particularly in dry forested ecosystems such as those found in the southwestern USA and elsewhere [17,39,67].

It is not entirely clear whether the factors that control low-severity fire can be inferred from studies of high-severity fire. Consequently, we suggest that our explicit attention to low-severity fire avoids ambiguity and potential misinterpretations that could arise from making inferences from high-severity fire studies. This is particularly important given that we focused on forests of the southwestern USA that historically experienced frequent surface fire prior to the late 19th century [4,5]. Moreover, our evaluation included four main drivers of low-severity fire (live fuel, topography, climate, and inter-annual climate variation), whereas most fire severity studies to date have included only one to three of these factors (e.g., [41,51,68]) (but see Parks et al. [62]). Lastly, many evaluations of high-severity fire included a limited number of fires (e.g., [51,69,70]), which potentially prevents generalizing their findings over broader regions; in contrast, our study included data from over 400 fires.

Live fuel was by far the most important variable group promoting low-severity fire (relative influence = 70.0%); Parks et al. [62] also found that fuel was most important in their evaluation of high-severity fire in the western USA. Other studies that used proxies for fuel (e.g., vegetation type or canopy cover) have also highlighted the influence of this factor in driving fire severity [71,72]. Moreover, we show that the probability of low-severity fire increased with decreasing levels of live fuel, as represented by EVI and NDMI (Figure 3). This result supports the findings of numerous studies based on field data [40,73], fire simulation modelling [74,75], and satellite-inferred severity metrics [59,76,77] that showed a reduction in fuel resulted in lower severity fire.

Year-of-fire climate (i.e., inter-annual climate variation) was the second most important variable group driving low-severity fire (relative influence = 28.6%). Keyser and Westerling [56], in their evaluation of high-severity fire, also highlighted the importance of climate variation. Importantly, our finding that the probability of low-severity fire increased with decreasing year-of-fire temperature and climatic moisture deficit is consistent with the findings of Abatzoglou et al. [55], who found a positive correlation between fire severity and year-of-fire fuel aridity. We find it notable that the climate metrics we used (departures from the mean value for the month of June, which are at a fairly coarse temporal resolution) exhibited a rather high relative influence. This suggests that near-term wildland fire forecasts, which currently address only area burned or number of large fires based on expected weather and other factors [78,79], could potentially forecast fire severity, thereby providing fire managers and others with a more complete prediction of the upcoming fire season.

Surprisingly, topography and climate (30-year normals representing spatial variability) had a negligible influence on the prevalence of low-severity fire (relative influence = 0.9% and 0.5%, respectively). This contrasts with a multitude of studies that showed topography is moderately to highly important in controlling fire severity (e.g., [41,51,68,72,80]). Likewise, recent studies conducted at scales ranging from individual fires to numerous fires across large regions have concluded that climate is related to fire severity [51,52,81]. We posit here, similar to Parks et al. [62], that topography and climate are indirect measures of fuel, and because we explicitly include fuel in our model, topography and climate are regarded as inconsequential. Indeed, Dillon et al. [41] acknowledged that topography was likely serving as a proxy for variation in fuel and other factors that were not accounted for in their study. Regardless, it is worth noting that Parks et al. [62], who evaluated high-severity fire, found substantial ecoregional variation in terms of the relative influence of topography and climate, suggesting that the findings presented here might not be generalizable to other regions.

The results of our study can be considered in relation to the growing body of literature pertaining to fire refugia [82–85]. Most fire refugia studies involve the study of unburned or low-severity remnants within a matrix of high-severity effects (e.g., [86]) or are focused on regions that are inherently characterized by mixed-severity and stand-replacing fire regimes (e.g., [87]). For the most part, these studies have not investigated those factors that create or promote the creation of fire refugia, but have instead focused on characterizing their prevalence and spatial patterns. This said, a limited number of studies have evaluated the factors promoting the creation of fire refugia; they found that topography and fire weather were important drivers [28,29]. Nevertheless, we suggest more research is needed to gain a better understanding of the factors that promote the creation of fire refugia and promote low-severity fire in general.

Producing statistical models of low-severity fire (or any severity fire) is challenging for several reasons. Remotely sensed metrics of fire severity are imperfect estimates of complex processes [88]. Nonetheless, such metrics are arguably the most consistent and appropriate for describing and analyzing fire severity over large landscapes and across multi-decadal timeframes. Furthermore, we used satellite indices to characterize fuel, but this approach generally describes live overstory vegetation and does not account for sub-canopy live and dead surface fuels that influence fire severity [89]. However, adequately characterizing live and dead sub-canopy fuel over large landscapes is difficult, if not impossible. Also, we used climate departures from the month with the highest average fire activity (June) to broadly characterize weather conditions conducive to fire. Fire severity, however, is known to vary with daily to hourly fluctuations in weather conditions [62,69]. Future investigations of low-severity fire could employ satellite fire detection data to infer the day that each pixel burned [90,91] and incorporate daily fire weather into their models (cf. [28,92]). Lastly, all else being equal, fire behavior and effects are different depending on the direction of fire spread (e.g., heading vs. flanking fire) [93], and at this time, we cannot capture this directional effect in our models.

5. Conclusions

Our study elucidates those conditions conducive to low-severity fire. Fuel and inter-annual climate variation (i.e., year-of-fire climate) were the dominant factors controlling the prevalence of low-severity fire, although the relative influence of fuel was ~2.4 times greater than that of climate variation. The probability of low-severity fire increased at lower levels of fuels and in years that were cooler and wetter than average. The influence of topography and climate (30-year normals representing a spatial gradient) was negligible. These findings support the notion that fuel treatments will likely increase the probability of low-severity fire [40,73,94]. Nevertheless, the influence inter-annual climate variation should not be discounted. Low-severity fire was more prevalent in cooler and wetter fire seasons (than average), which provides rationale for allowing more fires to burn (i.e., less aggressive fire suppression) in non-extreme years. These wildland fires are efficient means to reduce fuel loads, which has important consequences given that fuels are the prominent driver of high-severity fire [62]. Put another way, promoting low-severity fire in non-extreme years will reduce fuel loads and potentially decrease the probability of high-severity in fire extreme years.

It is recognized that low-severity fire consumes ladder and surface fuels [95,96] and reduces the prevalence of shade-tolerant trees in many cases [97]. These changes to fuels and the structure and composition of vegetation have important implications in terms of the behavior and effects of subsequent fires [17,19]. For example, a recent study concluded that sites with a restored fire regime were more likely to retain conifer trees and less likely to convert to non-forest during a subsequent extreme fire event [40]. Moreover, low-severity fire often reinforces a pattern of low-severity fire in subsequent fire events [18,59,98]. Other beneficial aspects of low-severity fire are also evident. For example, low-severity fire increases the ability of trees to defend against bark beetle attacks [99]. These examples illustrate that low-severity fire increases resilience to subsequent abiotic and biotic disturbance events and that managers could consider taking active measures to promote low-severity fire in regions dominated by dry conifer forest. Our findings provide land managers with general principles for promoting low-severity fire. As such, our study is both timely and relevant given the increasing desire to allow fire to burn to achieve restoration objectives [25,26,66] and the desire to avoid stand-replacing fire in dry forests in the southwestern USA [39,100].

Acknowledgments: We acknowledge funding from the Joint Fire Science Program under project 15-1-03-20 and from the National Fire Plan through the Rocky Mountain Research Station. We also thank two anonymous reviewers for providing comments that considerably improved our manuscript.

Author Contributions: S.A.P. and S.Z.D. conceived and designed the study; M.H.P. extracted and produced the dataset; S.A.P. analyzed the data; S.A.P. wrote the paper with input from the other authors.

Conflicts of Interest: The authors declare no conflict of interest.

References

1. Agee, J.K. *Fire Ecology of Pacific Northwest Forests*; Island Press: Washington, DC, USA, 1993.
2. Margolis, E.Q.; Malevich, S.B. Historical dominance of low-severity fire in dry and wet mixed-conifer forest habitats of the endangered terrestrial Jemez Mountains salamander (Plethodon neomexicanus). *For. Ecol. Manag.* **2016**, *375*, 12–26. [CrossRef]
3. Touchan, R.; Allen, C.D.; Swetnam, T.W. Fire history and climatic patterns in ponderosa pine and mixed-conifer forests of the Jemez Mountains, northern New Mexico. In *Fire Effects in Southwestern Forests: Proceedings of the Second La Mesa Fire Symposium*; Allen, C.D., Ed.; General Technical Report RM-GTR-286; USDA Forest Service: Fort Collins, CO, USA, 1996; pp. 33–46.
4. Baisan, C.H.; Swetnam, T.W. Fire history on a desert mountain range: Rincon Mountain Wilderness, Arizona, USA. *Can. J. For. Res.* **1990**, *20*, 1559–1569. [CrossRef]
5. Swetnam, T.W.; Baisan, C.H. Historical fire regime patterns in the southwestern United States since AD 1700. In *Fire Effects in Southwestern Forests: Proceedings of the Second La Mesa Fire Symposium*; Allen, C.D., Ed.; General Technical Report RM-GTR-286; USDA Forest Service: Fort Collins, CO, USA; 1996; pp. 11–32.
6. Fulé, P.Z.; Covington, W.W.; Moore, M.M. Determining reference conditions for ecosystem management of southwestern ponderosa pine forests. *Ecol. Appl.* **1997**, *7*, 895–908. [CrossRef]
7. Covington, W.W.; Moore, M.M. Southwestern ponderosa forest structure: Changes since Euro-American settlement. *J. For. Soc. Am. For.* **1994**, *92*, 39–47.
8. Swetnam, T.W.; Allen, C.D.; Betancourt, J.L. Applied historical ecology: Using the past to manage for the future. *Ecol. Appl. Ecol. Soc. Am.* **1999**, *9*, 1189–1206. [CrossRef]
9. Savage, M.; Mast, J.N. How resilient are southwestern ponderosa pine forests after crown fires? *Can. J. For. Res.* **2005**, *35*, 967–977. [CrossRef]
10. Williams, J. Exploring the onset of high-impact mega-fires through a forest land management prism. *For. Ecol. Manag.* **2013**, *294*, 4–10. [CrossRef]
11. Mallek, C.; Safford, H.; Viers, J.; Miller, J. Modern departures in fire severity and area vary by forest type, Sierra Nevada and southern Cascades, California, USA. *Ecosphere* **2013**, *4*, 1–28. [CrossRef]
12. O'Connor, C.D.; Falk, D.A.; Lynch, A.M.; Swetnam, T.W. Fire severity, size, and climate associations diverge from historical precedent along an ecological gradient in the Pinaleño Mountains, Arizona, USA. *For. Ecol. Manag.* **2014**, *329*, 264–278. [CrossRef]
13. Fornwalt, P.J.; Huckaby, L.S.; Alton, S.K.; Kaufmann, M.R.; Brown, P.M.; Cheng, A.S. Did the 2002 Hayman Fire, Colorado, USA, burn with uncharacteristic severity? *Fire Ecol.* **2016**, *12*, 117–132. [CrossRef]
14. Chambers, M.E.; Fornwalt, P.J.; Malone, S.L.; Battaglia, M.A. Patterns of conifer regeneration following high severity wildfire in ponderosa pine—Dominated forests of the Colorado Front Range. *For. Ecol. Manag.* **2016**, *378*, 57–67. [CrossRef]
15. Rother, M.T.; Veblen, T.T. Limited conifer regeneration following wildfires in dry ponderosa pine forests of the Colorado Front Range. *Ecosphere* **2016**, *7*. [CrossRef]
16. Stevens-Rumann, C.S.; Kemp, K.B.; Higuera, P.E.; Harvey, B.J.; Rother, M.T.; Donato, D.C.; Morgan, P.; Veblen, T.T. Evidence for declining forest resilience to wildfires under climate change. *Ecol. Lett.* **2018**, *21*, 243–252. [CrossRef] [PubMed]
17. Coop, J.D.; Parks, S.A.; Mcclernan, S.R.; Holsinger, L.M. Influences of prior wildfires on vegetation response to subsequent fire in a reburned southwestern landscape. *Ecol. Appl.* **2016**, *26*, 346–354. [CrossRef] [PubMed]

18. Coppoletta, M.; Merriam, K.E.; Collins, B.M. Post-fire vegetation and fuel development influences fire severity patterns in reburns. *Ecol. Appl.* **2016**, *26*, 686–699. [CrossRef] [PubMed]

19. Stevens-Rumann, C.; Morgan, P. Repeated wildfires alter forest recovery of mixed-conifer ecosystems. *Ecol. Appl.* **2016**, *26*, 1842–1853. [CrossRef] [PubMed]

20. Mast, J.N.; Fule, P.Z.; Moore, M.M.; Covington, W.W.; Waltz, A.E.M. Restoration of presettlement age structure of an Arizona ponderosa pine forest. *Ecol. Appl.* **1999**, *9*, 228–239. [CrossRef]

21. Brown, R.T.; Agee, J.K.; Franklin, J.F. Forest restoration and fire: Principles in the context of place. *Conserv. Biol.* **2004**, *18*, 903–912. [CrossRef]

22. Agee, J.K.; Skinner, C.N. Basic principles of forest fuel reduction treatments. *For. Ecol. Manag.* **2005**, *211*, 83–96. [CrossRef]

23. North, M.; Brough, A.; Long, J.; Collins, B.; Bowden, P.; Yasuda, D.; Miller, J.; Sugihara, N. Constraints on Mechanized Treatment Significantly Limit Mechanical Fuels Reduction Extent in the Sierra Nevada. *J. For.* **2015**, *113*, 40–48. [CrossRef]

24. Moritz, M.A.; Batllori, E.; Bradstock, R.A.; Gill, A.M.; Handmer, J.; Hessburg, P.F.; Leonard, J.; McCaffrey, S.; Odion, D.C.; Schoennagel, T.; et al. Learning to coexist with wildfire. *Nature* **2014**, *515*, 58. [CrossRef] [PubMed]

25. North, M.P.; Stephens, S.L.; Collins, B.M.; Agee, J.K.; Aplet, G.; Franklin, J.F.; Fulé, P.Z. Reform Forest Fire Management. *Science* **2015**, *349*, 1280–1281. [CrossRef] [PubMed]

26. Allen, C.D.; Savage, M.; Falk, D.A.; Suckling, K.F.; Swetnam, T.W.; Schulke, T.; Stacey, P.B.; Morgan, P.; Hoffman, M.; Klingel, J.T. Ecological restoration of southwestern ponderosa pine ecosystems: A broad perspective. *Ecol. Appl.* **2002**, *12*, 1418–1433. [CrossRef]

27. Zimmerman, T.; Frary, T.; Crook, S.; Fay, B.; Koppenol, P.; Lasko, R. Wildland fire use: Challenges associated with program management across multiple ownerships and land use situations. In *Fuels Manag How to Meas Success*; Andrews, P.L., Butl, B.W., Eds.; USDA For Serv Rocky Mt Res Station: Fort Collins, CO, USA, 2006; p. 809.

28. Krawchuk, M.A.; Haire, S.L.; Coop, J.; Parisien, M.-A.; Whitman, E.; Chong, G.; Miller, C. Topographic and fire weather controls of fire refugia in forested ecosystems of northwestern North America. *Ecosphere* **2016**, *7*. [CrossRef]

29. Camp, A.; Oliver, C.; Hessburg, P.; Everett, R. Predicting late-successional fire refugia pre-dating European settlement in the Wenatchee Mountains. *For. Ecol. Manag.* **1997**, *95*, 63–77. [CrossRef]

30. Parks, S.A.; Dillon, G.K.; Miller, C. A new metric for quantifying burn severity: The relativized burn ratio. *Remote Sens.* **2014**, *6*, 1827–1844. [CrossRef]

31. Rollins, M.G. LANDFIRE: A nationally consistent vegetation, wildland fire, and fuel assessment. *Int. J. Wildl. Fire* **2009**, *18*, 235–249. [CrossRef]

32. Olson, D.M.; Dinerstein, E. The Global 200: Priority ecoregions for global conservation. *Ann. Mo. Bot. Gard.* **2002**, *89*, 199–224. [CrossRef]

33. AdaptWest Project. Gridded Current and Projected Climate Data for North America at 1 km Resolution, Interpolated Using the ClimateNA v5.10 Software [Internet]. 2015. Available online: adaptwest.databasin.org (accessed on 2 April 2015).

34. Fick, S.E.; Hijmans, R.J. WorldClim 2: New 1-km spatial resolution climate surfaces for global land areas. *Int. J. Climatol.* **2017**, *37*, 4302–4315. [CrossRef]

35. Rollins, M.G.; Morgan, P.; Swetnam, T. Landscape-scale controls over 20th century fire occurrence in two large Rocky Mountain (USA) wilderness areas. *Landsc. Ecol.* **2002**, *17*, 539–557. [CrossRef]

36. Adams, D.K.; Comrie, A.C. The north American monsoon. *Bull. Am. Meteorol. Soc.* **1997**, *78*, 2197–2213. [CrossRef]

37. Holden, Z.A.; Morgan, P.; Evans, J.S. A predictive model of burn severity based on 20-year satellite-inferred burn severity data in a large southwestern US wilderness area. *For. Ecol. Manag.* **2009**, *258*, 2399–2406. [CrossRef]

38. Swetnam, T.W.; Dieterich, J.H. Fire history of ponderosa pine forests in the Gila Wilderness, New Mexico. In *Gen Tech Rep INT-GTR-182*; US Department of Agriculture, Forest Service, Intermountain Forest and Range Experiment Station: Ogden, UT, USA, 1985.

39. Savage, M.; Mast, J.N.; Feddema, J.J. Double whammy: High-severity fire and drought in ponderosa pine forests of the Southwest. *Can. J. For. Res.* **2013**, *43*, 570–583. [CrossRef]

40. Walker, R.B.; Coop, J.D.; Parks, S.A.; Trader, L. Fire regimes approaching historic norms reduce wildfire-facilitated conversion from forest to non-forest. *Ecosphere* **2018**, in press.

41. Dillon, G.K.; Holden, Z.A.; Morgan, P.; Crimmins, M.A.; Heyerdahl, E.K.; Luce, C.H. Both topography and climate affected forest and woodland burn severity in two regions of the western US, 1984 to 2006. *Ecosphere* **2011**, *2*. [CrossRef]

42. Eidenshink, J.C.; Schwind, B.; Brewer, K.; Zhu, Z.-L.; Quayle, B.; Howard, S.M. A project for monitoring trends in burn severity. *Fire Ecol.* **2007**, *3*, 3–21. [CrossRef]

43. Key, C.H. Ecological and sampling constraints on defining landscape fire severity. *Fire Ecol.* **2006**, *2*, 34–59. [CrossRef]

44. Pettorelli, N.; Vik, J.O.; Mysterud, A.; Gaillard, J.-M.; Tucker, C.J.; Stenseth, N.C. Using the satellite-derived NDVI to assess ecological responses to environmental change. *Trends Ecol. Evol.* **2005**, *20*, 503–510. [CrossRef] [PubMed]

45. Chu, T.; Guo, X.; Takeda, K. Temporal dependence of burn severity assessment in Siberian larch (Larix sibirica) forest of northern Mongolia using remotely sensed data. *Int. J. Wildl. Fire* **2016**, *25*, 685–698. [CrossRef]

46. McDonald, A.J.; Gemmell, F.M.; Lewis, P.E. Investigation of the utility of spectral vegetation indices for determining information on coniferous forests. *Remote Sens. Environ.* **1998**, *66*, 250–272. [CrossRef]

47. Huete, A.; Didan, K.; Miura, T.; Rodriguez, E.P.; Gao, X.; Ferreira, L.G. Overview of the radiometric and biophysical performance of the MODIS vegetation indices. *Remote Sens. Environ.* **2002**, *83*, 195–213. [CrossRef]

48. Evans, I.S. General geomorphometry, derivatives of altitude, and descriptive statistics. In *Spatial Analysis in Geomorphology*; 1972; pp. 17–90. Available online: https://books.google.com/books/about/Spatial_Analysis_in_Geomorphology.html?id=rvANAAAAQAAJ (accessed on 21 March 2018).

49. Flint, A.L.; Flint, L.E.; Hevesi, J.A.; Blainey, J.B. Fundamental concepts of recharge in the desert southwest: A regional modeling perspective. In *Groundwater Recharge in a Desert Environment: The Southwestern United States*; Hogan, J.F., Phillips, F.M., Scanlon, B.R., Eds.; 2004. Available online: https://agupubs.onlinelibrary.wiley.com/doi/book/10.1029/WS009 (accessed on 21 March 2018). [CrossRef]

50. Wang, T.; Hamann, A.; Spittlehouse, D.; Carroll, C. Locally downscaled and spatially customizable climate data for historical and future periods for North America. *PLoS ONE* **2016**, *11*, e0156720. [CrossRef] [PubMed]

51. Kane, V.R.; Cansler, C.A.; Povak, N.A.; Kane, J.T.; McGaughey, R.J.; Lutz, J.A.; Churchill, D.J.; North, M.P. Mixed severity fire effects within the Rim fire: Relative importance of local climate, fire weather, topography, and forest structure. *For. Ecol. Manag.* **2015**, *358*, 62–79. [CrossRef]

52. Parks, S.A.; Parisien, M.A.; Miller, C.; Dobrowski, S.Z. Fire activity and severity in the western US vary along proxy gradients representing fuel amount and fuel moisture. *PLoS ONE* **2014**, *9*, e99699. [CrossRef] [PubMed]

53. McKenzie, D.; Littell, J.S. Climate change and the eco-hydrology of fire: Will area burned increase in a warming western USA? *Ecol. Appl.* **2017**, *27*, 26–36. [CrossRef] [PubMed]

54. USDA Forest Service. MODIS Fire Detection GIS Data [Internet]. 2016. Available online: http://activefiremaps.fs.fed.us/gisdata.php (accessed on 3 January 2016).

55. Abatzoglou, J.T.; Kolden, C.A.; Williams, A.P.; Lutz, J.A.; Smith, A.M.S. Climatic influences on interannual variability in regional burn severity across western US forests. *Int. J. Wildl. Fire* **2017**, *26*, 269–275. [CrossRef]

56. Keyser, A.; Westerling, A. Climate drives inter-annual variability in probability of high severity fire occurrence in the western United States. *Environ. Res. Lett.* **2017**, *12*, 065003. [CrossRef]

57. Huang, C.; Goward, S.N.; Masek, J.G.; Thomas, N.; Zhu, Z.; Vogelmann, J.E. An automated approach for reconstructing recent forest disturbance history using dense Landsat time series stacks. *Remote Sens. Environ.* **2010**, *114*, 183–198. [CrossRef]

58. Stevens-Rumann, C.; Prichard, S.; Strand, E.; Morgan, P. Prior wildfires influence burn severity of subsequent large fires. *Can. J. For. Res.* **2016**, *46*, 1375–1385. [CrossRef]

59. Parks, S.A.; Miller, C.; Nelson, C.R.; Holden, Z.A. Previous Fires Moderate Burn Severity of Subsequent Wildland Fires in Two Large Western US Wilderness Areas. *Ecosystems* **2014**, *17*, 29–42. [CrossRef]

60. Kane, V.R.; Lutz, J.A.; Alina Cansler, C.; Povak, N.A.; Churchill, D.J.; Smith, D.F.; Kane, J.T.; North, M.P. Water balance and topography predict fire and forest structure patterns. *For. Ecol. Manag.* **2015**, *338*, 1–13. [CrossRef]

61. R Core Team. *R: A Language and Environment for Statistical Computing [Internet]*; R foundation for Statistical Computing: Vienna, Austria, 2016. Available online: https://www.r-project.org/ (accessed on 1 July 2017).

62. Parks, S.A.; Holsinger, L.M.; Panunto, M.H.; Jolly, W.M.; Dobrowski, S.Z.; Dillon, G.K. High-severity fire: Evaluating its key drivers and mapping its probability across western US forests. *Environ. Res. Lett.* **2018**, in press. [CrossRef]

63. Cansler, C.A.; McKenzie, D. Climate, fire size, and biophysical setting control fire severity and spatial pattern in the northern Cascade Range, USA. *Ecol. Appl. Ecol. Soc. Am.* **2014**, *24*, 1037–1056. [CrossRef]

64. Harvey, B.J.; Donato, D.C.; Turner, M.G. Drivers and trends in landscape patterns of stand-replacing fire in forests of the US Northern Rocky Mountains (1984–2010). *Landsc. Ecol.* **2016**, 1–17. [CrossRef]

65. North, M.; Collins, B.M.; Stephens, S. Using fire to increase the scale, benefits, and future maintenance of fuels treatments. *J. For. Soc. Am. For.* **2012**, *110*, 392–401. [CrossRef]

66. Stephens, S.L.; Collins, B.M.; Biber, E.; Fulé, P.Z. US federal fire and forest policy: Emphasizing resilience in dry forests. *Ecosphere* **2016**, *7*. [CrossRef]

67. Tepley, A.J.; Thompson, J.R.; Epstein, H.E.; Anderson-Teixeira, K.J. Vulnerability to forest loss through altered postfire recovery dynamics in a warming climate in the Klamath Mountains. *Glob. Chang. Biol.* **2017**, *23*, 4117–4132. [CrossRef] [PubMed]

68. Fang, L.; Yang, J.; Zu, J.; Li, G.; Zhang, J. Quantifying influences and relative importance of fire weather, topography, and vegetation on fire size and fire severity in a Chinese boreal forest landscape. *For. Ecol. Manag.* **2015**, *356*, 2–12. [CrossRef]

69. Lydersen, J.M.; Collins, B.M.; Brooks, M.L.; Matchett, J.R.; Shive, K.L.; Povak, N.A.; Kane, V.R.; Smith, D.F. Evidence of fuels management and fire weather influencing fire severity in an extreme fire event. *Ecol. Appl.* **2017**, *27*, 2013–2030. [CrossRef] [PubMed]

70. Harris, L.; Taylor, A.H. Topography, Fuels, and Fire Exclusion Drive Fire Severity of the Rim Fire in an Old-Growth Mixed-Conifer Forest, Yosemite National Park, USA. *Ecosystems* **2015**, *18*, 1192–1208. [CrossRef]

71. Fang, L.; Yang, J.; White, M.; Liu, Z. Predicting Potential Fire Severity Using Vegetation, Topography and Surface Moisture Availability in a Eurasian Boreal Forest Landscape. *Forests* **2018**, *9*, 130. [CrossRef]

72. Birch, D.S.; Morgan, P.; Kolden, C.A.; Abatzoglou, J.T.; Dillon, G.K.; Hudak, A.T.; Smith, A.M.S. Vegetation, topography and daily weather influenced burn severity in central Idaho and western Montana forests. *Ecosphere* **2015**, *6*. [CrossRef]

73. Kennedy, M.C.; Johnson, M.C. Fuel treatment prescriptions alter spatial patterns of fire severity around the wildland—Urban interface during the Wallow Fire, Arizona, USA. *For. Ecol. Manag.* **2014**, *318*, 122–132. [CrossRef]

74. Mitchell, S.R.; Harmon, M.E.; O'connell, K.E.B. Forest fuel reduction alters fire severity and long-term carbon storage in three Pacific Northwest ecosystems. *Ecol. Appl.* **2009**, *19*, 643–655. [CrossRef] [PubMed]

75. Ager, A.A.; Finney, M.A.; Kerns, B.K.; Maffei, H. Modeling wildfire risk to northern spotted owl (Strix occidentalis caurina) habitat in Central Oregon, USA. *For. Ecol. Manag.* **2007**, *246*, 45–56. [CrossRef]

76. Finney, M.A.; McHugh, C.W.; Grenfell, I.C. Stand- and landscape-level effects of prescribed burning on two Arizona wildfires. *Can. J. For. Res.* **2005**, *35*, 1714–1722. [CrossRef]

77. Wimberly, M.C.; Cochrane, M.A.; Baer, A.D.; Pabst, K. Assessing fuel treatment effectiveness using satellite imagery and spatial statistics. *Ecol. Appl.* **2009**, *19*, 1377–1384. [CrossRef] [PubMed]

78. Preisler, H.K.; Riley, K.L.; Stonesifer, C.S.; Calkin, D.E.; Jolly, W.M. Near-term probabilistic forecast of significant wildfire events for the Western United States. *Int. J. Wildl. Fire* **2016**, *25*, 1169–1180. [CrossRef]

79. Preisler, H.K.; Westerling, A.L.; Gebert, K.M.; Munoz-Arriola, F.; Holmes, T.P. Spatially explicit forecasts of large wildland fire probability and suppression costs for California. *Int. J. Wildl. Fire* **2011**, *20*, 508–517. [CrossRef]

80. Estes, B.L.; Knapp, E.E.; Skinner, C.N.; Miller, J.D.; Preisler, H.K. Factors influencing fire severity under moderate burning conditions in the Klamath Mountains, northern California, USA. *Ecosphere* **2017**, *8*. [CrossRef]

81. Parks, S.A.; Holsinger, L.M.; Miller, C.; Parisien, M.-A. Analog-based fire regime and vegetation shifts in mountainous regions of the western US. *Ecography* **2018**, in press. [CrossRef]

82. Kolden, C.A.; Lutz, J.A.; Key, C.H.; Kane, J.T.; van Wagtendonk, J.W. Mapped versus actual burned area within wildfire perimeters: Characterizing the unburned. *For. Ecol. Manag.* **2012**, *286*, 38–47. [CrossRef]

83. Kolden, C.A.; Abatzoglou, J.T.; Lutz, J.A.; Cansler, C.A.; Kane, J.T.; Van Wagtendonk, J.W.; Key, C.H. Climate contributors to forest mosaics: Ecological persistence following wildfire. *Northwest Sci.* **2015**, *89*, 219–238. [CrossRef]

84. Meddens, A.J.H.; Kolden, C.A.; Lutz, J.A. Detecting unburned areas within wildfire perimeters using Landsat and ancillary data across the northwestern United States. *Remote Sens. Environ.* **2016**, *186*, 275–285. [CrossRef]
85. Kolden, C.A.; Bleeker, T.M.; Smith, A.; Poulos, H.M.; Camp, A.E. Fire Effects on Historical Wildfire Refugia in Contemporary Wildfires. *Forests* **2017**, *8*, 400. [CrossRef]
86. Haire, S.L.; Coop, J.D.; Miller, C. Characterizing Spatial Neighborhoods of Refugia Following Large Fires in Northern New Mexico USA. *Land* **2017**, *6*, 19. [CrossRef]
87. Berry, L.E.; Driscoll, D.A.; Stein, J.A.; Blanchard, W.; Banks, S.C.; Bradstock, R.A.; Lindenmayer, D.B. Identifying the location of fire refuges in wet forest ecosystems. *Ecol. Appl.* **2015**, *25*, 2337–2348. [CrossRef] [PubMed]
88. Morgan, P.; Keane, R.E.; Dillon, G.K.; Jain, T.B.; Hudak, A.T.; Karau, E.C.; Sikkink, P.G.; Holden, Z.A.; Strand, E.K. Challenges of assessing fire and burn severity using field measures, remote sensing and modelling. *Int. J. Wildl. Fire* **2014**, *23*, 1045–1060. [CrossRef]
89. Reinhardt, E.D.; Keane, R.E.; Brown, J.K. Modeling fire effects. *Int. J. Wildl. Fire* **2001**, *10*, 373–380. [CrossRef]
90. Parks, S.A. Mapping day-of-burning with coarse-resolution satellite fire-detection data. *Int. J. Wildl. Fire* **2014**, *23*, 215–223. [CrossRef]
91. Veraverbeke, S.; Sedano, F.; Hook, S.J.; Randerson, J.T.; Jin, Y.; Rogers, B.M. Mapping the daily progression of large wildland fires using MODIS active fire data. *Int. J. Wildl. Fire* **2014**, *23*, 655–667. [CrossRef]
92. Holsinger, L.; Parks, S.A.; Miller, C. Weather, fuels, and topography impede wildland fire spread in western US landscapes. *For. Ecol. Manag.* **2016**, *380*. [CrossRef]
93. Finney, M.A. The challenge of quantitative risk analysis for wildland fire. *For. Ecol. Manag.* **2005**, *211*, 97–108. [CrossRef]
94. Safford, H.D.; Stevens, J.T.; Merriam, K.; Meyer, M.D.; Latimer, A.M. Fuel treatment effectiveness in California yellow pine and mixed conifer forests. *For. Ecol. Manag.* **2012**, *274*, 17–28. [CrossRef]
95. Ryan, K.C.; Knapp, E.E.; Varner, J.M. Prescribed fire in North American forests and woodlands: History, current practice, and challenges. *Front. Ecol. Environ.* **2013**, *11*, s1. [CrossRef]
96. Vaillant, N.M.; Fites-Kaufman, J.A.; Stephens, S.L. Effectiveness of prescribed fire as a fuel treatment in Californian coniferous forests. *Int. J. Wildl. Fire* **2009**, *18*, 165–175. [CrossRef]
97. Becker, K.M.L.; Lutz, J.A. Can low-severity fire reverse compositional change in montane forests of the Sierra Nevada, California, USA? *Ecosphere* **2016**, *7*. [CrossRef]
98. Collins, B.M.; Miller, J.D.; Thode, A.E.; Kelly, M.; van Wagtendonk, J.W.; Stephens, S.L. Interactions Among Wildland Fires in a Long-Established Sierra Nevada Natural Fire Area. *Ecosystems* **2009**, *12*, 114–128. [CrossRef]
99. Hood, S.; Sala, A.; Heyerdahl, E.K.; Boutin, M. Low-severity fire increases tree defense against bark beetle attacks. *Ecology* **2015**, *96*, 1846–1855. [CrossRef] [PubMed]
100. Fulé, P.Z.; Swetnam, T.W.; Brown, P.M.; Falk, D.A.; Peterson, D.L.; Allen, C.D.; Aplet, G.H.; Battaglia, M.A.; Binkley, D.; Farrsi, C.; et al. Unsupported inferences of high-severity fire in historical dry forests of the western United States: Response to Williams and Baker. *Glob. Ecol. Biogeogr.* **2014**, *23*, 825–830. [CrossRef]

Article

Can Land Management Buffer Impacts of Climate Changes and Altered Fire Regimes on Ecosystems of the Southwestern United States?

Rachel Loehman [1],*, Will Flatley [2], Lisa Holsinger [3] and Andrea Thode [4]

[1] US Geological Survey, Alaska Science Center, Anchorage, AK 99508, USA
[2] Department of Geography, University of Central Arkansas, Conway, AR 72035, USA; wflatley@uca.edu
[3] US Forest Service Rocky Mountain Research Station, Missoula, MT 59808, USA; lisamholsinger@fs.fed.us
[4] School of Forestry, Northern Arizona University, Flagstaff, AZ 86011, USA; andi.thode@nau.edu
*   Correspondence: rloehman@usgs.gov; Tel.: +1-907-786-7089

Received: 15 February 2018; Accepted: 4 April 2018; Published: 7 April 2018

Abstract: Climate changes and associated shifts in ecosystems and fire regimes present enormous challenges for the management of landscapes in the Southwestern US. A central question is whether management strategies can maintain or promote desired ecological conditions under projected future climates. We modeled wildfire and forest responses to climate changes and management activities using two ecosystem process models: FireBGCv2, simulated for the Jemez Mountains, New Mexico, and LANDIS-II, simulated for the Kaibab Plateau, Arizona. We modeled contemporary and two future climates—"Warm-Dry" (CCSM4 RCP 4.5) and "Hot-Arid" (HadGEM2ES RCP 8.5)—and four levels of management including fire suppression alone, a current treatment strategy, and two intensified treatment strategies. We found that Hot-Arid future climate resulted in a fundamental, persistent reorganization of ecosystems in both study areas, including biomass reduction, compositional shifts, and altered forest structure. Climate changes increased the potential for high-severity fire in the Jemez study area, but did not impact fire regime characteristics in the Kaibab. Intensified management treatments somewhat reduced wildfire frequency and severity; however, management strategies did not prevent the reorganization of forest ecosystems in either landscape. Our results suggest that novel approaches may be required to manage future forests for desired conditions.

Keywords: wildfire; climate change; management; resilience; modeling; southwest

1. Introduction

Gradual changes in landscape composition and structure are predicted with shifting climate patterns [1–5]. However, climate changes occur in the context of increased landscape disturbance that can catalyze abrupt changes in ecosystems [6–8]. In particular, global ecosystems are highly influenced by fire disturbance [9–12]. In fire-adapted, fire-prone systems, landscape patterns and vegetation distributions are determined primarily by reciprocal interactions with fire [13,14] and then by fine-scale interactions within and among species (e.g., competition and dispersal) and their surrounding environment (e.g., climate and edaphic conditions) [15].

In the southwestern US, fire regimes have been altered by land management and climate changes. A hundred-plus years of livestock grazing, logging, and fire exclusion have altered pre-European era fire frequencies, creating increased surface fuel loads, dense, fuel-rich forests, and reduced structural and spatial heterogeneity of vegetation, especially in dry conifer forests with frequent-fire regimes (typically, those with fire return intervals <35 years) [16–18]. Fires in these forests are likely to be more intense with larger patches of high-severity fire than occurred historically [19–23], reducing biodiversity, ecological function, and resilience [12,17]. Observed 20th and 21st century anthropogenic climate changes of warming temperatures and an earlier onset of snowmelt have increased the length of fire seasons and lowered fuel moistures, making large portions of the landscape flammable for longer periods of time [21,24], and widespread, regional fire years have been associated with prolonged droughts [13,25,26].

Understanding how to manage changing fire regimes and fuel conditions will be a central challenge for decades to come, as warmer and drier climates cause more frequent, more severe, and larger fires than occurred historically [9,10,27,28]. Wildfires that are large and severe, that overlap in space or time, or that are at or beyond the bounds of historical range of variability can abruptly reorganize ecosystems [11,29,30]. In the Southwest, this reorganization may represent a tipping point in which changing climate and disturbance processes create novel fuelscapes, thus setting the stage for future fire regimes that are significantly different from those that have existed in the past. These regime changes pose serious threats to ecosystem integrity and resilience [31].

Several recent papers have addressed the appropriateness and effectiveness of fire management and forest restoration activities under changing climates [32–35]. A common finding is the limited ability of current strategies to ameliorate undesired wildfire impacts in many ecological systems, particularly given the potentiating effects of warmer, drier climates on fire frequency and severity. Littell et al. 2009 [32], in an analysis of the relationship of climate and wildfire area burned in the western US, concluded that fuel treatments may mitigate wildfire vulnerability in fuel-limited systems, but that treatments may be less effective in systems where future fire patterns are influenced more by climate than by fuels. Stephens et al. 2013 [36] suggested that new strategies to mitigate and adapt to increased fire are needed to sustain forest landscapes (e.g., promote resilience), including the restoration of historical stand conditions in high frequency, low-to-moderate severity fire regimes, while allowing for shifts away from historical forest structure and composition in forests with low-frequency, high-severity fire regimes. Facilitating the adaptation of forests to changing climate and fire regimes may ultimately create more resilient systems as vegetation communities come into equilibrium with climate [35]. For example, Schoennagel et al. 2017 [33] indicated the importance of adaptive management approaches that include increased use of prescribed fire, much reduced fire suppression, and recognition of the limited ability of fuel treatments to alter regional fire patterns.

The need for a better understanding of the potential impacts of climate changes on ecosystems is reaching new levels of urgency. To guide management strategies, current, scientifically credible information on how landscapes will respond to the synergistic interactions of climate and disturbance processes is required [37]. Research is ongoing, but projections of future conditions are somewhat uncertain and rarely produce the level of accuracy and precision needed by resource managers [38]. A core question central to fire and ecosystem management in fire-prone ecosystems is whether fuels and fire management strategies can be designed to maintain or promote desired ecological conditions under projected future climate and fire regimes [33]. As part of a Joint Fire Science Program project funded to improve understanding of future ecosystem and fire regime dynamics and management impacts, we modeled a range of climate and land management scenarios using two spatially explicit, mechanistic ecosystem-fire models, FireBGCv2 and LANDIS-II. Our simulation landscapes are the Jemez Mountains in northern New Mexico, modeled with FireBGCv2 (hereafter FireBGCv2-Jemez), and the Kaibab Plateau in Arizona, modeled with LANDIS-II (hereafter LANDIS-II-Kaibab) (Figure 1). Models were previously parameterized for respective study areas, providing an opportunity to compare outcomes and evaluate the interactions of climate, fire, and management under a common set of climate and management scenarios. Modeled management strategies were collaboratively developed by southwestern US researchers and managers as plausible responses to changing forests and fire regimes. We used our models to address three related topics for ponderosa pine and dry mixed conifer forests in each simulation landscape: (1) Will future climate cause fundamental changes in forests and fire regimes? (2) Will current management approaches be effective in preventing fundamental changes in forests and fires under future climates? (3) Could shifting climate regimes require new management approaches, or can fundamental ecological characteristics of southwestern forests be preserved through an intensification of current strategies?

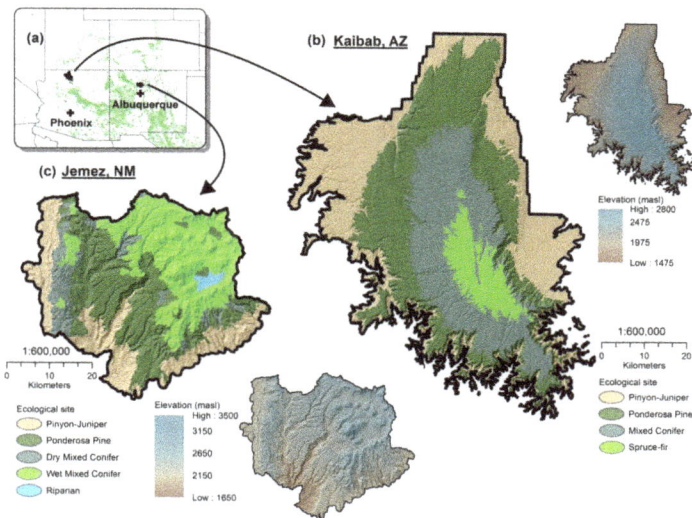

Figure 1. (a) Jemez, NM (FireBGCv2-Jemez) and Kaibab, AZ (LANDIS-II-Kaibab) study areas. Green shading denotes distribution of forests that historically experienced high frequency (≤35 Year Fire Return Interval) low- to mixed-severity fires (Fire Regime Group 1, LANDFIRE Program, Rollins 2009 [39]; (b) FireBGCv2-Jemez ecological setting and surface elevation; (c) LANDIS-II-Kaibab ecological setting and surface elevation.

2. Materials and Methods

2.1. Study Areas

The ~180,000 ha FireBGCv2-Jemez study area (Figure 1a,c) is a mainly forested, fire-adapted landscape of mesas and canyons (elevation range 1500–3500 meters above sea level (masl)). Landscape vegetation is strongly influenced by elevation and aspect, which are highly correlated with plant-available moisture [40]. Upper elevation (~2900 to 3500 masl), mesic forests contain Engelmann spruce (*Picea engelmanii*), corkbark fir (*Abies lasiocarpa var. arizonica*), and blue spruce (*P. pungens*); upper-middle elevation (~2300 to 2900 masl), dry mixed conifer forests consist of ponderosa pine (*Pinus ponderosa var. scopulorum*), Douglas-fir (*Pseudotsuga menziesii var. glauca*), white fir (*Abies concolor*), limber pine (*Pinus flexilis*), and southwestern white pine (*Pinus strobiformis*), with intermixed stands of aspen (*Populus tremuloides*); lower-middle elevations (~2100 to 2600 masl) are comprised of pure or nearly pure stands of ponderosa pine with Gambel oak (*Quercus gambelii*) understory; and dry woodlands of piñon pine (*Pinus edulis*) and juniper (primarily *Juniperus monosperma*) occur at lower elevations of 1500 to 2100 masl [40–42]. Regional climate is semi-arid with a bimodal precipitation pattern with peaks in winter (December–January) and summer (July–August; summer monsoon) [41]. Prior to persistent and substantial European settlement in the mid-1800s, much of the Jemez landscape was dominated by frequent, low-intensity surface fires with low overstory tree mortality, high understory tree mortality, and the regular consumption of surface fuels [41,43–45]. Large wildfires that have burned across the Jemez Mountains in the past five decades have included larger components of high-severity fire than occurred in the past, causing high levels of tree mortality, delayed forest regeneration, and erosion and other geomorphological changes [46,47].

The Kaibab Plateau (Figure 1a,b) is a broad, high elevation limestone plateau located in northern Arizona, encompassing 335,000 ha in Grand Canyon National Park (GCNP) and the North Kaibab Ranger District (NKRD) of the Kaibab National Forest. Elevation within the study landscape ranges from

1439 to 2830 masl, supporting a range of forest types that are distributed along a gradient of increasing moisture availability and decreasing temperature with increasing elevation. Forest types are similar to the Jemez landscape with low elevation piñon-juniper (primarily *J. osteosperma*), and Gambel oak, and mid elevation ponderosa pine. At mid to high elevations, dry mixed conifer and wet mixed conifer forests intermix and are composed of ponderosa pine, Douglas fir, white fir, blue spruce, Engelmann spruce, subalpine fir (*Abies lasiocarpa*), and aspen. Forests at the highest elevations are composed of Engelmann spruce, subalpine fir, and aspen [48,49]. Climate is similar to the Jemez landscape, with winter precipitation peaking slightly later (January–February) and lightning ignitions occurring most frequently in July [50,51]. Historically, frequent surface fires burned at intervals of about six to nine years in ponderosa pine and about seven to 31 years in dry mixed conifer forests [52,53]. Higher elevation forests experience less frequent, mixed- to high-severity fires [54,55]. The Kaibab Plateau has experienced large, high-severity fires in recent years [56].

2.2. Ecological Modeling

2.2.1. Firebgcv2-Jemez

FireBGCv2 (Fire BioGeoChemical model Version 2) is a spatially explicit, mechanistic ecosystem process model developed to evaluate interactions of climate, disturbance, and vegetation over long time scales [57–60]. Model details are described in Keane et al. 2011 [61] and in Supplement 1. Required model inputs are ecological site and stand maps, daily weather for ecological sites, fire regime and vegetation parameters, and initializing stand (plot) vegetation and fuels data. Ecological sites (ponderosa pine, 61,451 ha; wet mixed conifer, 60,795 ha; piñon-juniper, 41,009 ha; dry mixed conifer, 16,039 ha; and riparian, 1824 ha; Figure 1c) were mapped from the LANDFIRE environmental site potential (ESP) data layer [39], resampled to 90m, and generalized based on nearest neighbors to reduce fine-scale heterogeneity. Initial stand boundaries were defined using the LANDFIRE biophysical settings layer, and plot data from 84 plots collected across the Jemez landscape in 2012-2013 (for detailed field methods see [57,59]) or obtained from the Forest Inventory and Analysis Program (http://www.fia.fs.fed.us/) were assigned to stands based on similarities of dominant species, elevation, slope, and aspect. Weather data were obtained from the Jemez Springs National Climatic Data Center cooperative weather station (CO-OP ID 294369-2) (NCDC 2011) and were extrapolated across sites [62,63]. Historical fire frequency and size distributions were derived from fire history studies for southwestern ecosystems [18,64–72], including local studies [41,44,73]. Vegetation species parameters were gleaned from literature, previous FireBGCv2 projects [57–59,66,74], and field data. We adjusted fire size and frequency parameters until the model simulated landscape fire return intervals that were consistent with available fire history records [18,64–68,70–72,75]. We adjusted biological tree species parameters (e.g., shade tolerance, growing degree days, cone crop probability, bark thickness) until modeled spatial distributions and individual species basal area characteristics matched published estimates for southwest vegetation communities under non-managed conditions (e.g., without suppression, logging, or other activities) [69,76–78].

2.2.2. LANDIS-II-Kaibab

LANDIS-II is a spatially interactive, process-based landscape simulation model well suited for large spatial and temporal scales [79]. Specifics of the model structure, inputs, and validation for the Kaibab Plateau study landscape are described in Flatley and Fulé 2016 [80] and in Supplement 1. We used the Biomass Succession extension for LANDIS-II to model forest growth, competition, succession, and individual species response to climate change [81]. Inputs for the Biomass Succession extension, including species growth parameters across the varying environmental site conditions, climate scenarios, and time steps, were estimated using the Climate-Forest Vegetation Simulator [82]. The Dynamic Fire and Fuels (DFF) extension (v2.0) [83] simulated wildfire occurrence and spread according to site specific inputs of daily fire weather, ignition rates, and fire duration distributions. Daily fire weather data (ca. 1995–2013) was obtained from seven Remote Automated Weather Stations located within or adjacent to the study landscape (http://www.raws.dri.edu). The DFF extension simulated individual fires according to daily fire weather (for more detail see Supplement 1), then burned areas were aggregated within each 5-yr time step. Management actions were implemented with the Biomass Harvest extension (v2.1) [84], which reduces the biomass of cohorts according to species and age ranges specified in management prescriptions and alters fuel characteristics. We created initial forest conditions by preceding each model run with a 600-year spinup under contemporary climate conditions and historical fire frequencies, followed by 120 years of fire suppression and then biomass removal to simulate 20th century logging on forest service lands. We used a 1-ha cell resolution and each extension operated at a five-year time step.

2.3. Modeling Scenarios

We modeled 20 replicates of 100-year simulations for 12 factorial scenarios of climate (contemporary and two climate change factors, Figure 2) and management (wildfire suppression only, current management, and two intensified strategies). Contemporary climate in FireBGCv2 was a repeating loop of daily instrumental weather (1987–2006) from the Jemez NCDC station, extrapolated across sites [62,63]. For LANDIS-II-Kaibab, contemporary climate used to model forest growth and regeneration was based on downscaled climate normals from 1961–1990 [82]. Climate change factors spanned gradients of temperature and aridity projected for the southwestern US from Warm-Dry (based on the CCSM4 CMIP5 climate model [85] for the moderate emissions scenario RCP 4.5 [86,87]) to Hot-Arid (from the HadGEM2-ES CMIP5 climate model [88] for the high emissions scenario RCP 8.5). In FireBGCv2 data for both climate factors were acquired from the NASA Earth Exchange [89] for an 800-meter grid-cell coincident with the Jemez Springs weather station, then delta-downscaled for the period 2006-2009 by applying offsets derived from the slopes of linear regressions for individual seasons per weather year to the longest available Jemez station daily weather stream, 1954–2006. The delta method is a straightforward downscaling method that has a high level of climate realism desirable in future impact studies [90,91]. Future weather streams were then extrapolated to individual ecological sites as above. For LANDIS-II-Kaibab, climate change factors for vegetation dynamics were incorporated into C-FVS simulations. Site-specific growth and regeneration parameters for individual species were produced via downscaled climate surfaces available from the C-FVS webpage (http://charcoal.cnre.vt.edu/climate/customData/). Climate change factors for fire dynamics were incorporated into the daily fire weather data through the delta method. Adjusted fire weather data was then used to estimate fuel moisture inputs for fire modeling. Climate projections used for both the vegetation dynamics and fire dynamics were available at 30 year intervals, but we estimated intermediate values at 10 year intervals using linear interpolation (for more detail see Supplement 1).

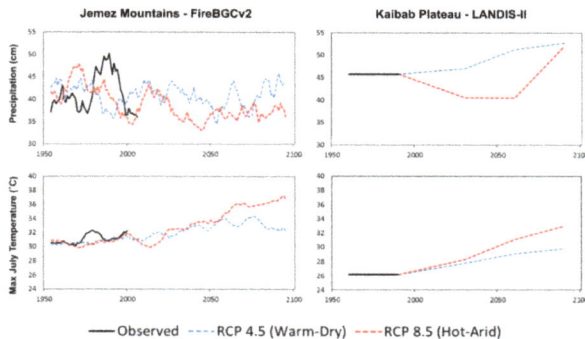

Figure 2. Annual precipitation (cm) and maximum July temperature (°C) for Contemporary (solid black lines), Warm-Dry (CMIP5/CCSM4, RCP 4.5, dashed blue lines), and Hot-Arid (CMIP5/ HadGEM2-ES, RCP 8.5, dashed red lines) climate factors for FireBGCv2-Jemez and LANDIS-II-Kaibab simulation models.

Management factors (Table S1) were derived from available prescriptions and burn plans for the study areas [16,92] refined with input from southwestern managers. All management factors included wildfire suppression at a level consistent with current policy and implementation, simulated as a randomly extinguished 90% of annual ignitions in FireBGCv2-Jemez or modeled as a function of the modern fire size distribution in LANDIS-II-Kaibab. We modeled treatments of Suppression Only, Business as Usual (BAU), representing current treatments in ponderosa pine and dry mixed conifer forests of the study regions, a three-fold annual increase over BAU treatment area (3xBAU), and a six-fold annual increase over BAU (6xBAU). The BAU treatment was applied to 1.5% of the areal extent of the ponderosa pine site and 1.5% of the dry mixed conifer site annually, with a combination of thinning followed by prescribed fire (0.75%) and prescribed fire only (0.75%) in each site, corresponding to a 67-year landscape treatment rotation. Treatment rotations for 3xBAU and 6x BAU were 22 years and 11 years, respectively, corresponding to larger annual treatment areas (Figures 3 and 4). Treatments were applied each year to stands that met specified criteria (Table S1), regardless of whether they were treated in previous years.

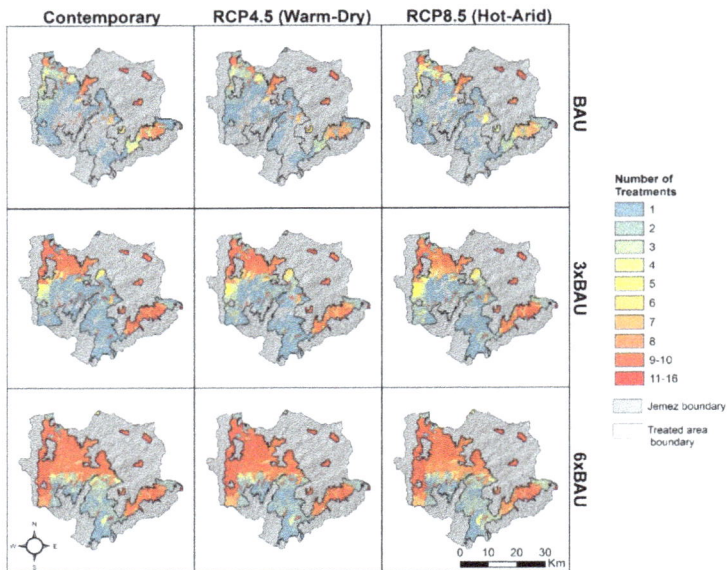

Figure 3. FireBGCv2-Jemez cumulative treated area and number of treatments per stand for the 100-year simulation period in managed forests of ponderosa pine and dry mixed conifer. All treatments are thinning followed by prescribed fire or prescribed fire only for business as usual (BAU) scenarios (1.5% treated annually, 76-year treatment rotation), three-fold increase in BAU (3xBAU, 22-year treatment rotation) scenarios, and six-fold increase in BAU (6xBAU, 11-year treatment rotation) scenarios for contemporary climate and two climate change factors.

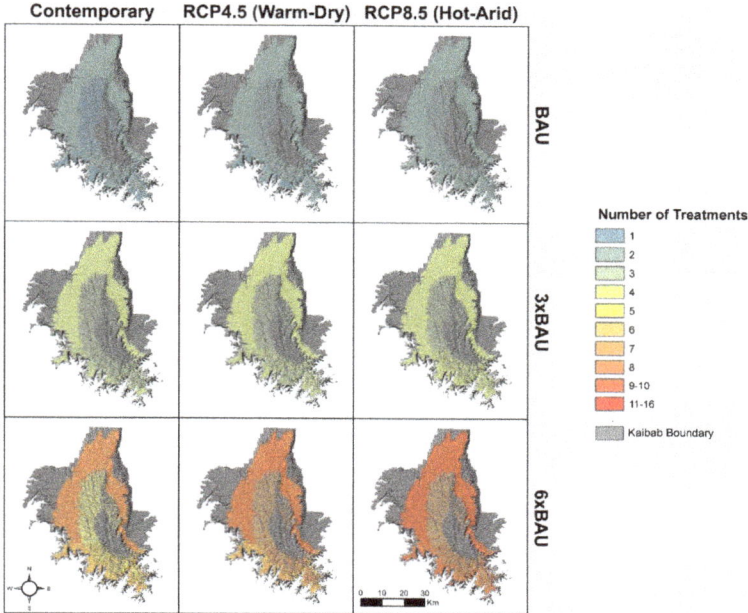

Figure 4. LANDIS-II-Kaibab cumulative treated area and number of treatments per stand for the 100-year simulation period for LANDIS-II-Kaibab in managed forests of ponderosa pine and mixed conifer. All treatments are thinning followed by prescribed fire or prescribed fire only for business as usual (BAU) scenarios (1.5% treated annually, 76-year treatment rotation), three-fold increase in BAU (3xBAU, 22-year treatment rotation) scenarios, and six-fold increase in BAU (6xBAU, 11-year treatment rotation) scenarios for contemporary climate and two climate change factors.

2.4. Model Responses and Analysis

We used a suite of fire regime and ecosystem metrics to evaluate the impacts of climate and management (Table 1), focusing our analysis on the combined 77,489 ha (FireBGCv2-Jemez) or 155,439 ha (LANDIS-II-Kaibab) area of ponderosa pine and mixed conifer sites in which management treatments were implemented. Variables were produced at annual time steps (FireBGCv2) or every five years (LANDIS-II), and were summarized across scenario replicates for each time step.

High-severity wildfires were identified as those for which tree mortality was greater than 70% of the pre-fire canopy (FireBGCv2-Jemez) [93] or as wildfires for which greater than 50% of the crown burned (LANDIS-II-Kaibab). For wildfire area burned (total wildfire area burned or high-severity wildfire area burned), we produced boxplots as measures of the central value (median) and variability (25th and 75th percentiles), calculated for the pool of all replicates and all years for each scenario (FireBGCv2) or all replicates (LANDIS-II). Maps of fire return intervals (FRI) show the mean point FRI across replicates, where FRI for each replicate is the number of simulation years/total number of simulated fires calculated for each pixel. Prescribed fires were not included in analyses. Ecosystem metrics, computed for individual stands within ecological sites, included biomass, vegetation composition calculated as dominant species by biomass, and species structural stage (FireBGCv2) or age class (LANDIS-II). We plotted proportional species composition, structural stage, and age class as mean occupancy per species, structural stage and age class across replicates, and basal area (BA, m^2/ha) or biomass (g/m^2) as time series plots of the replicate median and 25th and 75th percentiles of BA summed for all trees.

Table 1. Fire regime and ecosystem metrics tracked for FireBGCv2-Jemez and LANDIS-II-Kaibab models. Metrics are similar between models, but note differences in calculations of forest biomass, forest structure, and high-severity wildfire area burned that reflect differences in model mechanics.

	FireBGCv2-Jemez	LANDIS-II-Kaibab
	FIRE REGIME METRICS	
Point fire return interval	No. of simulation years/total number of wildfires per pixel	No. of simulation years/total number of wildfires per pixel
Area burned	Area of all wildfires (ha)	Area of all wildfires (ha)
High severity area burned	Area of all wildfires with tree mortality >70% (ha)	Area of all fires with >50% crown fraction burned (ha)
	ECOSYSTEM METRICS	
Vegetation composition	Proportional species biomass (%)	Proportional species biomass (%)
Forest structure	Proportional species structural stage [a] (%)	Proportional species age class [b] (%)
Forest production	Basal area by species (m^2/ha^{-1})	Biomass by species, g/m^{-2}

[a] FireBGCv2-Jemez structural stages correspond to the following diameter classes (cm): $2 \leq$ saplings ≤ 10, $10 <$ pole ≤ 23, $23 <$ mature ≤ 50, $50 <$ large ≤ 100, very large > 100; [b] LANDIS-II-Kaibab age classes correspond to the following: young 0 to 49 years, mid 50 to 99 years, and old > 100 years.

3. Results

3.1. Fire Regime Metrics

3.1.1. Fire Return Interval

FireBGCv2-Jemez fire return intervals decreased (wildfire frequency increased) for Warm-Dry and Hot-Arid climates relative to Contemporary climate for each management scenario (i.e., comparing climate impacts across all Suppression Only scenarios) (Figure 5). Fires were more frequent under progressively more severe drought conditions that increased the probability of ignition and fuel flammability, especially at lower elevations where fire return intervals were several decades shorter than for more moderate climates. As others have reported, fire suppression resulted in a fire deficit relative to historical reference conditions for dry southwestern forests [94,95], in which all simulated fire return intervals were substantially longer than reconstructed intervals of five to 12 years in ponderosa pine-dominated sites and 10 to 14 years in dry mixed conifer sites [41,43–45]. Management treatments had little effect on fire frequency for Contemporary climate (i.e., little difference between the Suppression Only scenario and BAU, 3xBAU, or 6xBAU) (Figure 5). Under Hot-Arid climate, fires were less frequent for 3xBAU and 6xBAU than for Suppression Only or BAU as the result of cumulative impacts of treatments and severe drought conditions on fuel availability, reducing fuel biomass to below the 0.05 kg/m^2 threshold required for successful fire ignition and spread in FireBGCv2.

LANDIS-II-Kaibab fire return intervals showed little to no change in response to climate, while management had a clear influence on fire frequency (Figure 6). Fires were most frequent for the Suppression Only management scenarios, with the shortest fire intervals of 30–40 years concentrated in large areas of contiguous low elevation ponderosa pine forest on the northern half of the plateau. Fuel treatments lengthened fire return intervals. The 3xBAU and 6xBAU treatments increased fire return intervals beyond 90 years for much of the study landscape. Due to the influence of fire suppression, fire return intervals under all scenarios (35–313 years) were longer than historical fire return intervals of six to nine years in ponderosa pine forests and seven to 31 years in mixed conifer forests [53,96].

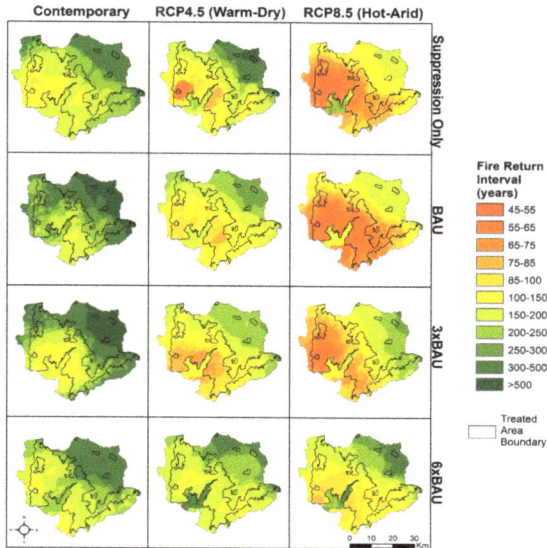

Figure 5. FireBGCv2-Jemez point fire return intervals (wildfires only; prescribed fires excluded). Scenarios are factorial combinations of management (Suppression Only; BAU, 76-year treatment rotation; 3xBAU, 22-year treatment rotation; 6xBAU, 11-year treatment rotation) and climate (Contemporary; Warm-Dry; Hot-Arid).

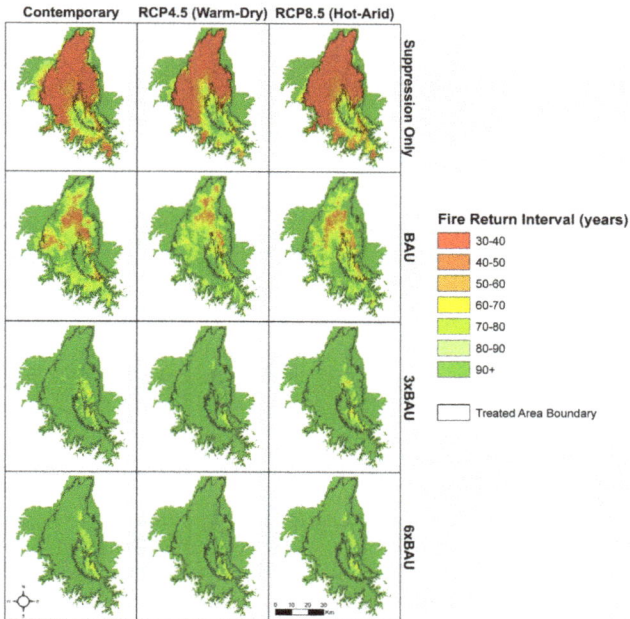

Figure 6. LANDIS-II-Kaibab fire return intervals (wildfires only; prescribed fires excluded). Scenarios are factorial combinations of management (Suppression Only; BAU, 76-year treatment rotation; 3xBAU, 22-year treatment rotation; 6xBAU, 11-year treatment rotation) and climate (Contemporary; Warm-Dry; Hot-Arid).

3.1.2. Wildfire Area Burned

FireBGCv2-Jemez annual area burned increased with Warm-Dry and Hot-Arid climates relative to Contemporary climate (Figure 7a, Table S2). Median annual area burned was a small portion of the total area of dry managed forest, ranging from less than 1000 ha (<1.5% of dry managed forest area) under Contemporary climate to over 2000 ha (>2.5%) for Hot-Arid climate, but among some replicate-years, about 3000 to 5000 ha (five to ten percent, Contemporary climate) or about 6000 to 15,000 ha (seven to 19 percent, Warm-Dry and Hot-Arid climates) of dry forested area burned annually, with outlier years in which much or all of the entire ~80,000 ha dry forested area burned (Table S2, Figure 7a). This outcome is consistent with projections of climate-driven increases in area burned in the western US, particularly those in which a small fraction of fires ("megafires") become very large despite fire suppression efforts [21,32,97,98]. Management impacts on annual area burned varied with climate (Figure 7a). For Contemporary and Warm-dry climates, treatments increased annual area burned proportionally to management intensity (Suppression Only, BAU, 3xBAU) because thinning and prescribed fire treatments produced fuels that easily carry fire; i.e., were a positive feedback to fire (but 6xBAU treatments were sufficiently frequent and extensive to limit burnable fuels and reduce area burned). For Hot-Arid climates, annual area burned decreased with management intensity (Suppression Only, BAU, 3xBAU, 6xBAU), because treatments in combination with altered climate suppressed fuel production.

LANDIS-II-Kaibab area burned did not differ between the Contemporary climate scenario and the Warm-Dry or Hot-Arid climates (Figure 8a, Table S3). The consistency in area burned for all climate factors within Suppression Only scenarios indicates that climate had minimal influence. As discussed above, wildfire was most prevalent in the Suppression Only scenarios, where median area burned ranged from 3600 to 4400 ha (2.3 to 2.9% of dry managed forests). The BAU treatments only slightly reduced area burned (2600 ha to 3200 ha) compared to Suppression Only. The enhanced treatment rates 3xBAU and 6xBAU notably reduced the average annual area burned (700 to 1200 ha and 500 to 700, respectively). The benefits of doubling the treatment rate from 3x to 6xBAU were limited, except for a reduction in the number of outlier years under the Hot-Arid scenario. However, exceptionally large fire years (outliers) may have outsized impacts on ecosystem function and recovery, justifying higher treatment rates.

3.1.3. High-Severity Wildfire Area Burned

FireBGCv2-Jemez median annual high-severity area burned (fires that resulted in >70% tree mortality in stands) was small (less than 300 ha, or less than one percent of the total area of dry managed forest), but in some replicate-years, as much as 1700 ha (about two percent, Contemporary climate) or 5000 ha (about seven percent, Hot-Arid climate) of dry forested area burned annually at high severity (Figure 7b, Table S2). Outlier years under Hot-Arid climates burned more than 75% of the dry forested area (60,000 ha) at high severity (Figure 7b), high-severity mega-fires that were much larger than those of the pre-European reference period [99]. There was no clear impact of management on median high-severity annual area burned; however, increasing treatment intensity reduced the upper range of high-severity burned area under Hot-Arid climate (Figure 7b). Management treatments had no effect on the extreme high-severity fire (outlier) years in the Hot-Arid climate scenarios.

Consistent with the previous results for the LANDIS-II-Kaibab simulations, management, not climate, was the primary influence on high-severity area burned. The area of high-severity fire declined considerably under the enhanced treatments (Figure 8b, Table S3). For the Hot-Arid climate and Suppression Only scenario, the high-severity fire rotation was 128 years for managed forest. The BAU, 3xBAU, and 6xBAU reduced this to 184, 526, and 3337 years, respectively. The rotation for the 3xBAU most closely aligns with a historical high-severity fire rotation of 528 years reconstructed in mixed conifer forests on the Kaibab Plateau [55]. Reductions in the area of high-severity fire could be particularly important in a changing climate, as forest turnover can be a catalyst for type changes [100,101].

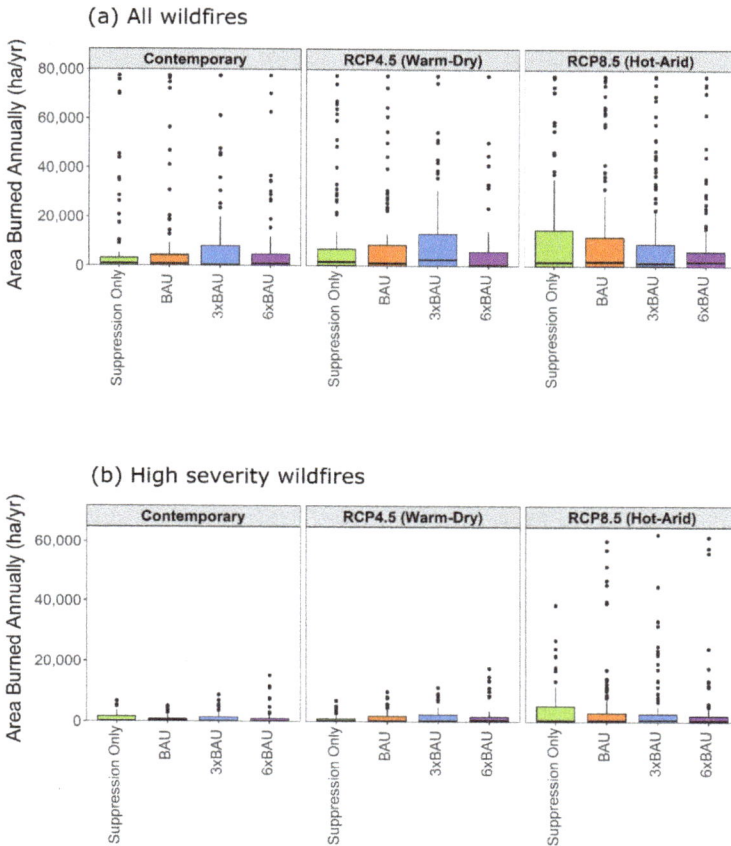

Figure 7. FireBGCv2-Jemez wildfire area burned annually (ha) in ponderosa pine and dry mixed conifer sites in (**a**) wildfires of all types and (**b**) high-severity wildfires (tree mortality >70%) for factorial combinations of management (Suppression Only; BAU, 76-year treatment rotation; 3xBAU, 22-year treatment rotation; 6xBAU, 11-year treatment rotation) and climate (contemporary; Warm-Dry; Hot-Arid). Boxplots show median, 25th, and 75th percentile wildfire area burned and outliers calculated for the pool of all replicates and all years for each scenario. The combined area of ponderosa pine and dry mixed conifer sites is 77,489 ha.

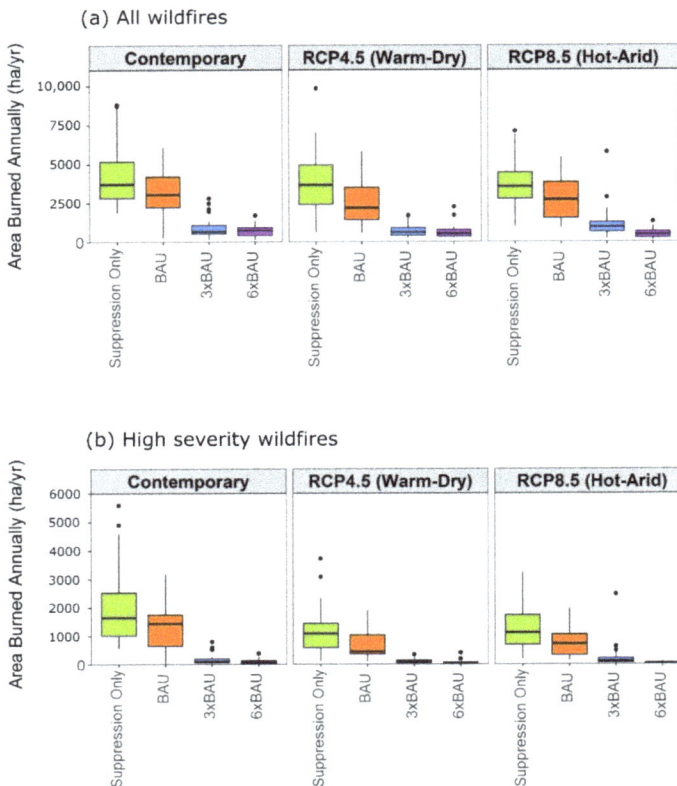

Figure 8. LANDIS-II-Kaibab wildfire area burned annually (ha) in ponderosa pine and dry mixed conifer sites for (**a**) wildfires of all types and (**b**) high-severity wildfires (>50% of crown burned) for factorial combinations of management (Suppression Only; BAU, 76-year treatment rotation; 3xBAU, 22-year treatment rotation; 6xBAU, 11-year treatment rotation) and climate (Contemporary; Warm-Dry; Hot-Arid). Boxplots show median area burned, 25th and 75th percentiles, and outliers among replicates and fire years for each scenario. The combined area of ponderosa pine and dry mixed conifer sites is 155,439 ha.

3.2. Ecosystem Responses

3.2.1. Forest Composition

For FireBGCv2-Jemez, Hot-Arid climate triggered a conversion of ponderosa pine forests to shrublands and woodlands dominated by Gambel oak, piñon pine, and juniper (Figure 9a). Conversion occurred ca AD 2075, corresponding to the hottest and driest period of future climate simulated in FireBGCv2-Jemez (Figure 2). Woodland expansion (but not shrub expansion) was somewhat mediated by management treatments: 3xBAU and 6xBAU reduced the proportion of piñon pine and juniper via thinning treatments that targeted smaller bole diameters characteristic of the species, and prescribed fire-caused tree mortality. In the dry mixed conifer site, where Gambel oak is not a significant component of the understory, aspen occupied an increasing proportion of the landscape as wildfires and area treated with prescribed fire increased (Figure 9b). This is consistent with aspen field studies documenting vigorous post-fire sprouting [41,102]; however, regeneration failed under Hot-Arid climate and co-occurring increased high severity burned area and more frequent fires.

Coniferous species composition was relatively stable in the dry mixed conifer site over time and among climate-management scenarios (Figure 9b) as compared with the ponderosa pine site (Figure 9a). More frequent prescribed fire rotations and larger treatment areas of 3xBAU and 6xBAU scenarios slightly decreased the amount of forest dominated by Douglas-fir in middle elevations in favor of ponderosa pine (Figure 9a), somewhat mitigating 20th century trends of infill by shade tolerant Douglas-fir facilitated by anthropogenic fire exclusion [19,77,103]. Treatments increased the amount of aspen in the dry mixed conifer site for contemporary and Warm-Dry climates (Figure 9b), especially by the mid-21st century, when intensified 22-year (3xBAU) or 11-year (6xBAU) rotations frequently retreated previously managed stands (Figure 3). Rapid (within five years) post-fire regeneration of aspen has been documented in southwestern conifer forests [104]; however, prescribed fires that stimulated aspen growth also reduced piñon pines, a fire-sensitive species that is usually killed by fire [105].

LANDIS-II-Kaibab compositional changes were minimal during the 21st century. Throughout the simulation period, ponderosa pine was the dominant species on sites initially classified as ponderosa pine forest, regardless of climate or management scenario (Figure 10a). Gambel oak was most common in ponderosa pine forests under the Suppression Only scenario, likely due to the prevalence of high-severity fire in the absence of treatments, while oak declined in response to the BAU and enhanced BAU treatments that reduced high-severity area burned. In contrast to the Jemez results, Gambel oak did not increase under climate change scenarios, actually decreasing in response to Hot-Arid climate conditions. Dry mixed conifer forests generally shifted towards dominance of ponderosa pine in the Suppression Only management scenario (more high-severity fire) or the climate change scenarios (Figure 10b). Ponderosa pine is the most fire tolerant and drought tolerant of the tree species in the dry mixed conifer forests on the Kaibab Plateau [54,106]. Ponderosa pine increased in dominance at the expense of white fir, aspen, and to a lesser extent, Douglas-fir, which is also relatively tolerant to fire and drought. Under the Hot-Arid climate, white fir declined initially but began to recover later in the century with the application of fuel treatments. This recovery is likely due to increased precipitation during the last decades of the century under the Hot-Arid climate scenario. Although, as modeled, LANDIS-II-Kaibab compositional changes were limited, they will likely be substantial over longer time periods, indicated by the declining regeneration probability of initial overstory species that occurred in all forest types. For example, piñon pine and juniper were much more viable than ponderosa pine in the ponderosa pine forest type by the end of the century under the Hot-Arid scenario. This suggests that, given time (more than a 100-year simulation period), the model would show the replacement of ponderosa pine with piñon-juniper vegetation, as in FireBGCv2-Jemez.

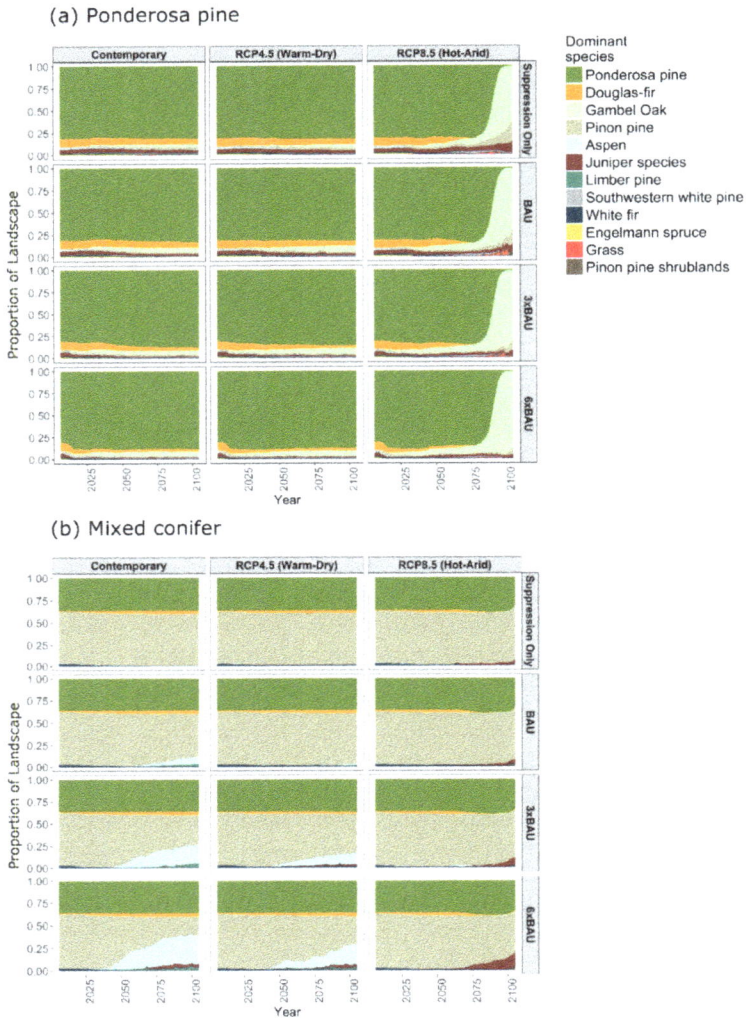

Figure 9. FireBGCv2-Jemez dominant vegetation by biomass for the ponderosa pine site (**a**) and dry mixed conifer site (**b**), represented as the proportional area within each site. Scenarios are factorial combinations of management (Suppression Only; BAU, 76-year treatment rotation; 3xBAU, 22-year treatment rotation; 6xBAU, 11-year treatment rotation) and climate (contemporary; Warm-Dry; Hot-Arid). The *x*-axes are simulation years and *y*-axes are proportional site area occupied by each species or life form.

(a) Ponderosa pine

(b) Mixed conifer

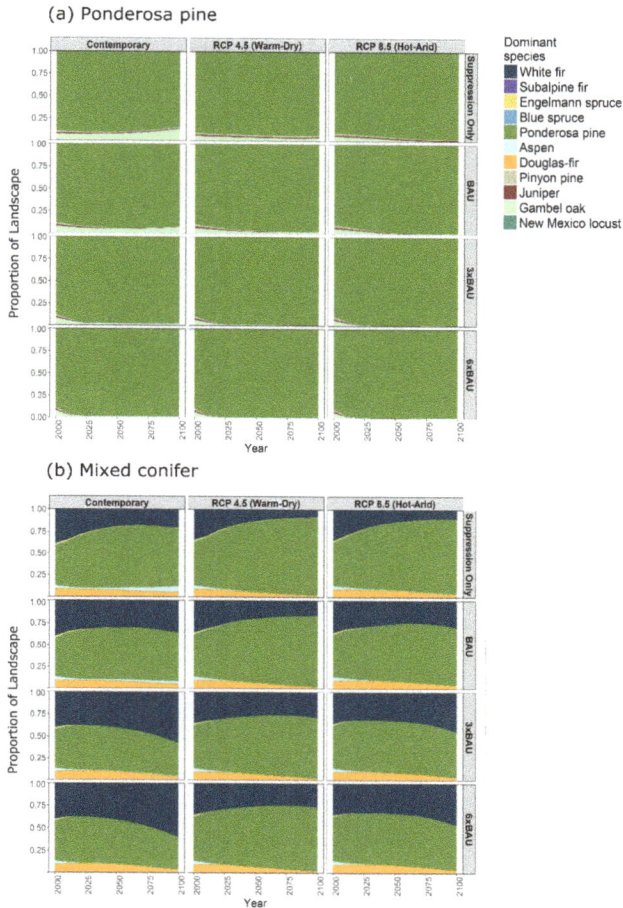

Figure 10. LANDIS-II-Kaibab dominant vegetation by biomass for the ponderosa pine site (**a**) and dry mixed conifer site (**b**), represented as the proportional area within each site. Scenarios are factorial combinations of management (Suppression Only; BAU, 76-year treatment rotation; 3xBAU, 22-year treatment rotation; 6xBAU, 11-year treatment rotation) and climate (Contemporary; Warm-Dry; Hot-Arid). The *x*-axes are simulation years and *y*-axes are proportional site area occupied by each species or life form.

3.2.2. Forest Structure

For FireBGCv2-Jemez, Hot-Arid climate triggered a transition of the ponderosa pine site to stands dominated by immature, sapling stage trees (2 to 10 cm DBH) ca AD 2075 (Figure 11a), corresponding to the hottest and driest period of future climate simulated in FireBGCv2-Jemez (Figure 2). We attribute the disappearance of larger-sized trees to drought- and heat-induced tree mortality, with some additional losses due to high-severity fire. Such impacts of climate change on tree mortality have been well-documented at regional to global scales [107–110]. Although early-successional forests occupy a large proportion of the landscape, they are a relatively minor biomass component when compared with Gambel oak (Figure 11a). For Contemporary and Warm-Dry climates, the distribution of structural stages in the ponderosa pine site was stable through the 100-year simulation period and consistent with reference conditions for uneven-aged forests with a mix of small to large size classes [16]. Trees were predominantly of a mature stage (23 to 50 cm DBH), with smaller proportions of saplings (2 to 10 cm DBH), pole-sized (10 to 23 cm DBH), and large (50 to 100 cm DBH) trees (Figure 11a). Saplings increased in the dry mixed conifer site under Warm-Dry and Hot-Arid climates ca. AD 2025 (Figure 11b), reflecting the increasing proportion of aspen (typically small in diameter [111]) and mortality and/or removal of trees that occurred with wildland fire and prescribed fires and fuels treatments.

LANDIS-II-Kaibab forest structure was impacted by declines in the regeneration of overstory species under the two climate change scenarios. Climate change driven regeneration failure was apparent from the age structure diagrams (Figure 12a,b), as proportional biomass shifted towards older cohorts during the middle of the century, when climate moved away from the regeneration niches of overstory species. Total biomass did not increase in older cohorts, but older trees gradually dominated the overall proportion of biomass due to a lack of regeneration moving new trees into younger age classes. The shift in age structure was less complete in the Warm-Dry scenario (Figure 12a,b) and in the dry mixed conifer stands for the Hot-Arid scenario (Figure 12b), where cooler and wetter conditions enabled some ponderosa pine regeneration.

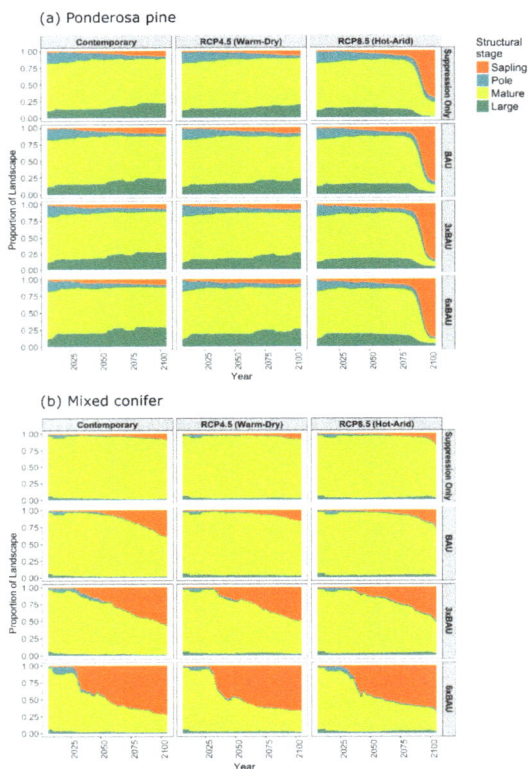

Figure 11. FireBGCv2-Jemez structural stage for the ponderosa pine site (**a**) and dry mixed conifer site (**b**). Scenarios are factorial combinations of management (Suppression Only; BAU, 76-year treatment rotation; 3xBAU, 22-year treatment rotation; 6xBAU, 11-year treatment rotation) and climate (Contemporary; Warm-Dry; Hot-Arid). Structural stages correspond to the following diameter classes (cm): $2 \leq$ saplings ≤ 10, $10 <$ pole ≤ 23, $23 <$ mature ≤ 50, $50 <$ large ≤ 100, very large > 100. The *x*-axes are simulation years and *y*-axes are proportional site area occupied by each structural stage.

(a) Ponderosa pine

(b) Mixed conifer

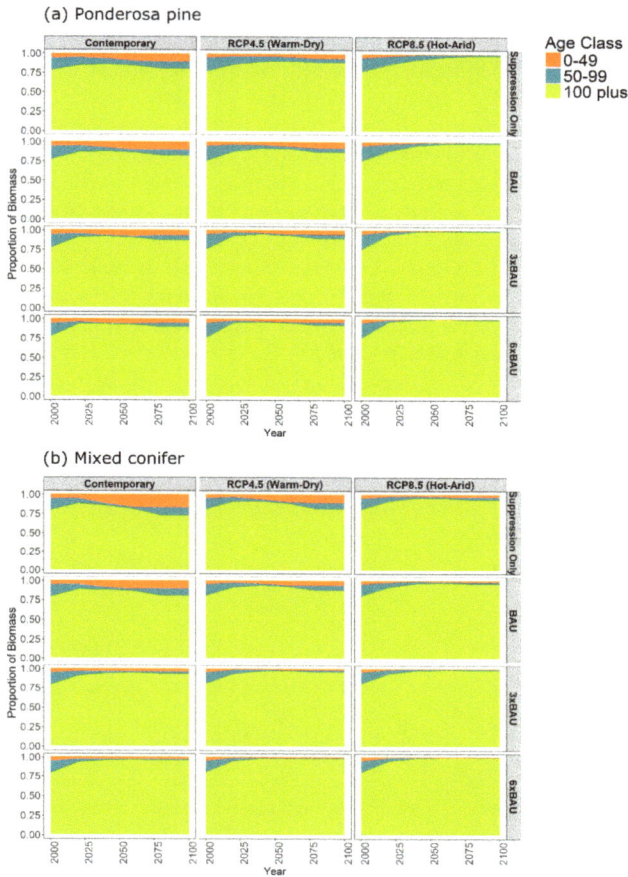

Figure 12. LANDIS-II-Kaibab proportion of biomass by age class for (**a**) ponderosa pine and (**b**) dry mixed conifer sites for factorial combinations of management (Suppression Only; BAU, 76-year treatment rotation; 3xBAU, 22-year treatment rotation; 6xBAU, 11-year treatment rotation) and climate (Contemporary; Warm-Dry; Hot-Arid). The *x*-axes are simulation years and *y*-axes are proportion of biomass in each age class. Age classes are young (0-49 years), mid (50-99 years), and old (100 plus years).

3.2.3. Forest Production

FireBGCv2-Jemez basal area (BA, sum of individual trees' cross-sectional area at 1.37m above ground) for ponderosa pine and dry mixed conifer sites was initially similar to measurements of contemporary, fire-excluded southwestern dry conifer forests [77] (Figure 13), around 20 m^2/ha. Climate changes and management treatments decreased forest biomass relative to initial conditions. Warm-Dry climate and any of the four management factors maintained basal area well above pre-Euroamerican era estimates of around 13 m^2/ha [16,77], although 3xBAU and 6xBAU treatments decreased basal area as compared with Suppression Only or BAU. Hot-Arid climate basal area decreased substantially, beginning early in the simulation period ca. AD 2025, with a further abrupt step change ca. AD 2075. By AD 2100, basal area was about 10 percent of its initial value, well below pre-settlement estimates. Loss of basal area occurred from a complex of ecological processes—tree mortality, regeneration failure, and compositional and structural shifts to shrublands or early successional forests—caused by climate stress, wildfires, management treatments, and changes in the distribution of bioclimatic space suitable for plant growth.

Biomass declines clearly illustrated the impact of climate on vegetation in the LANDIS-II-Kaibab landscape. Tree biomass decreased drastically under the Hot-Arid climate scenario (Figure 14). By the end of the century, average biomass in managed forests under the Hot-Arid scenario (3881 g/m^2) was well below historical estimates of pre-fire suppression biomass for ponderosa pine (9850 g/m^2) and dry mixed conifer (7460 g/m^2) on the Kaibab Plateau [52,112], indicating an overall forest decline and likely type change from forest to woodland or grassland. Biomass declines were driven by the failure of overstory species to regenerate under warmer and drier future climate conditions, as illustrated in the age structure shift to older cohorts. The absence of viable lower elevation species capable of replacing the declining overstory species delays biomass recovery and is also responsible for delays in compositional change. However, the drastic decline of forest biomass demonstrates that, despite the relatively static species composition, these forests are fundamentally altered by the Hot-Arid climate. Biomass declined slightly under the Warm-Dry scenario (7467 g/m^2) compared to Contemporary climate (9798 g/m^2), but remained within the historical pre-fire suppression estimates of biomass for ponderosa pine and dry mixed conifer forests referenced above. The enhanced treatment scenarios (3xBAU and 6xBAU) slightly increased the rate of biomass declines, suggesting that high rates of thinning and burning may accelerate forest decline under drastic climate shifts [113].

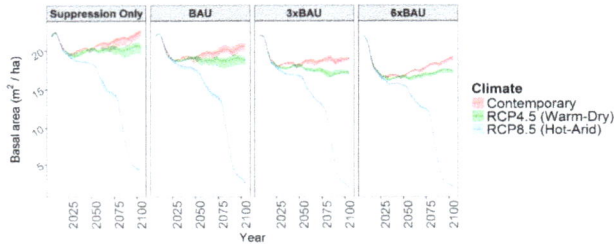

Figure 13. FireBGCv2-Jemez basal area (m^2/ha) of ponderosa pine and dry mixed conifer sites. Scenarios are factorial combinations of management (Suppression Only; BAU, 76-year treatment rotation; 3xBAU, 22-year treatment rotation; 6xBAU, 11-year treatment rotation) and climate (Contemporary; Warm-Dry; Hot-Arid). Envelopes show median (darker line) and 25th and 75th percentiles (lighter shading) among replicates for each scenario.

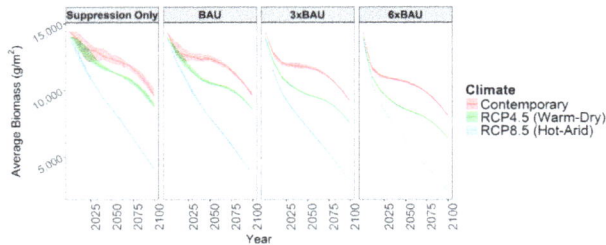

Figure 14. LANDIS-II-Kaibab average tree species biomass (g/m^2) for ponderosa pine and dry mixed conifer sites. Scenarios are factorial combinations of management (Suppression Only; BAU, 76-year treatment rotation; 3xBAU, 22-year treatment rotation; 6xBAU, 11-year treatment rotation) and climate (Contemporary; Warm-Dry; Hot-Arid). Envelopes show median (darker line) and 25th and 75th percentiles (lighter shading) among replicates for each scenario.

4. Discussion

4.1. Will Climate Changes Cause Fundamental Changes in Southwestern Fire Regimes and Forests?

Models did not produce consistent impacts of climate on fire regimes. For FireBGCv2-Jemez, wildfires were more frequent under Warm-Dry and Hot-Arid climates, and the Hot-Arid scenario resulted in an increase in median high-severity wildfire burned area and in episodic, high-severity "mega-fire" events. In contrast, climate had little impact on fire outcomes in LANDIS-II-Kaibab. Varying responses reflect fundamental differences in model mechanics (see below).

Our models were consistent in projecting a fundamental reorganization of ecosystem properties under the Hot-Arid climate scenario, and to a lesser extent, the Warm-Dry climate scenario. As modeled, the more extreme Hot-Arid climate is a "tipping point," driving biomass declines, shifts in forest structure, and compositional changes. Tipping points are critical thresholds at which even small perturbations radically and persistently reorganize system patterns or processes [114,115]. This fundamental reorganization of ecosystems aligns with previous modeling that incorporates climate change, fire, and vegetation interactions in temperate forests [57,116], as well as conceptual models that identify shrublands as an alternative, stable state in dry conifer-shrub ecosystems catalyzed by interacting anthropogenic stressors (e.g., climate changes and altered fire regimes) that push systems past a tipping point [117,118]. Modeled forest to shrubland transformations are aligned with field studies; for example, a recent study in the Jemez Mountains attributes the presence of Gambel oak shrubfields to high-severity wildfire disturbances [118]. Oak shrubfields, once established, can be highly resilient to subsequent high-severity fire events [45,117]. Ecotonal shifts between ponderosa pine forests and piñon-juniper woodlands have also been documented in the southwest, in response to a severe drought in the 1950s [119] and more recent drought conditions [120]. From our results, we infer that dry forests of the Jemez, NM and Kaibab, AZ ecosystems—and, by extension, other dry forest ecosystems in the southwestern US—may be vulnerable to a type change from forest to shrubland or grassland. Persistent changes in these systems may be driven by (1) shifts in fire regimes due to changes in climate and fuel availability, type, and structure; and (2) climate-driven regeneration failure as ecosystems depart from optimal conditions for overstory tree species.

Ponderosa pine forests of the Jemez Mountains were more substantially altered by climate and wildfire than dry mixed conifer forests, which contain species of varying drought resistance and physiological tolerance and therefore exhibited a more stable response to changing climate conditions. This outcome provides support for the hypothesis that species diversity promotes functional resilience to climate perturbations [121,122]. This conclusion is further supported by comparing responses between the two landscapes. In the Jemez Mountains, the ponderosa pine (Figure 9a) and dry mixed conifer (Figure 9b) sites both contain a significant component of lower-elevation piñon and juniper species, with mature individuals providing propagules as soon as climate conditions shift to favor these species. In contrast, piñon and juniper species are largely missing from ponderosa pine and dry mixed conifer sites on the Kaibab Plateau (Figure 10a,b); therefore, these species must encroach from lower elevations. This recruitment mechanism is also apparent in compositional shifts and biomass recovery that occurred more rapidly in the wet mixed conifer and spruce-fir forests as compared to ponderosa pine forests on the Kaibab Plateau (data not shown). Contemporary wet mixed conifer and spruce fir forests include ponderosa pine, which remains viable at these sites throughout the climate changes of the next century. In forests, types where none of the current overstory species remain viable, delayed uphill movement of lower elevation species will likely exacerbate and lengthen biomass declines and increase the potential for persistent type changes where shrub or cheatgrass communities establish and resist future invasion by tree species [118,123]. Because our models did not include insect disturbance, which has been a significant cause of recent tree mortality in the southwestern US and particularly in piñon pines in the southwest [124], the role of piñon pine in maintaining tree cover and initiating forest recovery could be unrealistic. We hypothesize that if insect disturbance were included in the models, differential piñon mortality would shift woodland dominance to juniper species, which may or may not play a similar ecological role in woodland systems [119,125].

4.2. Will Current Management Approaches be Effective in Preventing Departures under Future Climates?

Management treatments had little effect on ecosystem responses to climate change in FireBGCv2-Jemez and LANDIS-II-Kaibab. For both landscapes, the current management strategy (BAU scenarios) was consistently ineffective in preventing changes under future climate. Thinning and prescribed burning treatments at current application rates had little appreciable influence on area burned or high-severity area burned. At the stand scale, fuel treatments have been shown to be highly effective at reducing potential fire severity [126,127]. However, our modeling indicates that the current rate of application has little impact on wildfire outcomes at the landscape scale. The BAU scenarios were also ineffective in preventing biomass declines, shifts in age structure, and compositional changes under future climate. Targeted treatments may temporarily achieve objectives and protect high value landscape components. However, the central role of climate in driving forest changes through either mortality or regeneration failure suggests that the benefits of current treatments will likely be temporary.

4.3. Do Shifting Climate and Fire Regimes Require Novel Management Approaches?

Management strategies did not maintain current biomass, composition, or structure with changing climate. Although 3xBAU and 6xBAU treatments impacted wildfire regimes on the Kaibab landscape, they were ineffective at buffering or delaying the reorganization of forest ecosystems on either landscape. Intensified treatment scenarios (3xBAU and 6xBAU) slightly increased the rate of biomass declines, suggesting that high rates of thinning and burning may accelerate forest decline under drastic climate shifts [113]. The 6xBAU treatment strategy achieved treatment rotations of eleven years, the approximate historic fire rotation for ponderosa pine and dry mixed conifer forests, but increased areal extent and frequency of treatments alone was not an effective tool for preventing fundamental ecological shifts, especially under Hot-Arid climate. This result suggests that novel management approaches may be required to sustain forest landscapes or facilitate the adaptation of forests to changing climate and fire regimes [33,35,36]. For example, the planting of lower elevation species, adapted to warmer and drier conditions, could facilitate species migration, accelerating ecosystem recovery and reducing the depth of biomass declines [128,129].

4.4. Model Influences on Outcomes

Outcomes can differ across ecological models as the result of different model structures, initializing conditions, parameterization, and data inputs. Complex models are sensitive to a large number of factors (e.g., climate, weather, terrain) because they explicitly simulate the relevant underlying mechanistic processes. FireBGCv2 and LANDIS-II are highly complex models that mechanistically link climate, weather, and fuel patterns to fire frequency and area burned [83,130]. In FireBGCv2, fire regime characteristics are emergent model properties, determined by iterative climate and fire influences on fuel properties (type, amount, moisture) and live vegetation characteristics (e.g., bark thickness, canopy base height, DBH, vigor). Climate can serve as either a positive feedback to fire (e.g., warming temperatures increase landscape burnability by decreasing fuel moisture) or a negative feedback (e.g., drought conditions decrease burnabilty via reduction in biomass). In this study, the combination of larger areal extent of thinning and prescribed fire treatments, increased high-severity wildfire area burned, climate conditions unfavorable for tree recruitment and regeneration, and understory biomass production modeled for the Hot-Arid, 3xBAU and 6xBAU scenarios served as a negative feedback to fire frequency via reductions in fuel availability and connectivity [94,103,131]. In the LANDIS-II DFF extension, climate influences on area burned are partially restricted by the input of a user defined fire size or fire duration distribution. We used a fire duration distribution, which enables individual fires to increase in size in response to climate influences on fuel moisture and consequent fire spread rates. Therefore, individual fires should increase in size in response to a warmer and drier climate and potentially increase area burned. Yet, this model dynamic does not account for additional large fires that might be expected to escape suppression under a hotter and drier future. LANDIS-II simulations, in contrast to the FireBGCv2 simulations, also did not incorporate the direct influence of climate changes on the number of ignitions (e.g., fuel moisture influences on the likelihood of ignitions). However, the model does incorporate the indirect influence of climate-driven vegetation changes and subsequent changes in fuel conditions on the number of ignitions. Still, the missing link between direct impacts of climate on the number of ignitions represents an important limitation in the assessment of climate influences on future fire regimes in the LANDIS-II DFF extension as it is configured in this particular study, and it may explain the reduced sensitivity of the fire regime to climate. Ultimately, we did not produce consistent projections of future fire regimes, due to differences in model mechanics. However, despite their differences, models did produce consistent projections of substantially altered forest composition, structure, and biomass, suggesting a level of biotic reorganization that will affect management goals and strategies for southwestern landscapes.

Sources of uncertainty in model results come from input climate model data, which are inherently uncertain because climate change and the severity of its impacts depend on future emissions and mitigation measures. Additionally, global climate models (GCMs) may not accurately represent climate and weather at the regional and local scales, particularly concerning precipitation trends and climate oscillations that influence fire patterns [132–134]. The delta method used to locally downscale climate model inputs only accounts for changes to the mean climate signal, and not to shifts in synoptic-scale climate patterns outside of observed weather [90,135]. Finally, neither of our models included insects as a disturbance agent, although insect outbreaks are predicted to increase with warmer and drier climates [136,137]. Significant insect-caused tree mortality has already occurred in piñon pines in the Southwest [124], and insect disturbance, if modeled, could influence forest dynamics.

5. Conclusions

Although this study was not designed as a factorial model comparison (i.e., in which both models are developed for each landscape), similarities in the model results suggest common, regional-scale ecological inferences useful for understanding current and future climate, fire, and vegetation dynamics across southwestern forested ecosystems. Both models projected persistent compositional and structural changes in present-day dry conifer forests caused by climate changes and shifting fire patterns, particularly in ponderosa pine forests. We found resilience traits in dry mixed conifer forests in both model landscapes, and indications that ecotonal zones—for example, piñon-juniper ponderosa pine ecotone in the Jemez Mountains—can facilitate relatively rapid upslope movement of drought-adapted species into areas that have become too arid to support more mesic forests. Models produced dissimilar outcomes related to management and climate impacts on fire regimes, the result of inherent differences in model mechanics. However, both models captured cumulative, reciprocal interactions of climate, fires, and vegetation that highlight the complexity of fire-prone ecological systems in which key driving processes (e.g., climate) have both direct and indirect and short- and long-term influence on landscape patterns and processes.

Our results are compatible with recent papers that have identified the need for new strategies to promote the resilience of fire-prone forested ecosystems. Current and intensified management treatments simulated for FireBGCv2-Jemez and LANDIS-II-Kaibab did not prevent fundamental reorganization of the study landscapes under changing climates, suggesting that historical or present-day forest and fire regime characteristics may not be achievable management targets in the future. The design of novel management approaches will present two important challenges. First, it requires managers to reach a consensus on achievable objectives under future climate conditions, not based on historic reference conditions. Potential objectives could include the maintenance of functional types or ecosystem services, biomass conservation, carbon sequestration, the maintenance of key habitat types, or the conservation of species and genetic diversity [138]. Second, managers would need to begin implementing and experimenting with untested approaches that could produce unintended consequences. Modeling studies will be an important component of this process, helping to inform the selection of promising treatments and anticipate risks. However, ultimately, these approaches will require testing in actual landscapes, perhaps initially at smaller scales. This approach poses a difficult but critical path forward, requiring a dynamic, experimental land management framework that anticipates change, acknowledges that current systems have transformed or will transform away from historical references, and allows dynamic ecological processes to occur [139].

Supplementary Materials: The following are available online at http://www.mdpi.com/1999-4907/9/4/192/s1, Supplement 1: Methods: The FireBGCv2 and LANDIS-II models, Table S1: Treatment parameters for FireBGCv2-Jemez and LANDIS-II-Kaibab modeling simulations. Results: Table S2: FireBGCv2-Jemez wildfire area burned, Table S3: LANDIS-II-Kaibab wildfire area burned.

Acknowledgments: This research was funded by the Joint Fire Science Program (project 15-1-03-26, Landscape Impacts of Fire and Climate Change in the Southwest: A Science-Management Partnership, "SW FireCLIME"), the National Science Foundation Coupled Natural and Humans Systems Program (project 1114898, Jemez Fire and Humans in Resilient Ecosystems, "FHiRE"), the Arizona Technology and Research Initiative Fund, and the USGS Land Change Science Program. We also received support from the JFSP Southwest Fire Science Consortium. The authors gratefully acknowledge the contributions of many collaborators including Bob Keane, USFS Rocky Mountain Research Station, for FireBGCv2 development; Casey Teske and Windy Bunn, National Park Service, Shaula Headwall, US Fish and Wildlife Service, and Dennis Carril, Craig Wilcox, and David Robinson, US Forest Service, for guidance on and suggestions for management inputs; other members of the SW FireCLIME team for their project contributions; and the large number of participants at several science-management workshops. We thank Larissa Yocum Kent and two anonymous reviewers for their valuable comments and Mary Whalen (USGS) for graphics assistance. Any use of trade names is for descriptive purposes only and does not imply endorsement by the US Government.

Author Contributions: R.L., W.F., and A.T. conceived of and designed the experiments; L.H. and W.F. performed the experiments and produced summary data and graphics; R.L. and W.F. analyzed the data and wrote the paper.

Conflicts of Interest: The authors declare no conflict of interest.

References

1. Case, M.J.; Lawler, J.J. Integrating mechanistic and empirical model projections to assess climate impacts on tree species distributions in northwestern North America. *Glob. Chang. Biol.* **2017**, *23*, 2005–2015. [CrossRef] [PubMed]
2. Danby, R.K.; Hik, D.S. Variability, contingency and rapid change in recent subarctic alpine tree line dynamics. *J. Ecol.* **2007**, *95*, 352–363. [CrossRef]
3. Hamann, A.; Wang, T. Potential effects of climate change on ecosystem and tree species distribution in british columbia. *Ecology* **2006**, *87*, 2773–2786. [CrossRef]
4. Kelly, A.E.; Goulden, M.L. Rapid shifts in plant distribution with recent climate change. *Proc. Natl. Acad. Sci. USA* **2008**, *105*, 11823–11826. [CrossRef] [PubMed]
5. Rehfeldt, G.E.; Crookston, N.L.; Warwell, M.V.; Evans, J.S. Empirical analyses of plant-climate relationships for the western United States. *Int. J. Plant Sci.* **2006**, *167*, 1123–1150. [CrossRef]
6. Dale, V.H.; Joyce, L.A.; McNulty, S.; Neilson, R.P.; Ayres, M.P.; Flannigan, M.D.; Hanson, P.J.; Irland, L.C.; Lugo, A.E.; Peterson, C.J. Climate change and forest disturbances. *Bioscience* **2001**, *51*, 723–734. [CrossRef]
7. Flannigan, M.; Stocks, B.; Wotton, B. Climate change and forest fires. *Sci. Total Environ.* **2000**, *262*, 221–229. [CrossRef]
8. Gavin, D.G.; Hu, F.S. Spatial variation of climatic and non-climatic controls on species distribution: The range limit of tsuga heterophylla. *J. Biogeogr.* **2006**, *33*, 1384–1396. [CrossRef]
9. Bond, W.J.; Woodward, F.I.; Midgley, G.F. The global distribution of ecosystems in a world without fire. *New Phytol.* **2005**, *165*, 525–537. [CrossRef]
10. Krawchuk, M.A.; Moritz, M.A.; Parisien, M.A.; Van Dorn, J.; Hayhoe, K. Global pyrogeography: The current and future distribution of wildfire. *PLoS ONE* **2009**, *4*, e5102. [CrossRef]
11. Loehman, R.A.; Reinhardt, E.; Riley, K.L. Wildland fire emissions, carbon, and climate: Seeing the forest and the trees–a cross-scale assessment of wildfire and carbon dynamics in fire-prone, forested ecosystems. *For. Ecol. Manag.* **2014**, *317*, 9–19. [CrossRef]

12. Turner, M.G.; Romme, W.H. Landscape dynamics in crown fire ecosystems. *Landsc. Ecol.* **1994**, *9*, 59–77. [CrossRef]
13. Heyerdahl, E.K.; Morgan, P.; Riser, J.P. Multi-season climate synchronized historical fires in dry forests (1650–1900), northern rockies, USA. *Ecology* **2008**, *89*, 705–716. [CrossRef] [PubMed]
14. Whitlock, C.; Shafer, S.L.; Marlon, J. The role of climate and vegetation change in shaping past and future fire regimes in the northwestern US and the implications for ecosystem management. *For. Ecol. Manag.* **2003**, *178*, 5–21. [CrossRef]
15. Falk, D.A.; Miller, C.; McKenzie, D.; Black, A.E. Cross-scale analysis of fire regimes. *Ecosystems* **2007**, *10*, 809–823. [CrossRef]
16. Reynolds, R.T.; Meador, A.J.S.; Youtz, J.A.; Nicolet, T.; Matonis, M.S.; Jackson, P.L.; DeLorenzo, D.G.; Graves, A.D. *Restoring Composition and Structure in Southwestern Frequent-Fire Forests*; General Technical Report RMRS-GTR-310; USDA Forest Service, Rocky Mountain Research Station: Fort Collins, CO, USA, 2013.
17. Allen, C.D.; Savage, M.; Falk, D.A.; Suckling, K.F.; Swetnam, T.W.; Schulke, T.; Stacey, P.B.; Morgan, P.; Hoffman, M.; Klingel, J.T. Ecological restoration of southwestern ponderosa pine ecosystems: A broad perspective. *Ecol. Appl.* **2002**, *12*, 1418–1433. [CrossRef]
18. Covington, W.W.; Fule, P.Z.; Moore, M.M.; Hart, S.C.; Kolb, T.E.; Mast, J.N.; Sackett, S.S.; Wagner, M.R. Restoring ecosystem health in ponderosa pine forests of the Southwest. *J. For.* **1997**, *95*, 23–29.
19. Covington, W.W.; Moore, M.M. Southwestern ponderosa forest structure: Changes since euro-american settlement. *J. For.* **1994**, *92*, 39–47.
20. Steele, R.; Arno, S.F.; Geier-Hayes, K. Wildfire patterns change in central Idaho's ponderosa pine-Douglas-fir forest. *West. J. Appl. For.* **1986**, *1*, 16–18.
21. Westerling, A.L.; Hidalgo, H.G.; Cayan, D.R.; Swetnam, T.W. Warming and earlier spring increase western U.S. Forest wildfire activity. *Science* **2006**, *313*, 940–943. [CrossRef]
22. Swetnam, T.W.; Farella, J.; Roos, C.I.; Liebmann, M.J.; Falk, D.A.; Allen, C.D. Multiscale perspectives of fire, climate and humans in western North America and the Jemez mountains, USA. *Phil. Trans. R. Soc. B* **2016**, *371*, 20150168. [CrossRef] [PubMed]
23. Savage, M.; Mast, J.N. How resilient are southwestern ponderosa pine forests after crown fires? *Can. J. For. Res.* **2005**, *35*, 967–977. [CrossRef]
24. Miller, C.; Abatzoglou, J.; Brown, T.; Syphard, A.D. Wilderness fire management in a changing environment. In *The Landscape Ecology of Fire*; McKenzie, D., Miller, C., Falk, D.A., Eds.; Springer: Berlin, Germany, 2011; pp. 269–294.
25. Littell, J.S.; Peterson, D.L.; Riley, K.L.; Liu, Y.; Luce, C.H. A review of the relationships between drought and forest fire in the United States. *Glob. Chang. Biol.* **2016**, *22*, 2353–2369. [CrossRef] [PubMed]
26. Westerling, A.L.; Swetnam, T.W. Interannual to decadal drought and wildfire in the western United States. *EOS Trans. Am. Geophys. Union* **2003**, *84*, 545–555. [CrossRef]
27. Flannigan, M.D.; Amiro, B.D.; Logan, K.A.; Stocks, B.J.; Wotton, B.M. Forest fires and climate change in the 21st century. *Mitig. Adapt. Strateg. Glob. Chang.* **2006**, *11*, 847–859. [CrossRef]
28. Abatzoglou, J.T.; Williams, A.P. Impact of anthropogenic climate change on wildfire across western US forests. *Proc. Natl. Acad. Sci. USA* **2016**, *113*, 11770–11775. [CrossRef]
29. Drever, C.R.; Peterson, G.; Messier, C.; Bergeron, Y.; Flannigan, M. Can forest management based on natural disturbances maintain ecological resilience? *Can. J. For. Res.* **2006**, *36*, 2285–2299. [CrossRef]
30. Turner, M.G. Disturbance and landscape dynamics in a changing world 1. *Ecology* **2010**, *91*, 2833–2849. [CrossRef] [PubMed]
31. Falk, D.A. The resilience dilemma: Incorporating global change into ecosystem policy and management. *Ariz. State Law J.* **2016**, *48*, 145–156.
32. Littell, J.S.; McKenzie, D.; Peterson, D.L.; Westerling, A.L. Climate and wildfire area burned in western US ecoprovinces, 1916–2003. *Ecol. Appl.* **2009**, *19*, 1003–1021. [CrossRef] [PubMed]
33. Schoennagel, T.; Balch, J.K.; Brenkert-Smith, H.; Dennison, P.E.; Harvey, B.J.; Krawchuk, M.A.; Mietkiewicz, N.; Morgan, P.; Moritz, M.A.; Rasker, R. Adapt to more wildfire in western North American forests as climate changes. *Proc. Natl. Acad. Sci. USA* **2017**, *114*, 4582–4590. [CrossRef] [PubMed]
34. Stephens, S.L.; McIver, J.D.; Boerner, R.E.J.; Fettig, C.J.; Fontaine, J.B.; Hartsough, B.R.; Kennedy, P.L.; Schwilk, D.W. The effects of forest fuel-reduction treatments in the United States. *BioScience* **2012**, *62*, 549–560. [CrossRef]

35. Svenning, J.-C.; Sandel, B. Disequilibrium vegetation dynamics under future climate change. *Am. J. Bot.* **2013**, *100*, 1266–1286. [CrossRef] [PubMed]

36. Stephens, S.L.; Agee, J.K.; Fulé, P.; North, M.; Romme, W.; Swetnam, T.; Turner, M.G. Managing forests and fire in changing climates. *Science* **2013**, *342*, 41–42. [CrossRef] [PubMed]

37. Folke, C.; Carpenter, S.; Walker, B.; Scheffer, M.; Elmqvist, T.; Gunderson, L.; Holling, C.S. Regime shifts, resilience, and biodiversity in ecosystem management. *Annu. Rev. Ecol. Evol. Syst.* **2004**, *35*, 557–581. [CrossRef]

38. Millar, C.I.; Stephenson, N.L.; Stephens, S.L. Climate change and forests of the future: Managing in the face of uncertainty. *Ecol. Appl.* **2007**, *17*, 2145–2151. [CrossRef] [PubMed]

39. Rollins, M.G. Landfire: A nationally consistent vegetation, wildland fire, and fuel assessment. *Int. J. Wildl. Fire* **2009**, *18*, 235–249. [CrossRef]

40. Anderson, R.S.; Jass, R.B.; Toney, J.L.; Allen, C.D.; Cisneros-Dozal, L.M.; Hess, M.; Heikoop, J.; Fessenden, J. Development of the mixed conifer forest in northern New Mexico and its relationship to holocene environmental change. *Quat. Res.* **2008**, *69*, 263–275. [CrossRef]

41. Touchan, R.; Allen, C.D.; Swetnam, T.W. *Fire History and Climatic Patterns in Ponderosa Pine and Mixed-Conifer Forests of the Jemez Mountains, Northern New Mexico*; U.S. Department of Agriculture, Forest Service, Rocky Mountain Research Station: Fort Collins, CO, USA, 1996; pp. 33–46.

42. Muldavin, E.; Kennedy, A.; Jackson, C.; Neville, P.; Neville, T.; Schultz, K.; Reid, M. *Vegetation Classification and Map: Bandelier National Monument*; Natural Resource Technical Report NPS/SCPN/NRTR—2011/438; National Park Service: Fort Collins, CO, USA, 2011; p. 468.

43. Margolis, E.Q.; Malevich, S.B. Historical dominance of low-severity fire in dry and wet mixed-conifer forest habitats of the endangered terrestrial Jemez mountains salamander (*Plethodon neomexicanus*). *For. Ecol. Manag.* **2016**, *375*, 12–26. [CrossRef]

44. Swetnam, T.W.; Baisan, C.H. Historical fire regime patterns in the southwestern United States since ad 1700. In *Fire Effects in Southwestern Forests: Proceedings of the 2nd La Mesa Fire Symposium*; General Technical Report RM-GTR-286; Allen, C., Ed.; U.S. Department of Agriculture, Forest Service, Rocky Mountain Research Station: Fort Collins, CO, USA, 1996; pp. 11–32.

45. Swetnam, T.W.; Baisan, C.H. Tree-ring reconstructions of fire and climate history in the Sierra Nevada and southwestern United States. In *Fire and Climatic Change in Temperate Ecosystems of the Western Americas*; Veblen, T.T., Baker, W.L., Montenegro, G., Swetnam, T.W., Eds.; Springer: New York, NY, USA, 2003; Volume 160, pp. 158–195.

46. Allen, C.D. Interactions across spatial scales among forest dieback, fire, and erosion in northern New Mexico landscapes. *Ecosystems* **2007**, *10*, 797–808. [CrossRef]

47. Coop, J.D.; Parks, S.A.; McClernan, S.R.; Holsinger, L.M. Influences of prior wildfires on vegetation response to subsequent fire in a reburned southwestern landscape. *Ecol. Appl.* **2016**, *26*, 346–354. [CrossRef] [PubMed]

48. Vankat, J.L. Post-1935 changes in forest vegetation of Grand Canyon National Park, Arizona, USA: Part 1–ponderosa pine forest. *For. Ecol. Manag.* **2011**, *261*, 309–325. [CrossRef]

49. Vankat, J.L. Post-1935 changes in forest vegetation of Grand Canyon National Park, Arizona, USA: Part 2–mixed conifer, spruce-fir, and quaking aspen forests. *For. Ecol. Manag.* **2011**, *261*, 326–341. [CrossRef]

50. Bradshaw, L.; McCormick, E. *FireFamily Plus User's Guide, Version 2.0*; Gen. Tech. Rep. RMRS-GTR-67WWW; U.S. Department of Agriculture, Forest Service, Rocky Mountain Research Station: Ogden, UT, USA, 2000.

51. NCEI, N. *1981–2010 U.S. Climate Normals*; National Oceanic and Atmospheric Administration, National Centers for Environmental Information: Asheville, NC, USA, 2011.

52. Fulé, P.Z.; Crouse, J.E.; Heinlein, T.A.; Moore, M.M.; Covington, W.W.; Verkamp, G. Mixed-severity fire regime in a high-elevation forest of Grand Canyon, Arizona, USA. *Landsc. Ecol.* **2003**, *18*, 465–486. [CrossRef]

53. Fulé, P.Z.; Heinlein, T.A.; Covington, W.W.; Moore, M.M. Assessing fire regimes on Grand Canyon landscapes with fire-scar and fire-record data. *Int. J. Wildl. Fire* **2003**, *12*, 129–145. [CrossRef]

54. White, M.A.; Vankat, J.L. Middle and high elevation coniferous forest communities of the north rim region of Grand Canyon National Park, Arizona, USA. *Vegetatio* **1993**, *109*, 161–174. [CrossRef]

55. Yocom-Kent, L.L.; Fulé, P.Z.; Bunn, W.A.; Gdula, E.G. Historical high-severity fire patches in mixed-conifer forests. *Can. J. For. Res.* **2015**, *45*, 1587–1596. [CrossRef]

56. U.S. Forest Service. *Warm Fire Assessment: Post-Fire Conditions and Management Considerations*; U.S. Forest Service, North Kaibab Ranger District, Kaibab National Forest: Coconino County, AZ, USA, 2007.

57. Clark, J.A.; Loehman, R.A.; Keane, R.E. Climate changes and wildfire alter vegetation of Yellowstone National Park, but forest cover persists. *Ecosphere* **2017**, *8*. [CrossRef]
58. Holsinger, L.; Keane, R.E.; Isaak, D.J.; Eby, L.; Young, M.K. Relative effects of climate change and wildfires on stream temperatures: A simulation modeling approach in a Rocky Mountain watershed. *Clim. Chang.* **2014**, *124*, 191–206. [CrossRef]
59. Loehman, R.A.; Clark, J.A.; Keane, R.E. Modeling effects of climate change and fire management on western white pine (*Pinus monticola*) in the northern Rocky Mountains, USA. *Forests* **2011**, *2*, 832–860. [CrossRef]
60. Loehman, R.A.; Keane, R.E.; Holsinger, L.M.; Wu, Z. Interactions of landscape disturbances and climate change dictate ecological pattern and process: Spatial modeling of wildfire, insect, and disease dynamics under future climates. *Landsc. Ecol.* **2017**, *32*, 1447–1459. [CrossRef]
61. Keane, R.E.; Loehman, R.A.; Holsinger, L.M. *The FireBGCv2 Landscape Fire and Succession Model: A Research Simulation Platform for Exploring Fire and Vegetation Dynamics*; Gen. Tech. Rep. RMRS-GTR-255; U.S. Department of Agriculture, Forest Service, Rocky Mountain Research Station: Fort Collins, CO, USA, 2011; p. 137.
62. Thornton, P.E.; Running, S.W. An improved algorithm for estimating incident daily solar radiation from measurements of temperature, humidity, and precipitation. *Agric. For. Meteorol.* **1999**, *93*, 211–228. [CrossRef]
63. Hungerford, R.D.; Nemani, R.R.; Running, S.W.; Coughlan, J.C. *Mtclim: A Mountain Microclimate Simulation Model. Gen. Tech. Rep. Int-414*; U.S. Department of Agriculture, Forest Service, Intermountain Research Station: Ogden, UT, USA, 1989; p. 52.
64. Schussman, H.; Enquist, C.; List, M. *Historic Fire Return Intervals for Arizona and New Mexico: A Regional Perspective for Southwestern Land Managers*; The Nature Conservancy: Phoenix, AZ, USA, 2006.
65. Paysen, T.E.; Ansley, R.J.; Brown, J.K.; Gottfried, G.J.; Haase, S.M.; Harrington, M.G.; Narog, M.G.; Sackett, S.S.; Wilson, R.C. Fire in western shrubland, woodland, and grassland ecosystems. In *Wildland Fire in Ecosystems: Effects of Fire on Flora. Gen. Tech. Rep. RMRS-GTR-42-Vol 2*; U.S. Department of Agriculture, Forest Service: Ogden, UT, USA, 2000; Volume 2, pp. 121–159.
66. O'Connor, C.D.; Lynch, A.M.; Falk, D.A.; Swetnam, T.W. Post-fire forest dynamics and climate variability affect spatial and temporal properties of spruce beetle outbreaks on a sky island mountain range. *For. Ecol. Manag.* **2015**, *336*, 148–162. [CrossRef]
67. Margolis, E.Q. Fire regime shift linked to increased forest density in a piñon–juniper savanna landscape. *Int. J. Wildl. Fire* **2014**, *23*, 234–245. [CrossRef]
68. Huffman, D.W.; Fule, P.Z.; Pearson, K.M.; Crouse, J.E. Fire history of pinyon–juniper woodlands at upper ecotones with ponderosa pine forests in Arizona and New Mexico. *Can. J. For. Res.* **2008**, *38*, 2097–2108. [CrossRef]
69. Gottfried, G.J.; Swetnam, T.W.; Allen, C.D.; Betancourt, J.L.; Chung-MacCoubrey, A.L. Pinyon-juniper woodlands. In *Ecology, Diversity, and Sustainability of the Middle Rio Grande Basin*; U.S. Forest Service Technical Report; Finch, D.M., Tainter, J.A., Eds.; USDA Forest Service: Fort Collins, CO, USA, 1995; Volume RM-GTR-268, pp. 95–132.
70. Brown, P.M.; Kaye, M.W.; Huckaby, L.S.; Baisan, C.H. Fire history along environmental gradients in the sacramento mountains, New Mexico: Influences of local patterns and regional processes. *Ecoscience* **2001**, *8*, 115–126. [CrossRef]
71. Baisan, C.H.; Swetnam, T.W. Fire history on a desert mountain range: Rincon mountain wilderness, Arizona, USA. *Can. J. For. Res.* **1990**, *20*, 1559–1569. [CrossRef]
72. Abella, S.R.; Fulé, P.Z. *Fire Effects on Gambel Oak in Southwestern Ponderosa Pine-Oak Forests. USDA Forest Service Research Note RMRS-RN-34*; USDA Forest Service: Fort Collins, CO, USA, 2008; p. 6.
73. Allen, C.D.; Touchan, R.; Swetnam, T.W. Overview of fire history in the Jemez mountains, New Mexico. In Proceedings of the New Mexico Geological Society Forty-seventh Annual Field Conference, Albuquerque, NM, USA, 25–28 September 1996; Goff, F., Kues, B.S., Rogers, M.A., McFadden, L.D., Gardner, J.N., Eds.; pp. 35–36.
74. Keane, R.E.; Ryan, K.C.; Running, S.W. Simulating effects of fire on northern Rocky Mountain landscapes with the ecological process model Fire-BGC. *Tree Physiol.* **1996**, *16*, 319–331. [CrossRef] [PubMed]
75. Gottfried, G.J. Pinyon-juniper woodlands in the southwestern United States. In *Ecology and Management of Forests, Woodlands, and Shrublands in the Dryland Regions of the United States and Mexico: Perspectives for the 21st Century*; University of Arizona: Tucson, AZ, USA, 1999; pp. 53–68.

76. FEIS. *Fire Effects Information System [online]*; U.S. Department of Agriculture, Forest Service, Rocky Mountain Research Station, Fire Sciences Laboratory: Missoula, MT, USA, 2015.

77. Fulé, P.Z.; Covington, W.W.; Moore, M.M. Determining reference conditions for ecosystem management of southwestern ponderosa pine forests. *Ecol. Appl.* **1997**, *7*, 895–908. [CrossRef]

78. Chojnacky, D.C. *Volume equations for New Mexico's pinyon-juniper dryland forests*; Research Paper INT-471; Department of Agriculture, Forest Service, Intermountain Research Experiment Station: Ogden, UT, USA, 1994; 10p.

79. Scheller, R.M.; Domingo, J.B.; Sturtevant, B.R.; Williams, J.S.; Rudy, A.; Gustafson, E.J.; Mladenoff, D.J. Design, development, and application of LANDIS-II, a spatial landscape simulation model with flexible temporal and spatial resolution. *Ecol. Model.* **2007**, *201*, 409–419. [CrossRef]

80. Flatley, W.T.; Fulé, P.Z. Are historical fire regimes compatible with future climate? Implications for forest restoration. *Ecosphere* **2016**, *7*. [CrossRef]

81. Scheller, R.M.; Mladenoff, D.J. A forest growth and biomass module for a landscape simulation model, LANDIS: Design, validation, and application. *Ecol. Model.* **2004**, *180*, 211–229. [CrossRef]

82. Crookston, N.L.; Rehfeldt, G.E.; Dixon, G.E.; Weiskittel, A.R. Addressing climate 246 change in the forest vegetation simulator to assess impacts on landscape forest 247 dynamics. *For. Ecol. Manag.* **2010**, *260*, 1198–1211. [CrossRef]

83. Sturtevant, B.R.; Scheller, R.M.; Miranda, B.R.; Shinneman, D.; Syphard, A. Simulating dynamic and mixed-severity fire regimes: A process-based fire extension for LANDIS-ii. *Ecol. Model.* **2009**, *220*, 3380–3393. [CrossRef]

84. Gustafson, E.J.; Shifley, S.R.; Mladenoff, D.J.; Nimerfro, K.K.; He, H.S. Spatial simulation of forest succession and timber harvesting using LANDIS. *Can. J. For. Res.* **2000**, *30*, 32–43. [CrossRef]

85. Meehl, G.A.; Washington, W.M.; Arblaster, J.M.; Hu, A.; Teng, H.; Tebaldi, C.; Sanderson, B.N.; Lamarque, J.-F.; Conley, A.; Strand, W.G. Climate system response to external forcings and climate change projections in CCSM4. *J. Clim.* **2012**, *25*, 3661–3683. [CrossRef]

86. Van Vuuren, D.P.; Edmonds, J.; Kainuma, M.; Riahi, K.; Thomson, A.; Hibbard, K.; Hurtt, G.C.; Kram, T.; Krey, V.; Lamarque, J.-F. The representative concentration pathways: An overview. *Clim. Chang.* **2011**, *109*, 5. [CrossRef]

87. Collins, M.; Knutti, R.; Arblaster, J.; Dufresne, J.-L.; Fichefet, T.; Friedlingstein, P.; Gao, X.; Gutowski, W.; Johns, T.; Krinner, G. Long-term climate change: Projections, commitments and irreversibility. In *Climate Change 2013: The Physical Science Basis. Contribution of Working Group i to the Fifth Assessment Report of the Intergovernmental Panel on Climate Change*; Stocker, T.F., Qin, D., Plattner, G.-K., Tignor, M., Allen, S.K., Boschung, J., Nauels, A., Xia, Y., Bex, V., Midgley, P.M., Eds.; Cambridge University Press: Cambridge, UK; New York, NY, USA, 2013; pp. 1029–1136.

88. Jones, C.; Hughes, J.; Bellouin, N.; Hardiman, S.; Jones, G.; Knight, J.; Liddicoat, S.; O'Connor, F.; Andres, R.J.; Bell, C. The HadGEM2-ES implementation of CMIP5 centennial simulations. *Geosci. Model Dev.* **2011**, *4*, 543. [CrossRef]

89. Thrasher, B.; Xiong, J.; Wang, W.; Melton, F.; Michaelis, A.; Nemani, R. Downscaled climate projections suitable for resource management. *EOS Trans. Am. Geophys. Union* **2013**, *94*, 321–323. [CrossRef]

90. Ekström, M.; Grose, M.R.; Whetton, P.H. An appraisal of downscaling methods used in climate change research. *Wiley Interdiscip. Rev. Clim. Chang.* **2015**, *6*, 301–319. [CrossRef]

91. Snover, A.K.; Hamlet, A.F.; Lettenmaier, D.P. Climate-change scenarios for water planning studies: Pilot applications in the Pacific Northwest. *Bull. Am. Meteorol. Soc.* **2003**, *84*, 1513–1518.

92. U.S. Department of Agriculture, Forest Service (USFS). *Final Environmental Impact Statement for the Southwest Jemez Mountains Landscape Restoration Project, Santa Fe National Forest, Sandoval County, New Mexico*; U.S. Department of Agriculture, Forest Service: Washington, DC, USA, 2015.

93. Morgan, P.; Bunting, S.C.; Black, A.E.; Merrill, T.; Barrett, S. Fire regimes in the interior columbia river basin: Past and present. In *USDA Forest Service, Rocky Mountain Research Station Final Report for RJVA-INT-94913*; Fire Sciences Laboratory: Missoula, MT, USA, 1996.

94. Marlon, J.R.; Bartlein, P.J.; Gavin, D.G.; Long, C.J.; Anderson, R.S.; Briles, C.E.; Brown, K.J.; Colombaroli, D.; Hallett, D.J.; Power, M.J. Long-term perspective on wildfires in the western USA. *Proc. Natl. Acad. Sci. USA* **2012**, *109*, E535–E543. [CrossRef] [PubMed]

95. Parks, S.A.; Miller, C.; Parisien, M.-A.; Holsinger, L.M.; Dobrowski, S.Z.; Abatzoglou, J. Wildland fire deficit and surplus in the western United States, 1984–2012. *Ecosphere* **2015**, *6*, 1–13. [CrossRef]
96. Wolf, J.J.; Mast, J.N. Fire history of mixed-conifer forests on the north rim, Grand Canyon National Park, Arizona. *Phys. Geogr.* **1998**, *19*, 1–14.
97. Running, S.W. Is global warming causing more, larger wildfires? *Science* **2006**, *313*, 927–928. [CrossRef] [PubMed]
98. Adams, M.A. Mega-fires, tipping points and ecosystem services: Managing forests and woodlands in an uncertain future. *For. Ecol. Manag.* **2013**, *294*, 250–261. [CrossRef]
99. Fulé, P.Z.; Swetnam, T.W.; Brown, P.M.; Falk, D.A.; Peterson, D.L.; Allen, C.D.; Aplet, G.H.; Battaglia, M.A.; Binkley, D.; Farris, C. Unsupported inferences of high-severity fire in historical dry forests of the western United States: Response to Williams and Baker. *Glob. Ecol. Biogeogr.* **2014**, *23*, 825–830. [CrossRef]
100. Savage, M.; Mast, J.N.; Feddema, J.J. Double whammy: High-severity fire and drought in ponderosa pine forests of the Southwest. *Can. J. For. Res.* **2013**, *43*, 570–583. [CrossRef]
101. Millar, C.I.; Stephenson, N.L. Temperate forest health in an era of emerging megadisturbance. *Science* **2015**, *349*, 823–826. [CrossRef] [PubMed]
102. Jones, J.; DeByle, N.; Winokur, R. Aspen: Ecology and management in the western United States. In *USDA Forest Service, Rocky Mountain Forest and Range Experiment Station General Technical Report RM-119*; USDA Forest Service, Rocky Mountain Forest and Range Experiment Station: Fort Collins, CO, USA, 1985.
103. Schoennagel, T.; Veblen, T.T.; Romme, W.H. The interaction of fire, fuels, and climate across Rocky Mountain forests. *BioScience* **2004**, *54*, 661–676. [CrossRef]
104. Margolis, E.Q.; Swetnam, T.W.; Allen, C.D. Historical stand-replacing fire in upper montane forests of the Madrean Sky Islands and Mogollon Plateau, southwestern USA. *Fire Ecol.* **2011**, *7*, 88–107. [CrossRef]
105. Romme, W.H.; Allen, C.D.; Bailey, J.D.; Baker, W.L.; Bestelmeyer, B.T.; Brown, P.M.; Eisenhart, K.S.; Floyd, M.L.; Huffman, D.W.; Jacobs, B.F. Historical and modern disturbance regimes, stand structures, and landscape dynamics in pinon–juniper vegetation of the western United States. *Rangel. Ecol. Manag.* **2009**, *62*, 203–222. [CrossRef]
106. Ganey, J.L.; Vojta, S.C. Tree mortality in drought-stressed mixed-conifer and ponderosa pine forests, Arizona, USA. *For. Ecol. Manag.* **2011**, *261*, 162–168. [CrossRef]
107. Allen, C.D.; Macalady, A.K.; Chenchouni, H.; Bachelet, D.; McDowell, N.; Vennetier, M.; Kitzberger, T.; Rigling, A.; Breshears, D.D.; Hogg, E.T. A global overview of drought and heat-induced tree mortality reveals emerging climate change risks for forests. *For. Ecol. Manag.* **2010**, *259*, 660–684. [CrossRef]
108. Van Mantgem, P.J.; Stephenson, N.L.; Byrne, J.C.; Daniels, L.D.; Franklin, J.F.; Fule, P.Z.; Harmon, M.E.; Larson, A.J.; Smith, J.M.; Taylor, A.H.; et al. Widespread increase of tree mortality rates in the western United States. *Science* **2009**, *323*, 521–524. [CrossRef] [PubMed]
109. Williams, A.P.; Allen, C.D.; Macalady, A.K.; Griffin, D.; Woodhouse, C.A.; Meko, D.M.; Swetnam, T.W.; Rauscher, S.A.; Seager, R.; Grissino-Mayer, H.D. Temperature as a potent driver of regional forest drought stress and tree mortality. *Nat. Clim. Chang.* **2013**, *3*, 292–297. [CrossRef]
110. Williams, A.P.; Allen, C.D.; Millar, C.I.; Swetnam, T.W.; Michaelsen, J.; Still, C.J.; Leavitt, S.W. Forest responses to increasing aridity and warmth in the southwestern United States. *Proc. Natl. Acad. Sci. USA* **2010**, *107*, 21289–21294. [CrossRef] [PubMed]
111. Jones, J.R.; DeByle, N.V.; Winokur, R.P. Climates. General Technical Report RM-119. In *Aspen: Ecology and Management in the Western United States*; DeByle, N.V., Winokur, R.P., Eds.; USDA Forest Service Rocky Mountain Forest and Range Experiment Station: Fort Collins, CO, USA, 1985; pp. 57–64.
112. Fulé, P.Z.; Covington, W.W.; Moore, M.M.; Heinlein, T.A.; Waltz, A.E.M. Natural variability in forests of the Grand Canyon, USA. *J. Biogeogr.* **2002**, *29*, 31–47. [CrossRef]
113. Diggins, C.; Fulé, P.Z.; Kaye, J.P.; Covington, W.W. Future climate affects management strategies for maintaining forest restoration treatments. *Int. J. Wildl. Fire* **2010**, *19*, 903. [CrossRef]
114. Brook, B.W.; Ellis, E.C.; Perring, M.P.; Mackay, A.W.; Blomqvist, L. Does the terrestrial biosphere have planetary tipping points? *Trends Ecol. Evol.* **2013**, *28*, 396–401. [CrossRef]
115. Reyer, C.P.; Brouwers, N.; Rammig, A.; Brook, B.W.; Epila, J.; Grant, R.F.; Holmgren, M.; Langerwisch, F.; Leuzinger, S.; Lucht, W. Forest resilience and tipping points at different spatio-temporal scales: Approaches and challenges. *J. Ecol.* **2015**, *103*, 5–15. [CrossRef]

116. Scheller, R.M.; Kretchun, A.M.; Loudermilk, E.L.; Hurteau, M.D.; Weisberg, P.J.; Skinner, C. Interactions among fuel management, species composition, bark beetles, and climate change and the potential effects on forests of the lake tahoe basin. *Ecosystems* **2017**. [CrossRef]

117. Falk, D.A. Are Madrean ecosystems approaching tipping points? Anticipating interactions of landscape disturbance and climate change. In *Merging Science and Management in a Rapidly Changing World: Biodiversity and Management of the Madrean Archipelago;* Proceedings RMRS-P-67; Ffolliott, P.F., Gottfried, G., Gebow, B., Eds.; U.S. Department of Agriculture, Forest Service, Rocky Mountain Research Station: Fort Collins, CO, USA, 2013; pp. 40–47.

118. Guiterman, C.H.; Margolis, E.Q.; Allen, C.D.; Falk, D.A.; Swetnam, T.W. Long-term persistence and fire resilience of oak shrubfields in dry conifer forests of northern New Mexico. *Ecosystems* **2017**, 1–17. [CrossRef]

119. Allen, C.D.; Breshears, D.D. Drought-induced shift of a forest–woodland ecotone: Rapid landscape response to climate variation. *Proc. Natl. Acad. Sci. USA* **1998**, *95*, 14839–14842. [CrossRef]

120. Sanderson, C. Ecotone Conditions Along Piñon-Juniper and Ponderosa Pine Elevational Ranges, Jemez Mountains, NM. Master's Thesis, University of New Mexico, Albuquerque, NM, USA, 2015.

121. Hooper, D.U.; Chapin, F.S.; Ewel, J.J.; Hector, A.; Inchausti, P.; Lavorel, S.; Lawton, J.H.; Lodge, D.M.; Loreau, M.; Naeem, S.; et al. Effects of biodiversity on ecosystem functioning: A consensus of current knowledge. *Ecol. Monogr.* **2005**, *75*, 3–35. [CrossRef]

122. Díaz, S.; Cabido, M. Vive la difference: Plant functional diversity matters to ecosystem processes. *Trends Ecol. Evol.* **2001**, *16*, 646–655. [CrossRef]

123. McGlone, C.M.; Sieg, C.H.; Kolb, T.E. Invasion resistance and persistence: Established plants win, even with disturbance and high propagule pressure. *Biol. Invasions* **2011**, *13*, 291–304. [CrossRef]

124. Breshears, D.D.; Cobb, N.S.; Rich, P.M.; Price, K.P.; Allen, C.D.; Balice, R.G.; Romme, W.H.; Kastens, J.H.; Floyd, M.L.; Belnap, J.; et al. Regional vegetation die-off in response to global-change-type drought. *Proc. Natl. Acad. Sci. USA* **2005**, *102*, 15144–15148. [CrossRef]

125. Floyd, M.L.; Clifford, M.; Cobb, N.S.; Hanna, D.; Delph, R.; Ford, P.; Turner, D. Relationship of stand characteristics to drought-induced mortality in three southwestern piñon–juniper woodlands. *Ecol. Appl.* **2009**, *19*, 1223–1230. [CrossRef] [PubMed]

126. Stephens, S.L.; Moghaddas, J.J.; Edminster, C.; Fiedler, C.E.; Haase, S.; Harrington, M.; Keeley, J.E.; Knapp, E.E.; McIver, J.D.; Metlen, K.; et al. Fire treatment effects on vegetation structure, fuels, and potential fire severity in western U.S. Forests. *Ecol. Appl.* **2009**, *19*, 305–320. [CrossRef] [PubMed]

127. Fulé, P.Z.; Crouse, J.E.; Roccaforte, J.P.; Kalies, E.L. Do thinning and/or burning treatments in western USA ponderosa or Jeffrey pine-dominated forests help restore natural fire behavior? *For. Ecol. Manag.* **2012**, *269*, 68–81. [CrossRef]

128. Duveneck, M.J.; Scheller, R.M. Climate suitable planting as a strategy for maintaining forest productivity and functional diversity. *Ecol. Appl.* **2015**, *25*, 1653–1668. [CrossRef]

129. Williams, M.I.; Dumroese, R.K. Preparing for climate change: Forestry and assisted migration. *J. For.* **2013**, *111*, 287–297. [CrossRef]

130. Keane, R.E.; Cary, G.J.; Davies, I.D.; Flannigan, M.D.; Gardner, R.H.; Lavorel, S.; Lenihan, J.M.; Li, C.; Rupp, T.S. A classification of landscape fire succession models: Spatial simulations of fire and vegetation dynamics. *Ecol. Modell.* **2004**, *179*, 3–27. [CrossRef]

131. Ager, A.; Barros, A.; Preisler, H.; Day, M.; Spies, T.; Bailey, J.; Bolte, J. Effects of accelerated wildfire on future fire regimes and implications for the United States federal fire policy. *Ecol. Soc.* **2017**, *22*. [CrossRef]

132. McKenzie, D.; Gedalof, Z.E.; Peterson, D.L.; Mote, P. Climatic change, wildfire, and conservation. *Conserv. Biol.* **2004**, *18*, 890–902. [CrossRef]

133. Dominguez, F.; Cañon, J.; Valdes, J. Ipcc-ar4 climate simulations for the southwestern us: The importance of future enso projections. *Clim. Chang.* **2010**, *99*, 499–514. [CrossRef]

134. Geil, K.L.; Serra, Y.L.; Zeng, X. Assessment of CMIP5 model simulations of the North American monsoon system. *J. Clim.* **2013**, *26*, 8787–8801. [CrossRef]

135. Jardine, A.; Garfin, G.; Merideth, R.; Black, M.; LeRoy, S. *Assessment of climate change in the southwest united states: A report prepared for the national climate assessment;* Island Press: Washington DC, USA, 2013.

136. Bentz, B.J.; Régnière, J.; Fettig, C.J.; Hansen, E.M.; Hayes, J.L.; Hicke, J.A.; Kelsey, R.G.; Negrón, J.F.; Seybold, S.J. Climate change and bark beetles of the western United States and Canada: Direct and indirect effects. *BioScience* **2010**, *60*, 602–613. [CrossRef]

137. Creeden, E.P.; Hicke, J.A.; Buotte, P.C. Climate, weather, and recent mountain pine beetle outbreaks in the western United States. *For. Ecol. Manag.* **2014**, *312*, 239–251. [CrossRef]

138. Stephenson, N.L. Making the transition to the third era of natural resources management. *GWS J. Parks Prot. Areas Cult. Sites* **2014**, *31*, 227–235.

139. Seastedt, T.R.; Hobbs, R.J.; Suding, K.N. Management of novel ecosystems: Are novel approaches required? *Front. Ecol. Environ.* **2008**, *6*, 547–553. [CrossRef]

forests

Article

Exploring the Future of Fuel Loads in Tasmania, Australia: Shifts in Vegetation in Response to Changing Fire Weather, Productivity, and Fire Frequency [†]

Rebecca Mary Bernadette Harris [1,*], Tomas Remenyi [1], Paul Fox-Hughes [2], Peter Love [1] and Nathaniel L. Bindoff [1,3,4,5]

[1] Antarctic Climate and Ecosystems Cooperative Research Centre (ACE CRC), University of Tasmania, Hobart 7001, Australia; tom.remenyi@utas.edu.au (T.R.); p.t.love@utas.edu.au (P.L.); N.Bindoff@utas.edu.au (N.L.B.)
[2] Bureau of Meteorology, Hobart 7001, Australia; paul.fox-hughes@bom.gov.au
[3] Institute for Marine and Antarctic Studies (IMAS), University of Tasmania, Hobart 7001, Australia
[4] ARC Centre of Excellence for Climate Systems Science, University of Tasmania, Hobart 7001, Australia
[5] Centre for Australian Weather and Climate Research (CAWCR), Commonwealth Scientific and Industrial Research Organisation (CSIRO) Marine and Atmospheric Research, Hobart 7001, Australia
* Correspondence: rmharris@utas.edu.au; Tel.: +61-3-6226-2920
† This paper is an extended version of our paper published in Proceedings of The Modelling and Simulation Society of Australia and New Zealand (MSSANZ) conference 12 2017, pp. 1097–1103. ISBN: 978-0-9872143-7-9. http://www.mssanz.org.au/modsim2017/H10/harris.pdf.

Received: 23 February 2018; Accepted: 9 April 2018; Published: 16 April 2018

Abstract: Changes to the frequency of fire due to management decisions and climate change have the potential to affect the flammability of vegetation, with long-term effects on the vegetation structure and composition. Frequent fire in some vegetation types can lead to transformational change beyond which the vegetation type is radically altered. Such feedbacks limit our ability to project fuel loads under future climatic conditions or to consider the ecological tradeoffs associated with management burns. We present a "pathway modelling" approach to consider multiple transitional pathways that may occur under different fire frequencies. The model combines spatial layers representing current and future fire danger, biomass, flammability, and sensitivity to fire to assess potential future fire activity. The layers are derived from a dynamically downscaled regional climate model, attributes from a regional vegetation map, and information about fuel characteristics. Fire frequency is demonstrated to be an important factor influencing flammability and availability to burn and therefore an important determinant of future fire activity. Regional shifts in vegetation type occur in response to frequent fire, as the rate of change differs across vegetation type. Fire-sensitive vegetation types move towards drier, more fire-adapted vegetation quickly, as they may be irreversibly impacted by even a single fire, and require very long recovery times. Understanding the interaction between climate change and fire is important to identify appropriate management regimes to sustain fire-sensitive communities and maintain the distribution of broad vegetation types across the landscape.

Keywords: climate change; prescribed burning; vegetation change; climate adaptation

---

1. Introduction

One of the key determinants of fire activity is the available fuel load. The fuel load is influenced at the landscape scale by community structure and composition (e.g., grassland vs. forest), and at more local scales by fuel age, structure, and composition; rates of decomposition, and vegetation growth rates. Attempts to project future fire danger must therefore account for changes in vegetation growth and fuel dynamics under future climatic conditions. The challenges associated with quantifying these processes have been identified as a significant gap that limits our ability to project future fire danger [1].

Projecting future fire danger is further complicated by the interactions and feedbacks that exist between the fire regime, vegetation, climate, and human intervention [1]. The frequency of fire is an important aspect of the fire regime and is projected to increase under the warmer and drier climate conditions expected with ongoing climate change. In addition, prescribed burning (also referred to as management burning) is a major component of the fire regime in many parts of the world. Prescribed burning regimes are likely to change in the future, for several reasons. In the warmer and drier conditions that are projected under ongoing climate change, there may be reduced opportunities available for prescribed burning, as fire danger increases and the fire season starts earlier in the year [2–6]. At the same time, community attitudes may shift to demand more extensive and/or frequent prescribed burning to protect lives and property following destructive wildfires. Alternatively, support for prescribed burning may decline due to concerns about the health effects of smoke [7,8].

Changes to the frequency of fire due to such management decisions and climate change have the potential to affect the flammability of the vegetation, with long-term effects on the vegetation structure and composition. Frequent fire in some vegetation types can lead to transformational change when a threshold is crossed, beyond which the vegetation type is radically altered, and this is not always a gradual process. For example, in forests dominated by obligate seeders, increased frequency of intense fire can cause a state change from woodland to grassland [9]. In Tasmania, Australia, changes to anthropogenic burning have caused rainforest to shift to moorland and vice versa [10–12]. An increase in the frequency of prescribed burning may also increase flammability in some vegetation types [13,14]. In subalpine and alpine forests of southeastern Australia, for example, Zylstra [15] demonstrated that frequent burning (up to a 14-year cycle) led to changes in forest structure that more than doubled the average size of fires, which spread faster and were more difficult to suppress.

It is therefore of interest to explore future potential fire activity under different scenarios of fire frequency. While there are major impediments to projecting fuel loads under future climatic conditions, it is possible to project several important factors determining fire activity into the future. Values for future climate conditions, including fire weather, Soil Dryness Index, and productivity can be calculated from projections of future climate, available at increasingly fine resolution. For other ecological factors, the general trends expected under climate change, such as growth rates and time to maturity, can be estimated. This enables potential pathways of change to be identified, starting with the current flammability and sensitivity to fire of broad vegetation types.

We present an approach to identify the main drivers of change to potential fire activity under future climate change and explore potential pathways of change to broad vegetation types affecting flammability across the landscape. We illustrate the approach using data from Tasmania, Australia, but the method could be applied anywhere in the world. We use a "pathway modelling" approach to consider multiple transitional pathways that may occur under different fire frequencies. We do not include changes to the distribution of vegetation in response to changing climate suitability because we expect that such change will occur slowly over long timeframes for the main forest types in Tasmania. Since the dominant forest species that make up the bulk of the fuel load are long-lived and adapted to a broad range of climate conditions (as shown by their current broad geographical distributions), they are likely to persist, even if stressed, for much of the 100-year timeframe covered by the model.

While the model involves a considerable simplification of the real world of vegetation and fire at the landscape scale, the approach enables a range of plausible futures to be explored and provides a framework for considering the vegetation responses and feedbacks that may occur between fuel loads and fire weather in the future. It is not intended as a predictive model of vegetation flammability or spread under future conditions. Rather, it is a tool to explore the range of plausible futures arising from changing fire weather over time in combination with changes to the fire regime due to management decisions.

This article expands on research presented at the International Congress on Modelling and Simulation (MODSIM), December 2017, Hobart, Australia, and published in The Modelling and Simulation Society of Australia and New Zealand (MSSANZ) conference proceedings. A more detailed methodology is presented here, to enable the tool to be developed and applied elsewhere. Ways in which the tool could be applied to communicate and illustrate the potential consequences of changes of fire frequency and climate change are also more fully developed.

2. Materials and Methods

2.1. Representing the Response to Fire of Tasmanian Vegetation Communities

A regional vegetation map, TASVEG version 3.0 [16], was used to provide information about the existing composition, structure, flammability, and fire sensitivity of broad vegetation groups across the region. This provided the baseline for the potential response of the vegetation to changing fire weather, productivity (biomass), and fire frequency. TASVEG 3.0 provides a map of the Tasmanian vegetation at a resolution of 1:25,000, comprising 158 mapping units, most of which represent distinct vegetation communities. Associated with each mapping unit is detailed information about the composition, structure, and floristics of the unit [17], from which flammability and fire sensitivity categories have been derived based on the attributes of the common plant species [18]. We focus on the dominant plants because they are often the "fuel species" [19] that provide most of the biomass and determine the structure of the vegetation community. Changes to the distribution, abundance, or dominance of fuel species under altered fire regimes have the potential to set up positive or negative feedbacks.

The response of a community to fire is related to the flammability and sensitivity of the present vegetation type to fire. The fire attributes categories (24 categories) are groups of communities in the TASVEG vegetation map that have similar fire sensitivity and flammability characteristics. There are five fire sensitivity categories (low, moderate, high, very high, and extreme) which reflect the potential ecological impact of a single fire on a stand of vegetation (Table 1). Sensitivity to fire will determine the response of the vegetation to fire or, alternatively, its resilience to frequent burning. Sensitivity is influenced by the reproductive strategy of the dominant species (e.g., obligate seeders vs. resprouters, time to maturity) [20], which has been widely used to represent response to changing fire intervals, e.g., [21]. The four flammability categories (low, moderate, high, and very high) are based on knowledge of the dynamics of fuel dryness for each vegetation type, which affects how many days per year the vegetation type will burn (Table 1). The distribution of vegetation across Tasmania belonging to the flammability categories in TASVEG 3.0 is shown in Supplementary Material Figure S2.1.

We use a broad functional type approach to understand the effect of altered fire regimes across the landscape because it is independent of taxonomic identity and therefore focusses on process [22]. Different species assemblages are likely to have very similar fuel properties because fuel is strongly influenced by vegetation structure and spatial distribution [23].

Table 1. Fire sensitivity and flammability categories from Pyrke and Marsden-Smedley (2005) [18].

Fire Sensitivity Categories	
Extreme	Any fire will cause irreversible or very-long-term (>500 years) damage
Very high	A single fire will cause significant change to community structure for 50–100 years and will increase the probability of subsequent fires
High	At least 30 years between fires is required to maintain the defining species. Fire intervals greater than 80 years are required to reach mature stand structure

Table 1. *Cont.*

Moderate	At least 15 years between fires is required to maintain the defining species
Low	A single fire will generally not affect the vegetation, but repeated short intervals (i.e., <10 years) may cause long-term changes
Flammability Categories	
Very high	Will burn readily throughout the year even under mild weather conditions, except after recent rain (i.e., less than 2–7 days ago)
High	Will burn readily when fuels are dry enough (from late spring to early autumn) but will be too moist to burn for lengthy periods, particularly in winter
Moderate	Will only burn after extended periods without rain (i.e., 2 weeks or more), and in moderate or stronger wind conditions
Low	Will burn only after extended drought (i.e., 4 weeks or more without rain) and/or under severe fire weather conditions (i.e., Forest Fire Danger Index > 40)

2.2. Vegetation Pathways through Time

The model starts with broad vegetation type to determine the general transition pathway, but the rate at which change occurs is based on the attributes of the underlying mapping units. Transitional gradients, from wet forest types through to dry forests, woodlands, and grasslands, are followed, dependent on the fire frequency and the changing fire weather over time.

Eucalyptus forests, Non-eucalyptus forests (e.g., *Melaleuca* L., *Leptospermum* J.R. Forst. & G. Forst., *Acacia* Mill)., and Rainforests (e.g., *Athrotaxis* D. Don., *Nothofagus* Blume) follow different pathways, represented by a gradient of moisture and fire frequency. Figure 1 shows the general pathways followed in the model. The detailed steps for all vegetation types are shown in Supplementary Material Table S2.2. Subalpine and Alpine types are treated separately to reflect their higher sensitivities to fire. The pathways can be reversed under fire suppression scenarios except where site factors determine the present vegetation type. For example, grassland can move towards forest if fire is suppressed, and non-eucalypt wet forest may become drier in the future and with increased fire frequency. However, dry non-eucalypt forests cannot become wet forests because the current composition reflects the moisture of the site (e.g., *Allocasuarina* L.A.S. Johnson occurs on dry sites, *Acacia* on wet sites).

Different understorey types within the broad vegetation types reflect the fertility of the site, moisture, and fire history. We assigned broad understorey types to enable this to be incorporated into the transition pathway and influence the rate of change. We started all communities with the understorey it would have if it had been unburnt for long periods. We did not attempt to recreate the state of the vegetation in its current state, although this could be incorporated with further model development.

Fire sensitivity was changed at each time step to reflect any changes to vegetation type, based on the assumption that the vegetation community will shift in the direction of lower fire sensitivity (i.e., more fire adapted) if the fire interval is shorter than the interval that the original community requires to maintain the defining species. For example, a vegetation type in the Extreme category moves one step to the Very High category if the fire interval is less than 500 years, because any fire will cause either irreversible or very-long-term (>500 years) damage [18]. A fire-adapted community with Moderate fire sensitivity will move one step to the Low category if the fire interval is less than 15 years, and remain at Moderate if the fire interval is greater than 15 years, because vegetation communities in this category require at least 15 years between fires to maintain the defining species. Conversely, a grassland with Low fire sensitivity can move in the other direction if fire is excluded for more than 100 years, as the community shifts towards a more mesic vegetation type. Flammability was also updated at each time step to reflect any changes to vegetation type.

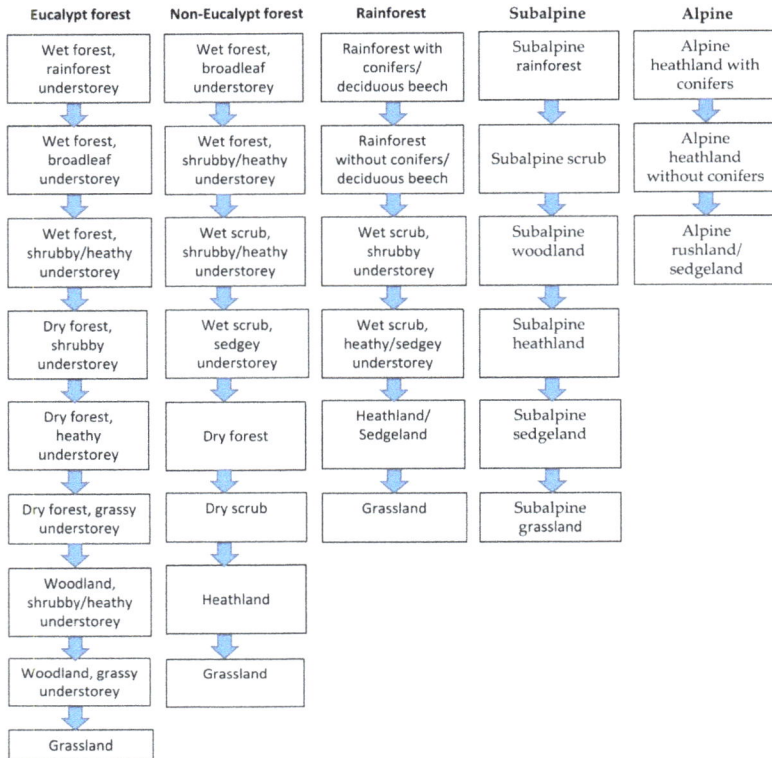

Eucalypt forest	Non-Eucalypt forest	Rainforest	Subalpine	Alpine
Wet forest, rainforest understorey	Wet forest, broadleaf understorey	Rainforest with conifers/ deciduous beech	Subalpine rainforest	Alpine heathland with conifers
Wet forest, broadleaf understorey	Wet forest, shrubby/heathy understorey	Rainforest without conifers/ deciduous beech	Subalpine scrub	Alpine heathland without conifers
Wet forest, shrubby/heathy understorey	Wet scrub, shrubby/heathy understorey	Wet scrub, shrubby understorey	Subalpine woodland	Alpine rushland/ sedgeland
Dry forest, shrubby understorey	Wet scrub, sedgey understorey	Wet scrub, heathy/sedgey understorey	Subalpine heathland	
Dry forest, heathy understorey	Dry forest	Heathland/ Sedgeland	Subalpine sedgeland	
Dry forest, grassy understorey	Dry scrub	Grassland	Subalpine grassland	
Woodland, shrubby/heathy understorey	Heathland			
Woodland, grassy understorey	Grassland			
Grassland				

Figure 1. The vegetation pathways followed in the model. Detailed steps with the associated TASVEG 3.0 vegetation codes representing the original vegetation types used to develop the pathway are shown in Supplementary Material Table S2.2.

2.3. Modelling Potential Fire Activity

The vegetation pathway model is based on the four-switch model [24], which describes fire activity in terms of four factors that must be fulfilled simultaneously (switched "on") for fire to occur. There must be fuel available (biomass); it must be dry enough to burn (availability to burn); weather conditions must be conducive to fire spread (fire weather); and there must be an ignition source (ignition).

Modelling "Potential Fire Activity" (PFA), the level of fire activity possible if an ignition source were present, is a two-step process. First, the broad vegetation type is determined for each cell for a particular time and fire interval. Then, the PFA is calculated at each grid cell (10 km) across Tasmania, Australia, using the appropriate attributes for that type, following the equation

$$\text{Potential Fire Activity (PFA)} = \text{Biomass} + \text{Availability to Burn} + \text{Fire Weather} \qquad (1)$$

where Biomass = (productivity $\times$ fuel load at time since fire); Availability to Burn = flammability of vegetation type at current Soil Dryness Index (SDI) $\times$ slope factor; and Fire Weather = Fire Danger Index (FFDI or MFDI, depending on vegetation type). Each term is described in detail below.

Potential Fire Activity was calculated for seven time periods (1961–1980, 1981–2000, 2001–2020, 2021–2040, 2041–2060, 2061–2080, 2081–2100) under a range of fire frequency scenarios. Mean values for each term were calculated for current and future time periods using a combination of climate layers (Productivity, SDI, Forest Fire Danger and Moorland Forest Danger Indices (FFDI and MFDI), attributes from TASVEG 3.0 [16], and information on fuel characteristics from the scientific literature. Climate indices used in the equation reflect the appropriate time period, so if the model is 50 years from 2000, then the climate layer at that time step is 2040–2060. The layers used to calculate each term and their relationship to each other are summarized in Figure 2.

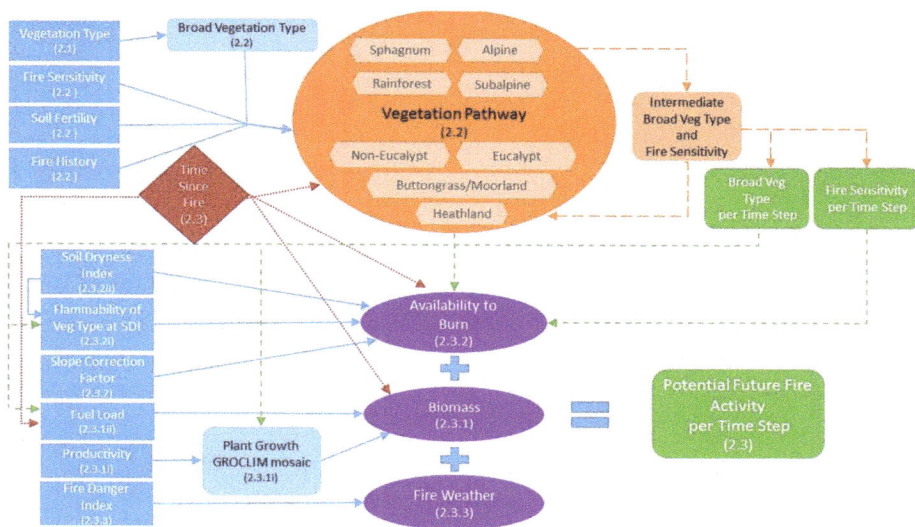

Figure 2. The components of the vegetation pathway model. Blue boxes are inputs. Light-blue boxes are derived products. Orange components represent the different vegetation pathways followed over time. Purple boxes represent the "switches" calculated. Green boxes are outputs reflecting changes to vegetation over time. 'Broad Vegetation Type per timestep' is used to define the vegetation conditions and estimate the Potential Fire Activity at each timestep. Numbers in brackets refer to the section within the text within which the term is described in detail.

All climate-driven layers were calculated from the output of a dynamically downscaled regional climate model (The Commonwealth Scientific and Industrial Research Organisation's Conformal Cubic Atmospheric Model (CCAM)), at a resolution of ~10 km. Downscaling of six global climate models from the Climate Model Intercomparison Project 3 (CMIP3) was carried out by the Climate Futures for Tasmania (CFT) project. Details of the climate modelling can be found in Corney et al. [25]. All projections and layers are available through the Tasmanian Partnership for Advanced Computing (TPAC) portal (http://portal.sf.utas.edu.au/thredds/catalog.html).

2.3.1. Biomass

There are two components to the Biomass term: (i) productivity and (ii) the fuel load.

(i)   Productivity

The GROCLIM submodel from the ANUCLIM model [26] was used to generate an index of relative potential plant growth, based on plant growth response to light, temperature, and water regimes under current and future climate conditions. The growth index does not represent actual biomass production, but describes plant production potential across the landscape.

Annual mean growth indices were computed for three thermal types using a parabolic thermal response curve, which is most appropriate for the C3 photosynthetic pathway, and the following optimum temperature and thermal ranges. The C3-Mesotherm plant type has a relatively broad range of growing temperatures (3–36 °C) with an optimum temperature of 19 °C, and is most applicable to temperate species. The C3-Microtherm plant type has a range of growing temperatures from 0 to 20 °C with an optimum temperature of 10 °C, so is most applicable to conifers and cool to cold temperate climate plants. An additional GROCLIM index was customized to represent forest growth based on the known thermal requirements of a *Eucalyptus* L'Hér. species (*E. globulus*: Minimum temperature (T_{min}) 8 °C, Maximum temperature (T_{max}) 40 °C, Optimum temperature (T_{opt}) 16 °C) [27]. The growth index is a dimensionless index with a scale of zero to one, where plant growth is minimal or nonexistent below a growth index value of 0.2. This was scaled (by multiplying by 1.6), so that productivity could increase or decrease under future conditions of temperature and rainfall.

A composite GROCLIM layer was calculated for each time period, with the appropriate GROCLIM thermal types applied to each broad vegetation type, as follows. The C3-Microtherm index was used for all areas with alpine and subalpine vegetation types; Buttongrass because it generally occurs at altitudes greater than 600 m in colder regions; and Rainforest with conifers or deciduous beech, because of the presence of conifers. The index based on the *Eucalyptus* thermal type was applied to all areas with Eucalypt forest types, and the C3-Mesotherm index was applied to all other regions to incorporate the broad thermal range of temperate plants in general.

(ii)   Fuel Load

The fuel load (tonnes per hectare) was calculated for each broad vegetation type at each time step, using Olson's model of fuel accumulation [28]:

$$\text{Biomass (of fuel)} = L \times (1 - \exp(-k \times A)) \tag{2}$$

where L represents the carrying capacity (or maximum fuel load), k is the growth rate (or decomposition rate) and A is age (or time since fire).

Values for carrying capacity and growth rate in Tasmanian vegetation types were decided upon after consultation with the literature and fire ecologists (Jon Marsden-Smedley, Dave Taylor, personal communication), and resulted in the accumulation curves shown in Supplementary Material Figure S2.3. The value for the TASVEG type that made up the greatest area of each Broad Vegetation Type was used.

2.3.2. Availability to Burn

This term of the Potential Fire Activity equation incorporates (i) the Flammability of the vegetation type and (ii) a measure of fuel dryness, the Soil Dryness Index (SDI).

(i)   Flammability

The four flammability categories (low, moderate, high, and very high) are based on knowledge of the dynamics of fuel dryness for each vegetation type, which affects how many days per year the vegetation type will burn. The categories are defined in Table 1. As with fire sensitivity, the flammability category was changed at each time step to reflect any changes to vegetation type.

(ii)   Soil Dryness Index (SDI)

The Soil Dryness Index (SDI) [29] is a measure of soil moisture which is used as an index of fuel moisture and relative flammability of different vegetation types (values from [30]) (Table S2.4, Supplementary Material). An overview of the SDI and its strengths and weaknesses can be found in Marsden-Smedley (2009) [30]. The mean SDI from the CFT projections was calculated at each time period, and determined the flammability of each vegetation type at that time.

In addition, a slope correction factor was applied to each pixel to capture the effect of slope on fuel preheating and wind speed. Slopes of >30% were weighted by 10; slopes of 21–30% by 5; slopes of 16–20% by 3; and slopes of 0–10% were weighted by 1. This follows the BRAM–Bushfire Risk Assessment Model (Parks and Wildlife Service) which is used operationally in Tasmania.

2.3.3. Fire Weather (FFDI, MFDI)

Two different fire danger indices were used to indicate fire weather at each time period. Both indices incorporate surface air temperature, relative humidity, and wind speed, combined with an estimate of fuel dryness (Drought Factor, based on Soil Dryness Index and recent precipitation) to give an index of daily fire danger. The Moorland Fire Danger Index (MFDI) [31] was used for areas with Buttongrass Moorland, Sphagnum, and Sedgeland vegetation. This index is better suited to moorlands and other types where soil dryness has less of an influence on fire behavior. The annual cumulative MacArthur's Forest Fire Danger Index (FFDI) [32] was applied for all other vegetation types.

2.4. Fire Frequency

We explored the effect of different fire frequencies on the potential fire activity and flammability of vegetation across Tasmania. The climate-driven layers (productivity, SDI, and fire weather) were updated to reflect the changing climate over time, and the vegetation type shifted along the appropriate pathway when the fire frequency was above the threshold for each type. Values for the time between fires, or fire interval, required for recovery were based on available literature (e.g., Table 2).

Table 2. Fire sensitivity categories used in the vegetation model [18].

	Frequency of Fire above Which Community Does Not Survive (No. of Fires per 100 Years)	Time between Fires below Which Community Does Not Survive (Stand Replacement)	Time between Fires below Which Community Will Change Gradually
Extreme (E)	1	300	500
Very High (VH)	1	50	100
High (H)	3	30	80
Moderate (M)	6	-	15
Low (L)	-	2–5	10

3. Results

3.1. Impact of Fire Frequency on Vegetation Type

Very frequent fire, with only 4 years between fires, results in a shift towards drier vegetation types across the state (Figure 3). Regions with fire-sensitive vegetation are highlighted, as the vegetation shifts quickly at high fire frequency. For example, the wet sclerophyll forests with rainforest or broadleaved understoreys in southern Tasmania (shown in orange) quickly move towards dry forest types. In contrast, at very long fire intervals (or low fire frequency), which would occur if fire were actively suppressed, some vegetation types could potentially transition towards different vegetation types (Figure 4). For instance, if fire were suppressed in native grasslands, there would be a shift towards woodland vegetation as trees establish in the absence of frequent fire. Buttongrass moorland transitions to a woody vegetation type (mauve to pink) if fire is suppressed and the fire interval is longer than 30 years.

Such changes have the potential to affect the statewide distribution of structural types (Figure 5). If fires were to occur every two years for a period of 15 years, only grasslands and dry forests would remain, and many areas, such as alpine areas and sphagnum, would become bare ground (a category used to indicate when the limits of adaptability have been exceeded and no vegetation is able to establish). As the fire interval increases (e.g., to 7 years), there is less of an impact on the fire-adapted vegetation types such as grasslands and woodlands, but there is still an increase in their area as the more mesic vegetation types transition towards grassland and woodland. The area of forest appears stable at these fire frequencies, but there is a shift towards dry forest, away from wet eucalypt and non-eucalypt forests. The current area of woodlands can be sustained into the future at fire intervals above 16 years. At longer intervals, the area increases, as grasslands transition into woodlands when fire is suppressed. The dry eucalypt forests and woodland types in which prescribed burning is currently carried out are sustained at a 10-year fire interval (Figure 6). The transitions are seen as a series of steps in the output, reflecting the threshold values used in the model.

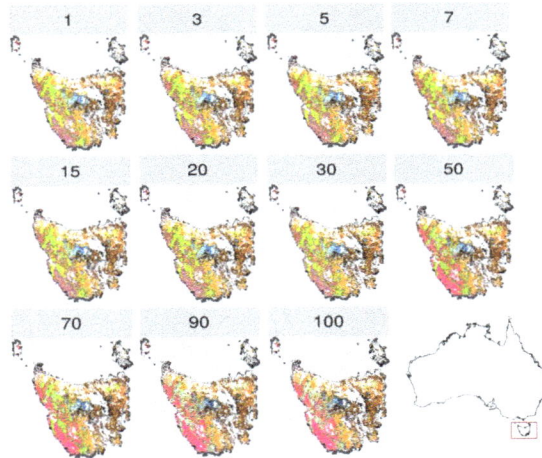

Figure 3. Impact of frequent fire (every 4 years) on vegetation type across Tasmania, incorporating annual layers from the Climate Futures for Tasmania projections. The numbers above each map refer to the number of years from 2000. Colors represent different vegetation types, as follows: Light blues, Subalpine vegetation; Dark blue, Alpine vegetation; Purple, Buttongrass; Oranges, Wet sclerophyll; Dark Orange to Brown, Dry sclerophyll; Greys, Woodland; Reds, Non-eucalypt wet forests; Dark Purple, Non-eucalypt dry forests, Light to Dark Greens, Rainforest.

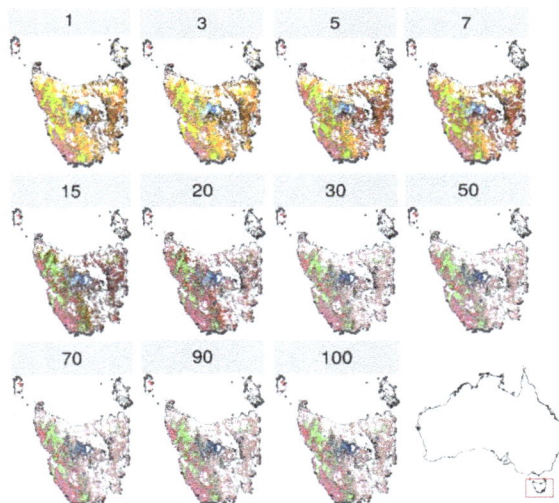

Figure 4. Impact of infrequent fire (every 35 years) on vegetation type across Tasmania. The numbers above each map refer to the number of years from 2000. Colors represent different vegetation types, as follows: Light blues, Subalpine vegetation; Dark blue, Alpine vegetation; Purple, Buttongrass; Oranges, Wet sclerophyll; Dark Orange to Brown, Dry sclerophyll; Greys, Woodland; Reds, Non-eucalypt wet forests; Dark Purple, Non-eucalypt dry forests, Light to Dark Greens, Rainforest.

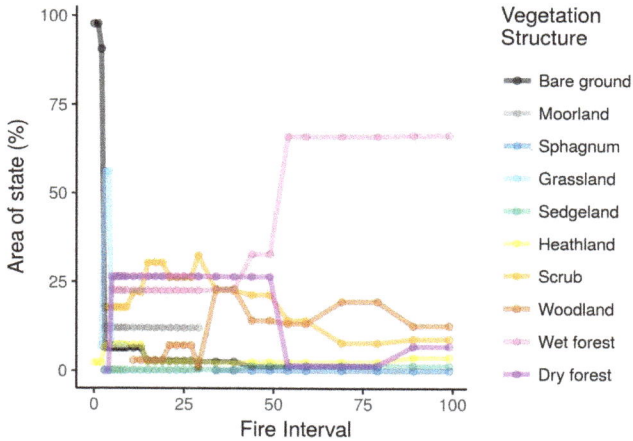

Figure 5. The change in the area of Tasmania covered by broad vegetation structural types after 100 years of burns at a range of fire intervals. The area at fire interval 0 corresponds to the distribution of vegetation types in the year 2000.

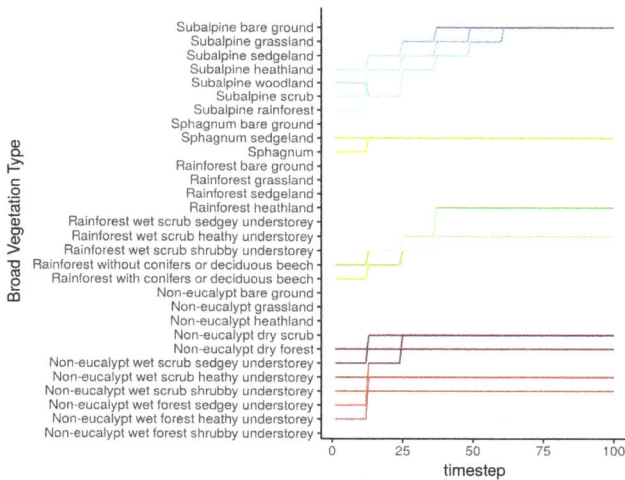

Figure 6. The impact of a ten-year fire interval on vegetation types across Tasmania over a period of 100 years, beginning in 2000. Transitions were constrained along vegetation pathways, indicated by different colors, with each type having different tolerances to fire frequency.

3.2. Impact of Fire Frequency on Future Potential Fire Activity

Fire frequency has a greater impact on future Potential Fire Activity (PFA) than climate change over the coming decades. With very high fire frequency (fire interval of 1–2 years), the future Potential Fire Activity is very low because all vegetation is pushed towards the bare ground state in the model over time (Figure 7). Beyond three-yearly intervals, the more frequent the fire, the lower the distribution and the peak of the state-wide PFA. The highest PFA values are all fire intervals greater than 30 years, reflecting the contribution of fuel accumulation and carrying capacity to fire activity.

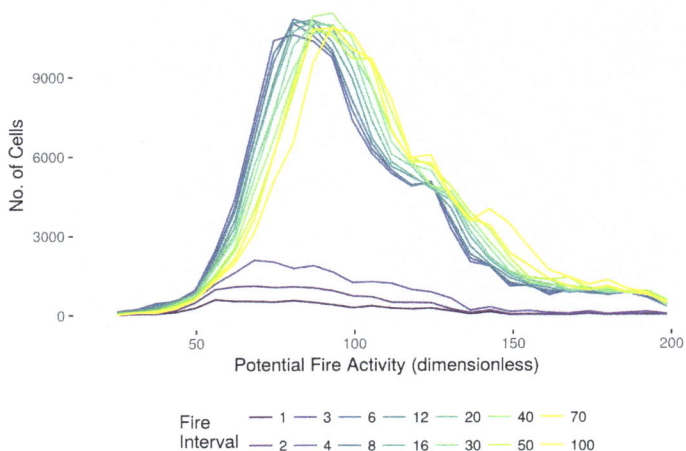

Figure 7. Impact of fire interval on future Potential Fire Activity across Tasmania.

3.3. Impact of Fire Frequency on Treatability

Vegetation that requires fire management to fulfil operational requirements is referred to as "treatable". At fire intervals of between 15 and 50 years, there is little change in the percentage treatability across Tasmania from the baseline to the end of the century (Figure 8). Intervals of less than 5 years maintain the highest proportion of vegetation requiring fuel management. The drop in treatability at very high frequencies reflects the shift to the bare ground state of fire-sensitive vegetation types over time. At very long fire intervals, the percentage of treatable vegetation drops over time because in the absence of fire, vegetation transitions to wetter forest types.

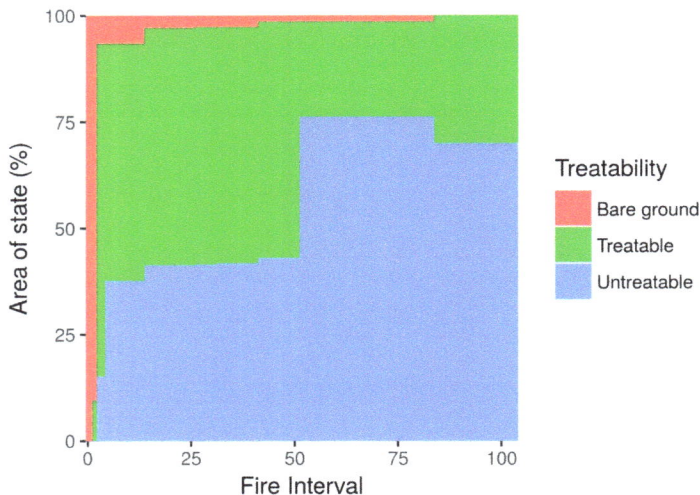

Figure 8. The impact of fire interval on the area of the state requiring fuel management.

4. Discussion

Future fire danger is projected to increase substantially under ongoing climate change [2,33]. More frequent bushfires can therefore be expected, leading to a greater need for prescribed burning to reduce bushfire risks. However, trade-offs will occur between fuel reduction and vegetation transitions in response to more frequent fire. The vegetation pathway model is a tool to illustrate the potential impacts of a dryer and warmer future climate in combination with management decisions about the frequency of prescribed burning. Within the model, ecological theory is translated into visualizations and summaries of potential landscape-scale change, to consider the impact of fire frequency on vegetation type, potential future fire activity, and the consequences of such changes for the proportion of a region that will require fuel management.

Currently, prescribed burning is applied in Tasmania for fuel reduction, ecological management, and weed control purposes [30]. Each of these objectives requires different intensities and frequencies of burning, which also vary in different vegetation types. In asset protection zones, fires of sufficient intensity are required to reduce the fuel load while ensuring that safety standards are not compromised and fires can be contained. Except in cases where the asset is a fire-sensitive species or community, there is a trade-off in these zones between fire risk reduction and ecological impacts. Broad-scale fuel management is then applied in strategic management zones to increase the potential to suppress bushfires and reduce wildfire size, whilst aiming to minimize adverse impacts on other values. In ecological management zones, fires may be suppressed, or prescribed burning applied at a range of intensities and frequencies appropriate for the target species or community, so that a mosaic of burnt and unburnt areas is maintained. The aim in these zones is that no vegetation transition should occur as a result of fire management. The Tasmanian operational guidelines for asset protection zones recommend fire frequencies of between 4 and 10 years in dry forests and scrub, and the exclusion of prescribed burning from the wet forests, alpine areas, and other fire-sensitive vegetation communities [18]. The results presented here therefore do not represent the current management approach to prescribed burning. Instead, we illustrate the potential impact of a range of scenarios of fire frequency.

Fire frequency has a substantial impact on the future Potential Fire Activity (PFA) relative to the impact of the changing climate over the coming decades. While the climate is projected to become warmer and drier over time, leading to higher fire danger, fire frequency is the dominant driver of future fire activity because of the feedback between fire and flammability in drier, fire-adapted vegetation types. Frequent fire has the potential to lead to shifts in vegetation type, away from mesic, fire-sensitive types, towards drier, more fire-adapted vegetation. The rate of change differs across the vegetation types, with some fire-sensitive communities irreversibly impacted by even a single fire, and requiring very long recovery times. For example, rainforest communities with conifers may never recover after a fire, as *Athrotaxis* is an extremely slow growing and very long-lived tree that is killed by fire. In such communities, there is a positive feedback where fires promote vegetation that is more flammable, increasing the risk of fire. In contrast, fire-adapted vegetation such as dry eucalypt forests recover relatively quickly after fire, and are only impacted by very frequent fires.

Vegetation types such as alpine and subalpine heathland and grasslands and rainforests are considered "untreatable" and excluded from fuel management because their sensitivity to fire would result in the loss of fire-sensitive species and long-term changes to their composition. All other vegetation types are considered "treatable", requiring consideration of prescribed burning to fulfil operational requirements. The percentage of land that is treatable therefore has important implications for resource allocation and planning. Fire intervals of less than 5 years maintain the highest proportion of vegetation requiring fuel management across Tasmania. A drop in treatability over time can be achieved either by maintaining very high frequencies (less than 3 years), which result in the shift to the bare ground state of fire-sensitive vegetation types over time, or very long fire intervals (more than 30 years), because in the absence of fire, vegetation transitions to wetter forest types in the model. However, the latter is an unrealistic scenario requiring active fire suppression in the very long term. The challenge of suppressing fires even at relatively small scales is already becoming evident in Tasmania, as shown by the impact of recent fires in The Tasmanian Wilderness World Heritage Area. This area contains the core refugium of the paleo-endemic conifer *Athrotaxis cupressoides* (*A. cupressoides*) D. Don, a species restricted to cool, wet climates and fire-free environments. Following an extremely hot and dry summer in 2015/2016, a lightning storm ignited numerous fires which burnt large stands of *A. cupressoides* [34,35]. Recovery is unlikely because of the species' slow growth and limited seedling establishment and the positive feedback between fire and flammability discussed above. Fire suppression is likely to become increasingly difficult in the future as fire danger increases, the fire season becomes longer [33], and the window available for prescribed burning narrows under ongoing climate change [36].

We have presented results for the state of Tasmania, but similar assessments could be generated for any subregion and will reflect the different vegetation types within the region of interest (for example, see Supplementary Material, Figure S2.5, for results for each of the Bureau of Meteorology forecast districts). The percentage of vegetation requiring fuel treatment is likely to differ across different districts depending on the vegetation types present, as they follow different transition pathways. Further exploration of the changes within a region, or particular forest, would be useful to inform conversations about the range of possible futures under different fuel management strategies.

The pathway approach is a useful tool for assisting community adaptation, by illustrating the potential impacts of a dryer and warmer future in combination with decisions seeking to manage fire risk in the future. Change over time under different scenarios of fire frequency can be spatially represented to show the shifts in vegetation type across the landscape and, hence, flammability. Maps can be used to show the distribution of the different vegetation types across the landscape, and how this changes at different fire intervals. The regions with the most fire-sensitive vegetation types and, therefore, greatest potential for vegetation transitions can be highlighted in this way to improve understanding of the tradeoffs between conservation, flammability, and fuel management.

The model involves a considerable simplification of the real world of vegetation and fire at the landscape scale. Flammability and fire sensitivity, for instance, are categorized into four and five classes, respectively. We have based these classes on available research in Tasmania, but any number of classes could be incorporated. Recently logged wet eucalypt forest and rainforest, for example, might be better represented by an additional flammability class, because the increased exposure of the understorey to insolation and altered floristics leads to higher flammability compared to undisturbed forests. More refined categories, based on understandings of the many fuel characteristics that influence fire, could be incorporated to make the model more regionally specific.

There are several factors influencing fire activity that could be included in the model with further development. Aspect could be incorporated to consider its influence on fire intensity and frequency, through temperature and drying effects and differential fire spread. MODIS (Moderate Resolution Imaging Spectroradiometer) canopy cover class could be used to distinguish different canopy cover within the broad vegetation types and within mapping units (e.g., "forest" and "woodland" canopy structure; recently logged or cleared vegetation). However, some factors such as forest growth under elevated CO_2 are unable to be projected into the future because of lack of knowledge or complex interactions and feedbacks (summarized in [1]). Other improvements would require targeted empirical research to increase understanding of fire intensity across vegetation types; improve accumulation curves across a range of vegetation and geological types; and include variations in ignitions and fire size, informed by fire history. These developments would better reflect the mosaic of burnt and unburnt areas that is maintained by the fire agencies within the different management zones (e.g., asset protection, ecological, and strategic management zones). Further work is also necessary to incorporate changes to vegetation composition over time due to changing climate suitability. While we expect that vegetation change will occur slowly over long timeframes in the forest types because of the longevity of the dominant species, vegetation change may be more rapid in the alpine and subalpine regions, where the suitable climate is projected to constrict over the coming decades as Tasmania becomes

Forests **2018**, *9*, 210

warmer and drier. Additionally, extreme events such as heatwaves and droughts may cause sudden shifts in the vegetation in these regions, where the dominant species are less resilient to extremely high temperatures and/or low moisture conditions [35]. Similarly, the distribution of vegetation types in which structurally important species have particular climatic requirements (e.g., *Athrotaxis*) may change over time.

The transitions in the model are based on the assumption of low to moderate fire intensity such as might be applied in asset protection zones and in some areas within strategic management zones. They do not capture the impact of very high-intensity, high-severity fires that can sometimes lead to immediate change after a single fire. The occurrence of such fires is likely to increase in the future as lightning ignitions and fire danger increase. Additionally, we have assumed that vegetation responses will remain the same under future climate conditions, but this may not be the case as vegetation becomes stressed by ongoing climate changes such as droughts. For example, dry eucalypt forests, which in the past have recovered relatively quickly after fire, may become more vulnerable to transition due to the cumulative effect of drought.

5. Conclusions

Fire frequency has a large impact on future fire activity relative to the impact of the changing climate over the coming decades. Frequent fire has the potential to lead to shifts in vegetation type, away from mesic, fire-sensitive types, towards drier, more fire-adapted vegetation. This leads to a positive feedback between fire and flammability in drier, fire-adapted vegetation types. The rate of change differs across vegetation types, leading to changes in vegetation structure and flammability at the landscape scale. The pathway model consolidates current understanding in the field into an interactive framework, enabling plausible futures to be explored. It could be used as a tool in community adaptation, to frame potential futures and identify the consequences of decisions seeking to manage fire risk in the future. Change over time, under different management regimes (frequency of prescribed burning), can be spatially represented to show the shifts in vegetation types across the landscape.

Supplementary Materials: The following are available online at http://www.mdpi.com/1999-4907/9/4/210/s1. Figure S2.1: The distribution of vegetation across Tasmania belonging to the flammability categories in TASVEG 3.0, Table S2.2. The vegetation pathways followed in the model, Figure S2.3: Fuel load versus time since fire in the broad vegetation types, Table S2.3: The vegetation types in the model associated with each fuel accumulation curve; Table S2.4, Flammability at different levels of Soil Dryness Index (SDI), Figure S2.5: Impact of fire interval on Potential Future Fire Activity in the Bureau of Meteorology forecast districts.

Acknowledgments: This work was funded by the National Bushfire Mitigation—Tasmanian Grants Program (NBMP). Rebecca Harris was supported in part by a Humboldt Research Fellowship. Jon Marsden-Smedley and Dave Taylor provided guidance and data for fuel accumulation curves and vegetation attributes. Sandra Whight and Paul Black supported the concept and gave valuable insights into the operational implications of prescribed burning regimes in Tasmania. Jayne Balmer (Department of Primary Industries, Parks, Water and Environment) gave ecological advice that helped in translating the TASVEG types into the model.

Author Contributions: R.M.B.H. and T.R. conceived and designed the approach and analyses; T.R., R.M.B.H. and P.L. developed the model; R.M.B.H. and T.R. analyzed the data; all authors contributed to the concept and model development; R.M.B.H. and T.R. wrote the paper, with contributions from P.F.-H., P.L. and N.L.B.

Conflicts of Interest: The authors declare no conflict of interest. The funding sponsors had no role in the design of the study; in the collection, analyses, or interpretation of data; in the writing of the manuscript, and in the decision to publish the results.

References

1.   Harris, R.M.B.; Remenyi, T.; Williamson, G.; Bindoff, N.L.; Bowman, D. Climate–vegetation–fire interactions and feedbacks: Major barrier or trivial detail in projecting the future of the earth system? *Wiley Interdiscip. Rev. Clim. Chang.* **2016**. [CrossRef]

2. Fox-Hughes, P.; Harris, R.M.; Lee, G.; Grose, M.; Bindoff, N.L. Future fire danger climatology for Tasmania, Australia, using a dynamically downscaled regional climate model. *Int. J. Wildland Fire* **2014**, *23*, 309–321. [CrossRef]
3. Flannigan, M.; Cantin, A.S.; de Groot, W.J.; Wotton, M.; Newbery, A.; Gowman, L.M. Global wildland fire season severity in the 21st century. *For. Ecol. Manag.* **2013**, *294*, 54–61. [CrossRef]
4. Liu, Y.Q.; Stanturf, J.; Goodrick, S. Trends in global wildfire potential in a changing climate. *For. Ecol. Manag.* **2010**, *259*, 685–697. [CrossRef]
5. Westerling, A.L.; Hidalgo, H.G.; Cayan, D.R.; Swetnam, T.W. Warming and earlier spring increase western us forest wildfire activity. *Science* **2006**, *313*, 940–943. [CrossRef] [PubMed]
6. Jolly, W.M.; Cochrane, M.A.; Freeborn, P.H.; Holden, Z.A.; Brown, T.J.; Williamson, G.J.; Bowman, D. Climate-induced variations in global wildfire danger from 1979 to 2013. *Nat. Commun.* **2015**. [CrossRef] [PubMed]
7. Johnston, F.; Bowman, D. Bushfire smoke: An exemplar of coupled human and natural systems. *Geogr. Res.* **2014**, *52*, 45–54. [CrossRef]
8. Haikerwal, A.; Reisen, F.; Sim, M.R.; Abramson, M.J.; Meyer, C.P.; Johnston, F.H.; Dennekamp, M. Impact of smoke from prescribed burning: Is it a public health concern? *J. Air Waste Manag. Assoc.* **2015**, *65*, 592–598. [CrossRef] [PubMed]
9. Bowman, D.M.J.S.; Murphy, B.P.; Neyland, D.L.J.; Williamson, G.J.; Prior, L.D. Abrupt fire regime change may cause landscape-wide loss of mature obligate seeder forests. *Glob. Chang. Biol.* **2014**, *20*, 1008–1015. [CrossRef] [PubMed]
10. Di Folco, M.B.; Kirkpatrick, J.B. Organic soils provide evidence of spatial variation in human-induced vegetation change following European occupation of Tasmania. *J. Biogeogr.* **2013**, *40*, 197–205. [CrossRef]
11. Fletcher, M.S.; Thomas, I. A Holocene record of sea level, vegetation, people and fire from western Tasmania, Australia. *Holocene* **2010**, *20*, 351–361. [CrossRef]
12. Fletcher, M.S.; Thomas, I. The origin and temporal development of an ancient cultural landscape. *J. Biogeogr.* **2010**, *37*, 2183–2196. [CrossRef]
13. Fernandes, P.M.; Botelho, H.S. A review of prescribed burning effectiveness in fire hazard reduction. *Int. J. Wildland Fire* **2003**, *12*, 117–128. [CrossRef]
14. Lindenmayer, D.B.; Hobbs, R.J.; Likens, G.E.; Krebs, C.J.; Banks, S.C. Newly discovered landscape traps produce regime shifts in wet forests. *Proc. Natl. Acad. Sci. USA* **2011**, *108*, 15887–15891. [CrossRef] [PubMed]
15. Zylstra, P. The historical influence of fire on the flammability of subalpine Snowgum forest and woodland. *Vic. Nat.* **2013**, *130*, 232–239.
16. Tasmanian Department of Primary Industries, Parks, Water and Environment. *Tasmanian Vegetation Monitoring and Mapping Program*; TASVEG 3.0; Resource Management and Conservation Division, Department of Primary Industries, Parks, Water and Environment: Hobart, Australia, 2013.
17. Harris, S.; Kitchener, A. *From Forest to Fjaeldmark: Descriptions of Tasmania's Vegetation*; Department of Primary Industries, Parks, Water and Environment, Printing Authority of Tasmania: Hobart, Australia, 2005.
18. Pyrke, A.F.; Marsden-Smedley, J.B. Fire-attributes categories, fire sensitivity, and flammability of Tasmanian vegetation communities. *TasForests* **2005**, *16*, 35–46.
19. Gill, A.M.; Woinarski, J.C.Z.; York, A. *Australia's Biodiversity—Responses to Fire: Plants, Birds and Invertebrates*; Department of Environment and Heritage: Canberra, Australia, 1999; p. 206.
20. Noble, I.R.; Slatyer, R.O. The use of vital attributes to predict successional changes in plant communities subject to recurrent disturbances. *Vegetatio* **1980**, *43*, 5–21. [CrossRef]
21. Hammill, K.; Penman, T.; Bradstock, R. Responses of resilience traits to gradients of temperature, rainfall and fire frequency in fire-prone, Australian forests: Potential consequences of climate change. *Plant Ecol.* **2016**, *217*, 725–741. [CrossRef]
22. Cary, G.J.; Bradstock, R.A.; Gill, A.M.; Williams, R.J. Global change and fire regimes in Australia. In *Flammable Australia: Fire Regimes, Biodiversity and Ecosystems in a Changing World*; Bradstock, R.A., Gill, A.M., Williams, R.J., Eds.; CSIRO Publishing: Clayton, Australia, 2012; pp. 149–169.
23. Bradstock, R.A.; Cary, G.J.; Davies, I.; Lindenmayer, D.B.; Price, O.F.; Williams, R.J. Wildfires, fuel treatment and risk mitigation in Australian eucalypt forests: Insights from landscape-scale simulation. *J. Environ. Manag.* **2012**, *105*, 66–75. [CrossRef] [PubMed]

24. Bradstock, R.A. A biogeographic model of fire regimes in Australia: Current and future implications. *Glob. Ecol. Biogeogr.* **2010**, *19*, 145–158. [CrossRef]

25. Corney, S.P.; Katzfey, J.J.; McGregor, J.L.; Grose, M.R.; Bennett, J.C.; White, C.J.; Holz, G.K.; Gaynor, S.M.; Bindoff, N.L. *Climate Futures for Tasmania: Climate Modelling Technical Report*; Antarctic Climate & Ecosystems Cooperative Research Centre: Hobart, Australia, 2010.

26. Hutchinson, M.F. *ANUCLIM*; Version 6.1; Fenner School of Environment and Society, Australian National University: Canberra, Australia, 2011.

27. Sands, P.J.; Landsberg, J.J. Parameterisation of 3-PG for plantation grown *Eucalyptus globulus. For. Ecol. Manag.* **2002**, *163*, 273–292. [CrossRef]

28. Olson, J.S. Energy storage and balance of producers and decomposers in ecological systems. *Ecology* **1963**, *34*, 322–331. [CrossRef]

29. Mount, A.B. *The Derivation and Testing of a Soil Dryness Index Using Run-Off Data*; Bulletin 4; Forestry Commission: Hobart, Australia, 1972.

30. Marsden-Smedley, J.B. *Planned Burning in Tasmania: Operational Guidelines and Review of Current Knowledge*; Fire Management Section, Parks and Wildlife Service, Department of Primary Industries, Parks, Water and the Environment: Hobart, Australia, 2009.

31. Marsden-Smedley, J.B.; Rudman, T.; Pyrke, A.; Catchpole, W.R. Buttongrass moorland fire-behaviour prediction and management. *TasForests* **1999**, *11*, 87–99.

32. McArthur, A.G. Fire behaviour in eucalypt forests. In *Forestry and Timber Bureau Leaflet 107*; Forestry and Timber Bureau: Canberra, Australia, 1967.

33. Holz, A.; Wood, S.W.; Veblen, T.T.; Bowman, D.M.J.S. Effects of high-severity fire drove the population collapse of the subalpine Tasmanian endemic conifer *Athrotaxis cupressoides. Glob. Chang. Biol.* **2015**, *21*, 445–458. [CrossRef] [PubMed]

34. Harris, R.M.B.; Beaumont, L.J.; Vance, T.R.; Tozer, C.; Remenyi, T.A.; Perkins-Kirkpatrick, S.E.; Mitchell, P.J.; Nicotra, A.B.; McGregor, S.; Andrew, N.R.; et al. Biological responses to the press and pulse of climate trends and extreme events. *Nat. Clim. Chang.* **2018**, in press.

35. Fox-Hughes, P.; Harris, R.M.B.; Lee, G.; Jabour, J.; Grose, M.R.; Remenyi, T.A.; Bindoff, N.L. *Climate Futures for Tasmania Future Fire Danger: The Summary and the Technical Report*; Antarctic Climate & Ecosystems Cooperative Research Centre: Hobart, Australia, 2015.

36. Harris, R.M.B.; Remenyi, T.; Fox-Hughes, P.; Love, P.; Phillips, H.E.; Bindoff, N.L. An assessment of the viability of prescribed burning as a management tool under a changing climate: a Tasmanian case study. In Proceedings of the Bushfire and Natural Hazards CRC & AFAC Conference, Melbourne, Australia, 4 September 2017; Rumsewicz, M., Ed.; Bushfire and Natural Hazards CRC: Melbourne, Australia, 2017.

Article

Composition and Structure of Forest Fire Refugia: What Are the Ecosystem Legacies across Burned Landscapes?

Garrett W. Meigs *and Meg A. Krawchuk

Department of Forest Ecosystems and Society, College of Forestry, Oregon State University, 321 Richardson Hall, Corvallis, OR 97331, USA; meg.krawchuk@oregonstate.edu
* Correspondence: gmeigs@gmail.com; Tel.: +1-541-737-2244

Received: 13 April 2018; Accepted: 27 April 2018; Published: 2 May 2018

Abstract: Locations within forest fires that remain unburned or burn at low severity—known as fire refugia—are important components of contemporary burn mosaics, but their composition and structure at regional scales are poorly understood. Focusing on recent, large wildfires across the US Pacific Northwest (Oregon and Washington), our research objectives are to (1) classify fire refugia and burn severity based on relativized spectral change in Landsat time series; (2) quantify the pre-fire composition and structure of mapped fire refugia; (3) in forested areas, assess the relative abundance of fire refugia and other burn severity classes across forest composition and structure types. We analyzed a random sample of 99 recent fires in forest-dominated landscapes from 2004 to 2015 that collectively encompassed 612,629 ha. Across the region, fire refugia extent was substantial but variable from year to year, with an annual mean of 38% of fire extent and range of 15–60%. Overall, 85% of total fire extent was forested, with the other 15% being non-forest. In comparison, 31% of fire refugia extent was non-forest prior to the most recent fire, highlighting that mapped refugia do not necessarily contain tree-based ecosystem legacies. The most prevalent non-forest cover types in refugia were vegetated: shrub (40%), herbaceous (33%), and crops (18%). In forested areas, the relative abundance of fire refugia varied widely among pre-fire forest types (20–70%) and structural conditions (23–55%). Consistent with fire regime theory, fire refugia and high burn severity areas were inversely proportional. Our findings underscore that researchers, managers, and other stakeholders should interpret burn severity maps through the lens of pre-fire land cover, especially given the increasing importance of fire and fire refugia under global change.

Keywords: biological legacies; burn severity; disturbance; forest composition and structure; land cover; US Pacific Northwest; pyrogeography; refugia; resilience; wildfire

1. Introduction

Wildland fire is a pervasive ecological disturbance process that interacts with and shapes landscape patterns throughout the world. In forest ecosystems, large wildfire perimeters encompass a variety of land cover types, including forest, non-forest, and unvegetated areas, and the interaction of fuels, weather, and topography results in patchy burn severity mosaics that range from high severity (i.e., large ecological change such as complete tree mortality) to low severity (i.e., little or no ecological change) [1–3]. Land managers, scientists and policy makers increasingly rely on remotely sensed burn severity maps to characterize and interpret these fire effects at landscape scales [4–6]. Fire refugia, defined here following Krawchuk et al. [7] as places that burn less frequently or severely than the surrounding landscape, have become a topic of increasing interest, particularly in the context of global change and conservation of broader refugia [8–10]. Fire refugia represent ecosystem legacies that can perform important ecological functions, such as protecting fire-sensitive flora and fauna and providing propagules for the regeneration of more severely burned locations (e.g., [11–14]). In this way, the resistance of fire refugia may confer resilience to landscapes that will be increasingly important given projections of increasing fire activity due to climate warming and land use [15–17]. Although previous studies in western North American forests have used satellite imagery to quantify the distribution and abundance of fire refugia [5,18,19] or their predictability [7], very little is known about the composition and structure of these areas. Because forest-dominated landscapes can include diverse forest and non-forest conditions, quantifying the variability of fire refugia across heterogeneous regions is essential to evaluate assumptions regarding their ecological functions and to support ecosystem management. Our study develops new approaches to quantify and characterize the composition and structure of fire refugia at landscape and regional scales with detailed ecological resolution.

Studies to date typically characterize forest fire refugia with two basic approaches, either with intensive field observations at a limited number of locations or across extensive landscapes and regions without specific information on local conditions. Researchers have conducted field-based assessments in different geographical settings, including Australia (e.g., [12,20,21]) and western North America (e.g., [7–9,22]), typically with the goal of understanding conditions that give rise to fire refugia over long time scales and for specific types of organisms or species. For example, Camp et al. [8] used inventory plots to assess the composition and structure of forest sites that had not burned as frequently or severely as adjacent forests in Washington, USA, associating refugia with late-successional characteristics, including fire-intolerant species, old trees, multi-layered canopies, and downed coarse wood. These refugia contained abundant fuel for a subsequent fire event, leading to marginally higher overstory tree mortality in refugial than in non-refugial sites and demonstrating that late-successional refugia are dynamic [9]. These and other local-scale studies (e.g., [23,24]) raise questions about the persistence and sustainability of fire refugia under global change, but they are not designed to quantify fire refugia composition and structure at broader scales.

In contrast to field-based studies, landscape and regional assessments have leveraged spatially and temporally extensive satellite imagery to map and identify refugia locations within fire perimeters as areas that remain unburned or burn with low severity. These locations, typically defined by low spectral change between pre- and post-fire images, appear to be more abundant than previously thought (e.g., 20% of fire perimeters [18]). However, these remotely mapped refugia likely include a variety of forest and non-forest areas, with associated variation in ecosystem functions and management significance [19]. Although forest composition and structure vary widely across landscapes and regions (e.g., [25,26]), satellite-based studies typically have not characterized the types and structures of forested and non-forested conditions within mapped fire refugia. Here, we focus on recent fire events, using Landsat-based change detection and existing maps to identify fire refugia as areas experiencing minimal spectral change within generally forested landscapes. We recognize that these *recent forest fire refugia* represent only one characterization of refugia, but such areas are important to forest and fire managers, many of whom utilize Landsat-based burn severity maps as a primary tool to assess fire effects and implement post-fire management activities.

Forest ecosystems contain a variety of compositional and structural conditions that influence fire behavior, fire effects (i.e., burn severity), and post-fire ecosystem responses at multiple spatiotemporal scales. Forest composition is associated with fire regime attributes (i.e., fire frequency, burn severity) that vary from frequent, low-severity fire to infrequent, high-severity fire [1,3]. Due to inherent differences in fire tolerance, fire refugia are more likely to contain particular species, such as thick-barked Douglas-fir (*Pseudotsuga menziesii* [Mirb.] Franco) and mature ponderosa pine (*Pinus ponderosa* Lawson & C. Lawson) that can tolerate surface fires. Similarly, forest structure influences fire behavior, burn severity, and the capacity to form refugia, for instance in open forests with limited surface and ladder fuels and associated crown fire potential [1]. Structure also is important for wildlife habitat and ecosystem resilience, and structural complexity is a vital attribute of natural forests that both influences and emerges from disturbance dynamics [26,27]. Robust data on pre-fire forest composition and structure are critical for understanding the ecosystem legacies [13] and ecological memory [28] associated with wildfires, but previous studies have not quantified these attributes within fire refugia across heterogeneous forested regions.

In addition to trees, forest landscapes typically include non-forest vegetation and unvegetated conditions within and among forested areas. Although they may represent a relatively small portion of forest landscapes at any given time, non-forest areas—including grasslands, shrublands, alpine zones, and unvegetated environments—directly and indirectly influence the patterns and processes of tree-dominated areas [29,30]. As such, non-forest areas influence both the conceptualization and management of forest fire refugia. Whereas unvegetated areas could provide fuel breaks adjacent to forested refugia, non-forest vegetation could serve as a vector of surface fire within and among forested areas (e.g., dry herbaceous vegetation). Non-forest vegetation also responds differently to fire than forests, including lower absolute or relative biomass loss and more rapid regeneration [31].

From the perspective of satellite remote sensing, pre-fire biomass and post-fire vegetation growth also are important factors influencing spectral change and associated burn severity maps. In forested areas, open forests have less biomass and canopy cover to lose than closed-canopy forests, which translates to lower capacity for absolute spectral change and highlights the value of relativized indices that account for pre-fire spectral reflectance (e.g., RdNBR [32]). Low biomass and rapid post-fire vegetation response in non-forested areas also may contribute to lower remotely sensed estimates of burn severity in non-forested than in forested areas because locations with lower woody biomass tend to exhibit lower absolute spectral differences that can attenuate rapidly [32–34]. In addition, despite the key role that non-forested areas play in fire behavior and effects, standard burn severity mapping approaches have been developed in forested areas [35,36]. For instance, in the western United States, the Monitoring Trends in Burn Severity program (MTBS; https://mtbs.gov) maintains a widely used fire perimeter and burn severity database. Importantly, although MTBS provides absolute, relative, and classified burn severity maps, as well as pre- and post-fire Landsat imagery, the MTBS approach does not directly account for different pre-fire land cover types, particularly non-forest areas. Moreover, the MTBS classified burn severity maps are based on an absolute change metric, dNBR, rather than relativized change. These limitations could lead to the misinterpretation of burn severity, especially regarding the quantity and quality of forest fire refugia.

The goal of this study is to quantify and describe the composition and structure of contemporary fire refugia across the US Pacific Northwest (Oregon and Washington, hereafter "PNW"). Increases in wildfire activity and novel region-wide vegetation and disturbance maps provide an unprecedented opportunity to investigate fire refugia across numerous fire events spanning a variety of pre-fire conditions. The advent of Landsat time-series approaches for disturbance mapping across landscape and regional scales (e.g., [37]) and the availability of annualized vegetation maps (e.g., [38]) make it possible to address fundamental questions about the composition and structure of fire refugia while also evaluating mapping tools for scientists, forest managers, and policy makers. By developing and exploring classified burn severity maps similar to widely used databases (e.g., MTBS), we seek to reveal conditions within mapped refugia that map users might otherwise overlook, even when accounting for pre-fire variability with relativized spectral indices. The specific objectives of this study are to:

1. Classify fire refugia and burn severity based on relativized spectral change in Landsat time series and previously published tree mortality thresholds [6].
2. Quantify the pre-fire composition and structure of mapped fire refugia, including forested, non-forested, and unvegetated conditions.
3. In forested areas, assess the relative abundance of fire refugia and other burn severity classes across forest composition and structure types.

2. Materials and Methods

2.1. Overview of Approach

We selected a random, representative sample of recent large fire events in forest-dominated landscapes of the PNW. We then developed burn severity and fire refugia maps using Landsat time series and relationships between relative spectral change and field-based estimates of tree mortality published by Reilly et al. [6]. Next, we overlaid the burn severity maps with existing land cover and vegetation maps representing pre-fire conditions, which also were developed in part with Landsat imagery, thereby enabling a relatively fine-resolution analysis (30-m grain). Our primary focus was to describe fire refugia at the low end of the burn severity gradient, but our third objective evaluates refugia and other severity classes across variable forest compositional and structural conditions (Figure 1). Although we do not assess individual fires in our quantitative analyses, we illustrate the fine spatial patterning of our landscape maps—including refugia, burn severity, land cover, and forest conditions—for a representative large fire, the Table Mountain Complex (Figures 2–4). This 2012 event was part of the broader Wenatchee Complex studied by Kolden et al. [9].

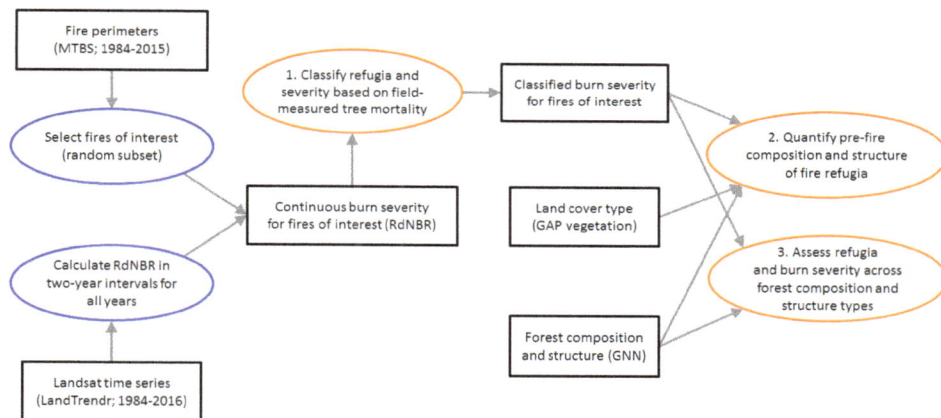

Figure 1. Overview of key spatial datasets (black), processing steps (blue), and objectives (orange). See Section 2 for fire selection criteria. Data sources and references: MTBS: https://mtbs.gov; LandTrendr: Kennedy et al. [37]; RdNBR: Miller and Thode [32]; field-measured tree mortality: Reilly et al. [6]; GAP land cover: https://gapanalysis.usgs.gov; GNN based on Ohmann et al. [38].

2.2. Study Area and Fires of Interest

Conifer forests are widespread across the PNW region, and their composition, structure, and productivity vary across gradients of climate, topography, soil parent material, disturbance regime, and management history [39–41] (Figures S1 and S2). Precipitation and temperature regimes differ across forested ecoregions of the PNW, but a common climatic feature is low summer precipitation [39] conducive to fire and other disturbances (e.g., [41,42]). From west to east, important conifer forest types and tree species are encompassed by broad ecoregions (Figure 2, Supplemental Figure S1) [39,43,44]. Relatively moist forests occur primarily in the Coast Range and West Cascades and are dominated by Douglas-fir and western hemlock (*Tsuga heterophylla* [Raf.] Sarg.). Subalpine forests occupy multiple ecoregions, especially higher elevations in the Cascade Range and inland mountain ranges, featuring subalpine fir (*Abies lasiocarpa* [Hook.] Nutt.), lodgepole pine (*Pinus contorta* Douglas ex Loudon), and mountain hemlock (*Tsuga mertensiana* [Bong.] Carrière). Mixed-conifer forests occur in portions of all ecoregions except for the Coast Range and feature grand fir (*Abies grandis* [Douglas ex D. Don] Lindl.), western larch (*Larix occidentalis* Nutt.), ponderosa pine, and Douglas-fir. Ponderosa pine forests and woodlands and western juniper (*Juniperus occidentalis* Hook.) woodlands occur primarily in the East Cascades and Blue Mountains. Broadleaf trees intermix with conifer forests in riparian areas and in the mixed forests of the southwest portion of the region (e.g., Klamath Mountains; Figure 2, Supplemental Figure S1).

Across the region, forested areas intermix with non-forest and unvegetated land cover types. Non-forest vegetation types above treeline include alpine meadows, and non-forest vegetation types below treeline include sagebrush-steppe shrublands and herbaceous vegetation (e.g., grasslands, meadows). Important unvegetated conditions include barren areas, high alpine environments, open water, and developed land [39,43].

In general, PNW forests occupy relatively remote, mountainous areas managed primarily by US federal agencies for multiple resource objectives. These landscapes have experienced dramatic land-use changes, including widespread logging, grazing, fire exclusion, and associated fuel accumulations [40]. In turn, land use and climate change have contributed to recent increases in the activity of fire and other disturbances [40,42,45]. Given the widespread extent of similar geographic conditions and anthropogenic pressures, PNW forests and their recent fire dynamics are broadly representative of contemporary forest disturbance regimes in western North America.

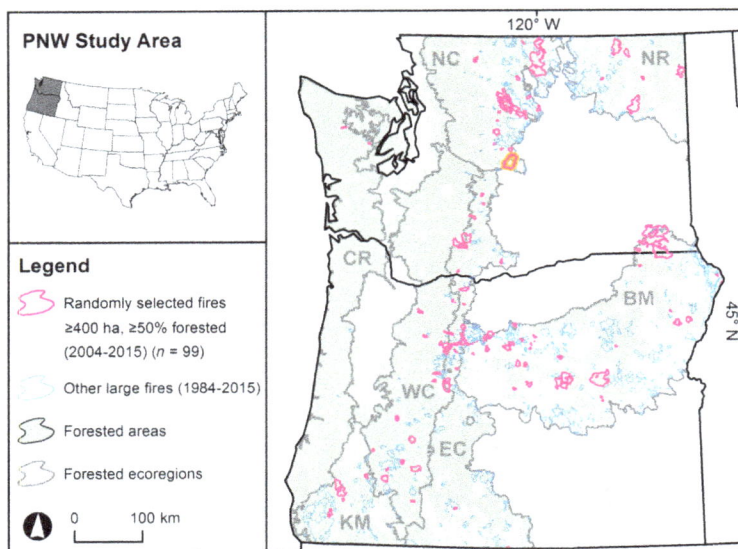

Figure 2. Study area and fires of interest across Oregon and Washington. Pink polygons are the randomly selected portions of large wildfires (≥400 ha) with ≥50% forest cover that burned only once from 2004 to 2015 (*n* = 99). Study fires occurred primarily east of the Cascade Range. Dark blue polygons are all MTBS fires that burned between 1984 and 2015 (level three [44]). The orange perimeter indicates location of the Table Mountain Complex (Figures 3 and 4). Oregon and Washington encompass ca. 40 M ha total and 20 M ha of forest (light green areas indicate forest cover [38]). Ecoregion abbreviations: NC: North Cascades; NR: Northern Rockies; CR: Coast Range; BM: Blue Mountains; WC: West Cascades; EC: East Cascades; KM: Klamath Mountains. We assess only the portions of ecoregions within Oregon and Washington.

We examined the distribution of fire refugia and their pre-fire composition and structure across recent large fire events. We acquired a database of large fire perimeters (≥400 ha) from the MTBS archive (available online: https://mtbs.gov) and identified fires across Oregon and Washington with the following criteria. We first selected fires with ≥50% forest cover by applying a regional forest mask (30 m grain [38]). We then selected fires after 2003 due to the timing of available land cover maps to assess pre-fire conditions (described below). Finally, to avoid the confounding effects of reburn we retained only those portions of fire polygons that burned once since 1985, excluding locations burned more than once. We also excluded burned fragments <400 ha that resulted from these geospatial processing steps. Within this subset, we removed fire events that were on the edge of the PNW study area (*n* = 6), were not classified as wildfires (*n* = 4), and had duplicate entries in the MTBS database (*n* = 2). These criteria yielded 172 distinct fire events that occurred between 2004 and 2015, from which we randomly selected 99 for this analysis (Figure 2, Supplemental Table S1). We manually reviewed this random selection to identify scanline errors from the Landsat 7 sensor, which could introduce errors into refugia maps, but none were apparent in our dataset.

2.3. Burn Severity and Fire Refugia Mapping

We mapped burn severity and fire refugia across the selected fires using regional mosaics of Landsat spectral change, following methods developed by Meigs et al. [46] and Reilly et al. [6] to analyze fire effects across numerous fires in heterogeneous conditions. Landsat imagery was pre-processed (atmospheric correction, cloud masking) and processed using temporal segmentation according to LandTrendr change detection algorithms, which are described in detail by Kennedy et al. [37]. Briefly, LandTrendr segmentation identifies vegetation disturbance and recovery by distilling an often-noisy annual time series into a simplified set of segments and vertices to capture the salient features of spectral trajectories while omitting most false changes [37,45]. Rather than applying disturbance estimates directly from LandTrendr outputs, we compiled annual Landsat time series of the normalized burn ratio (NBR) spectral vegetation index, which combines near-infrared and mid-infrared wavelengths of the Landsat TM/ETM+ sensor and is sensitive to forest vegetation change [32,37]. These NBR time series were centered around the median date of the Landsat stacks (generally 1 August) at the pixel scale, which reduces seasonal variability associated with phenology and sun angles. This process resulted in annual mosaics of NBR covering the full study area, which we then combined with MTBS fire perimeters to produce consistent burn severity maps across all study fires.

Specifically, for each fire perimeter, we computed the relative differenced normalized burn ratio (RdNBR [32]) in two-year intervals to ensure pre- and post-fire coverage for all pixels within a given fire event [46]. By capturing the relative change in dominant vegetation, RdNBR is appropriate for assessing fire effects across numerous events spanning heterogeneous pre-fire conditions [32,47]. Although Landsat spectral indices such as RdNBR have inherent limitations and do not capture very fine-scale fire effects and responses (e.g., tree charring, forest floor combustion, or post-fire regeneration [48,49]), they provide a spatially and temporally consistent metric of burn severity for landscape and regional analysis of fires since 1985. Moreover, the NBR index is at the core of many current fire monitoring protocols (e.g., MTBS [35,36]), and our aim was to characterize areas that fire researchers and managers might identify as fire refugia using these protocols and data.

After clipping the regional RdNBR mosaics within the fires of interest, our next step was to classify the continuous RdNBR maps to specific burn severity categories based on previous field-based estimates of tree mortality (Figure 1). Specifically, we used an equation developed by Reilly et al. [6] that relates RdNBR to relative tree mortality observed at US federal forest inventory plots in the Current Vegetation Survey across the PNW [50]:

$$y = 134.87 + 259.38x + 567.68x^2 \tag{1}$$

where *y* is continuous RdNBR and *x* is the percent basal area mortality estimated from changes in live tree basal area before and after fire at 304 inventory locations. We designated five burn severity classes corresponding to distinct ranges of basal area (BA) mortality. In addition to the low- (<25%

BA mortality), moderate- (>25–75%), and high-severity (>75–100%) classes applied by Reilly et al. [6], we added very low/unchanged (0–10% BA mortality) and very high-severity (>90–100%) classes to further resolve the two ends of the severity gradient. See Reilly et al. [6] for further details on the burn severity classification and field validation.

We defined fire refugia as all pixels within the very low/unchanged class. Recognizing the challenges inherent in remote sensing of fire effects at the low end of the burn severity spectrum [18], our goal was not to distinguish truly unburned areas. Rather, we assumed that pixels with ≥90% estimated tree survival within the first year post-fire include both unburned and lightly burned conditions that are difficult to distinguish remotely. Although these forests are not necessarily unburned, they experienced less severe fire effects than the rest of the burned landscape [7]. Additionally, we recognize that this classification approach based on basal area does not translate directly to locations without trees. Our mapped refugia represent locations with minimal spectral change regardless of tree cover, however, and we distinguish non-forest areas with ancillary spatial datasets (described below). Overall, these areas are conceptually and quantitatively similar to the lowest-severity category in the classified burn severity maps from MTBS ("Unburned to low"; Figure 3), which are based on absolute spectral change (dNBR) and do not integrate a forest mask.

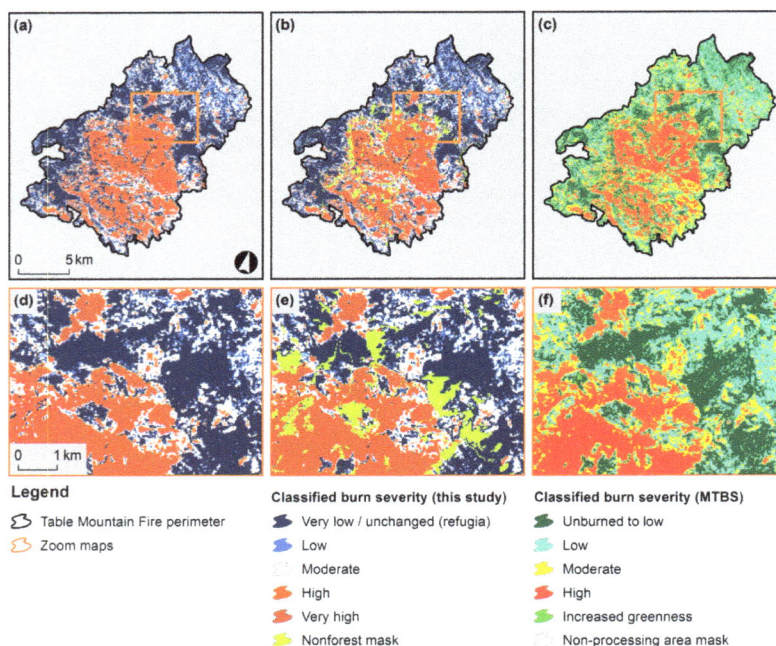

Figure 3. Spatial patterns of burn severity mosaic, refugia, and non-forest areas across the 2012 Table Mountain Complex. Fire location is indicated in Figure 1. Burn severity classes in this study (**a**,**b**,**d**,**e**) are based on Landsat time series, RdNBR, and field-based tree mortality estimates (see Section 2). MTBS severity classes (**c**,**f**) are based on dNBR protocols described by Eidenshink et al. [35] and exhibit similar spatial patterns, particularly the lowest- and highest-severity classes. According to our severity maps, non-forest conditions (non-forest mask) accounted for 31% of refugia extent across all fires and 10% of refugia extent across the Table Mountain Complex. Zoom maps (**d**–**f**) show how non-forest conditions are more prevalent in some refugia areas. MTBS: Monitoring Trends in Burn Severity; https://mtbs.gov.

2.4. Geospatial Overlay Analysis

Our final analytical step was to overlay the classified burn severity maps with land cover and vegetation data available for the study area (Figure 1). We assessed land cover types, including forest vegetation, non-forest vegetation, and unvegetated conditions with spatial data from the Gap Analysis Program (GAP; available online: https://gapanalysis.usgs.gov/). We used a map of terrestrial ecological systems, which represent groups of biological communities that occur within landscapes with similar ecological processes, substrates, and/or environmental gradients [51]. We combined the level three ecological system types into a simplified set of land cover types based on the ecological system descriptions and metadata (Table 1, Figure 4, Supplemental Table S2). This map reflects conditions existing in the year 2001, when the first generation of the US National Land Cover Database was developed, thereby providing information on land cover prior to our fires of interest.

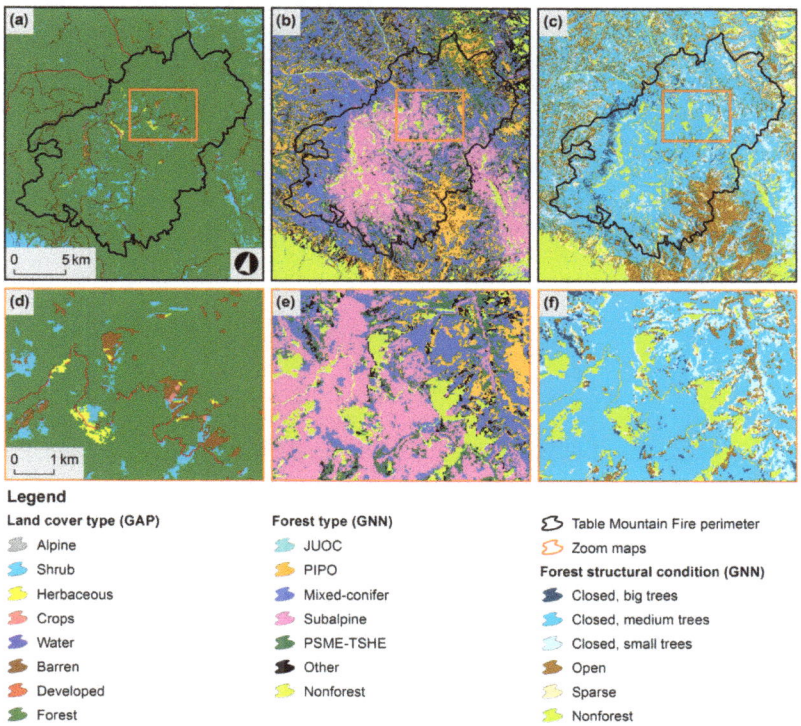

Figure 4. Spatial patterns of pre-fire land cover (**a,d**), forest type (**b,e**), and forest structural condition (**c,f**) across the 2012 Table Mountain Complex. Fire location is indicated in Figure 1. The 2003 GNN-based forest maps (**b–e,f**) illustrate more variability in forest type across this plateau landscape than in forest structure, which was generally closed-canopy forest dominated by medium trees (see Section 2 for classification details). Zoom maps (**d–f**) show how fine-grained variability of pre-fire conditions. Data sources and references: GAP (Gap Analysis Program) land cover: https://gapanalysis.usgs.gov; GNN (gradient nearest-neighbor imputation) based on Ohmann et al. [38]. See Supplemental Figures S1 and S2 for distribution of forest type and structural condition across the study area.

For forested areas identified with the GAP data, we assessed pre-fire (2003) forest composition and structure using annualized maps derived from gradient nearest-neighbor imputation (GNN [38,52]).

GNN maps integrate data from federal forest inventory plots (n ≈ 17,000), key spatial predictors, and Landsat time series to impute plot-level attributes for all forested pixels across the PNW [38]. The GNN imputation is based on Euclidean distance in a multivariate space defined by the predictor variables and derived from canonical correspondence analysis [53,54]. GNN maps include numerous plot variables (available online: https://lemma.forestry.oregonstate.edu/data), and we selected a subset of forest composition and structure variables for our analysis (Table 2). Similar to the GAP land cover types, we combined GNN forest types into a more constrained set applicable to forest vegetation across the PNW based on dominant tree species basal area (Table 3, Figure 4, Supplemental Table S3, Supplemental Figure S1). We combined GNN forest structural conditions into five classes based on live tree canopy cover and tree size (Figure 4, Supplemental Figure S2) [25,43]. Specifically, the sparse and open forest structure classes had canopy cover <10% and 10–40%, respectively, and closed forest structure classes had canopy cover >40% in three size classes based on dominant tree quadratic mean diameter (small: <25 cm QMD, medium: 25–50 cm QMD, large: >50 cm QMD). QMD is a standard metric of average tree size in forestry that gives greater weight to larger trees influencing basal area [55].

We deliberately chose GNN attributes spanning a variety of compositional and structural dimensions, recognizing that the GAP and GNN spatial datasets and variables have distinct strengths, weaknesses, and sources of uncertainty. Because our goal was to describe pre-fire conditions within mapped fire refugia, we focus primarily on relative rather than absolute differences among land cover and forest conditions. We present results from analyses across all fires and years combined to provide a regional perspective on conditions in fire refugia. For the Table Mountain Complex that we use as an example to illustrate our concepts at a landscape event scale, we also show the standard MTBS severity classes to compare with our burn severity maps, both with and without the 30 m grain forest mask (Figure 3).

Table 1. Land cover types across study fires according to GAP analysis data.

Land Cover	Extent (Total ha)	Extent (% of Total)	Extent (Refugia ha)	Extent (% of Refugia)
Forest	519,391	84.8	157,386	69.4
Non-forest total	93,238	15.2	69,413	30.6
Non-forest vegetation	87,426	14.3	65,689	29.0
Alpine	3905	0.6	2784	1.2
Shrub	38,951	6.4	27,498	12.1
Herbaceous	31,398	5.1	22,926	10.1
Crops	13,172	2.2	12,481	5.5
Unvegetated	5812	0.9	3724	1.6
Water	439	0.1	326	0.1
Barren	2279	0.4	1689	0.7
Developed	3094	0.5	1709	0.8
Total	612,629	100.0	226,798	100.0

Notes: See Figure 5 for example of landscape spatial pattern and Figure 6 for distribution among burn severity classes. Refugia areas are the lowest burn severity class (very low/unchanged).

Table 2. Gradient nearest-neighbor (GNN) variables included in spatial analysis.

Variable	Units	Description
Forest type	categorical	Forest type, which describes dominant tree species (based on basal area) of current vegetation; simplified to general types (Table 3).
Structural condition	categorical	Structural condition based on size class and cover class (O'Neil et al. 2001)
Live tree basal area	$m^2\,ha^{-1}$	Basal area of live trees ≥2.5 cm DBH

Table 2. *Cont.*

Variable	Units	Description
Live tree density	stems ha^{-1}	Density of live trees $\geq$2.5 cm DBH
Tree age	years	Basal area weighted stand age based on field recorded or modeled ages of dominant and codominant trees
Quadratic mean diameter of dominant trees	cm	[a] Quadratic mean diameter (QMD) in centimeters of trees whose heights are in the top 25% of all tree heights on the plot
Diameter diversity index	H′	[b] Diameter diversity index (DDI): a measure of stand structural complexity, based on tree densities in different diameter classes

Notes: GNN analysis imputes inventory plot data to forested pixels [38]. Full list of mapped variables available online (https://lemma.forestry.oregonstate.edu/data/structure-maps). [a] QMD of the upper quartile indicates the average size of dominant overstory trees. QMD can be calculated as the square root of the arithmetic mean of squared diameters or based on basal area and tree number [55]. [b] DDI is based on the number of live trees in four standardized tree size classes, and higher values correspond to higher levels of structural complexity.

Table 3. Forest types across study fires according to GNN data [38].

Forest Type	Extent (Total ha)	Extent (% of Forested Total)	Extent (Refugia ha)	Extent (% of Refugia)
Other	45,524	8.8	21,176	13.5
PSME-TSHE	91,234	17.6	25,980	16.5
Subalpine	133,311	25.7	26,207	16.7
Mixed-conifer	153,763	29.6	46,208	29.4
PIPO	79,818	15.4	26,786	17.0
JUOC	15,741	3.0	11,030	7.0
Forested total	519,391	100.0	157,387	100.0

Notes: See Figure 4 for landscape spatial pattern and Figure 7 for distribution among burn severity classes. Species codes: PSME-TSHE = Douglas-fir-western hemlock; PIPO = ponderosa pine; JUOC = western juniper. Other species include miscellaneous conifers (7.3%) deciduous hardwoods (1.5%).

3. Results

3.1. Classification of Fire Refugia and Burn Severity in Recent Forest Fires

The randomly selected fires occurred primarily east of the crest of the Cascade Range, consistent with the spatial distribution of fires during the entire Landsat era (1984–2015) (Figure 2). Our random subset of fires exhibited the same temporal pattern as the general population of large fires during the study period (2004–2015) (Figure 5a). Total annual fire extent typically was below 50,000 ha but was punctuated by two episodic fire years (2006, 2015; Figure 5a). The cumulative extent of the study fires, which included only those locations that burned once, was 612,629 ha over the 12-year study period, equivalent to a mean of 51,052 ha per year.

The burn severity classes we derived based on relative tree basal area mortality corresponded to five ranges of RdNBR (Table 4). Overall, three burn severity classes accounted for the vast majority of fire extent; very low/unchanged was 37%, moderate was 30%, and very high was 18% of total extent (Table 4). Refugia areas (very low/unchanged severity class) were extensive but varied widely from fire to fire and year to year (interannual mean: 38%; range: 15–60%) (Figure 5b). The spatial distribution of refugia varied within fires, as illustrated by the Table Mountain Complex (Figure 3). The Table Mountain example also shows how burn severity distributions were similar between our Landsat-based maps and the standard classified severity maps from MTBS (Figure 3).

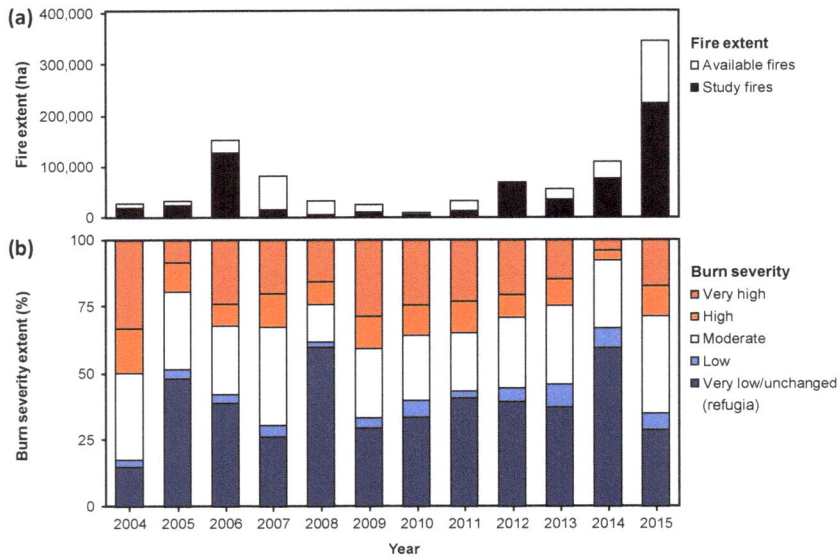

Figure 5. (**a**) Temporal patterns of study fires (*n* = 99) and available fires matching study criteria (*n* = 172). (**b**) Relative distribution of burn severity classes for study fires across all land cover types. The study fires exhibited the same temporal pattern as the available fires (see spatial pattern in Figure 1). The refugia class (very low/unchanged) was extensive but varied widely from year to year (mean ± SD: 38.1 ± 13.2%). Burn severity classes are based on the relationship between tree basal area mortality at federal inventory plots and Landsat spectral change (RdNBR; Reilly et al. [6]).

Table 4. RdNBR values, tree mortality ranges from forest inventory data, and extent of severity classes across study fires.

Burn Severity Class	RdNBR Value	Basal Area Mortality (%)	Extent (ha)	Extent (%)
Very low/unchanged (refugia)	≤166.48	0–10	226,798	37
Low	>166.48–235.20	>10–25	32,645	5
Moderate	>235.20–648.73	>25–75	185,957	30
High	>648.73–828.13	>75–90	58,287	10
Very high	>828.13	90–100	108,943	18

Notes: See Section 2 for burn severity classification equation between RdNBR and basal area mortality (adapted from Reilly et al. [6]).

3.2. Composition and Structure of Fire Refugia

Across the study fires, forests were the most extensive land cover type (Table 1, Figure 6). Total fire extent was 85% forested and 15% non-forested (Table 1). In refugia areas, however, the non-forested component was substantially higher (31%) (Table 1, Figure 6). Across all burn severity classes, the most prevalent vegetated non-forest cover types were shrub (42%), herbaceous (34%), and crops (14%), cumulatively representing 90% of non-forest areas (Figure 6). Within the refugia class, these cover types exhibited a similar distribution, with shrub (40%), herbaceous (33%), and crops (18%) accounting for 91% of non-forest areas (Figure 6). Unvegetated areas cumulatively represented 1.6% of refugia areas and 6% of non-forest extent (3% developed [including roads], 2% barren, 0.5% water) (Table 1).

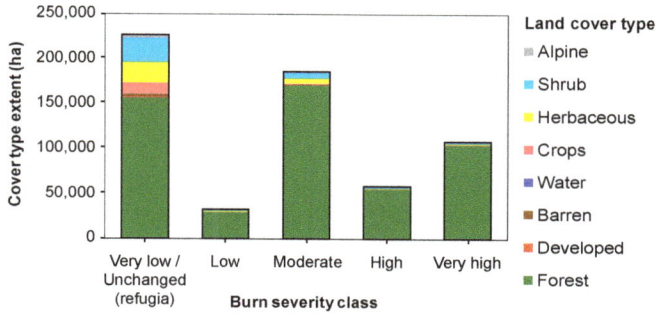

Figure 6. Pre-fire land cover across study fires according to GAP analysis data (https://gapanalysis. usgs.gov). Although these fires were predominantly forested (85%), a substantial portion of the refugia class was non-forested (31%). In addition, most of the non-forest extent (74%) was in refugia areas, and the most prevalent cover types in refugia were shrub (40%), herbaceous (33%), and crops (18%).

In forested areas, fire refugia extent varied with pre-fire forest composition. Mixed-conifer forests in relatively dry parts of the region were the most extensive forest type and contained the most refugia, covering 46,000 ha (Figure 7a). Refugia extent was similar in the Douglas-fir/western hemlock, subalpine, ponderosa pine, and other forest types, with each forest type covering approximately 25,000 ha (Figure 7a). Western juniper woodland was the least extensive forest type and contained the lowest refugia extent, covering 11,000 ha (Figure 7a). As demonstrated by the Table Mountain landscape, pre-fire forest types were intermixed but changed with increasing elevation, with ponderosa pine transitioning into mixed-conifer and subalpine forests (Figure 4b,e).

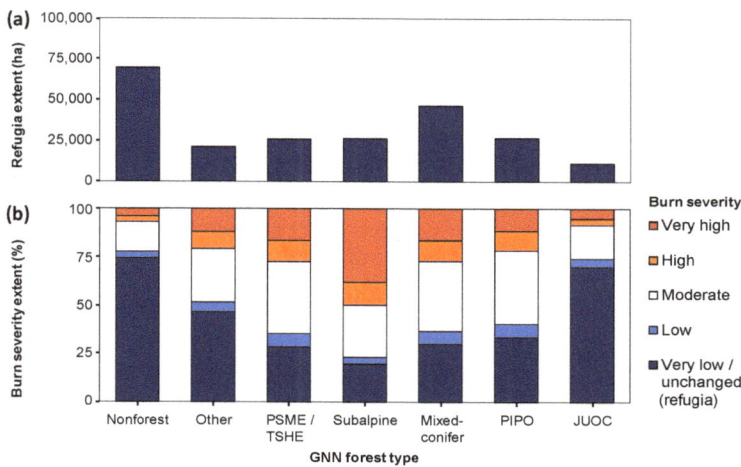

Figure 7. Forest composition of fire refugia in terms of extent of refugia (**a**) and relative distribution of other burn severity classes (**b**). Mixed-conifer forests in relatively dry parts of the region were the most extensive forest type and contained the most refugia (**a**). The percentage of refugia was lowest in subalpine forests and highest in juniper woodlands (**b**). Pre-fire forest types are consolidated into general forest types, ordered from west to east, and are based on live basal area of dominant tree species according to 2003 GNN maps [38]. See Section 2 for details regarding burn severity and forest-type classification and Figures 3 and 4 for landscape spatial patterns. We include non-forested areas for reference but do not interpret the severity classes in direct comparison with the forested areas.

Fire refugia extent also varied with pre-fire forest structure. Closed forests (>40% canopy cover) dominated by medium trees (dominant tree QMD of 25–50 cm) contained the most refugia, encompassing 53,000 ha (Figure 8a). Open forests also contained substantial refugia (44,000 ha), followed by closed forests with small trees (27,000 ha), sparse forests (17,000 ha), and closed forests with large trees (16,000 ha; Figure 8a). As illustrated by the Table Mountain landscape, forest structural conditions varied with elevation but to a lesser degree than forest types (Figure 4c,f). Non-forest areas contained a substantial number of locations identified as refugia based on spectral change alone, representing 69,000 ha (Figures 7a and 8a), although such areas are qualitatively different from forest fire refugia.

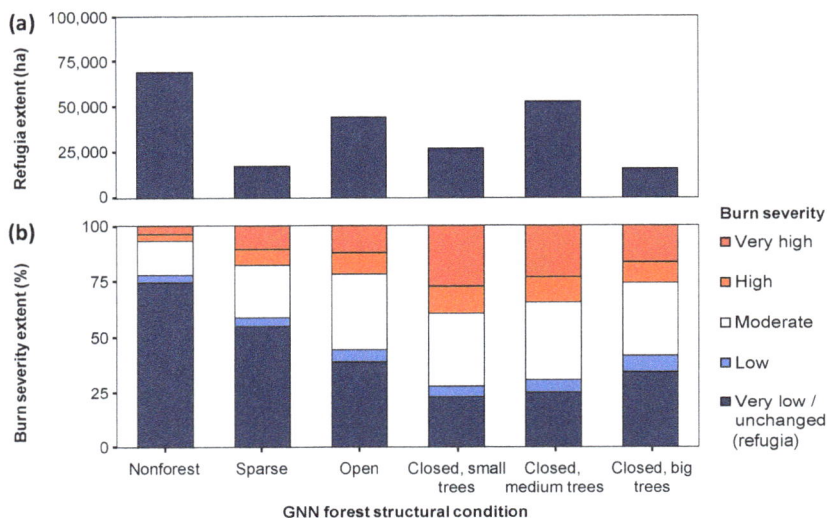

Figure 8. Forest structure of fire refugia in terms of extent of refugia (**a**) and relative distribution of other burn severity classes (**b**). Closed forests (>40% canopy cover) dominated by medium trees (dominant tree diameter 25–50 cm) were the most extensive structural class and contained the most refugia (**a**). The percentage of refugia generally declined with increasing tree cover and size but then increased in closed forests with large trees. Pre-fire structural conditions are based on live tree canopy cover and size classes according to 2003 GNN maps [38]. Structure classes are arranged in increasing order of tree cover and size. See Section 2 for details regarding burn severity and structure classification and Figures 3 and 4 for landscape spatial patterns. We include non-forested areas for reference but do not interpret the severity classes in direct comparison with the forested areas.

3.3. Fire Refugia and Burn Severity across Forest Composition and Structure Types

Fire refugia and the other burn severity classes were not evenly distributed among forest types, and refugia were generally most abundant where high- and very high-severity fire were least abundant (and vice versa; Figure 7b). The relative abundance of fire refugia ranged from 20% of fire extent in subalpine forests to 70% in juniper woodlands (Figure 7b). Conversely, the relative abundance of very high-severity fire ranged from 5% of fire extent in juniper woodlands to 38% in subalpine forests (Figure 7b). The Douglas-fir/western hemlock, mixed-conifer, and ponderosa pine forests exhibited very similar amounts of the lowest and highest burn severity classes, with refugia ranging from 28% to 34% and very high severity ranging from 11% to 16% of fire extent (Figure 7b).

As with forest composition, fire refugia and burn severity classes varied among forest structural conditions (Figure 8b). In general, the relative abundance of refugia was lower in settings with moderate tree cover and size. Importantly, however, refugia abundance was higher in closed forests with large trees than in closed forests with medium trees (Figure 8b). Refugia areas ranged from 23% of fire extent in closed forests dominated by small trees to 55% in sparse forests (Figure 8b). In contrast, very high-severity areas ranged from 11% of fire extent in sparse forests to 27% in closed forests with small trees (Figure 8b). The other three forest structural conditions were intermediate in their distributions of burn severity classes. Closed forests dominated by medium trees were similar to closed forests with small trees, and closed forests with big trees were similar to open forests (Figure 8b). For the continuous structural variables, refugia tended to have lower live tree basal area and density, while very high-severity fire occurred in forests with higher live basal area and density (Table 5). Similarly, refugia tended to exhibit lower pre-fire tree age, quadratic mean diameter (an indicator of dominant tree size), and structural complexity (based on tree diameter distributions) than areas experiencing very high burn severity (Table 5).

Table 5. GNN structure variables across study fires.

Variable (Units)	Statistic	Very Low/Unchanged (Refugia)	Low	Moderate	High	Very High
		Burn Severity				
Live tree basal area	mean	16.63	25.60	24.98	26.91	32.49
($m^2\ ha^{-1}$)	SD	19.17	19.00	18.05	17.98	18.52
Live tree density	mean	589.96	904.37	951.25	1094.74	1385.24
(stems ha^{-1})	SD	904.28	1014.74	1099.79	1230.42	1277.28
Tree age	mean	80.98	108.15	109.02	112.68	123.21
(year)	SD	72.29	63.84	60.37	58.17	56.44
Quadratic mean diameter	mean	12.51	17.22	16.99	16.87	16.52
(QMD; cm)	SD	11.11	9.83	9.33	8.92	8.43
Diameter diversity index	mean	2.40	3.56	3.49	3.59	3.80
(DDI; H')	SD	2.29	2.08	1.95	1.87	1.82

Notes: See Table 3 for descriptions of GNN variables including QMD and DDI.

4. Discussion

4.1. Composition and Structure of Forest Fire Refugia across the US Pacific Northwest

This study elucidates substantial variability in the composition and structure of fire refugia across forested ecosystems of the PNW study area, underscoring the need to account for pre-fire forest and non-forest conditions when creating and interpreting burn severity maps. In many cases, our analyses support the common intuition that fire refugia identified in classified severity maps (such as MTBS) broadly capture forests that experience minimal fire effects. These forested fire refugia vary in forest type and structural condition, demonstrating a range of forested conditions that will influence the transmission of ecological memory from the pre- to post-fire environment (i.e., information and material legacies [28]). However, non-forest vegetation accounted for a substantial component of mapped refugia, highlighting the importance of these areas both for ecosystem functions and mapping applications. Unvegetated conditions within mapped fire refugia were relatively rare in our study fires, but they may contribute disproportionately to landscape fire patterns if they influence the distribution of fire refugia in adjacent vegetated areas (e.g., by acting as fuel breaks). Overall, our assessment illustrates that the ecological role of fire refugia depends on site-specific pre-fire conditions, as well as the broader burn severity mosaic. As such, ecological interpretation of burn severity maps generated according to Landsat spectral change requires users to leverage additional datasets, such as the regional land cover and forest maps used here, to refine fire refugia assessments to specific ecosystems of interest and to characterize ecosystem legacies more comprehensively.

In addition to characterizing important variability of fire refugia, our study quantifies general ranges of conditions where fire refugia occur across the PNW study area. For example, although the extent and proportion of mapped refugia varied from year to year, refugia were widespread across burned areas, averaging 38% of mapped fire extent annually. Indeed, refugia were relatively extensive even in the forest type with the lowest percentage of refugia, subalpine fir. Although subalpine forests typically are characterized by infrequent, high-severity fire, our analyses identify one fifth of subalpine forests as unburned or low-severity refugia and one half experiencing <75% basal area mortality. The prevalence of non-stand-replacing fire in the forest type with the most severe fire effects, coupled with less extensive but notable high-severity conditions in the other forest types, supports increasing recognition of the importance of mixed-severity fire regimes [3,29]. The substantial extent of fire refugia across recent fires highlights that pre-fire conditions persist in many cases, despite concerns about increasing fire activity [15,17,25]. In addition, because the percentage of fire refugia was lower in forest compositional and structural conditions with a higher percentage of high-severity fire and vice versa, our findings demonstrate relative differences in fire effects among forest types that are consistent with expectations from fire history studies and fire regime theory [1,6,30].

As expected in these generally forested locations, forests were the dominant land cover type overall (85%) and in refugia areas (69%). However, our study also indicates that the nature of post-fire ecosystem legacies and potential functions of fire refugia depends on specific forest conditions in the pre-fire landscape. For example, the post-fire trajectory of a refugia site with surviving dense, large trees will be very different from a sparsely forested or unvegetated site. Locations with abundant overstory trees likely will function as forest refugia with live tree legacies (i.e., seed sources [11]) and fauna source populations [12]. These ecological functions are particularly important for refugia sites adjacent to high-severity areas and in cases where drought conditions hinder seedling establishment [11,14]. Another key function for forested refugia is the provision of critical habitat for forest specialists both during and following fire [12,20,23]. In western portions of the Pacific Northwest region, because late-successional and old-growth forests provide nesting and roosting habitat for the Northern Spotted Owl (*Strix occidentalis caurina* Merriam), fire refugia in closed canopy forest with large trees represent an especially vital subset of refugia for this and other vulnerable species.

Although the majority of mapped refugia were forested prior to the most recent fires, non-forest conditions represented 31% of refugia extent, a considerable component of burned landscapes with distinct implications for forest ecosystems and fire dynamics. For example, locations with non-forest vegetation prior to fire likely contain shrub and herbaceous communities that contribute to heterogeneity in both the pre- and post-fire landscape, providing habitat for early-successional species that might otherwise require stand-replacing disturbance. Non-forest vegetation also may respond rapidly following fire [31,33], increasing surface fuel connectivity and potential exposure of forest refugia to future fires, at least where herbaceous grasslands interface with forests. In contrast, unvegetated non-forest conditions like rocky slopes in barren and alpine locations may protect adjacent forested areas from fire via fuel breaks despite not harboring surviving trees themselves. The different ways that non-forest cover types intermix, and potentially influence, forest fire refugia within generally forested ecosystems highlights the need to account for the diversity of land cover types and spatial complexity of burn severity mosaics in fire assessments.

4.2. Implications for Fire Refugia Research, Monitoring, and Management

This study describes previously undocumented variability in remotely mapped fire refugia across a heterogeneous region and numerous fire events, suggesting several avenues for future research. Finer-resolution analyses are possible in both forest and non-forest areas, including examination of more specific forest types, forest structural conditions, or non-forest land cover types. Such assessments could be particularly fruitful at sub-regional scales, especially where detailed pre-fire field data are available within specific landscapes or land-management units (i.e., National Forests). The landscape-scale maps of the Table Mountain Fire (Figures 3 and 4) illustrate important pixel- and stand-scale variation that future studies could integrate further with intensive field surveys (e.g., [8,9]). In addition to assessing composition and structure as separate components of forest ecosystems, future work could explore the interactions of composition and structure, identifying, for instance, the structural conditions more conducive to fire refugia in forest types with the least amount of refugia. Additional studies also could investigate the variability of post-fire forest and non-forest conditions in order to document the influence of pre-fire heterogeneity on post-fire heterogeneity and ecosystem responses, building on recent analyses of refugia spatial patterns. For example, Meddens et al. [19] determined that refugia patch size varies with land cover type and topography (i.e., larger patch size in flatter locations with sparse vegetation). Finally, subsequent work could focus on statistical modeling of the environmental controls underpinning the predictability and persistence of fire refugia (e.g., [7,24]), as well as how fire refugia might overlap with hydrological and climate refugia (e.g., [10]). In anticipation of these prospects for further inquiry, the current study provides more detailed ecological resolution than previous efforts for a regional sample of large fires spanning a broad range of environmental settings. The opportunity to conduct this type of assessment will only increase with ongoing improvements in fire [49], vegetation [52], and land cover [56] mapping.

Our findings have immediate applications for the development and interpretation of refugia and burn severity maps. Specifically, this study underscores that the same estimate of spectral change (or lack thereof) can mean very differ things in forested and non-forested areas with differing composition and structure [32,34]. Categorical maps amplify this potential ambiguity because burn severity classes necessarily include a range of change values. As such, map users should exercise caution when interpreting burn severity products, particularly classified maps at the low end of the severity gradient in environments with a substantial non-forest component. If one's primary interest is forest applications, rather than assuming that tree-based thresholds are applicable throughout fire perimeters, a prudent approach would be to apply a robust forest mask and assess only those fire events occurring after the forest mask imagery date (as in this study). This principle applies whether the refugia maps are based on two-image Landsat change detection with dNBR (an absolute change index, as in the MTBS classification) or Landsat time-series change detection with RdNBR (a relative change index, as in our classification). A related implication of this finding is that current off-the-shelf approaches (e.g., MTBS) overestimate the extent of forest fire refugia if an appropriate forest mask is not incorporated (e.g., Figure 3).

Finally, this study has direct implications for fire and forest management in the PNW region and other temperate forests with abundant wildfires. First, our findings suggest that land managers explicitly consider the pre-fire variability of burned areas when developing and applying burn severity maps, post-fire management activities, and ecosystem service assessments (e.g., [2]). The estimated difference between conditions before and after fire—whether spectral or field-based—is only one piece of the fire effects puzzle. The full picture of burn severity and ecosystem response to fire depends on pre-fire conditions, short-term fire effects, and post-fire vegetation trajectories [4,31]. Second, the substantial extent and variability of non-forest vegetation within fire refugia warrant special management attention and coordination with non-forest specialists. Because we assessed only those fire events with >50% forest cover, the prevalence of non-forest areas is higher across the broader PNW [19] and western North America. Accordingly, monitoring and management activities should integrate pre-fire land cover and other ancillary spatial data to characterize contemporary burn mosaics more comprehensively. Third, the high variability within mapped refugia locations confirms the value of developing clear terminology and conceptual frameworks for fire refugia [57], especially in the context of broader discussions of refugia conservation [10].

5. Conclusions

As fire activity continues to increase due to changing climate and land use [15–17], the topic of fire refugia will become increasingly important in ecosystems throughout the world. In the Pacific Northwest, fire extent has increased dramatically in recent years, although the proportion of

Forests **2018**, *9*, 243

different burn severity classes has remained relatively consistent [6,19]. This study develops a new approach to map and describe forest fire refugia and overlays those refugia with readily available land cover and vegetation maps, illustrating that not all fire refugia are equivalent. The variability and potential interactions of land cover types, forest types, and forest structural conditions demonstrate the importance of understanding the full range of pre-fire conditions in burned areas. Our findings also underscore that burn severity map users should be careful in their assumptions when identifying potential forest fire refugia with satellite imagery because non-forest and sparsely forested areas can represent a considerable percentage of locations experiencing minimal spectral change, which could result in the overestimation of functional forest fire refugia. Future research, monitoring, and management activities could further elucidate the patterns and ecological functions of fire refugia, as well as strategies to increase the capacity of refugia to enhance forest resistance and resilience in fire-prone landscapes.

Supplementary Materials: The following are available online at http://www.mdpi.com/1999-4907/9/5/243/s1: Figure S1: Forest composition across the Pacific Northwest study area based on GNN data; Figure S2: Forest structure across the Pacific Northwest study area based on GNN data; Table S1: Attributes of selected fire events; Table S2: Land cover types from GAP data; Table S3: Forest types from GNN data.

Author Contributions: G.W.M. and M.A.K. conceived and designed the study; G.W.M. prepared and analyzed the spatial data; G.W.M. and M.A.K. wrote the paper.

Funding: This research was funded by Oregon State University and the USFS Rocky Mountain Research Station under 16-JV-11221639-101.

Acknowledgments: We are thankful for assistance with data analysis and interpretation from William Downing, Christopher Dunn, Matthew Gregory, and Matthew Reilly. We appreciate thought-provoking feedback from Geneva Chong, Jonathan Coop, Sandra Haire, Carol Miller, Marc-André Parisien, Marie-Pierre Rogeau, Ryan Walker, Ellen Whitman, and the Fierylabs at Oregon State University. We acknowledge constructive comments from two anonymous reviewers.

Conflicts of Interest: The authors declare no conflict of interest.

References

1. Agee, J.K. The landscape ecology of western forest fire regimes. *Northwest Sci.* **1998**, *72*, 24–34.
2. Meigs, G.W.; Donato, D.C.; Campbell, J.L.; Martin, J.G.; Law, B.E. Forest fire impacts on carbon uptake, storage, and emission: The role of burn severity in the Eastern Cascades, Oregon. *Ecosystems* **2009**, *12*, 1246–1267. [CrossRef]
3. Halofsky, J.; Donato, D.; Hibbs, D.; Campbell, J.; Cannon, M.D.; Fontaine, J.; Thompson, J.R.; Anthony, R.; Bormann, B.; Kayes, L. Mixed-severity fire regimes: Lessons and hypotheses from the Klamath-Siskiyou ecoregion. *Ecosphere* **2011**, *2*, 1–19. [CrossRef]
4. Lentile, L.B.; Holden, Z.A.; Smith, A.M.S.; Falkowski, M.J.; Hudak, A.T.; Morgan, P.; Lewis, S.A.; Gessler, P.E.; Benson, N.C. Remote sensing techniques to assess active fire characteristics and post-fire effects. *Int. J. Wildland Fire* **2006**, *15*, 319–345. [CrossRef]
5. Kolden, C.A.; Lutz, J.A.; Key, C.H.; Kane, J.T.; van Wagtendonk, J.W. Mapped versus actual burned area within wildfire perimeters: Characterizing the unburned. *For. Ecol. Manag.* **2012**, *286*, 38–47. [CrossRef]
6. Reilly, M.J.; Dunn, C.J.; Meigs, G.W.; Spies, T.A.; Kennedy, R.E.; Bailey, J.D.; Briggs, K. Contemporary patterns of fire extent and severity in forests of the Pacific Northwest, USA (1985–2010). *Ecosphere* **2017**, *8*, 1–28. [CrossRef]
7. Krawchuk, M.A.; Haire, S.L.; Coop, J.; Parisien, M.A.; Whitman, E.; Chong, G.; Miller, C. Topographic and fire weather controls of fire refugia in forested ecosystems of northwestern North America. *Ecosphere* **2016**, *7*, 1–18. [CrossRef]
8. Camp, A.; Oliver, C.; Hessburg, P.; Everett, R. Predicting late-successional fire refugia pre-dating European settlement in the Wenatchee Mountains. *For. Ecol. Manag.* **1997**, *95*, 63–77. [CrossRef]
9. Kolden, C.A.; Bleeker, T.M.; Smith, A.; Poulos, H.M.; Camp, A.E. Fire effects on historical wildfire refugia in contemporary wildfires. *Forests* **2017**, *8*, 400. [CrossRef]

10. Morelli, T.L.; Daly, C.; Dobrowski, S.Z.; Dulen, D.M.; Ebersole, J.L.; Jackson, S.T.; Lundquist, J.D.; Millar, C.I.; Maher, S.P.; Monahan, W.B. Managing climate change refugia for climate adaptation. *PLoS ONE* **2016**, *11*, 1–17. [CrossRef] [PubMed]

11. Haire, S.L.; McGarigal, K. Effects of landscape patterns of fire severity on regenerating ponderosa pine forests (Pinus ponderosa) in New Mexico and Arizona, USA. *Landsc. Ecol.* **2010**, *25*, 1055–1069. [CrossRef]

12. Robinson, N.M.; Leonard, S.W.; Ritchie, E.G.; Bassett, M.; Chia, E.K.; Buckingham, S.; Gibb, H.; Bennett, A.F.; Clarke, M.F. Refuges for fauna in fire-prone landscapes: Their ecological function and importance. *J. Appl. Ecol.* **2013**, *50*, 1321–1329. [CrossRef]

13. Jõgiste, K.; Korjus, H.; Stanturf, J.A.; Frelich, L.E.; Baders, E.; Donis, J.; Jansons, A.; Kangur, A.; Köster, K.; Laarmann, D. Hemiboreal forest: Natural disturbances and the importance of ecosystem legacies to management. *Ecosphere* **2017**, *8*, 1–20. [CrossRef]

14. Stevens-Rumann, C.S.; Kemp, K.B.; Higuera, P.E.; Harvey, B.J.; Rother, M.T.; Donato, D.C.; Morgan, P.; Veblen, T.T. Evidence for declining forest resilience to wildfires under climate change. *Ecol. Lett.* **2017**, *21*, 243–252. [CrossRef] [PubMed]

15. Moritz, M.A.; Parisien, M.-A.; Batllori, E.; Krawchuk, M.A.; Van Dorn, J.; Ganz, D.J.; Hayhoe, K. Climate change and disruptions to global fire activity. *Ecosphere* **2012**, *3*, 1–22. [CrossRef]

16. North, M.; Stephens, S.; Collins, B.; Agee, J.; Aplet, G.; Franklin, J.; Fulé, P. Reform forest fire management. *Science* **2015**, *349*, 1280–1281. [CrossRef] [PubMed]

17. Abatzoglou, J.T.; Williams, A.P. Impact of anthropogenic climate change on wildfire across western us forests. *Proc. Natl. Acad. Sci. USA* **2016**, *113*, 11770–11775. [CrossRef] [PubMed]

18. Meddens, A.J.H.; Kolden, C.A.; Lutz, J.A. Detecting unburned areas within wildfire perimeters using Landsat and ancillary data across the northwestern united states. *Remote Sens. Environ.* **2016**, *186*, 275–285. [CrossRef]

19. Meddens, A.J.H.; Kolden, C.A.; Lutz, J.A.; Abatzoglou, J.T.; Hudak, A.T. Spatiotemporal patterns of unburned areas within fire perimeters in the northwestern United States from 1984 to 2014. *Ecosphere* **2018**, *9*, 1–16. [CrossRef]

20. Banks, S.C.; Dujardin, M.; McBurney, L.; Blair, D.; Barker, M.; Lindenmayer, D.B. Starting points for small mammal population recovery after wildfire: Recolonisation or residual populations? *Oikos* **2011**, *120*, 26–37. [CrossRef]

21. Wood, S.W.; Murphy, B.P.; Bowman, D.M.J.S. Firescape ecology: How topography determines the contrasting distribution of fire and rain forest in the south-west of the Tasmanian wilderness world heritage area. *J. Biogeogr.* **2011**, *38*, 1807–1820. [CrossRef]

22. Keeton, W.S.; Franklin, J.F. Fire-related landform associations of remnant old-growth trees in the southern Washington Cascade Range. *Can. J. For. Res.* **2004**, *34*, 2371–2381. [CrossRef]

23. Hylander, K.; Johnson, S. In situ survival of forest bryophytes in small-scale refugia after an intense forest fire. *J. Veg. Sci.* **2010**, *21*, 1099–1109. [CrossRef]

24. Ouarmim, S.; Paradis, L.; Asselin, H.; Bergeron, Y.; Ali, A.A.; Hély, C. Burning potential of fire refuges in the boreal mixedwood forest. *Forests* **2016**, *7*, 246. [CrossRef]

25. Reilly, M.J.; Elia, M.; Spies, T.A.; Gregory, M.J.; Sanesi, G.; Lafortezza, R. Cumulative effects of wildfires on forest dynamics in the eastern Cascade Mountains, USA. *Ecol. Appl.* **2018**, *28*, 291–308. [CrossRef] [PubMed]

26. Reilly, M.J.; Spies, T.A. Regional variation in stand structure and development in forests of Oregon, Washington, and inland Northern California. *Ecosphere* **2015**, *6*, 1–27. [CrossRef]

27. Meigs, G.W.; Morrissey, R.C.; Bače, R.; Chaskovskyy, O.; Čada, V.; Després, T.; Donato, D.C.; Janda, P.; Lábusová, J.; Seedre, M.; et al. More ways than one: Mixed-severity disturbance regimes foster structural complexity via multiple developmental pathways. *For. Ecol. Manag.* **2017**, *406*, 410–426. [CrossRef]

28. Johnstone, J.F.; Allen, C.D.; Franklin, J.F.; Frelich, L.E.; Harvey, B.J.; Higuera, P.E.; Mack, M.C.; Meentemeyer, R.K.; Metz, M.R.; Perry, G.L. Changing disturbance regimes, ecological memory, and forest resilience. *Front. Ecol. Environ.* **2016**, *14*, 369–378. [CrossRef]

29. Hessburg, P.F.; Spies, T.A.; Perry, D.A.; Skinner, C.N.; Taylor, A.H.; Brown, P.M.; Stephens, S.L.; Larson, A.J.; Churchill, D.J.; Povak, N.A. Tamm review: Management of mixed-severity fire regime forests in Oregon, Washington, and Northern California. *For. Ecol. Manag.* **2016**, *366*, 221–250. [CrossRef]

30. Miller, R.F.; Rose, J.A. Fire history and western juniper encroachment in sagebrush steppe. *J. Range Manag.* **1999**, *52*, 550–559. [CrossRef]

31. Keeley, J.E. Fire intensity, fire severity and burn severity: A brief review and suggested usage. *Int. J. Wildland Fire* **2009**, *18*, 116–126. [CrossRef]

32. Miller, J.D.; Thode, A.E. Quantifying burn severity in a heterogeneous landscape with a relative version of the delta normalized burn ratio (dNBR). *Remote Sens. Environ.* **2007**, *109*, 66–80. [CrossRef]

33. Keeley, J.E.; Brennan, T.; Pfaff, A.H. Fire severity and ecosytem responses following crown fires in California shrublands. *Ecol. Appl.* **2008**, *18*, 1530–1546. [CrossRef] [PubMed]

34. Strand, E.K.; Bunting, S.C.; Keefe, R.F. Influence of wildland fire along a successional gradient in sagebrush steppe and western juniper woodlands. *Rangel. Ecol. Manag.* **2013**, *66*, 667–679. [CrossRef]

35. Eidenshink, J.; Schwind, B.; Brewer, K.; Zhu, Z.L.; Quayle, B.; Howard, S. A project for monitoring trends in burn severity. *Fire Ecol.* **2007**, *3*, 3–21. [CrossRef]

36. Key, C.H.; Benson, N.C. Landscape assessment: Ground measure of severity, the composite burn index; and remote sensing of severity, the normalized burn ratio. In *FIREMON: Fire Effects Monitoring and Inventory System*; General Technical Report RMRS-GTR-164-CD; USDA Forest Service: Fort Collins, CO, USA, 2006; pp. 1–55.

37. Kennedy, R.E.; Yang, Z.G.; Cohen, W.B. Detecting trends in forest disturbance and recovery using yearly Landsat time series: 1. LandTrendr—Temporal segmentation algorithms. *Remote Sens. Environ.* **2010**, *114*, 2897–2910. [CrossRef]

38. Ohmann, J.L.; Gregory, M.J.; Roberts, H.M.; Cohen, W.B.; Kennedy, R.E.; Yang, Z. Mapping change of older forest with nearest-neighbor imputation and Landsat time-series. *For. Ecol. Manag.* **2012**, *272*, 13–25. [CrossRef]

39. Franklin, J.F.; Dyrness, C.T. *Natural Vegetation of Oregon and Washington*; General Technical Report PNW-GTR-8; USDA Forest Service: Portland, OR, USA, 1973; pp. 1–452.

40. Hessburg, P.F.; Smith, B.G.; Salter, R.B.; Ottmar, R.D.; Alvarado, E. Recent changes (1930s–1990s) in spatial patterns of interior northwest forests, USA. *For. Ecol. Manag.* **2000**, *136*, 53–83. [CrossRef]

41. Meigs, G.W.; Campbell, J.L.; Zald, H.S.J.; Bailey, J.D.; Shaw, D.C.; Kennedy, R.E. Does wildfire likelihood increase following insect outbreaks in conifer forests? *Ecosphere* **2015**, *6*, 1–24. [CrossRef]

42. Littell, J.S.; Oneil, E.E.; McKenzie, D.; Hicke, J.A.; Lutz, J.A.; Norheim, R.A.; Elsner, M.M. Forest ecosystems, disturbance, and climatic change in Washington state, USA. *Clim. Chang.* **2010**, *102*, 129–158. [CrossRef]

43. O'Neil, T.A.; Bettinger, K.A.; Vander Heyden, M.; Marcot, B.; Barrett, C.; Mellen, T.K.; Vanderhaegen, W.M.; Johnson, D.H.; Doran, P.J.; Wunder, L. Structural conditions and habitat elements of Oregon and Washington. In *Wildlife Habitats and Relationships in Oregon and Washington*; OSU Press: Corvallis, OR, USA, 2001; pp. 115–139.

44. Omernik, J.M. Ecoregions of the conterminous United States. Map (scale 1:7,500,000). *Ann. Assoc. Am. Geogr.* **1987**, *77*, 118–125. [CrossRef]

45. Meigs, G.W.; Kennedy, R.E.; Gray, A.N.; Gregory, M.J. Spatiotemporal dynamics of recent mountain pine beetle and western spruce budworm outbreaks across the Pacific Northwest Region, USA. *For. Ecol. Manag.* **2015**, *339*, 71–86. [CrossRef]

46. Meigs, G.W.; Zald, H.S.; Campbell, J.L.; Keeton, W.S.; Kennedy, R.E. Do insect outbreaks reduce the severity of subsequent forest fires? *Environ. Res. Lett.* **2016**, *11*, 1–10. [CrossRef]

47. Cansler, C.A.; McKenzie, D. Climate, fire size, and biophysical setting control fire severity and spatial pattern in the northern Cascade Range, USA. *Ecol. Appl.* **2014**, *24*, 1037–1056. [CrossRef] [PubMed]

48. Harvey, B.J.; Donato, D.C.; Turner, M.G. Recent mountain pine beetle outbreaks, wildfire severity, and postfire tree regeneration in the US Northern Rockies. *Proc. Natl. Acad. Sci. USA* **2014**, *111*, 15120–15125. [CrossRef] [PubMed]

49. Parks, S.A.; Dillon, G.K.; Miller, C. A new metric for quantifying burn severity: The relativized burn ratio. *Remote Sens.* **2014**, *6*, 1827–1844. [CrossRef]

50. Max, T.A.; Schreuder, H.T.; Hazard, J.W.; Oswald, D.D.; Teply, J.; Alegria, J. *The Pacific Northwest Region Vegetation and Inventory Monitoring System*; Research Paper PNW-RP-493; USDA Forest Service: Portland, OR, USA, 1996; pp. 1–22.

51. Comer, P.; Faber-Langendoen, D.; Evans, R.; Gawler, S.; Josse, C.; Kittel, G.; Menard, S.; Pyne, M.; Reid, M.; Schulz, K. *Ecological Systems of the United States: A Working Classification of US Terrestrial Systems*; NatureServe: Arlington, VA, USA, 2003; pp. 1–75.

52. Kennedy, R.E.; Ohmann, J.; Gregory, M.; Roberts, H.; Yang, Z.; Bell, D.M.; Kane, V.; Hughes, M.J.; Cohen, W.B.; Powell, S. An empirical, integrated forest biomass monitoring system. *Environ. Res. Lett.* **2018**, *13*, 1–10. [CrossRef]

53. Ohmann, J.L.; Gregory, M.J. Predictive mapping of forest composition and structure with direct gradient analysis and nearest-neighbor imputation in coastal Oregon, USA. *Can. J. For. Res.* **2002**, *32*, 725–741. [CrossRef]

54. Ter Braak, C.J. Canonical correspondence analysis: A new eigenvector technique for multivariate direct gradient analysis. *Ecology* **1986**, *67*, 1167–1179. [CrossRef]

55. Curtis, R.O.; Marshall, D.D. Why quadratic mean diameter? *West J. Appl. For.* **2000**, *15*, 137–139.

56. Soulard, C.E.; Acevedo, W.; Stehman, S.V. Removing rural roads from the National Land Cover Database to create improved urban maps for the United States, 1992 to 2011. *Photogramm. Eng. Remote Sens.* **2018**, *84*, 101–109.

57. Meddens, A.J.H.; Kolden, C.A.; Lutz, J.A.; Smith, A.M.S.; Cansler, C.A.; Abatzoglou, J.T.; Meigs, G.W.; Downing, W.M.; Krawchuk, M.A. Fire refugia: What are they and why do they matter for global change? *BioScience* **2018**, in review.

forests

Article

Spruce Budworm (*Choristoneura fumiferana* Clem.) Defoliation Promotes Vertical Fuel Continuity in Ontario's Boreal Mixedwood Forest

Graham A. Watt [1,*], Richard A. Fleming [2], Sandy M. Smith [3] and Marie-Josée Fortin [4]

[1]   Faculty of Forestry, University of Toronto, 33 Willcocks St., Toronto, ON M5S 3B3, Canada
[2]   Natural Resources Canada, Canadian Forest Service, Great Lakes Forestry Centre, Sault Ste. Marie, ON P6A 2E5, Canada; Rich.Fleming@canada.ca
[3]   Faculty of Forestry, University of Toronto, 33 Willcocks St., Toronto, ON M5S 3B3, Canada; s.smith.a@utoronto.ca
[4]   Department of Ecology & Evolutionary Biology, University of Toronto, 25 Willcocks Street, Toronto, ON M5S 3B2, Canada; mariejosee.fortin@utoronto.ca
*   Correspondence: gwatt14@schulich.yorku.ca; Tel.: +1-416-978-5480

Received: 15 February 2018; Accepted: 6 May 2018; Published: 9 May 2018

Abstract: Spruce budworm, *Choristoneura fumiferana* (Clem.), defoliation has been shown to affect the occurrence of crown fire in Ontario, highlighting the need to better understand the driving factors of this effect on forest structure, including changes in fuel loading, type and position. Here, we investigate five boreal mixedwood sites within four zones that experienced different durations of continuous defoliation by spruce budworm in northeastern Ontario. Duration of defoliation had significant effects on vertical stand components, namely, host overstory to host understory crown overlap, host overstory and host understory crown to downed woody debris overlap, and downed woody debris height and quantity. Vertical stand components tended to increase with the duration of continuous defoliation, with the highest vertical fuel continuity occurring after 16 years of continuous defoliation. Such increases in the vertical spatial continuity of fuels may be a key reason for the greater percentage of area burned in those forests which have recently sustained a spruce budworm outbreak.

Keywords: spruce budworm defoliation; vertical fuel continuity; crown fire; forest fire management; forest structure; natural disturbance; insect outbreak; boreal mixedwood forest; interaction

1. Introduction

The spruce budworm, *Choristoneura fumiferana* (Clem.), periodically erupts into large-scale outbreaks during which its favoured host species, balsam fir (*Abies balsamea*), and to a lesser extent white spruce (*Picea glauca*) and black spruce (*Picea mariana*), are defoliated and killed [1]. This insect is a major biotic disturbance throughout the boreal forest in the province of Ontario, Canada. During its last major outbreak (1977–1987), the spruce budworm defoliated roughly 20 million hectares in Ontario [1] and depleted an average of 35 million cubic meters of host tree wood volume annually in Canada [2]. Research points to a large-scale outbreak cycle of variable length averaging about 35 years [3] and lasting between five and fifteen years [2]. Historical records showed that in Ontario the last major outbreak's peak in defoliated area occurred around 1980 [1].

Severe and repeated spruce budworm defoliation is a major stress on forest stands [4] and can begin the process of stand breakdown. Defoliation usually begins at the top of the tree and becomes more extensive as budworm populations grow, with severity a function of the number of larvae feeding on the individual tree [5]. Several years of defoliation can remove the majority of foliage from crown branches [5], initially causing growth reduction, but eventually leading to tree crown mortality termed "top-kill" [5]. If only a few years of defoliation occur, a tree may completely recover; however, after five years of severe defoliation, full-tree mortality is likely to begin [6]. Partially eaten needles of live trees and needles from dying trees may be held in the tree crown by silk produced by budworms for feeding shelters. Reduced structural integrity in a dying or dead tree at first leaves the crown and branches susceptible to breakage from environmental stressors such as wind [7], gradually extending to the entire bole. After this, broken tree components may begin to accumulate suspended in the lower canopy or on the forest floor. The finer diameter elements of this suspended biomass are referred to as ladder fuels (i.e., biomass found above fuel accumulated on the forest floor and below the overstory crown) due to their ability to carry fire vertically into and above the overstory canopy [8]. As tree mortality and windthrow occur, canopy gaps allow previously suppressed trees to reach the canopy. The resulting canopy form will vary in initial composition, and in mixedwood forests a multilevel canopy can be expected [7]. Variable tree mortality rates among spruce budworm host tree species [5] provide a vertical distribution of biomass in differential amounts and stages of decomposition [9]. High levels of tree mortality, top-kill, and an abundance of dead leaf material in the canopy and on the ground represent the drastic shifts in distribution of flammable fuel types and loads that result from defoliation.

Changes in forest structure and the build-up of downed woody debris have been postulated to affect forest fire hazard [10], and historical records show the likelihood of large (>200 ha) fires to be greater during the short 'window of opportunity' following spruce budworm defoliation [11]. Study suggests that the potential for crown fire is likely greatest five to eight years following complete spruce budworm-caused stand mortality, largely due to stand breakdown and the accumulation of surface fuel [10]. After spring flush, the transition of surface fire to crown fire is inhibited by the moist deciduous and herbaceous vegetation layer up to four to five years after stand mortality, after which surface fuel accumulation overcomes this inhibition, making the height of surface fuels with respect to the herbaceous vegetation layer important. Previous work has shown that crown fire occurs disproportionately more often three to nine years following the end of spruce budworm defoliation in Ontario's boreal forest, and that there is clear geographic variation in this 'window of opportunity': Fire begins later and lasts longer in western than in eastern parts of the region, with the difference being attributed to drier climates in the West [11]. The effect of a drier climate likely impacts fuel accumulation and decomposition directly, but may also operate indirectly on spruce budworm population dynamics and host tree composition (i.e., greater ratio of white spruce to balsam fir in the western compared to the eastern region). Subsequent analysis has shown that lagged spruce budworm defoliation (8–10 years) increases the probability of fire ignition, confirming previous conclusions [12].

Vertical fuel continuity describes the vertical spatial distribution of fuels within a stand. Limited attention has been placed on understanding the vertical position of fuels within the fuel ladder and surface stratums, as well as on how the amount and position of those fuels change, with respect to the duration of spruce budworm defoliation. Smaller vertical gaps between surface, ladder, and crown fuels indicate a more vertically continuous fuel. An understanding of such structural changes due to spruce budworm defoliation at the stand-level may point to one of the key driving factors behind landscape-scale observations [11,13], and help highlight the areas where risk of a large forest fire is elevated due to spruce budworm defoliation.

In Ontario, historical fire records show that the vast majority of forest fires burn relatively small areas. Stocks (2018, personal communication) deduced that most of these small fires burn along the surface and only rarely establish themselves in the crown. While relatively few forest fires involve the crown in a substantial way, those few fires that do cause the vast majority of the total area burned [14]. Studies suggest that vertical fuel continuity is an important factor in allowing what began as a surface fire to establish itself in the crown [10,15].

Here, we focus on factors that may contribute to the spruce budworm–crown fire interaction at the stand level. Specifically, we begin to quantify how an increase in duration of defoliation by spruce budworm alters fuel loads and leads to the re-arrangement of fuels in ways that can impact a stand's propensity to support a crown fire.

2. Materials and Methods

2.1. Study Area

The study took place in a forested area over 150 km to the northeast of Warren, Ontario (i.e., referred to as 'Warren'), identified based on the duration of past spruce budworm defoliation and general forest characteristics. Spruce budworm defoliation was established using the Forest Insect and Disease Survey (FIDS) of the Canadian Forest Service (CFS) and the OMNR, which provided aerial mapping of moderate–severe defoliation between 1941 and 2009. Moderate–severe defoliation was considered here to be the loss of 40–100% of new foliage [16].

From the aerially mapped dataset, four zones of continuous, moderate–severe spruce budworm defoliation were identified, corresponding to zero (i.e., no mapped defoliation), four (2004–2007), eight (2000–2007), and sixteen (1993–2008) years. Each zone of defoliation was a minimum of 30 km from the next. The previous defoliation event experienced by each zone occurred from 1973 to 1984, respectively nineteen, fifteen, and eight years prior to the defoliation identified in the most recent zones of continuous defoliation. Forest characteristics were established using Forest Resource Inventory (FRI) data provided by the Ontario Ministry of Natural Resources (OMNR) (Table 1). These data were collected in 1989 (three years prior to the commencement of the most recent period of spruce budworm defoliation), and showed the vegetation type to be boreal mixedwood composed of host (i.e., balsam fir, white spruce, black spruce) and non-host (i.e., sugar maple (*Acer saccharum*), trembling aspen (*Populus tremuloides*), white cedar (*Thuja occidentalis*), white birch (*Betula papyrifera*) and yellow birch (*Betula alleghaniensis*)) trees. According to the FRI data, mean dominant (or co-dominant) tree age of the stands selected in the defoliation zones ranged from 68 to 111 years and from 12 to 18 m in height. Mean host composition ranged from 30% to 50%.

2.2. Site Selection

Earlier work [11] focused on the time lag between the end of spruce budworm defoliation and the subsequent occurrence of a large fire in that area, without regard to the length of the defoliation period. Here, we chose zones that had been continuously defoliated for various lengths of time in order to measure the effect of defoliation duration on subsequent fuel characteristics that influence a stand's ability to sustain a crown fire. Within each defoliation zone (i.e., areas defoliated continuously for zero, four, eight, and sixteen years according to the FIDS defoliation maps), five sites representing independent stands were randomly selected for data collection. These sites were measured in July 2011. We ensured that each of the sites had a minimum host tree composition of 20%, and dominant or co-dominant tree age of 50 years as defined by the FRI data collected from the study area in 1989 (Table 1). The sites also fell within Ontario's Intensive Fire Management Zone, and did not show signs of recent fire. We included five more sites per zone to estimate the frequencies with which dead, top-killed, and topped trees occurred. These sites were measured in May 2011. Latitudinal and longitudinal coordinates were recorded (Table 1) using a handheld global positioning system (GPS) with an accuracy varying according to signal quality (i.e., error < 15 m). Coordinates corresponded to a randomly selected point at 0 m on a 30-m transect that ran as a North–South transect, and was used for data collection.

Table 1. Geographic locations and Forest Resource Inventory (FRI) data pertaining to the 20 sites inventoried in July 2011. Sites were located within boreal mixedwood forest continuously defoliated by spruce budworm for 0, 4, 8, and 16 years from 1993–2008. FRI data for these sites were collected in 1989 by the Ontario Ministry of Natural Resources (OMNR), three years prior to the start of the defoliation period. Stocking is a relative measure of stand density against the reference level for the dominant and co-dominant species. Height and Age pertain to the dominant and co-dominant trees of the leading species in the stand, averaged over the stand. The reported value of a given metric is represented by Value, while standard error of the mean is represented by SEM. Host trees include balsam fir, white spruce, and black spruce.

		July		Forest Resource Inventory Data (1989)							
		Location		Structure				Stocking (0–1)		Composition	
		Coordinate		Height (m)		Age (Years)				Host (%)	
Continuous Defoliation	Site	N (°)	W (°)	Value	SEM	Value	SEM	Value	SEM	Value	SEM
(Years)	(#)										
0	1	46 29 52.9999	79 09 39.5012	21		90		0.8		20	
0	2	46 32 25.9689	79 09 09.1302	19		110		0.7		40	
0	3	46 32 26.2641	79 08 48.2574	15		90		0.5		80	
0	4	46 33 53.6288	79 09 29.5137	14		90		0.7		50	
0	5	46 33 50.3132	79 09 34.8001	11		95		0.7		30	
Mean				16	±2	95	±4	0.7	±0.0	44	±10
4	6	46 47 49.6223	79 48 08.4162	13		110		0.9		50	
4	7	46 37 35.6161	79 48 04.3366	13		150		0.5		20	
4	8	46 45 36.4558	79 48 25.0054	11		103		0.4		20	
4	9	46 44 40.7008	79 49 16.8733	16		95		0.7		30	
4	10	46 44 35.7205	79 49 09.5751	16		95		0.7		30	
Mean				14	±1	111	±10	0.6	±0.1	30	±5
8	11	46 37 58.8879	80 00 18.5172	22		85		0.8		30	
8	12	46 37 18.0186	80 00 38.8205	21		80		0.0		50	
8	13	46 37 02.8668	80 00 26.1413	20		80		0.8		30	
8	14	46 35 50.8079	80 00 26.7027	13		65		0.7		60	
8	15	46 35 34.8847	80 00 32.9673	15		62		0.9		30	
Mean				18	±2	74	±5	0.6	±0.2	40	±6
16	16	46 30 32.7797	80 16 03.1156	13		75		0.6		40	
16	17	46 29 50.5380	80 15 14.1664	7		50		0.5		50	
16	18	46 30 34.4934	80 16 09.8633	13		70		0.5		60	
16	19	46 30 50.4948	80 16 03.5893	13		75		0.6		70	
16	20	46 31 56.0937	80 76 04.3965	14		70		0.5		50	
Mean				12	±1	68	±5	0.5	±0.0	20	
										50	±8

2.3. Data Collection

For the sites measured in each defoliation zone, forest structural data were collected from three points along a 30-m transect (0 m, 15 m, 30 m). Sampling methods followed those previously established [17] and used the point-centered quarter method [18] to inventory host overstory and understory trees. Species type (i.e., non-host, balsam fir, white spruce, black spruce) was also identified. The five sites measured in July were the basis for all analyses with the exception of the inventory of the state of trees. An additional five sites per defoliation zone, those measured in May, were included for that analysis. The vertical forest characteristics measured near Warren, and discussed throughout this study, are presented in Figure 1.

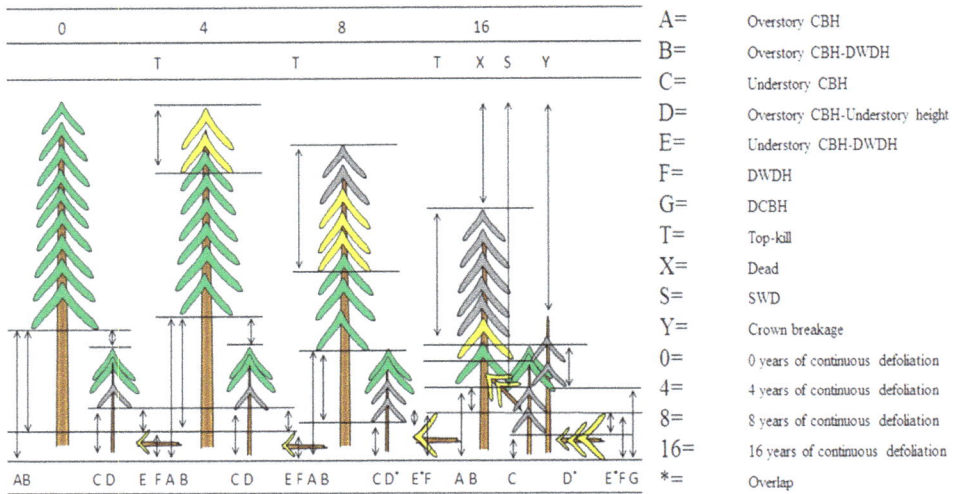

A=	Overstory CBH
B=	Overstory CBH-DWDH
C=	Understory CBH
D=	Overstory CBH-Understory height
E=	Understory CBH-DWDH
F=	DWDH
G=	DCBH
T=	Top-kill
X=	Dead
S=	SWD
Y=	Crown breakage
0=	0 years of continuous defoliation
4=	4 years of continuous defoliation
8=	8 years of continuous defoliation
16=	16 years of continuous defoliation
*=	Overlap

Figure 1. Schematic representation of vertical forest characteristics in boreal forest defoliated by spruce budworm. Represented are host overstory and understory crown base height (CBH), host understory height, top-kill, crown breakage, dead crown base height (DCBH), downed woody debris height (DWDH), suspended woody debris (SWD) and related vertical spacing. Note: The length of top-kill (T) was not measured in this study.

Data collected for all host trees included: the distance from the transect reference point, diameter at breast height (DBH), height (H), live crown base height (LCBH), dead crown base height (DCBH), and crown base height (CBH). Distance from the transect and DBH were measured for host overstory trees having a DBH $\geq$ 3.0 cm. Height was measured using a clinometer for the overstory tree in the first quadrant at each point, and all other tree heights were visually estimated by reference to the measured tree. LCBH was defined here as the lowest height of live foliage and DCBH as the lowest height of dead branches, dead foliage, and lichen visually estimated to sustain vertical fire propagation. Both LCBH and DCBH were measured using a tape if possible, or otherwise visually estimated. Visual estimation of base heights was used when the tape had reached the horizontal plane of the base height from an unobstructed location, in instances when there was a vertical obstruction. In this study, we define crown base height (CBH) as a combination of live and dead crown heights, where live crown base heights were included for live trees, and dead crown base heights for dead trees.

The status of host overstory trees was also assessed. Defoliation was visually estimated as the percentage per branch of absent needles compared to the undefoliated state, then averaged for the entire tree. While this gives a robust estimate of crown foliar fuel load, it differs from the FIDS' method of measurement (i.e., percentage of new foliage lost), and may provide a higher estimate of defoliation.

The health of each tree was assessed following an existing classification system [19] and categorized as either: (i) "Affected" including all dead trees (i.e., stages 3, 4, 6, 7), topped trees (i.e., stage 2), and trees experiencing top-kill (i.e., stage 2); (ii) "Dead" including all dead (i.e., stages 3, 4, 6, 7) and topped trees (i.e., stage 2); and (iii) "Top-kill" trees including only those experiencing top-kill (i.e., stage 2). The categories were not mutually exclusive. The occurrence of suspended woody debris (SWD), fallen branches or upper stems caught in lower stand branches, was recorded for each host tree at the five sites per defoliation zone measured in July.

Measurement of the surface region was important to understand the lower section of the vertical fuel continuum. Downed woody debris height (DWDH) and the maximum height of the herbaceous plant matter were measured at 3-m intervals along the first 15 m of each 30-m transect. DWDH was defined here as the height above ground of the highest branch, bole or foliage that had fallen to the ground. A count of downed woody debris with a diameter of 7.0 cm or greater that crossed the transect was recorded following the line-intersect method [20]. Pieces assessed as sound (i.e., class 1,2) were recorded [19].

2.4. Statistical Methods

The experimental design was stratified, with balanced, random sampling within strata. Data analysis involved comparison of mean forest structural characteristics among treatment groups (i.e., defoliation zones) to determine if any significant differences existed. A one-way analysis of variance (ANOVA) and then a Tukey's post-hoc test were used to compare differences in means for the following variables: host overstory tree composition and dominant tree age obtained from FRI data. Yet some variables were not normally distributed and did not have equal variance. For these variables (i.e., stand density, defoliation extent, occurrence of affected trees, occurrence of suspended woody debris, down woody debris height, vertical forest structure, and the overlap of key forest components) a non-parametric one-way comparison of means was performed using the Kruskal-Wallis rank sum test. A post-hoc, pairwise comparison using the Wilcoxon rank sum test was used to test for differences among means. To describe the occurrence of a trend categorically, contingency tables were created and tested with a Fisher's exact test (labelled as "Fisher's test").

Linear regression was used to test for any significant trend with increased duration of defoliation. Dependent variables tested by regression described vertical forest structure and the overlap of key forest components. These dependent variables were normally distributed and of equal variance, and included: (a) host overstory to host understory crown overlap (host overstory CBH − host understory H); (b) host overstory and host understory crown to downed woody debris overlap (host overstory CBH − host understory H + host understory CBH − DWDH); (c) host overstory to downed woody debris overlap (host overstory CBH − DWDH); and d) host understory to downed woody debris overlap (host understory DCBH − DWDH). The independent variable was years of continuous defoliation.

3. Results

3.1. Forest Characteristics and Defoliation Duration

FRI data collected in 1989 showed general stand characteristics for the study sites, three years prior to the start of the most recent period of spruce budworm defoliation. Host overstory species composition did not significantly differ among defoliation zones ($F_{3,16} = 1.149$, $p = 0.640$). However, sites that later endured four years of continuous defoliation consisted of a significantly lower composition of balsam fir (Kruskal-Wallis H: 10.480, df: 3, n: 5, $p = 0.013$) than those that endured sixteen years. The zone of four years of continuous defoliation was also significantly greater in age than zones of eight and sixteen years of continuous defoliation (Kruskal-Wallis H: 15.387, df: 3, n: 5, $p = 0.002$ and Wilcoxon rank sum test; $p < 0.05$). Stocking did not significantly differ among defoliation zones (Kruskal-Wallis H: 4.075, df: 3, n: 5, $p = 0.253$). Dominant tree height prior to sixteen years of continuous defoliation was significantly less than eight years ($F_{3,16}$ = 3.255, p = 0.049 and Tukey's post-hoc test; $p < 0.05$).

Post-defoliation, stands were of similar density and dominated by host tree species (Table 2). Mean stand density and associated standard error of the mean (SEM) was 3321 stems/ha ± 103 stems/ha with a minimum of 50% host overstory composition. While the balsam fir component of host overstory trees varied among defoliation zones (Kruskal-Wallis H: 8.368, df: 7, p = 0.039), the overall density of balsam fir trees in each defoliation zone was not significantly different and averaged 1680 stems/ha ± 371 stems/ha (SEM). Independent (i.e., July) and pooled (i.e., May and July) stand densities were not significantly different (Kruskal-Wallis H: 6.69, df: 7, p = 0.46). Host understory trees were primarily balsam fir and to a lesser extent, white spruce. Mean host understory density was 903 (±174) stems per hectare, 2348 (±1947) stems per hectare, 1844 (±425) stems per hectare and 3624 (±854) stems per hectare in the zones of zero, four, eight and sixteen years of continuous defoliation, respectively.

The severity of spruce budworm defoliation and impact to stand structure tended to increase with duration of defoliation. Defoliation was greater in zones of four, eight and sixteen years of continuous defoliation than zero years (18% ± 3% SEM), with 50% ± 4% SEM, 52% ± 5% SEM, 63% ± 4% SEM respectively (Kruskal-Wallis H: 71.930, df: 3, n: 60, $p < 0.001$ and Wilcoxon rank sum test; $p < 0.05$) with zone sixteen showing the greatest defoliation (Wilcoxon rank sum test; $p < 0.05$). The occurrence of trees showing symptoms of spruce budworm defoliation (i.e., affected trees) tended to increase to eight years of continuous defoliation (Figure 2) and was significantly greater than after zero and four years (Kruskal-Wallis H: 22.194, df: 3, n: 10, $p < 0.001$ and Wilcoxon rank sum test; $p < 0.05$). The percentage of dead and topped trees followed a similar trend (Dead Kruskal-Wallis H: 21.305, df: 3, n: 10, $p < 0.001$, Wilcoxon rank sum test; $p < 0.05$ and Topped Kruskal-Wallis H: 20.929, df: 3, n: 10, $p < 0.001$ and Wilcoxon rank sum test; $p < 0.05$). Top-kill peaked after sixteen years of continuous defoliation, and was significantly greater than after zero and four years (Kruskal-Wallis H: 11.98, df: 3, n: 10, $p < 0.01$ and Wilcoxon rank sum test; $p < 0.05$).

Table 2. Overstory tree (DBH ≥ 3 cm) density and composition in zones of boreal mixedwood forest continuously defoliated by spruce budworm for zero, four, eight and sixteen years from 1993–2008 near Warren, Ontario. Forest characteristics were measured at five sites within each defoliation zone in May 2011. The same measurements were repeated at independent sites in July 2011. The pooled data (i.e., ten sites per defoliation zone) are shown separately from the data collected in July 2011. Standard error of the mean is represented by SEM. Host trees include balsam fir, white spruce and black spruce.

Continuous Defoliation	n	Density (Stems/ha)		Composition (%)			
(Years)	(# of Sites)	Overall	SEM	Host (of Total)	SEM (of Total)	Balsam Fir (of Host)	SEM (of Host)
Data from July 2011							
0	5	3142	±702	72	±31	93	±3
4	5	2664	±412	51	±29	57	±12
8	5	2847	±337	94	±24	48	±16
16	5	4630	±280	90	±40	67	±18
Pooled data from May and July 2011							
0	10	3203	±720	72	±37	-	-
4	10	3242	±744	59	±33	-	-
8	10	2895	±392	94	±53	-	-
16	10	3793	±1056	87	±48	-	-

A spike in suspended and downed woody debris reflected the occurrence of dead and topped trees. The quantity of suspended woody debris after sixteen years was significantly different from all other years in both the host overstory and host understory (Overstory Kruskal-Wallis H: 9.78, df: 3, n: 5, p = 0.02, Wilcoxon rank sum test; $p < 0.05$ and Understory Kruskal-Wallis H: 11.95, df: 3, n: 5, $p < 0.01$, Wilcoxon rank sum test; $p < 0.05$). The height of downed woody debris was significantly greater after eight and sixteen years than zero years (Kruskal-Wallis *H*: 22.169, df: 3, *n*: 25, *p* < 0.001, Wilcoxon rank sum test; *p* < 0.05; Figure 3) and surpassed the height of the herbaceous layer after four and eight years (Kruskal-Wallis *H*: 11.22, df: 3, *n*: 25, *p* = 0.01, Wilcoxon rank sum test; *p* < 0.05).

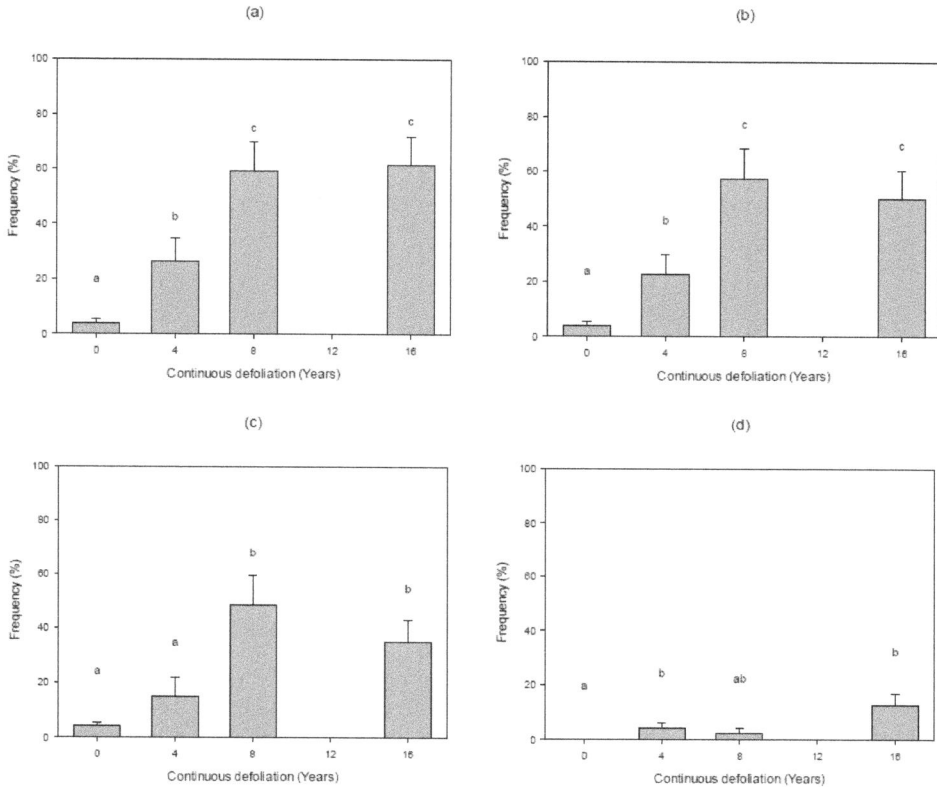

Figure 2. Mean percentage of host overstory trees (*n* = 10) with given trait by years of continuous spruce budworm defoliation from 1993–2008 in boreal mixedwood forest near Warren, Ontario: (**a**) Affected (all dead trees, topped trees and trees experiencing top-kill); (**b**) Dead (all dead trees); (**c**) Topped (all topped trees); and (**d**) Top-kill. Among treatments, means with the same letters were not significantly different (Wilcoxon rank sum test; *p* < 0.05). Error bars show standard error of the mean.

(a)   (b)

(c)   (d)

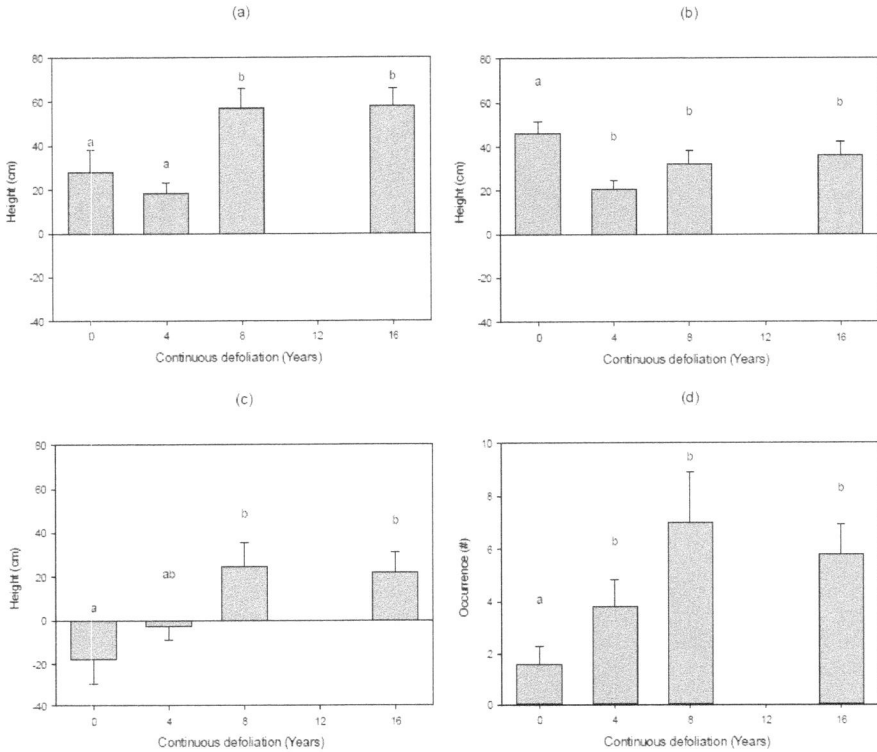

Figure 3. Mean surface biomass height by years of continuous spruce budworm defoliation from 1993–2008 in boreal mixedwood forest near Warren, Ontario: (**a**) Downed woody debris height (DWDH; *n* = 25); (**b**) Herbaceous height (*n* = 25); (**c**) Difference between DWDH and herbaceous height (*n* = 25); and (**d**) Downed woody debris (DWD) occurrence ≥ 7.0 cm (*n* = 5). Means with the same letters among treatments were not significantly different (Wilcoxon rank sum test; $p < 0.05$). Error bars show standard error of the mean.

3.2. Vertical Fuel Continuity

Aspects of vertical forest structure responded to variation in defoliation duration (Figure 4). Defoliation duration significantly predicted the observed decrease in host overstory height ($R^2 = 0.33$, $F_{3,17} = 8.84$, $p < 0.01$) and diameter at breast height ($R^2 = 0.30$, $F_{3,17} = 7.81$, $p = 0.01$). After eight and sixteen years, top height and crown base height were significantly lower than after zero and four years (Kruskal-Wallis *H*: 21.469, df: 3, *n*: 60, $p < 0.001$, Wilcoxon rank sum test; $p < 0.05$ and Kruskal-Wallis *H*: 24.351, df: 3, *n*: 60, $p < 0.001$, Wilcoxon rank sum test; $p < 0.05$, respectively). There was no significant difference in host understory heights.

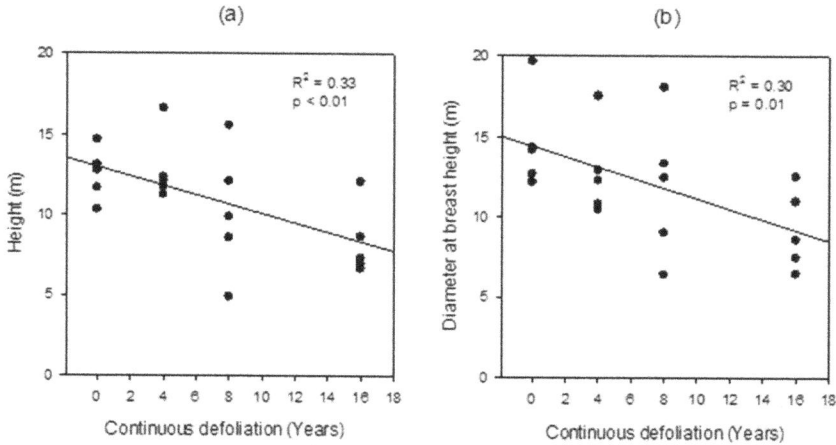

Figure 4. Linear regression of host overstory tree structure (dependent variables) by years of continuous spruce budworm defoliation (independent variable) from 1993–2008 in boreal mixedwood forest near Warren, Ontario: (**a**) Host tree height (Ht); and (**b**) Diameter at breast height (DBH). Linear regression showed years of continuous defoliation significantly predicted host tree height ($R^2 = 0.33$, $F_{3,17} = 8.84$, $p < 0.01$) and diameter at breast height ($R^2 = 0.30$, $F_{3,17} = 7.81$, $p = 0.01$). The equations of the regression lines were Ht = −0.29 (Continuous defoliation) +13.00 and DBH = −0.33 (Continuous defoliation) +14.41, respectively. The coefficient estimates with SEM for host tree height and diameter at breast height were −0.29 ± 0.10, 13.00 ± 0.89 and −0.33 ± 0.12, 14.41 ± 1.07, respectively.

Vertical fuel continuity among defoliation zones was assessed by determining the overlap of key stand components (Figure 5). The gap between host overstory and host understory crowns decreased with increasing years of defoliation such that after sixteen years of continuous defoliation, the vertical components overlapped. Overlap of host overstory and host understory crowns (i.e., the ladder region) occurred in all sites (Fisher's test $p < 0.01$) and in 49 of 60 quadrants (Fisher's test $p < 0.001$) after sixteen years. The gap between host tree crowns and downed woody debris was smaller after eight and sixteen years of continuous defoliation than zero and four years for both the host overstory (Kruskal-Wallis H: 38.003, df: 3, n: 60, $p < 0.001$, Wilcoxon rank sum test; $p < 0.05$) and host understory (Kruskal-Wallis H: 19.951, df: 3, n: 60, $p < 0.001$, Wilcoxon rank sum test; $p < 0.05$). Host understory dead crowns overlapped downed woody debris height after eight and sixteen years of defoliation (Kruskal-Wallis H: 122.202, df: 3, n: 60, $p < 0.001$, Wilcoxon rank sum test; $p < 0.05$). These results were supported by regression analysis (Table 3 and Figure 6).

Table 3. Estimated linear model parameters of the space between measured stand structural components (dependent variables) by years of continuous spruce budworm defoliation (independent variable) from 1993–2008 in boreal mixedwood forest near Warren, Ontario: Host overstory crown and downed woody debris (Overstory CBH – DWDH); host overstory crown, host understory crown and downed woody debris (Overstory CBH – Understory H + Understory CBH – DWDH); host overstory and understory crowns (Overstory CBH – Understory H); and host understory dead crown base and downed woody debris (Understory DCBH – DWDH). Standard error of the mean is represented by SEM.

Dependent Variable	Parameter	Estimate	R^2	SEM	*t* stat	*p*
Overstory CBH – DWDH (m)	Intercept	3.698		0.544	6.803	<0.001
Overstory CBH – DWDH (m)	Slope	−0.140	0.237	0.059	−2.364	0.030
Overstory CBH – Understory H + Understory CBH - DWDH (m)	Intercept	2.153		0.438	4.916	<0.001
Overstory CBH – Understory H + Understory CBH- DWDH (m)	Slope	−0.107	0.218	0.048	−2.240	0.038
Overstory CBH – Understory H (m)	Intercept	1.594		0.541	2.948	0.009
Overstory CBH – Understory H (m)	Slope	−0.126	0.203	0.059	−2.140	0.046
Understory DCBH – DWDH (m)	Intercept	−0.037		0.075	−0.499	0.624
Understory DCBH – DWDH (m)	Slope	−0.030	0.428	0.008	−3.669	0.002

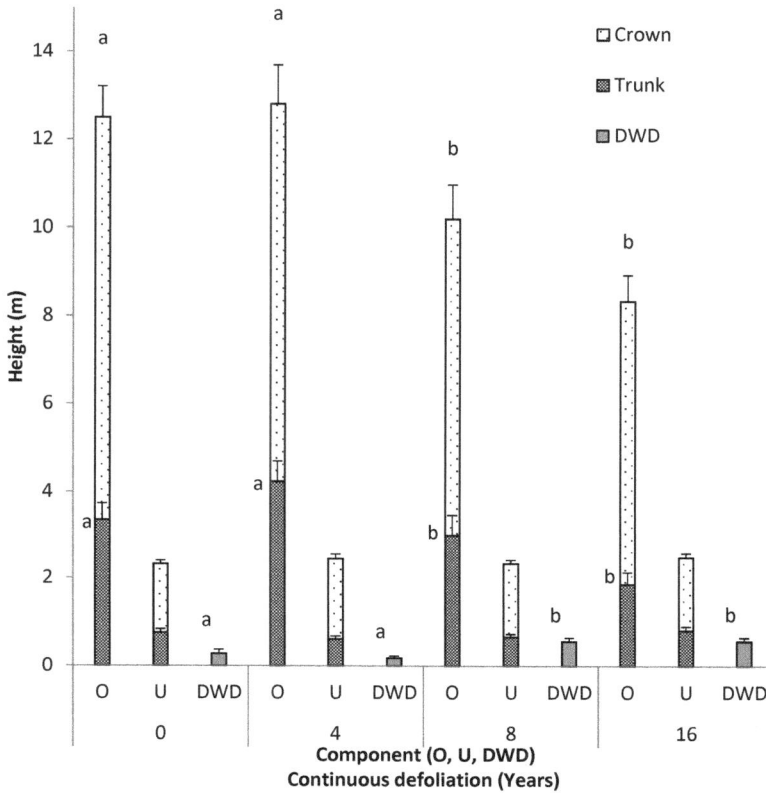

Figure 5. Vertical forest structure showing mean top height (top of crown component; *n* = 60), crown base height (bottom of crown component and top of trunk component; *n* = 60), downed woody debris height (DWDH; *n* = 25) by years of continuous spruce budworm defoliation from 1993–2008 in boreal mixedwood forest near Warren, Ontario for host overstory (O) trees, host understory (U) trees and downed woody debris (DWD). Among treatments, means with the same letters were not significantly different (Wilcoxon rank sum test; *p* < 0.05). Error bars show standard error of the mean. Note: Eight to sixteen years on *x*-axis is eight years.

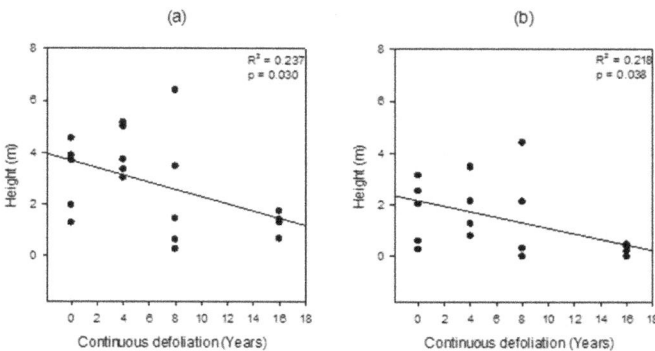

Figure 6. *Cont.*

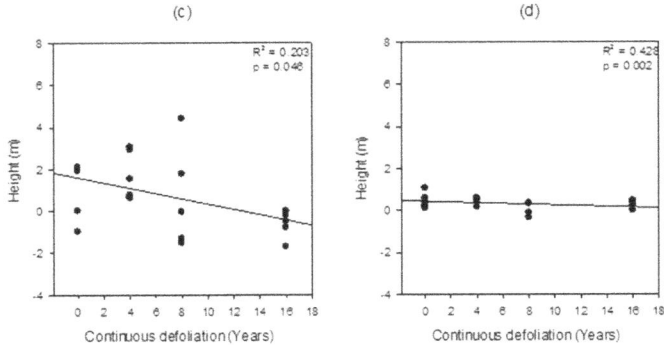

Figure 6. The distance between selected vertical stand components (dependent variables) by years of continuous spruce budworm defoliation (independent variable) from 1993–2008 in boreal mixedwood forest near Warren, Ontario: (**a**) Host overstory crown and downed woody debris (Overstory CBH − DWDH); (**b**) Host overstory crown, host understory crown and downed woody debris (Overstory CBH − Understory H + Understory CBH − DWDH); (**c**) Host overstory and understory crowns (Overstory CBH − Understory H); and (**d**) Host understory dead crown and downed woody debris (Understory DCBH − DWDH). For all four categories of number of years of continuous defoliation, *n* = 5.

4. Discussion

4.1. Forest Characteristics and Defoliation Duration

Tree mortality can be expected to begin after four to five years of moderate–severe spruce budworm defoliation, but will vary with forest composition and age [6]. After the onset of spruce budworm defoliation, stressed trees begin to experience crown breakage and windthrow, and the amount of freshly fallen, woody debris begins to accumulate above normal background levels [10]. We observed such accumulation of downed woody debris as the frequency of stressed, dying and dead trees increased with the duration of defoliation. This suggests that spruce budworm defoliation drove stand breakdown.

The accumulation of downed woody debris has implications for surface fire intensity. In our study, downed woody debris with a diameter $\geq$ 7.0 cm was used as a surrogate for understanding, in a relative sense, the amount of finer branch woody debris present (i.e., the material < 1.0 cm in diameter that would be consumed in flaming) for each defoliation zone. The build-up of downed woody debris in the larger size class (i.e., $\geq$7.0 cm) with the duration of defoliation suggests a greater presence of fuels consumed in flaming. Such an increase in available surface fuel loads would increase surface fire intensity [15]. However, it is important to be cautious here as smaller-diameter fuels may accumulate faster and decompose sooner than larger diameter fuels. As has been shown, the rapid decomposition of finer branch woody debris may limit any increase in the likelihood of fire despite the gradual accumulation of larger-diameter fuels following spruce budworm-caused stand mortality in wet regions of eastern Canada [21]. Thus, varying rates of decomposition can limit the accuracy of this relative comparison.

As a result of crown breakage and mortality, downed woody debris increased in both quantity and height with the duration of defoliation. After eight years of defoliation, downed woody debris height surpassed that of the herbaceous layer. This is important because the high-moisture leaves of the herbaceous layer may have a dampening effect on crown fire initiation. In a previous study, such a phenomenon was suggested as a possible contributing factor in the limited crowning observed during summer months for spruce budworm-defoliated stands [10]. However, when downed woody debris surpasses the herbaceous layer, as we observed, the dampening effect may be overcome.

The increase in suspended woody debris with the increase in the duration of continuous defoliation may also be attributed to the stand degradation caused by defoliation. Dispersing spruce budworm moths have been found to vertically drop from the horizontal air column [22]. Therefore, trees that reach higher into the canopy may be easy targets for oviposition by dispersing moths, making the upper canopy a likely site of oviposition. The abundance of eggs and the close proximity to present-year needles [23] may also make the upper canopy a favored feeding area for larvae. The preference for spruce budworm to defoliate mature, and taller, host trees [5] may have opened the canopy through defoliation, crown breakage, and the mortality of taller trees. Such an opening could promote the recruitment of immature, shade-tolerant spruce and balsam fir that were suppressed in the understory prior to disturbance [24,25], and explain the dense layer of immature host overstory trees (3–9 cm in DBH) and host understory trees (3624 stems per hectare) observed after 16 years of continuous defoliation. This layer may have also offered greater opportunity for the suspension of woody debris with a net-like effect as crown breakage progressed. The spike in suspended woody debris after 16 years of defoliation may also have been due to the greater composition of younger balsam fir and white spruce initially present, whose branches may have caught the falling trees, crowns, and branches of the mature generation that stood above.

4.2. Vertical Fuel Continuity

The accumulation of biomass in the surface and ladder fuel regions of a forest has implications for the transfer of fire from the surface to the crown region [15]. Greater fuel loads on the surface support higher energy surface fires, and consequently longer flames that are able to reach higher and approach the canopy [15]. While the increase in fuel quantity may have an even greater effect on fire behaviour, fuels above the surface help propagate surface flames upward towards the canopy, but at relatively lower energy levels. A continuous connection of fuel from the surface to the ladder to the crown region of a forest would therefore facilitate the climb of lower energy flames. Where overlaps occur (e.g., the host understory dead crown sat below the high point of downed woody debris), continuity will be enhanced, and where gaps occur (e.g., the host understory dead crown sat above the high point of downed woody debris), continuity will be reduced.

Fuel continuity increased with the duration of continuous defoliation. After 16 years, host overstory crowns overlapped host understory crowns, and host understory dead crowns overlapped downed woody debris. The reduced tendency for individual trees to shed their lowest branches as stands open and the recruitment of younger trees into the developing overstory gaps as crown breakage progresses may have contributed to the reduction in both host overstory height and crown base height. The defoliation of larger host trees likely deposited increased downed woody debris in close proximity and likely increased the probability that under these trees the downed woody debris height was greater than the height of the herbaceous layer. The combined effect here would suggest an increase in the probability that a surface fire would reach the crown of the larger host trees where additional continuity exists.

Previous study has found that large fires (>200 ha) were more likely to occur during a window of opportunity three to nine years following the cessation of spruce budworm defoliation, without regard to the length of the defoliation period [11]. The data in our study were gathered four and three years following the end of defoliation periods lasting zero, four, eight and sixteen years, which aligns closely with the beginning of the proposed 'window of opportunity' [11]. The increase in the parameters measured (i.e., vertical continuity) with progressive annual defoliation suggests that consideration of defoliation length may provide further resolution to these findings. More specifically, stand breakdown, surface fuel accumulation and vertical continuity were greater after sixteen years than four years of continuous defoliation. While the likelihood of large fires may increase three years after any length of defoliation, the likelihood of large fires occurring within the window of opportunity may be greater if the period of defoliation is longer. Determining the impact of the parameters measured here on the likelihood that a surface fire will transition to a crown fire presents a valuable area of further investigation.

4.3. Limitations

This was an unplanned 'silvicultural experiment'. We were unable to direct where and when the treatment (i.e., budworm defoliation) was applied, so some initial zone differences are to be expected. We were also unable to measure the same stands/trees prior to and after defoliation (over the 20-year period from 1989 to 2011), and therefore we cannot be certain that the changes in structure were due only to spruce budworm defoliation, and not also to, for example, site quality or host tree age. We believe that using the FRI data for our initial measurement of pre-defoliation (1989) site quality and stand age amplified any apparent differences among the zones, and that the difference would have been much less if we were able to make pre-defoliation measurements in 1989.

The scope of this study was to compare differences in forest structure among defoliation zones, while also limiting the variation of forest characteristics prior to defoliation. Hence, the primary criterion used to select the sample sites was 'years of moderate–severe continuous defoliation', according to the FIDS defoliation maps. The five sites selected within each zone of continuous defoliation were otherwise randomly selected with several constraints including: a minimum host tree composition of 20% and a minimum dominant or co-dominant tree age of 50 years, as defined by the FRI data collected in 1989.

The FRI is an effective tool to describe general stand characteristics. For example, described metrics (e.g., stand height and age) pertain to the dominant and co-dominant trees of the leading species in the stand, averaged over the stand. However, in the boreal mixedwood forest investigated here, host trees were not necessarily the dominant or co-dominant species, and the structural metrics in Table 1 do not necessarily describe the structural characteristics of the host trees in the stand (i.e., those trees defoliated by spruce budworm). Nonetheless, the FRI data were still used to estimate stands that were similar in structure prior to defoliation because, given the resources of this study, it was not feasible to measure the forest structure of sites in both 1989 (pre-defoliation) and 2011 (post-defoliation), nor ensure those sites would endure spruce budworm defoliation. Given that it was not possible to measure the same sites before and after defoliation, it could be true that to some extent the observed differences among defoliation zones (e.g., stand height and vertical fuel continuity) existed prior to defoliation. This may suggest, for example, that vertical fuel continuity, as shown in Figures 4–6, is greater in stands that later experience longer durations of defoliation.

It has long been recognized that spruce budworm preferentially attacks over-mature, dominant and co-dominant host trees and that, consequently, outbreaks can have major impacts on stand ages and tree heights of the host species [26]. In fact, our measurements suggest structural changes that would reduce tree height and stand age. Differences in stand structure were particularly evident in sites defoliated continuously for eight and sixteen years compared to those defoliated for zero and four years. For example, significant differences were found in the occurrence of top-killed, topped and dead host overstory trees, host overstory tree height, and downed woody debris height. These characteristics are related in that as the tallest trees die or lose their tops after top-kill, stand height declines and downed woody debris increases. Given such observations, we believe that the duration of spruce budworm defoliation does increase stand breakdown and vertical fuel continuity, as detailed in this study.

5. Conclusions

Vertical fuel continuity increased with the duration of defoliation. Further investigation into the changes in vertical forest structure by continuous spruce budworm defoliation, specifically those associated with crown fire initiation and spread probability, will advance a more mechanistic and causal understanding of the process underlying the patterns revealed by landscape-scale analyses of the interaction between spruce budworm and crown fire [11,13]. Surface, ladder, and crown fuel properties measured in this study should be compared with those of other studies (i.e., [10]), and if possible, combined with an estimate of the impact of vertical fuel continuity on crown fire. This will provide further insight into landscape-scale trends regarding the likelihood of large fires in forest following spruce budworm defoliation [11,13].

Author Contributions: G.A.W., R.A.F., S.M.S. and M.-J.F. conceived and designed the experiments; G.A.W. performed the experiments; G.A.W., R.A.F. and M.-J.F. analyzed the data; G.A.W. wrote the paper.

Acknowledgments: Sincere thank you to Mike Wotton, Brian Stocks, Jean-Noel Candau, and Bill DeGroot for valuable consultation. Thank you to the CFS (Ron Fournier) and the OMNR (Taylor Scarr) for providing historical spruce budworm data. We also thank the Ontario Ministry of Natural Resources CRA (SPA 90198 and 90170) for financial support. We would like to dedicate this paper to acknowledge the substantial contribution made by the late Jan Volney to the study of spruce budworm outbreaks.

Conflicts of Interest: The authors declare no conflict of interest.

References

1. Candau, J.; Fleming, R.; Hopkin, A. Spatiotemporal patterns of large-scale defoliation caused by the spruce budworm in Ontario since 1941. *Can. J. For. Res.* **1998**, *28*, 1733–1741. [CrossRef]
2. Fleming, R.A. Climate change and insect disturbance regimes in Canada's boreal forests. *World Resour. Rev.* **2000**, *12*, 520–554.
3. Volney, W.J.; Fleming, R. Climate change and impacts of boreal forest insects. *Agric. Ecosyst. Environ.* **2000**, *82*, 283–294. [CrossRef]
4. Brassard, B.; Chen, H. Stand structural dynamics of North American boreal forests. *Crit. Rev. Plant. Sci.* **2006**, *25*, 115–137. [CrossRef]
5. MacLean, D.A. Effects of spruce budworm outbreaks on the productivity and stability of balsam fir forests. *For. Chron.* **1984**, *60*, 273–279. [CrossRef]
6. MacLean, D.A. Vulnerability of fir-spruce stands during uncontrolled spruce budworm outbreaks: A review and discussion. *For. Chron.* **1980**, *56*, 213–221. [CrossRef]
7. Taylor, S.; MacLean, D. Legacy of insect defoliators: Increased wind-related mortality two decades after a spruce budworm outbreak. *For. Sci.* **2009**, *55*, 256–267.
8. Menning, K.; Stephens, S. Fire climbing in the forest: A semiqualitative approach to assessing ladder fuel hazards. *West. J. Appl. For.* **2007**, *22*, 88–93.
9. Taylor, S.; MacLean, D. Dead wood dynamics in declining balsam fir and spruce stands in New Brunswick, Canada. *Can. J. For. Res.* **2007**, *37*, 750–762. [CrossRef]
10. Stocks, B.J. Fire potential in the spruce budworm-damaged forests of Ontario. *For. Chron.* **1987**, *63*, 8–14. [CrossRef]
11. Fleming, R.; Candau, J.; McAlpine, R. Landscape-scale analysis of interaction between insect defoliation and forest fire in central Canada. *Clim. Chang.* **2002**, *55*, 251–272. [CrossRef]
12. James, P.M.; Robert, L.E.; Wotton, B.M.; Martell, D.L.; Fleming, R.A. Lagged cumulative spruce budworm defoliation affects the risk of fire ignition in Ontario, Canada. *Ecol. Appl.* **2017**, *27*, 532–544. [CrossRef] [PubMed]
13. Candau, J.; Fleming, R.; Wang, X. Ecoregional patterns of spruce budworm-wildfire interactions in central Canada's forests. *Forests* **2018**, *9*, 137. [CrossRef]
14. Stocks, B.J. The extent and impact of forest fires in northern circumpolar countries. In *Global Biomass Burning: Atmospheric, Climatic, and Biospheric Implications*; Levine, J.S., Ed.; MIT Press: Cambridge, MA, USA, 1991; pp. 197–202.
15. Van Wagner, C.E. Conditions for the start and spread of crown fire. *Can. J. For. Res.* **1977**, *7*, 23–34. [CrossRef]
16. Scarr, T.; Ryall, K.; Hodge, P. (Eds.) *Forest Health Conditions in Ontario, 2011*; Queen's Printer for Ontario: Sault Ste. Marie, ON, Canada, 2012; pp. 1–100.
17. Lavoie, N.; Alexander, M.; Macdonald, S. Photo guide for qualitatively assessing the characteristics of forest fuels in a jack pine—Black spruce chronosequence in the Northwest Territories. *Nat. Res. Can.* **2010**, 1–51. Available online: http://cfs.nrcan.gc.ca/publications/download-pdf/31785 (accessed on 19 April 2013).
18. Cottam, G.; Curtis, J. The use of distance measures in phytosociological sampling. *Ecology* **1956**, *37*, 451–460. [CrossRef]

19. Maser, C.; Anderson, R.G.; Cromack, K.; Williams, J.T.; Martin, R.E. Dead and down woody material. In *Wildlife Habitats in Managed Forests: The Blue Mountains of Oregon and Washington*; Thomas, J.W., Ed.; USDA: Portland, OR, USA, 1979; Volume 553, pp. 78–95.

20. Van Wagner, C.E. The line intersect method in forest fuel sampling. *For. Sci.* **1968**, *14*, 20–26.

21. Péch, G. Fire hazard in budworm-killed balsam fir stands on Cape Breton Highlands. *For. Chron.* **1993**, *69*, 178–186. [CrossRef]

22. Greenbank, D.O.; Schaefer, G.W.; Rainey, R.C. Spruce budworm (Lepidoptera: Tortricidae) moth flight and dispersal: New understanding from canopy observations, radar, and aircraft. *Mem. Entomol. Soc. Can.* **1980**, *110*, 1–49. [CrossRef]

23. Fleming, R.; Piene, H. Spruce budworm defoliation and growth loss in young balsam fir: Cohort models of needlefall schedules for spaced trees. *For. Sci.* **1992**, *38*, 287–304.

24. Bergeron, Y. Species and stand dynamics in the mixed woods of Quebec's southern boreal forest. *Ecology* **2000**, *81*, 1500–1516. [CrossRef]

25. Bouchard, M.; Kneeshaw, D.; Bergeron, Y. Mortality and stand renewal patterns following the last spruce budworm outbreak in mixed forests of western Quebec. *For. Ecol. Manag.* **2005**, *204*, 297–313. [CrossRef]

26. MacLean, D.A. Impact of insect outbreaks on tree mortality, productivity, and stand development. *Can. Entomol.* **2015**, *148*, S138–S159. [CrossRef]

![forests logo] *forests*

MDPI

Article

Deforestation-Induced Fragmentation Increases Forest Fire Occurrence in Central Brazilian Amazonia

Celso H. L. Silva Junior [1,*], Luiz E. O. C. Aragão [1,2], Marisa G. Fonseca [1], Catherine T. Almeida [1], Laura B. Vedovato [1,2] and Liana O. Anderson [3]

[1] Tropical Ecosystems and Environmental Sciences Laboratory (TREES), Remote Sensing Division, National Institute for Space Research-INPE, São José dos Campos 12227-010, SP, Brazil; luiz.aragao@inpe.br (L.E.O.C.A.); marisa_fonseca@yahoo.com.br (M.G.F.); cathe.torres@gmail.com (C.T.A.); lauravedovato2@gmail.com (L.B.V.)
[2] College of Life and Environmental Sciences, University of Exeter, Exeter EX4 4RJ, UK
[3] National Center for Monitoring and Early Warning of Natural Disasters-CEMADEN, São José dos Campos 12247-016, SP, Brazil; liana.anderson@cemaden.gov.br
* Correspondence: celso.junior@inpe.br; Tel.: +55-12-3208-6425

Received: 22 January 2018; Accepted: 1 May 2018; Published: 1 June 2018

Abstract: Amazonia is home to more than half of the world's remaining tropical forests, playing a key role as reservoirs of carbon and biodiversity. However, whether at a slower or faster pace, continued deforestation causes forest fragmentation in this region. Thus, understanding the relationship between forest fragmentation and fire incidence and intensity in this region is critical. Here, we use MODIS Active Fire Product (MCD14ML, Collection 6) as a proxy of forest fire incidence and intensity (measured as Fire Radiative Power—FRP), and the Brazilian official Land-use and Land-cover Map to understand the relationship among deforestation, fragmentation, and forest fire on a deforestation frontier in the Brazilian Amazonia. Our results showed that forest fire incidence and intensity vary with levels of habitat loss and forest fragmentation. About 95% of active fires and the most intense ones (FRP > 500 megawatts) were found in the first kilometre from the edges in forest areas. Changes made in 2012 in the Brazilian main law regulating the conservation of forests within private properties reduced the obligation to recover illegally deforested areas, thus allowing for the maintenance of fragmented areas in the Brazilian Amazonia. Our results reinforce the need to guarantee low levels of fragmentation in the Brazilian Amazonia in order to avoid the degradation of its forests by fire and the related carbon emissions.

Keywords: remote sensing; MODIS; Amazonian forests; Brazilian Forest Code; edge effects

1. Introduction

Tropical forests are globally important reservoirs of carbon (C) and biodiversity [1–3]. Vegetation in this region stores between 350–600 Pg C [3–7], while the atmosphere stores about 750 Pg C [8]. The loss of these C stocks due to deforestation and forest degradation is estimated to be approximately 1.1 Pg C·year^{-1} [9–11]. Amazonia, specifically, is home to more than half of the world's remaining rainforest areas [12]. However, in the Brazilian Amazonia, intense land-use and land-cover changes and forest degradation threaten the forest structure, biodiversity, and ecological functions [13].

The intense occupation of Brazilian Amazonia from the 70s [14], aiming to expand agricultural and livestock activities and to increase the wood supply, besides a general lack of enforcement of environmental laws, caused the dramatic increase of deforestation rates, reaching a peak of 27,772 km^2 in 2004 [15,16]. After 2005, a steep decrease in deforestation rates was observed, which can be attributed to a combination of factors, including governmental enforcement of environmental laws, restrictions on access to credit, expansion of protected areas, and civil society interventions in the soy and beef supply chains [16]. Nonetheless, the deforestation rate increased markedly in 2015 and 2016 [15] (24% and 27% in relation to the previous year, respectively), raising concerns that the recent weakening of environmental-protection policies could be already reversing the Brazilian progress in reducing the Amazonian forest destruction.

Whether at a slower or faster pace, continued deforestation cumulatively causes forest habitat loss, altering habitat configuration, such as the change in spatial arrangement of the remaining habitat through forest fragmentation. Metrics of habitat configuration, such as the number and mean size of forest patches and edge length covary with habitat amount. Understanding these relationships is important to correctly interpret the effects of habitat fragmentation on tropical forests [17]. Following Farhig (2003) [18], the mean patch size of remaining forests is expected to linearly decrease with the reduction in habitat amount, while both the number of patches and the total edge are expected to rise up to a certain threshold of habitat loss and then decrease with increasing deforestation.

Forest edges resulting from landscape fragmentation are highly fire-prone due to increased dryness, higher fuel load compared to forest interior and proximity to ignition sources from adjacent management areas [19–24]. Fragmentation and its resulting edge effects may act synergistically with the ongoing large-scale changes in climate and fire regimes, threatening the Amazonian forest ecological integrity [13,25].

Much of the literature on the effects of habitat loss and changes in habitat configuration has focused on biodiversity maintenance and population persistence. Studies concerning the effect of habitat loss and configuration on forest fires incidence and intensity at the landscape scale are rare in the Brazilian Amazonia, especially in active deforestation frontiers, where the interactions between deforestation, forest fragmentation and fire are evident. In other regions of the Amazon Basin, some authors have demonstrated a positive response of fire incidence and intensity to increased fragmentation and forest edges in the landscape [20–22,26,27].

In Brazil, the Forest Code (Federal Law 12.727/2012) is the main national law that is regulating the conservation of forests within private properties [28]. This law determines that, within the Amazon Biome, at least 80% of each rural property should not be deforested in order to ensure the sustainable use of natural resources, assisting in the conservation and rehabilitation of ecological processes, promoting the conservation of biodiversity, as well as the shelter and the protection of wildlife and native flora. The question of whether such a high level of habitat maintenance is necessary to reduce fire incidence in the region, however, has not been directly addressed yet.

To fill this gap, we relate, for the first time, habitat configuration metrics with fire incidence and intensity in an active Brazilian Amazonia deforestation frontier, aiming to identify the relationships between forest fragmentation and fire on the landscape scale. To achieve this, we address the following question: What is the relationship between habitat loss and measures of habitat configuration, and their implications for fire incidence and intensity in a central Amazonian landscape?

2. Study Area

Our study site is located in the northern region of Novo Progresso municipality, State of Pará, Central Brazilian Amazonia, with an area of 30,000 km^2 (3×10^6 ha) (Figure 1), which approximately corresponds to the area of Belgium. This region is known as a frontier of deforestation because of high rates of deforestation in the last 10 years. The vegetation is predominantly composed of the Dense Ombrophilous Forest, with trees that can reach heights up to 50 m [29].

The initial occupation of this area was associated with governmental settlement projects and the construction of road infrastructure, mainly the construction of BR-163 highway [30]. During the 70s and 80s, a spontaneous colonization phenomenon occurred in the region, which was characterized by the occupation of land by small subsistence farmers and gold miners [30]. There are three main deforestation patterns that are present in the study area (i) fishbone, associated with settlements,

(ii) rectangular patches, related to large rural properties, and (iii) stem of the rose pattern that is associated with mining areas, mainly in BR-163 [31].

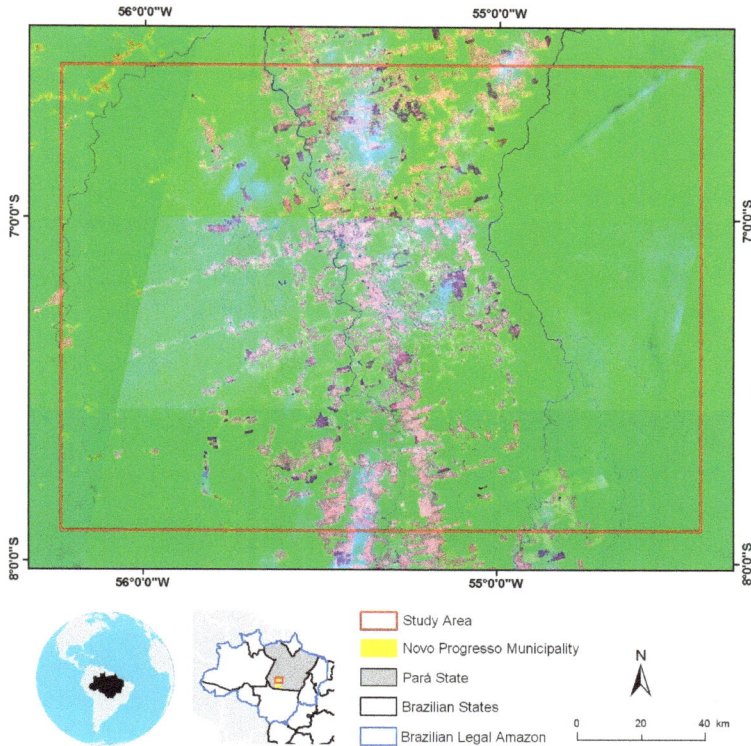

Figure 1. Location map of the study area. On the main map, in green are the old-growth and secondary forest areas, in magenta the productive lands and in purple the burned areas. Composition of Landsat 8 images (Operational Land Imager (OLI) sensor) for the dry season of the year 2014 (RGB composite: Shortwave Infrared 1 in Red, Near Infrared in Green and Red in Blue).

3. Datasets

3.1. Forest Cover Map

Land-use and land-cover data were obtained from the Amazonia Land-use Land-cover Monitoring Project (TerraClass Project/INPE) [32]. We used data for the year 2014, which corresponds to the last year of available mapping.

The TerraClass Project data are the result of a combination of deforestation data from the Brazilian Amazonia Deforestation Monitoring Project (PRODES/INPE) [15] and the land use classification based on orbital images from Landsat, Terra/Aqua, and SPOT-5 satellites.

We regrouped the original classes of the TerraClass Project into two new classes: Forest Cover and Deforested Areas (Table 1). In order to eliminate natural edges in the analyses, we jointed the areas of Cerrado (Brazilian Savannas) and water bodies to the Forest Cover class.

Table 1. Regroups of the original classes of the Amazonia Land-use Land-cover Monitoring Project (TerraClass Project) to obtain the forest cover map.

Original Classes	New Classes
Forest, Secondary Forest, Cerrado (Brazilian Savanna) and Hydrography	Forest Cover
Annual Crops, Urban area, Deforestation in 2014, Mining, Mosaic of Uses, Others, Pasture with exposed soil, Herbaceous Pastures, Shrubby Pasture and Regeneration with Pasture	Deforested Areas

3.2. Active Fire Data

Active fire data were obtained for the period between January and December 2014 from the Fire Information for Resource Management System (FIRMS). These data are derived from the MODIS Active Fire Product (MCD14ML, Collection 6) [33], adjusted to 1 km of spatial resolution. To generate the product, a contextual algorithm compares the daily data of the medium and thermal infrared bands with reference data (without thermal anomalies). Subsequently, false detections are rejected by examining the brightness temperature of the neighbouring pixels [34].

Fire Radiative Power (FRP) values are considered to be an indicator of fire intensity (given in Megawatts or MW) and they are commonly related to the amount of biomass that was consumed during the fire, where the higher the FRP value, the greater is the amount of biomass consumed [35].

During 2014, the number of detected active fires ($N = 35,873$) in Pará State was near the average from 1999 to 2017 ($N = 32,602$) [36] and the year presented a normal climatology (Figure S1) [37].

4. Methods

4.1. Landscape, Fire Incidence and Fire Intensity Metrics

Firstly, we use the forest cover map to calculate landscape metrics using the LecoS plug-in (version 2.0.7, Landscape Ecology Statistics, University of Évora, Évora, Portugal) [38] implemented in the QGIS software (version 2.18, Long-term Release (LTR), QGIS Development Team, https://qgis.org/en/site/) [39]. These metrics and its modifications are commonly used in the literature for analysis that is related to forest fires [26,40] and are based from the Fragstats software (University of Massachusetts, Amherst, MA, USA) [41].

For our analysis, we used 300 grid cells of 10 km by 10 km. This spatial resolution satisfactorily captures the different patterns of fragmentation in our study area. According to Saito et al. [42] the size of the cells do not statistically affect the results of the landscape metrics, and the user then chooses the size of the cells based on the phenomenon and scale analysed. The following metrics were adopted (Table 2): (1) Habitat Loss (percentage of deforestation), (2) Edges Proportion, (3) Number of Forest Patches, and (4) Mean Forest Patch Area.

Then, for each cell, two metrics were calculated for the active fire data. The first metric was the Fire Density (FD, as a proxy of fire incidence), which corresponds to the cumulative number of active fires in 2014 that occurred within forest areas in each cell divided by the total forest in that cell. The second metric was the FRP Mean (as a proxy of fire intensity), which was calculated by averaging the FRP values of active fires that were falling within the forest areas in each cell.

Table 2. Landscape metrics used and their respective descriptions.

Landscape Metric	Abbreviation	Equation	Description
Habitat Loss	HL	$\frac{\sum_{j=1}^{n} a_{ij}}{A} \times 100$	The sum of all deforested areas within a cell, divided by total cell area, and multiplied by 100 (to convert to a percentage). The final unit is given in percentage (%). Where a_{ij} is the area (km^2) of patch ij, and A is total cell area (km^2).
Edges Proportion	EP	$\frac{\sum_{k=1}^{n} e_{ik}}{\sum_{j=1}^{n} a_{ij}}$	The sum of the lengths of all forest edge segments within a cell, divided by total area of all forest patches. The final unit is given in kilometres of edge per square kilometres of forest (km·km^{-2}). Where e_{ik} is the total length (km) of edge in patch i, and a_{ij} is the area (km^2) of patch ij.
Number of Forest Patches	NFP	n_i	The number of forest patches within a cell (n_i).
Mean Forest Patch Area	MFPA	$\frac{\sum_{j=1}^{n} a_{ij}}{n_i}$	The mean area of all forest patches in each cell. The final unit is given in square kilometres (km^2). Where a_{ij} is the area (km^2) of patch ij, and n_i is the total of patches within a cell.

4.2. Statistical Analyzes

To evaluate the relationship among the variables (Fire Density, FRP Mean, and landscape metrics), we fitted curves using LOESS Regression (Locally Weighted Scatterplot Smoothing—LOESS), which is a form of local regression model [43,44]. This method is a non-parametric strategy for fitting a smooth curve to data, where noisy data values, sparse data points, or weak interrelationships interfere with your ability to see a line of best fit [45]. We used the span 0.75 (default setting) in LOESS Regression analyses.

In order to verify the existence of significant differences in the incidence and the intensity of fire as a function of the landscape metrics, we used the Kruskal-Wallis non-parametric test. This test is equivalent to Analysis of Variance (ANOVA), which compares three or more groups to test the hypothesis that they have the same distribution [46–48]. To identify how the analysed variables differ, a paired posthoc test was performed. To perform the posthoc test, we use the Fisher's least significant difference criterion with Bonferroni adjustment methods correction [49]. For all of the tests, the significance level of 95% (p-value < 0.05) was adopted.

We use the R software (version 3.4.4, https://www.r-project.org/) for all analysis [50]. For LOESS Regression, we use the "loess" native function [51]. In the Kruskal-Wallis test, we use the "agricolae" package [52].

We also separated and quantified active fires and the respective FRP values at three edge distances (1 km, 2 km, and greater than 2 km), both within forest areas (hereafter referred as edge of forest cover) and out of forest areas (hereafter referred as edge of deforested areas). Additionally, we calculated the percentage of active fires per FRP intervals, as suggested by Armenteras et al. [26]: $\leq$50 MW, 50 to $\leq$500 MW, 500 to $\leq$1000 MW, and >1000 MW.

5. Results

5.1. Relationship between Habitat Loss and Measures of Habitat Configuration

Our results showed that the analysed landscape metrics exhibited different relationships with habitat loss (HL, Figure 2). The number of forest patches (NFP), as well as its variance, increases with HL until it reaches 70%, which is the maximum level of deforestation within a grid cell that is found in the study area (Figure 2a). The mean forest patch area (MFPA) decreases sharply between 0 and 10% of HL and continues to decrease smoothly from about 10% to 70% of HL, with a lower variance in the larger HL values (Figure 2b). Similarly to NFP, EP and its variance increase with HL, mostly from 20% of HL onwards (Figure 2c).

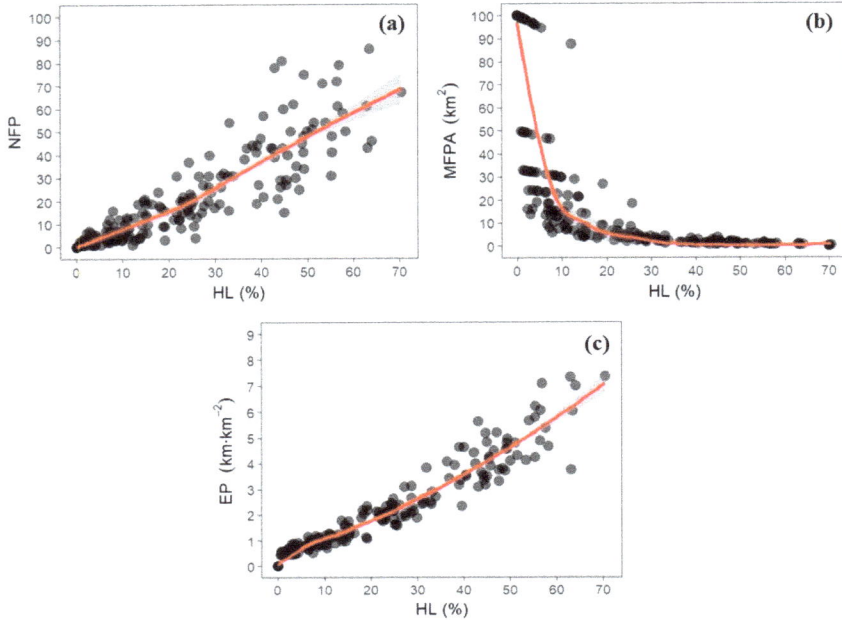

Figure 2. Landscape metrics as a function of Habitat Loss (HL): (**a**) relationship between Habitat Loss and Number of Forest Patches (NFP); (**b**) relationship between Habitat Loss and Mean of Forest Patches Areas (MFPA); and, (**c**) relationship between Habitat Loss and Edges Proportion (EP). Shaded areas represent 95% confidence intervals. The missing confidence intervals in some regions of the graphs are the result of the dispersion in the data at the upper end of the distribution.

The Kruskal-Wallis (KW) test showed that the NFP (KW = 196.04; p-value < 0.05; Figure S2a) and the EP (KW = 205.07; p-value < 0.05; Figure S2c) were significantly lower only in the interval between 0–20% of HL, while the MFPA (KW = 201.38; p-value < 0.05; Figure S2b) was significantly higher in the same interval.

5.2. Relationship between Habitat Configuration and Fire Incidence and Intensity

Fire density (FD) increased with habitat loss (HL), with greater variability in the higher levels of deforestation (Figure 3a). Furthermore, the FD increased until NFP reaches ~35 per grid cell, and then stabilized (Figure 3b). The FD decreased sharply up to 25 km^2 of MFPA, tending to zero after that. On the other hand, the FD increased up to 5 km·km^{-2} of EP, after which it plateaus.

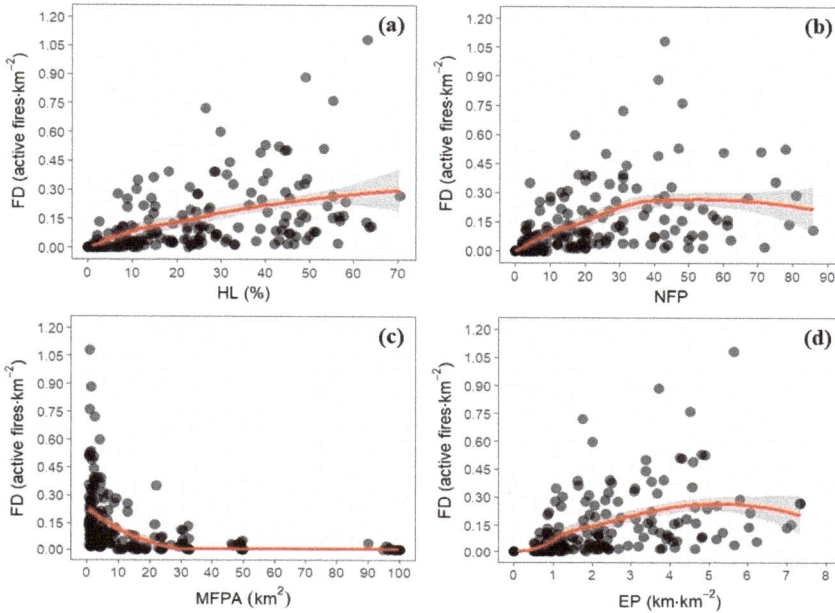

Figure 3. Fire Density (FD) as a function of (**a**) Habitat Loss (HL); (**b**) Number of Forest Patches (NFP); (**c**) Mean Forest Patches areas (MFPA); and, (**d**) Edges Proportion (EP). Shaded areas represent 95% confidence intervals. The missing confidence intervals in some regions of the graphs are the result of the dispersion in the data at the upper end of the distribution.

The Kruskal-Wallis (KW) test showed that FD was significantly lower only in the interval between 0–10% of the HL (KW = 191.76; *p*-value < 0.05; Figure S3a), between 0–10 NFP (KW = 180.68; *p*-value < 0.05; Figure S3b), between 90–100 km^2 of MFPA (KW = 224.86; *p*-value < 0.05; Figure S3c), and finally, between 0–1 km·km^{-2} of EP (KW = 166.82; *p*-value < 0.05; Figure S3d).

The fragmentation effect on the fire intensity, as measured by the Mean FRP, is presented in Figure 4. The Mean FRP increased until ~35% of HL and then decreased until the higher registered levels of HL (Figure 4a). The Mean FRP increased with the increase in the NFP up to 25, but decreased smoothly from about 25 to 80 forest patches (Figure 4b). A tendency of decrease in the Mean FRP was registered as the MFPA increases up to 50 km^2. On the other hand, the Mean FRP increased with the increase of the EP up to 3 km·m^{-2}, with a subsequent decrease up to 7.5 km·m^{-2}.

The Kruskal-Wallis test indicated that forest fire intensity (measured as mean FRP) was significantly lower at the lowest levels of fragmentation: 0–10% of HL (KW = 162.90; *p*-value < 0.05; Figure S4a), between 0–10 NFP (KW = 145.49; *p*-value < 0.05; Figure S4b), between 90–100 km^2 of MFPA (KW = 204.28; *p*-value < 0.05; Figure S4c) and between 0–1 km·km^{-2} of EP (KW = 121.89; *p*-value < 0.05; Figure S4d).

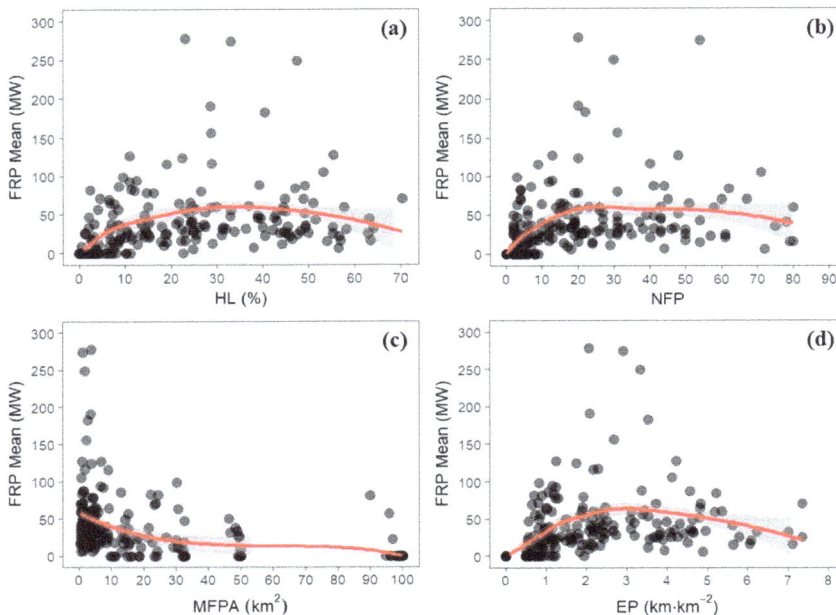

Figure 4. Mean Fire Radiative Power (FRP) as a function of (**a**) Habitat Loss; (**b**) Number of Forest Patches (NFP); (**c**) Mean of Forest Patches Areas (MFPA); and, (**d**) Edges Proportion (EP). Shaded areas represent 95% confidence intervals. The missing confidence intervals in some regions of the graphs are the result of the dispersion in the data at the upper end of the distribution.

Most of the active fires detected were located within 1 km from the forest edges (Table 3), corresponding to 95% and 98% of fires occurring in forest and deforested areas, respectively.

Table 3. Total of active fires per edge distance.

Class Cover	Edge Distance	Total Number of Active Fires	% of the Total
Forest Cover	>3 km	10	0.62
	2 km	66	4.07
	1 km	1546	95.31
Deforested Areas	1 km	2477	98.92
	2 km	27	1.08
	>3 km *	0	0

* No active fires were observed.

Most active fires were classified as low intensity (FRP less than 50 MW), representing between 70% and 90% of the total of active fires analysed for each edge distance (Table 4). Between 10 and 28% of the total active fires were in the 50–500 MW intensity category. The few observed higher intensities of active fires (FRP greater than 500 MW) were located in the first kilometre from the forest edges only. Corroborating the previous evidence, the Kruskal-Wallis test showed a significant difference between the FRP values for the different edge distances in the forest areas (KW = 6.95; p-value < 0.05; Figure S5a), where the highest FRP values were only observed in the first kilometre from the forest edges. For the deforested areas, no significant difference was observed (KW = 2.99; p-value > 0.05; Figure S5b).

Table 4. Percentage of Fire Radiative Power (FRP) per edges distance interval and fire intensity class.

Class Cover	Edge Distance	Class of FRP (%)			
		<50 MW	50–500 MW	500–1000 MW	>1000 MW
Forest Cover	>3 km	90.00	10.00	0	0
	2 km	75.76	24.24	0	0
	1 km	70.63	28.01	0.97	0.39
Deforested Areas	1 km	74.44	24.34	0.93	0.28
	2 km	74.07	25.93	0	0
	>3 km *	0	0	0	0

* No active fires were observed.

6. Discussion

6.1. Relationship between Habitat Loss and Measures of Habitat Configuration

Due to the complexity of anthropic actions in the Amazon region, deforestation occurs in different patterns, resulting in different spatial configurations of patches and forest edges [18,31,53]. Here, we show that in Central Amazonia, the NFP increases as deforestation progresses to levels that are up to 70% of HL. The increasing number of forest patches and its variability with increasing habitat loss is similar to the one found by Oliveira Filho and Metzger [54] for the "fishbone" fragmentation pattern. This relationship was also found by Villard and Metzger [17] in simulated landscapes. Although the maximum HL that was observed in our study area was 70%, the NFP should necessarily decrease at some point as deforestation approaches the 100% level. According to the literature review that was carried out by Fahrig [18], the number of forest patches is expected to increase up to a certain degree of deforestation (~80% of habitat loss), and decrease in the lower levels of habitat amount.

The non-linear relationship between the MFPA and HL that was found in our study area differed from the one that was previously presented by Fahrig [18] in a global study (meta-analysis) for real landscapes. However, the pattern found here is similar to that documented by Oliveira Filho and Metzger [54] in real and simulated landscapes in the Brazilian Amazonia. According to Oliveira Filho and Metzger [54], this response pattern is usually associated with the "fishbone" fragmentation pattern and small settlements, as they produce small patches that are close to each other, which is similar to our study area.

The theoretical model proposed by Fahrig [18] describes a significant increase in the total edges up to 50% of habitat removal level, tending progressively to zero after this threshold. However, in our study area, there was no reduction in EP up to at least 70% of HL, indicating a greater inflection point than that observed by Fahrig [18]. The same pattern was observed by Numata et al. [55] when analysing the forest fragmentation in old deforestation frontiers in the state of Rondônia (Brazilian Amazonia) with different patterns and levels of deforestation, and by Laurance et al. [56] when simulating the deforestation scenario for the same state. This pattern occurs over time as the habitat loss progresses to intermediate levels, increasing the number of forest patches, and consequently the density of forest edges. On the other hand, when forest removal approaches 100%, the number of forest patches and total area are reduced dramatically, resulting in a lower edge density in the landscape [18,57].

6.2. Relationship between Habitat Configuration and Fire Incidence and Intensity

Our results suggest that the landscape structure partly explains the variation of fire incidence and intensity in forest areas, which is similar to the results that were found by Armenteras et al. [25] in the Colombian Amazon. More fragmented landscapes, with smaller patches and a greater proportion of edges, tend to be more vulnerable to fire than landscapes with continuous and intact forests. The effect of fragmentation on the incidence and intensity of fire that was observed here is likely a result of changes in the original structural configuration of the forest, which changes the mass and energy balance. Fragmented forests tend to be drier than a continuous forest cover, due to the lower humidity retention, higher temperature, and the greater exposure to dry air masses and winds [58]. This dry condition causes a higher tree mortality (generally large trees) [59], resulting in a large amount of fuel load available (dead biomass), which increases the susceptibility of forest to fire [60].

Although fragmentation makes forests more susceptible to fire, the occurrence of fire is conditioned to the presence of ignition sources. In Amazonia, these sources are mostly associated with the escape of fire from newly deforested areas (Appendix A, Figure A1b), or from the management of agricultural and pasture areas (Figure A1c) [23,61,62]. This explains the observed variation in fire occurrence and intensity at different levels of landscape fragmentation in our results. This issue becomes even clearer when we observe that over than 95% of the active fires that occurred in the first kilometre from the edge, in both forested and deforested areas, indicating the escape of fires into forests. We verified that fire penetrates forest areas up to a distance of 3 km, which corroborates other studies that were carried out in the Amazon region [20,22,26,27,63]. All active fires of higher intensity (FRP above 500 MW) occurred in the first kilometre in the forest areas, with a significant difference when compared to the other edge distances. This can be explained by the greater amount of fuel available, due to the high rate of trees mortality that is closer to the forest edges [59].

The great variability in the incidence and intensity of fire observed at different levels of fragmentation in our results are likely related to the combined existence of ignition sources and fuel availability in the landscape. Conversely, it is important to note that our results are based on a year that is considered to be normal from the point of view of the amount of rainfall (Figure S1b). Thus, the effects of fragmentation on fire incidence and intensity can be more significant during drought years [25,37], thus increasing carbon emissions into the atmosphere [37,64]. This scenario is worrying since the occurrence of extreme droughts events have become increasingly frequent in Amazonia, and fire occurrence is predicted to increase in the region due to climate and land use change synergies [65–67].

6.3. Implications of the Effect of Fragmentation on Fire Occurrence in Amazonia for the Brazilian Forest Code

Land use regulation is a critical component of forest governance and conservation strategies [68]. In Brazil, the Brazilian Forest Code (BFC) is the main law for regulating land use with the objective of conserving native vegetation. Two instruments of this legislation are highlighted, the first is the Legal Reserve (LR), which requires the maintenance of at least 80% of intact forest areas on private properties in the Amazon biome; and, the other is the Permanent Preservation Area (PPA), which includes both Riparian Preservation Areas (RPA) that protect riverside forest buffers and Hilltop Preservation Areas in high elevations and steep slopes [69].

Our results showed that forest removal values limited by 20% guarantee a smaller number of patches (0–20 patches per 100 km^{-2}) with larger average areas (90–100 km^2) and a lower proportion of forest edges (0–2 km·km^{-2}) in relation to higher levels of habitat loss. This HL threshold coincides with values where the incidence and intensity of fire are significantly smaller when compared to the other levels of HL. The susceptibility of the landscape to forest fires clearly increases with greater HL. Therefore, maintaining native vegetation in at least 80% of the rural properties area, as prescribed in the LR definition for the Amazon biome, allow for low levels of fire incidence, even if the ignition sources are present. Regions with a lower proportion of forest cover are clearly more susceptible to forest degradation due to fire, unless appropriate prevention and management techniques are applied.

In 2012, the BFC was reviewed, and based on our results we argue that some of the current BFC rules for LR and PPA areas can contribute to increasing fire incidence and intensity in the Amazon region, since they substituted some instruments established in the previous version of the law. The most worrying from a conservation point of view is that "small" properties (from 40 ha to 440 ha depending on the region) were exempted from recovering areas of LR that were deforested illegally before 2008. Furthermore, the vegetation of PPA within a property is now considered to be part of the LR, while before the law's modification, the PPA and the LR areas were computed separately, as they serve to different conservation purposes. Additionally, the requirements for the restoration of PPA and the maintenance of LR were reduced. The LR requirement for 80% intact forest was reduced to 50% when (1) the proportion of conservation areas and indigenous territories within Amazonian municipalities is equal to or higher than 50% or (2) conservation areas and indigenous territories represent 65% of the state territory. These legal modifications together reduced the country's "forest debt" by 58% [69], which may allow for the maintenance of the fragmentation of Amazonian landscapes, keeping them susceptible to the occurrence of fire, as we demonstrated in our results.

Another legal modification allowed the rural owner who has forest liabilities to compensate for it in other properties that were located anywhere in the same biome. Given the vast extent of Brazilian biomes, this implies that an owner may compensate for an illegally deforested area by restoring another over 3000 km away. Such restoration effort, if undertaken in a region where forest cover is already well preserved, would not recover the landscape structure and local environmental services where it is needed most. Thus, the displacement of restoration efforts from highly fragmented to more preserved areas would make the former regions more susceptible to the incidence of fire.

According to the BFC, economic exploitation is allowed in the LR areas, including the collection of non-timber forest products (fruits, vines, leaves, and seeds) and the commercial and non-commercial selective extraction of wood. The sustainable economic exploitation of the forest is important for the rural owner as a source of income, thus avoiding the deforestation of the LR areas. However, good forest management practices should be applied. Selective logging can increase the forest susceptibility to fire [70] due the canopy damage [71–74], which allows for the penetration of solar radiation, raising the temperature, and decreasing the humidity within the forest. These microclimate changes that are associated with the greater amount of dead biomass are caused mainly by the logging operations [75], thus resulting in more severe fires [76,77].

This whole context is worrisome since the main sources of fire ignition in the Amazonia are related to the management of adjacent agricultural and livestock areas. The flexibilization of the Forest Code in comparison to its predecessor allowed for the maintenance of extensive fragmented areas, mainly in the region of the deforestation arc, where there are intense anthropic activities [53], and therefore abundant ignition sources.

7. Conclusions

We conclude that the susceptibility of the landscape to forest fires increases at the beginning of the deforestation process. In general, our results reinforce the need to guarantee low levels of fragmentation in the Brazilian Amazonia in order to avoid the degradation of its forests by fire and the related carbon emissions [37,64]. Future work could examine whether the relations that were found here are kept or modified during extreme drought events.

The reduction of forest liabilities resulting from the last modification of the forest code increases the probability of occurrence of forest degradation by fire since it allows the existence of areas with less than 80% of forest cover, contributing to the maintenance of high levels of fragmentation.

We anticipate that forest degradation by fire will continue to increase in the region, especially in light of the mentioned environmental law relaxation and its synergistic effects with climate change. All of this can affect efforts to Reduce Emissions from Deforestation and Forest Degradation (REDD). Therefore, actions to prevent and manage forest fires are necessary, mostly for the properties where forest liabilities exist and are compensated in other regions.

Supplementary Materials: The following are available online at http://www.mdpi.com/1999-4907/9/6/305/s1, Figure S1: (**a**) Seasonal rainfall pattern (the vertical black lines are the standard deviations). (**b**) Normalized rainfall anomalies (1998-2014), Figure S2: Boxplot of the habitat loss (HL) intervals for the number of forest patches, mean of forest patches areas and edges proportion. Figure S3: Boxplot of the fire density for the habitat loss intervals, number of forest patches, mean of forest patches areas and edges proportion. Figure S4: Boxplot of the Fire Radiative Power (FRP) for the habitat loss intervals number of forest patches, mean of forest patches areas and edges proportion. Figure S5: Boxplot of Fire Radiative Power (FRP) for different distances from the edges in forest areas and in deforested areas.

Author Contributions: C.H.L.S.J. and L.E.O.C.A. led in the design of the experiment. C.H.L.S.J. performed data analysis. C.H.L.S.J., L.E.O.C.A., M.G.F., C.T.A., L.B.V. and L.O.A. interpreted the results. C.H.L.S.J., M.G.F. and C.T.A. wrote the paper with significant contributions from all authors.

Acknowledgments: We gratefully acknowledge the Brazilian Federal Agency for Support and Evaluation of Graduate Education (CAPES) and National Council for Scientific and Technological Development (CNPq) agencies for providing research fellowships and support this work (CAPES Post-doctoral fellowship to M.G.F, CNPq-305054/2016-3, CNPq-458022/2013-6 and CNPq-300504/2016-0). L.O.A. acknowledge the productivity scholarship from CNPq (CNPq-309247/2016-0).

Conflicts of Interest: The authors declare no conflict of interest.

Appendix A

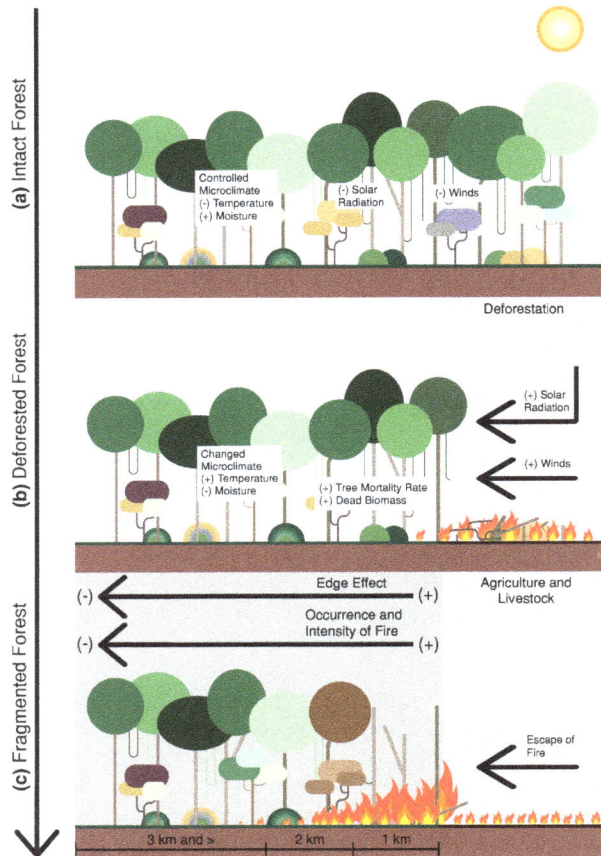

Figure A1. Graphic summary of the main results found in this paper. (**a**) Intact forest, with controlled microclimate, less penetration of solar radiation and action of the winds; (**b**) Deforested forest, resulting in a changed microclimate (higher temperature and lower humidity due to greater penetrability of solar radiation and wind action) and higher mortality rate of trees near the edges, resulting in a greater amount of available fuel material; (**c**) Fragmented forest, more susceptible to the occurrence of fire (more intense near the forest edge) due to the edge effect and fire escape from the agriculture and livestock management areas.

References

1. Sullivan, M.J.P.; Talbot, J.; Lewis, S.L.; Phillips, O.L.; Qie, L.; Begne, S.K.; Chave, J.; Cuni-Sanchez, A.; Hubau, W.; Lopez-Gonzalez, G.; et al. Diversity and carbon storage across the tropical forest biome. *Sci. Rep.* **2017**, *7*, 39102. [CrossRef] [PubMed]
2. Bonan, G.B. Forests and Climate Change: Forcings, Feedbacks, and the Climate Benefits of Forests. *Science* **2008**, *320*, 1444–1449. [CrossRef] [PubMed]
3. Baccini, A.; Goetz, S.J.; Walker, W.S.; Laporte, N.T.; Sun, M.; Sulla-Menashe, D.; Hackler, J.; Beck, P.S.A.; Dubayah, R.; Friedl, M.A.; et al. Estimated carbon dioxide emissions from tropical deforestation improved by carbon-density maps. *Nat. Clim. Chang.* **2012**, *2*, 182–185. [CrossRef]
4. Houghton, R.A.; Hall, F.; Goetz, S.J. Importance of biomass in the global carbon cycle. *J. Geophys. Res. Biogeosci.* **2009**, *114*. [CrossRef]
5. Pan, Y.; Birdsey, R.A.; Fang, J.; Houghton, R.; Kauppi, P.E.; Kurz, W.A.; Phillips, O.L.; Shvidenko, A.; Lewis, S.L.; Canadell, J.G.; et al. A Large and Persistent Carbon Sink in the World's Forests. *Science* **2011**, *333*, 988–993. [CrossRef] [PubMed]
6. Ciais, P.; Sabine, C.; Bala, G.; Bopp, L.; Brovkin, V.; Canadell, J.; Chhabra, A.; Defries, R.; Galloway, J.; Heimann, M.; et al. Carbon and Other Biogeochemical Cycles. In *Climate Change 2013—The Physical Science Basis*; Intergovernmental Panel on Climate Change, Ed.; Cambridge University Press: Cambridge, UK, 2013; Volume 9781107057, pp. 465–570. ISBN 9781107415324.
7. Saatchi, S.S.; Harris, N.L.; Brown, S.; Lefsky, M.; Mitchard, E.T.A; Salas, W.; Zutta, B.R.; Buermann, W.; Lewis, S.L.; Hagen, S.; et al. Benchmark map of forest carbon stocks in tropical regions across three continents. *Proc. Natl. Acad. Sci. USA* **2011**, *108*, 9899–9904. [CrossRef] [PubMed]
8. Grace, J. Understanding and managing the global carbon cycle. *J. Ecol.* **2004**, *92*, 189–202. [CrossRef]
9. Grace, J.; Mitchard, E.; Gloor, E. Perturbations in the carbon budget of the tropics. *Glob. Chang. Biol.* **2014**, 3238–3255. [CrossRef]
10. Malhi, Y. The carbon balance of tropical forest regions, 1990–2005. *Curr. Opin. Environ. Sustain.* **2010**, *2*, 237–244. [CrossRef]
11. Houghton, R.A.; House, J.I.; Pongratz, J.; van der Werf, G.R.; DeFries, R.S.; Hansen, M.C.; Le Quéré, C.; Ramankutty, N. Carbon emissions from land use and land-cover change. *Biogeosciences* **2012**, *9*, 5125–5142. [CrossRef]
12. Capobianco, J.P.R. *Biodiversidade na Amazônia Brasileira: Avaliação e Ações Prioritárias Para a Conservação, Uso sustentável e Repartição de Benefícios*; Instituto Socioambiental: São Paulo, Brazil, 2001; ISBN 8574480525.
13. Coe, M.T.; Marthews, T.R.; Costa, M.H.; Galbraith, D.R.; Greenglass, N.L.; Imbuzeiro, H.M.A; Levine, N.M.; Malhi, Y.; Moorcroft, P.R.; Muza, M.N.; et al. Deforestation and climate feedbacks threaten the ecological integrity of south-southeastern Amazonia. *Philos. Trans. R. Soc. Lond. B Biol. Sci.* **2013**, *368*, 20120155. [CrossRef] [PubMed]
14. Fearnside, P.M. Desmatamento na Amazônia brasileira: História, índices e conseqüências. *Megadiversidade* **2005**, *1*, 113–123. [CrossRef]
15. Instituto Nacional de Pesquisas Espaciais. Monitoramento da Floresta Amazônica Brasileira por Satélite. Available online: http://www.obt.inpe.br/prodes/ (accessed on 1 January 2018).
16. Nepstad, D.; McGrath, D.; Stickler, C.; Alencar, A.; Azevedo, A.; Swette, B.; Bezerra, T.; DiGiano, M.; Shimada, J.; da Motta, R.S.; et al. Slowing Amazon deforestation through public policy and interventions in beef and soy supply chains. *Science* **2014**, *344*, 1118–1123. [CrossRef] [PubMed]
17. Villard, M.-A.; Metzger, J.P. Beyond the fragmentation debate: A conceptual model to predict when habitat configuration really matters. *J. Appl. Ecol.* **2014**, *51*, 309–318. [CrossRef]
18. Fahrig, L. Effects of Habitat Fragmentation on Biodiversity. *Annu. Rev. Ecol. Evol. Syst.* **2003**, *34*, 487–515. [CrossRef]
19. Laurance, W.F.; Williamson, G.B. Positive Feedbacks among Forest Fragmentation, Drought, and Climate Change in the Amazon. *Conserv. Biol.* **2001**, *15*, 1529–1535. [CrossRef]
20. Cochrane, M.A. Synergistic interactions between habitat fragmentation and fire in evergreen tropical forests. *Conserv. Biol.* **2001**, *15*, 1515–1521. [CrossRef]
21. Alencar, A.A.C.; Solórzano, L.A.; Nepstad, D.C. Modeling forest understory fires in an Eastern Amazonian Landscape. *Ecol. Appl.* **2004**, *14*, 139–149. [CrossRef]

22. Cochrane, M.A.; Laurance, W.F. Fire as a large-scale edge effect in Amazonian forests. *J. Trop. Ecol.* **2002**, *18*, 311–325. [CrossRef]

23. Cano-Crespo, A.; Oliveira, P.J.C.; Boit, A.; Cardoso, M.; Thonicke, K. Forest edge burning in the Brazilian Amazon promoted by escaping fires from managed pastures. *J. Geophys. Res. Biogeosci.* **2015**, *120*, 2095–2107. [CrossRef]

24. Aragão, L.E.O.C.; Shimabukuro, Y.E. The incidence of fire in Amazonian forests with implications for REDD. *Science* **2010**, *328*, 1275–1278. [CrossRef] [PubMed]

25. Aragão, L.E.O.C.; Malhi, Y.; Roman-Cuesta, R.M.; Saatchi, S.; Anderson, L.O.; Shimabukuro, Y.E. Spatial patterns and fire response of recent Amazonian droughts. *Geophys. Res. Lett.* **2007**, *34*, L07701. [CrossRef]

26. Armenteras, D.; González, T.M.; Retana, J. Forest fragmentation and edge influence on fire occurrence and intensity under different management types in Amazon forests. *Biol. Conserv.* **2013**, *159*, 73–79. [CrossRef]

27. Armenteras, D.; Barreto, J.S.; Tabor, K.; Molowny-Horas, R.; Retana, J. Changing patterns of fire occurrence in proximity to forest edges, roads and rivers between NW Amazonian countries. *Biogeosciences* **2017**, *14*, 2755–2765. [CrossRef]

28. LEI No. 12.727, DE 17 DE OUTUBRO DE 2012. Available online: http://www.planalto.gov.br/ccivil_03/_ato2011-2014/2012/lei/l12727.htm (accessed on 10 January 2018).

29. Vieira, S.; de Camargo, P.B.; Selhorst, D.; da Silva, R.; Hutyra, L.; Chambers, J.Q.; Brown, I.F.; Higuchi, N.; dos Santos, J.; Wofsy, S.C.; et al. Forest structure and carbon dynamics in Amazonian tropical rain forests. *Oecologia* **2004**, *140*, 468–479. [CrossRef] [PubMed]

30. Pinheiro, T.F.; Escada, M.I.S.; Valeriano, D.M.; Hostert, P.; Gollnow, F.; Müller, H. Forest Degradation Associated with Logging Frontier Expansion in the Amazon: The BR-163 Region in Southwestern Pará, Brazil. *Earth Interact.* **2016**, *20*, 1–26. [CrossRef]

31. Arima, E.Y.; Walker, R.T.; Perz, S.; Souza, C. Explaining the fragmentation in the Brazilian Amazonian forest. *J. Land Use Sci.* **2016**, *11*, 257–277. [CrossRef]

32. De Almeida, C.A.; Coutinho, A.C.; Esquerdo, J.C.D.M.; Adami, M.; Venturi, A.; Diniz, C.G.; Dessay, N.; Durieux, L.; Gomes, A.R. High spatial resolution land use and land cover mapping of the Brazilian Legal Amazon in 2008 using Landsat-5/TM and MODIS data. *Acta Amaz.* **2016**, *46*, 291–302. [CrossRef]

33. Giglio, L.; Schroeder, W.; Justice, C.O. The collection 6 MODIS active fire detection algorithm and fire products. *Remote Sens. Environ.* **2016**, *178*, 31–41. [CrossRef]

34. Giglio, L.; Descloitres, J.; Justice, C.O.; Kaufman, Y.J. An Enhanced Contextual Fire Detection Algorithm for MODIS. *Remote Sens. Environ.* **2003**, *87*, 273–282. [CrossRef]

35. Wooster, M.J.; Roberts, G.; Perry, G.L.W.; Kaufman, Y.J. Retrieval of biomass combustion rates and totals from fire radiative power observations: FRP derivation and calibration relationships between biomass consumption and fire radiative energy release. *J. Geophys. Res.* **2005**, *110*, D24311. [CrossRef]

36. Instituto Nacional de Pesquisas Espaciais. Monitoramento de Queimadas. Available online: http://www.inpe.br/queimadas/portal (accessed on 1 January 2018).

37. Aragão, L.E.O.C.; Anderson, L.O.; Fonseca, M.G.; Rosan, T.M.; Vedovato, L.B.; Wagner, F.H.; Silva, C.V.J.; Silva Junior, C.H.L.; Arai, E.; Aguiar, A.P.; et al. 21st Century drought-related fires counteract the decline of Amazon deforestation carbon emissions. *Nat. Commun.* **2018**, *9*, 536. [CrossRef] [PubMed]

38. Jung, M. LecoS—A python plugin for automated landscape ecology analysis. *Ecol. Inform.* **2016**, *31*, 18–21. [CrossRef]

39. QGIS Development Team. QGIS Geographic Information System. Available online: http://qgis.osgeo.org (accessed on 22 June 2016).

40. Hayes, J.J.; Robeson, S.M. Relationships between fire severity and post-fire landscape pattern following a large mixed-severity fire in the Valle Vidal, New Mexico, USA. *For. Ecol. Manag.* **2011**, *261*, 1392–1400. [CrossRef]

41. McGarigal, K. Fragstats Help. Available online: http://www.umass.edu/landeco/research/fragstats/documents/fragstats.help.4.2.pdf (accessed on 21 April 2015).

42. Saito, É.A.; Fonseca, L.M.G.; Escada, M.I.S.; Korting, T.S. Efeitos da mudança de escala em padrões de desmatamento na Amazônia. *Rev. Bras. Cartogr.* **2011**, *63*, 401–414.

43. Cleveland, W.S.; Grosse, E.; Shyu, W.M. Local regression models. In *Statistical Models in S*; Chambers, J.M., Hastie, T.J., Eds.; Chapman and Hall: New York, NY, USA, 1992; pp. 309–376.

44. Cleveland, W.S.; Loader, C. Smoothing by Local Regression: Principles and Methods. In *Statistical Theory and Computational Aspects of Smoothing*; Härdle, W., Schimek, M.G., Eds.; Physica-Verlag: Heidelberg, Germany, 1996.

45. Tate, N.J.; Brunsdon, C.; Charlton, M.; Fotheringham, A.S.; Jarvis, C.H. Smoothing/filtering LiDAR digital surface models. Experiments with loess regression and discrete wavelets. *J. Geogr. Syst.* **2005**, *7*, 273–290. [CrossRef]

46. Gibbons, J.D.; Chakraborti, S. Nonparametric Statistical Inference. In *International Encyclopedia of Statistical Science*; Lovric, M., Ed.; Springer: Berlin/Heidelberg, Germany, 2011; pp. 977–979.

47. Hettmansperger, T.P.; McKean, J.W. *Robust Nonparametric Statistical Methods*, 2nd ed.; CRC Press: Boca Raton, FL, USA, 2010; ISBN 9781439809082.

48. Bonnini, S.; Corain, L.; Marozzi, M.; Salmaso, L. Nonparametric Hypothesis Testing. In *Wiley Series in Probability and Statistics*; John Wiley & Sons: Chichester, UK, 2014; ISBN 9781118763490.

49. Conover, W.J. *Practical Nonparametric Statistics*, 3rd ed.; John Wiley & Sons: Hoboken, NJ, USA, 1999; ISBN 978-0471160687.

50. R Core Team. R: A Language and Environment for Statistical Computing. Available online: Https://www.r-project.org/ (accessed on 1 January 2018).

51. Ripley, B.D. Local Polynomial Regression Fitting. Available online: http://stat.ethz.ch/R-manual/R-devel/library/stats/html/loess.html (accessed on 1 January 2018).

52. De Mendiburu, F. Statistical Procedures for Agricultural Research. Available online: https://cran.r-project.org/web/packages/agricolae/agricolae.pdf (accessed on 1 January 2017).

53. Vedovato, L.B.; Fonseca, M.G.; Arai, E.; Anderson, L.O.; Aragão, L.E.O.C. The extent of 2014 forest fragmentation in the Brazilian Amazon. *Reg. Environ. Chang.* **2016**, *16*, 2485–2490. [CrossRef]

54. De Filho, F.J.B.O.; Metzger, J.P. Thresholds in landscape structure for three common deforestation patterns in the Brazilian Amazon. *Landsc. Ecol.* **2006**, *21*, 1061–1073. [CrossRef]

55. Numata, I.; Cochrane, M.A.; Roberts, D.A.; Soares, J.V.; Souza, C.M.; Sales, M.H. Biomass collapse and carbon emissions from forest fragmentation in the Brazilian Amazon. *J. Geophys. Res.* **2010**, *115*, G03027. [CrossRef]

56. Laurance, W.F.; Laurance, S.G.; Delamonica, P. Tropical forest fragmentation and greenhouse gas emissions. *For. Ecol. Manag.* **1998**, *110*, 173–180. [CrossRef]

57. Liu, Z.; He, C.; Wu, J. The Relationship between Habitat Loss and Fragmentation during Urbanization: An Empirical Evaluation from 16 World Cities. *PLoS ONE* **2016**, *11*, e0154613. [CrossRef] [PubMed]

58. Cochrane, M.A.; Laurance, W.F. Synergisms among Fire, Land Use, and Climate Change in the Amazon. *AMBIO A J. Hum. Environ.* **2008**, *37*, 522–527. [CrossRef]

59. Laurance, W.F.; Camargo, J.L.C.; Fearnside, P.M.; Lovejoy, T.E.; Williamson, G.B.; Mesquita, R.C.G.; Meyer, C.F.J.; Bobrowiec, P.E.D.; Laurance, S.G.W. An Amazonian rainforest and its fragments as a laboratory of global change. *Biol. Rev.* **2018**, *93*, 223–247. [CrossRef] [PubMed]

60. Berenguer, E.; Ferreira, J.; Gardner, T.A.; Aragão, L.E.O.C.; De Camargo, P.B.; Cerri, C.E.; Durigan, M.; De Oliveira, R.C.; Vieira, I.C.G.; Barlow, J. A large-scale field assessment of carbon stocks in human-modified tropical forests. *Glob. Chang. Biol.* **2014**, *20*, 3713–3726. [CrossRef] [PubMed]

61. Aragão, L.E.O.C.; Malhi, Y.; Barbier, N.; Lima, A.A.; Shimabukuro, Y.; Anderson, L.; Saatchi, S. Interactions between rainfall, deforestation and fires during recent years in the Brazilian Amazonia. *Philos. Trans. R. Soc. Lond. B Biol. Sci.* **2008**, *363*, 1779–1785. [CrossRef] [PubMed]

62. Rosan, T.M.; Anderson, L.O.; Vedovato, L. Assessing the Origin of Hot Pixels in Extreme Climate Years in the Brazilian Amazon. *Rev. Bras. Cartogr.* **2017**, *69*, 731–741.

63. Briant, G.; Gond, V.; Laurance, S.G.W. Habitat fragmentation and the desiccation of forest canopies: A case study from eastern Amazonia. *Biol. Conserv.* **2010**, *143*, 2763–2769. [CrossRef]

64. Anderson, L.O.; Aragão, L.E.O.C.; Gloor, M.; Arai, E.; Adami, M.; Saatchi, S.S.; Malhi, Y.; Shimabukuro, Y.E.; Barlow, J.; Berenguer, E.; et al. Disentangling the contribution of multiple land covers to fire-mediated carbon emissions in Amazonia during the 2010 drought. *Glob. Biogeochem. Cycles* **2015**, *29*, 1739–1753. [CrossRef] [PubMed]

65. Marengo, J.A.; Espinoza, J.C. Extreme seasonal droughts and floods in Amazonia: Causes, trends and impacts. *Int. J. Climatol.* **2016**, *36*, 1033–1050. [CrossRef]

66. Malhi, Y.; Roberts, J.T.; Betts, R.A.; Killeen, T.J.; Li, W.; Nobre, C.A. Climate Change, Deforestation, and the Fate of the Amazon. *Science* **2008**, *319*, 169–172. [CrossRef] [PubMed]

67. Le Page, Y.; Morton, D.; Hartin, C.; Bond-Lamberty, B.; Pereira, J.M.C.; Hurtt, G.; Asrar, G. Synergy between land use and climate change increases future fire risk in Amazon forests. *Earth Syst. Dyn.* **2017**, *8*, 1237–1246. [CrossRef]

68. Stickler, C.M.; Nepstad, D.C.; Azevedo, A.A.; McGrath, D.G. Defending public interests in private lands: Compliance, costs and potential environmental consequences of the Brazilian Forest Code in Mato Grosso. *Philos. Trans. R. Soc. B Biol. Sci.* **2013**, *368*, 20120160. [CrossRef] [PubMed]

69. Soares-Filho, B.; Rajao, R.; Macedo, M.; Carneiro, A.; Costa, W.; Coe, M.; Rodrigues, H.; Alencar, A. Cracking Brazil's Forest Code. *Science.* **2014**, *344*, 363–364. [CrossRef] [PubMed]

70. Holdsworth, A.R.; Uhl, C. Fire in Amazonian selectively logged rain forest and the potential for fire reduction. *Ecol. Appl.* **1997**, *7*, 713–725. [CrossRef]

71. Pereira, R.; Zweede, J.; Asner, G.P.; Keller, M. Forest canopy damage and recovery in reduced-impact and conventional selective logging in eastern Para, Brazil. *For. Ecol. Manag.* **2002**, *168*, 77–89. [CrossRef]

72. Asner, G.P.; Broadbent, E.N.; Oliveira, P.J.C.; Keller, M.; Knapp, D.E.; Silva, J.N.M. Condition and fate of logged forests in the Brazilian Amazon. *Proc. Natl. Acad. Sci. USA* **2006**, *103*, 12947–12950. [CrossRef] [PubMed]

73. Verissimo, A.; Barreto, P.; Mattos, M.; Tarifa, R.; Uhl, C. Logging impacts and prospects for sustainable forest management in an old Amazonian frontier: The case of Paragominas. *For. Ecol. Manag.* **1992**, *55*, 169–199. [CrossRef]

74. Uhl, C.; Vieira, I.C.G. Ecological Impacts of Selective Logging in the Brazilian Amazon: A Case Study from the Paragominas Region of the State of Para. *Biotropica* **1989**, *21*, 98. [CrossRef]

75. Uhl, C.; Barreto, P.; Vidal, E.; Amaral, P.; Barros, A.C.; Souza, C.; Johns, J.; Gerwing, J. Natural Resource Management in the Brazilian Amazon. *Bioscience* **1997**, *47*, 160–168. [CrossRef]

76. Gerwing, J.J. Degradation of forests through logging and fire in the eastern Brazilian Amazon. *For. Ecol. Manag.* **2002**, *157*, 131–141. [CrossRef]

77. Siegert, F.; Ruecker, G.; Hinrichs, A.; Hoffmann, A.A. Increased damage from fires in logged forests during droughts caused by El Niño. *Nature* **2001**, *414*, 437–440. [CrossRef] [PubMed]

MDPI
St. Alban-Anlage 66
4052 Basel
Switzerland
Tel. +41 61 683 77 34
Fax +41 61 302 89 18
www.mdpi.com

Forests Editorial Office
E-mail: forests@mdpi.com
www.mdpi.com/journal/forests

www.ingramcontent.com/pod-product-compliance
Lightning Source LLC
Chambersburg PA
CBHW051713210326
41597CB00032B/5460